Human Centric Integrative Lighting

Human Centric Integrative Lighting

Technology, Perception, Non-visual Effects

Tran Quoc Khanh
Peter Bodrogi
Trinh Quang Vinh

WILEY-VCH

Authors

Prof. Dr.-Ing. Tran Quoc Khanh
Technische Universität Darmstadt
Laboratory for Adaptive Lighting
Systems and Visual Processing
Hochschulstraße 4A
64289 Darmstadt
Germany

Dr.-Ing. Peter Bodrogi
ERCO GmbH
Brockhauser Weg 80-82
58507 Lüdenscheid
Germany

Dr.-Ing. Trinh Quang Vinh
Technische Universität Darmstadt
Laboratory for Adaptive Lighting
Systems and Visual Processing
Hochschulstraße 4A
64289 Darmstadt
Germany

Cover © Perry Mastrovito/Getty Images

Library of Congress Card No.: applied for

British Library Cataloguing-in-Publication Data
A catalogue record for this book is available from the British Library.

Bibliographic information published by the Deutsche Nationalbibliothek
The Deutsche Nationalbibliothek lists this publication in the Deutsche Nationalbibliografie; detailed bibliographic data are available on the Internet at <http://dnb.d-nb.de>.

Print ISBN: 978-3-527-41400-0
ePDF ISBN: 978-3-527-83150-0
ePub ISBN: 978-3-527-83152-4
oBook ISBN: 978-3-527-83151-7

Cover Design SCHULZ Grafik-Design
Typesetting Straive, Chennai, India

Contents

Preface

Since the introduction of LED technology at the turn of the century and the discovery of a new light-sensitive receptor, whose signals play a key role in the control of human biorhythms, there has been a new dynamic in colour and light quality research, coupled with intensive research in the fields of neurobiology, sleep research, psychology, human medicine and occupational science, with the aim of qualitatively and quantitatively exploring the relationship between radiation properties and mechanisms of action in the brain as well as in physical organs.

Based on the conviction that such valuable scientific findings, which will certainly continue to be collected over the next few years, should already be implemented in lighting design, lighting product development, and lighting planning, the authors saw the need to combine and bundle these scientific findings with technological and systems engineering advances, such as the Internet of Things, networking of communication systems, and sensor systems. This book is aimed at scientists, lighting designers, lighting engineers, psychologists, engineers, as well as students.

The first German language book was written in an unexpectedly difficult time with the COVID-19 pandemic in winter 2020–2021. The present English language book is an extended and updated version of the German one written in the not-less-difficult year 2022. The authors would like to thank the scientific staff of the Technical University of Darmstadt, especially S. Beck, J. Klabes, Dr. S. Babilon, and B. Zandi for the literature collection. The authors would like to thank the research assistants S. Klir and S. Benkner, whose lecture slides on *Smart Lighting* formed part of the content of Chapter 11 for the students of lighting technology. The authors would like to thank W. Truong, who is doing his doctorate at the TU Darmstadt at this time, for jointly developing the concepts of measuring circadian-effective metrics and for the joint publication with the main author of this book on the health of early-shift workers, the results of which were included in a section of this book.

Darmstadt
25 January 2023

Tran Quoc Khanh
Peter Bodrogi
Trinh Quang Vinh

Acknowledgements

The authors would also like to express their sincere thanks to the two lighting experts P. Dehoff (Dornbirn) and J. Minnerup (Arnsberg) for numerous discussions and for sending the literature on lighting quality.

The authors would also like to thank their families for their moral support before and during the phase of writing a book with a complex content in a time of many crises in the world.

1

Introduction and Motivations

1.1 Introduction: A Historical Review. Current Issues

The International Commission on Illumination (CIE) defined the 2° standard colour matching functions in September 1931, more than 90 years before the publication of this book: the so-called $\bar{x}(\lambda)$, $\bar{y}(\lambda)$, $\bar{z}(\lambda)$ functions for visual colour matching with a visual field of view of 1–4°. This made it possible to calculate the tristimulus values and chromaticity coordinates for any colour stimulus of any spectral composition in the visible wavelength range, to characterise *perceived colours* and communicate them in scientific and industrial processes. In addition, from the end of the nineteenth century until today, the following steps of development have taken place in lighting technology:

(a) *In lighting technology and photometry*: from the end of the nineteenth century until around the 1980s, some parts of the world experienced a steadily growing development of the industrialisation process (e.g. mechanical engineering, shipbuilding, electrical engineering, and construction), so that research in lighting technology concentrated on formulating the requirements for workplaces in offices and manufacturing on the basis of physiological visual performance such as contrast perception ability, visual acuity, reading speed, or error rate of the work performed, using the parameters *illuminance* (in lux) or *luminance* (in $cd\,m^{-2}$) as a basis. The results of research in this field up to the end of the twentieth century formed the basis for today's international and national lighting standards [1–3].

(b) *In light source technology*: from 1879 to 1999, there were several important developmental steps from incandescent lamps to high-pressure discharge lamps, halogen incandescent lamps, three-band fluorescent lamps, and compact fluorescent lamps (see Table 1.1). From 1994 to the present day, light source technology has undergone enormous progress with the new development of high-power LEDs. The luminous efficacy of white LEDs exceeds the values of commonly used discharge lamps (e.g. T5 lamps, Cosmopolis lamps, and HMI lamps). The high luminous efficacy of the LEDs, rated according to the $V(\lambda)$-function for daytime vision, contributes positively to the worldwide effort to save energy and protect the environment.

Human Centric Integrative Lighting: Technology, Perception, Non-Visual Effects, First Edition.
Tran Quoc Khanh, Peter Bodrogi, and Trinh Quang Vinh.

Table 1.1 Major milestones in the development of light source technology.

Year	Contents
1854	Goebel: Light bulb with bamboo fibre
1879	Edison: Incandescent lamp with carbon filament
1900	Cooper, Hewitt: Patent on mercury vapour lamp
1906	Introduction of the tungsten metal filament lamp with nitrogen filling
1934	Introduction of the low-pressure discharge lamp with phosphors
1959–1960	Introduction of the tungsten halogen lamp
1971	Fluorescent lamps with a three-band concept
1980	Introduction of the CFL-i (energy-saving) lamp
1994	White LED based on InGaN material

Source: TU Darmstadt.

(c) *In CIE – colorimetry*: the history of CIE – colorimetry is characterised by the constant efforts to define perceptual colour attributes (brightness, lightness, hue, chroma, and saturation) and to arrange them in a perceptually equidistant colour space. If these perceptually equidistant colour spaces are created, the colour differences between different colours can be calculated there and used for industrial quality control. One benefit of accurate colour difference calculation is the colour rendering index. This task was carried out at several levels of knowledge over the last decades (see Table 1.2 as well as [6, 7]). The research results of colour science have been used in the colour industry (display technology, film technology, printing technology, and textile industry), and more intensively since about 2010 in lighting technology and light source technology (LED, OLED).

Since the beginning of the twenty-first century, some development trends relevant to lighting technology have intensified as follows:

- Societies in large parts of the world (North America, Europe, China, Japan, and South-East Asia) have been oriented towards information technology. The way of working, the work processes (day and night rhythms), as well as the work equipment (monitors, data, and display devices), have reached a new quality. In addition to quality features such as illuminance or uniformity and glare, other discussions about light and health, well-being during office work, stress reduction, and increased concentration through lighting have been added.
- The previous light source technologies had the decisive disadvantage that the spectrum and colour of the lamps could only be varied to a small extent. Today's high-power and mid-power LEDs with their high luminous efficacies and with a few advantages such as dimmability, controllability, and integrability also have the great advantage that they can be formed from coloured and white LEDs into a *lighting system* of variable spectral composition (chromaticity, colour temperature). The *dynamic light* formed in this way enhances the colour

Table 1.2 Major milestones in the development of CIE colorimetry.

Year	Contents
Year	Contents
1931	Definition of the 2° – standard colour matching functions $\bar{x}(\lambda)$, $\bar{y}(\lambda)$, and $\bar{z}(\lambda)$
1960	Definition of the UCS diagram (*Uniform Colour Scale*)
1964	Definition of the CIE 1964 ($U^*V^*W^*$) colour space
1964	Definition of the colour matching functions $\overline{x_{10}}(\lambda)$, $\overline{y_{10}}(\lambda)$, and $\overline{z_{10}}(\lambda)$ for 10° field of view
1976	Definition of the 2 colour spaces CIE $L^*a^*b^*$ (or CIELAB) and CIE $L^*u^*v^*$ (or CIELUV)
2004	CIE Publication: a colour appearance model for colour management systems: CIECAM02, Publication No. 159 (Vienna: Central Bureau of the CIE, 2004) [4]
2006	Definition of the CAM02-UCS colour space based on the colour appearance model CIECAM02 [5]

Source: TU Darmstadt.

and light quality of *interior lighting*, for the evaluation of which non-visual, colour-technological, and photometric approaches are now increasingly expected to come into play.

This makes it clear that the three important components of lighting technology (photometry, colorimetry, and light source technology) should be used much more intensively and closely together in current and future research for the evaluation of the colour and light quality of workplaces and in the lighting industry for the development of new lighting products. In addition – in the period between 2000 and today – the *non-visual effects of light* have been investigated by various international research groups. Despite numerous efforts in the experimental field, these findings are only partially implemented in the practice of lighting product development and lighting design in a comprehensible and interpretable way.

According to the above considerations, the authors elaborated on the present book to answer the following questions:

1. How do the visual and non-visual mechanisms in the brain and in other physiological areas function during night-time hours and during the day?
2. Which influencing parameters and which initial parameters with which metrics in the physiological and psychological – emotional area are decisive for the description of the subjective and objective characteristics of health, well-being, and work performance of human beings? To what extent can scientists and product developers control these parameters – according to the findings to date? Where is there still a need for research?
3. What findings are known so far about the effects of light at night? The focus will be on the relationship between irradiation and its effects on humans.
4. What findings have been made so far for the time during the day? Can some of them be scientifically established to the extent that the long-awaited

recommendation values for lighting designs as well as for the development of intelligent lighting products can be put up for discussion?

5. How to record, measure, and interpret the visual and non-visual parameters of lighting with daylight and electric light? Such measurements should be carried out not only with laboratory measuring devices but also with portable, inexpensive, and accurate measuring units to plausibly record the effects of light in alternating and dynamically changing workplaces and places where people stay.

Derived from these important questions, this book (except for this chapter) is divided into the thematic blocks summarised in Figure 1.1.

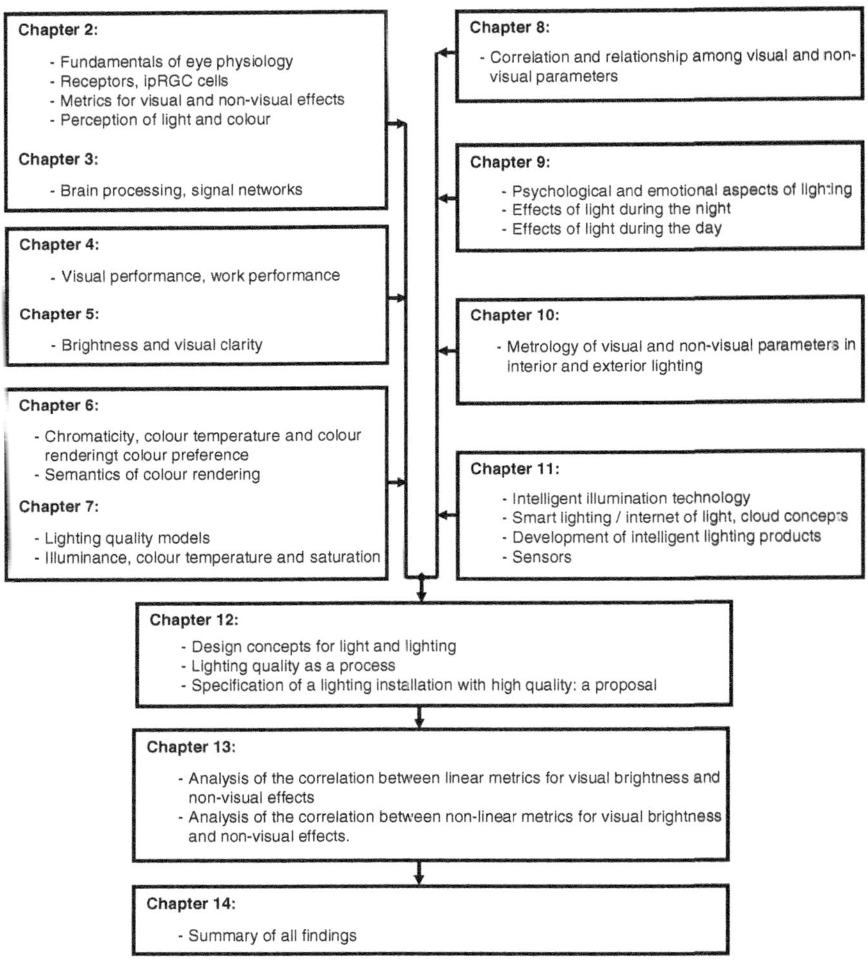

Figure 1.1 Contents of the present book. Source: TU Darmstadt.

References

1 DIN EN 12464 (2011). Licht und Beleuchtung, Teil 1: Beleuchtung von Arbeitsstätten in Innenräumen (Light and Lighting, Part 1: Lighting of Indoor Workplaces), German version EN 12464-1:2021). Beuth, Berlin.
2 Schmits, P.W. (2002). Innenraumbeleuchtung – Tendenzen und Reaktionen, Tagung Licht (Interior Lighting – Trends and Reactions, Conference Light), Maastricht, 34–47.
3 CIE (1981). An analytical model for describing the influence of lighting parameters upon visual performance. CIE Publication No. 19/2.1.
4 Commission Internationale de l'Eclairage (2004). A colour appearance model for color management systems: CIECAM02 CIE TC8-01. Technical Report. CIE Publication No. 159. https://cie.co.at/publications/colour-appearance-model-colour-management-systems-ciecam02.
5 Luo, M.R., Cui, G., and Li, C. (2006). Uniform colour spaces based on CIECAM02 colour appearance model. *Color Res. Appl.* 31: 320–330.
6 CIE (2004). Colorimetry. C. I. E. "Report No: CIE Pub No 15." *Vienna: CIE Central Bureau.* https://cie.co.at/publications/colorimetry-3rd-edition.
7 Commission Internationale de l'Eclairage (1995). Method of measuring and specifying colour rendering properties of light sources. Technical Report. CIE 13.3-1995.

2

Fundamentals of Lighting Technology – Basic Visual and Non-visual Aspects

This chapter summarises the basics of modern lighting technology (including colorimetry), with special emphasis on the connection between the visual and non-visual components of the human visual system to understand the most important aspects of Human-Centric Lighting. Section 2.1 deals with the structure of the eye and the retina to illustrate the mechanisms of visual and non-visual signal processing. Section 2.2 presents the basic photometric and colorimetric parameters and the modelling of colour appearance. The basics of non-visual aspects such as melatonin suppression at night, their modelling as well as the spectral activity functions and the metrics for the consideration of non-visual influencing variables are discussed in Section 2.3.

2.1 The Human Visual System. Visual and Non-visual Signal Processing

To understand the basic concepts of Human-Centric Lighting, the structure, and functioning of the human eye, including the retina (the '*biological image receptor*'), should first be understood. Figure 2.1 illustrates the structure of the human eye.

As can be seen in Figure 2.1, the human eye is an ellipsoid with an average length of about 26 mm and a diameter of about 24 mm. The eye is rotated in all directions with the help of eye muscles. The outer layer is called the sclera. The sclera continues in front as a transparent cornea. The choroid supplies the retina with oxygen and nourishment. The retina is the photoreceptive layer of the eye, containing both the photoreceptors and those cells that pre-process the signals from the photoreceptors through neuronal connections (ganglion cells, amacrine and bipolar cells, and horizontal cells), see Figure 2.2.

The vitreous body is responsible for maintaining the ellipsoidal shape of the eye. It consists of a suspension of water (98%) and hyaluronic acid (2%). The optical system of the human eye is a complex, slightly decentered lens system that projects an inverted and reduced image of the surroundings onto the retina. The cornea, anterior chamber, and iris form the front part of this optical system, followed by the posterior chamber and the biconvex lens of the eye. The lens is held in place by the zonula fibres.

Human Centric Integrative Lighting: Technology, Perception, Non-Visual Effects, First Edition.
Tran Quoc Khanh, Peter Bodrogi, and Trinh Quang Vinh.

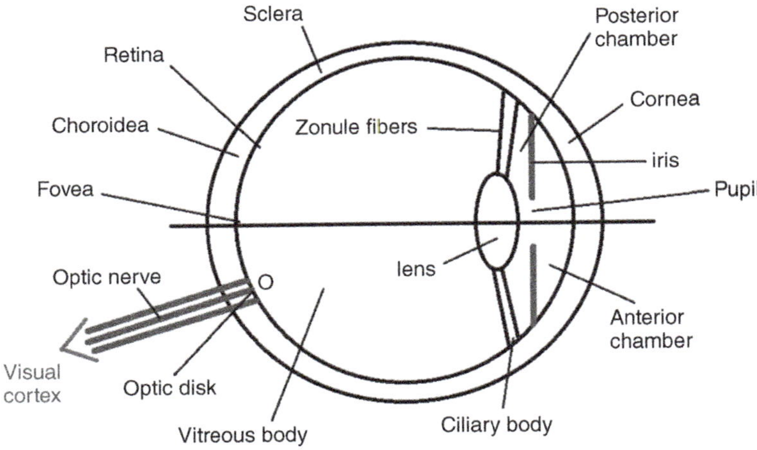

Figure 2.1 Structure of the human eye. O (optic disk): blind spot – the point where the optic nerve crosses the eye and transmits the pre-processed neuronal signals from the retina towards the visual cortex. Source: Reproduced with permission from Wiley-VCH [1].

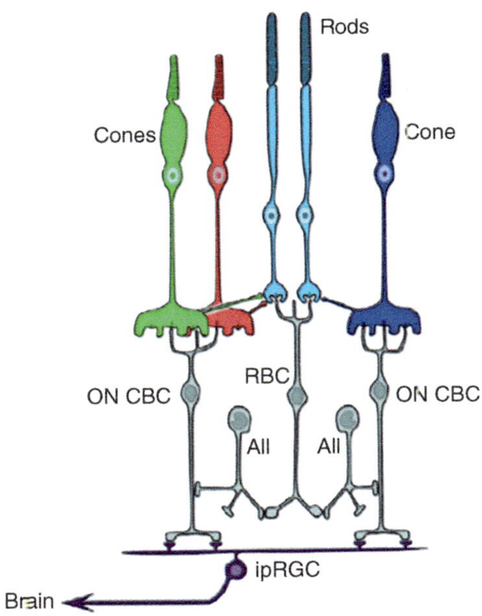

Figure 2.2 Schematic representation of the retinal circuits in humans. Photoreceptor connections via the cone bipolar cells (ON CBCs), amacrine cells (AII), and rod bipolar cells (RBCs), ipRGC: intrinsically photosensitive retinal ganglion cell with the pigment melanopsin. Source: Reproduced from Lucas et al. [2] with permission from Trends in Neurosciences.

By tensing the ciliary muscles, the focal length of the lens can be changed. The angle of vision intersects the retina at the *fovea centralis*, the site of the sharpest vision. The most important optical parameters of the components of the ocular medium include the refractive indices (which typically range from 1.33 to 1.43) and the spectral transmission factors. All parameters vary considerably in different individuals and are subject to significant changes with age. In particular, accommodation, visual acuity, and pupillary responses are impaired with age. The spectral transmission of the ocular media decreases significantly with age, especially for short wavelengths (see Section 12.6.1 for details).

After light rays reach the retina, they pass through the retinal layers and, in the central retina, also through the so-called *macula lutea* (a yellow pigment layer that protects the central retina) before reaching the photoreceptors, which are located on the back of the retina. The blind spot is the location (labelled O in Figures 2.1 and 2.3) where the optic nerve passes through the eye. The retina is blind at the O spot because the density of rods and cones is zero there.

The retina (a layer with an average thickness of 250 μm) is part of the optical system of the eye and with its photoreceptor structure also part of the visual brain. The retina contains a complex cell layer with two types of photoreceptors, rods and cones. Both the rod and cone receptors are connected to the nerve fibres of the optic nerve via a complex network of the pre-processing cells mentioned above, which calculates further neuronal signals from the receptor signals. The retina contains about 6.5 million cones and 110–125 million rods, while the number of nerve fibres is about 1 million. The density of the rods and cones varies and depends on the position of the retina (see Figure 2.3).

There is also a third type of photosensitive cell, the so-called ipRGC (*intrinsic photosensitive retinal ganglion cell*, which contains the pigment melanopsin), which is responsible for regulating the circadian rhythm, see Section 2.3. Figure 10.3 illustrates the distribution of ipRGCs on the retina. According to the illustration

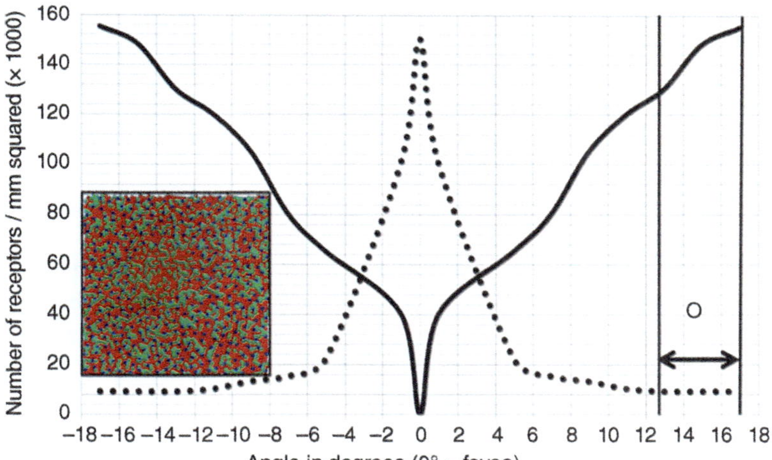

Figure 2.3 Rod density (solid curve) and cone density (dots) as a function of retinal position (abscissa: in degrees) drawn according to Oesterberg's data [3]. O: Blind spot; inset diagram: cone mosaic of the rod-free inner fovea with an extension of approx. 1.25 °, i.e. approx. 350 μm. Red dots: long-wavelength-sensitive cone photoreceptors (L-cones). Green dots: medium-wavelength-sensitive cones (M-cones). Blue dots: short-wavelength sensitive cones (S-cones). Source: Reproduced with permission from Wiley-VCH [1], except for the inset diagram. Source of inset diagram: Figure 1.1 from Sharpe, L. T., Stockman, A., Jägle, H., & Nathans, J. (1999), Opsin genes, cone photopigments, colour vision, and colourblindness, pp. 3–51 in [4]. Reproduced with permission from Cambridge University Press.

in Figure 2.2, which reflects current knowledge of neurophysiology, the signals from the cones and rods are also transmitted not only to the 'normal' ganglion cells but also to the *photosensitive* ganglion cells ipRGCs. The signals of the ipRGCs in turn flow into both *visual* and *non-visual signal processing channels*, whereby the visual and non-visual channels are strongly interconnected in the human brain, see Figure 2.2, Figure 10.6 as well as Chapter 3 of this book.

Due to the photopigment melanopsin, the ipRGCs (which represent only approx. 1–5% of the retinal ganglion cells [2]) are *directly sensitive to* light and can therefore also respond to light in isolation from the rest of the retina [2]. They are connected *in situ* to the rod and cone photoreceptors via the retinal circuits, see Figure 2.2. As a result, the ipRGC signal can be influenced by both intrinsic melanopsin photoreception and extrinsic rod and cone signals [2], see also Figure 2.13. Figure 2.3 shows rod and cone density as a function of retinal position, while the inset diagram of Figure 2.3 shows the long, medium, short (LMS) cone mosaic (which is similar to a digital camera) of the rod-free inner fovea.

As can be seen in Figure 2.3, there are no receptors at the site of the blind spot because the optic nerve exits the eye at this point (labelled O). The fovea is in the centre of the *macula lutea* region. A characteristic value to show the diameter of the fovea is 1.5 mm, which corresponds to a visual angle of about 5°. The fovea is responsible for the best visual acuity due to the high cone receptor density, see the cone density maximum in Figure 2.3. Outside the fovea, cone diameter increases to about 4.5 μm, cone density decreases, and rod density (diameter of rods: 2 μm) increases to reach a rod density maximum at about 18–20°.

Rods are responsible for night vision, also called *scotopic vision*, at a luminance lower than 0.001 cd m^{-2}. Rods are more sensitive than cones, but they become completely inactive above about 60–100 cd m^{-2}. Cones are responsible for daytime vision or *photopic vision* (at a luminance of about 10 cd m^{-2} or higher). The transitional area between scotopic (rod) vision and photopic (cone) vision is called the twilight or *mesopic area*, where both the rods and cones are active. Acceptable *colour quality* (see Section 6.7) can only be expected in the photopic range.

In addition to pupil contraction, the transition between rod and cone vision in the mesopic area represents a second important adaptation mechanism of the human visual system to changing light conditions (so-called *adaptation*). There is a third adaptation mechanism, the *gain control* of the receptor signals.

Photoreceptors contain pigments (opsins, certain types of proteins) that change their structure when they absorb photons and generate neural signals that are pre-processed by the horizontal, amacrine, bipolar, and ganglion cells of the retina to provide neural signals for later processing through the various visual and non-visual channels of the visual system. There are three types of cones (see the cones labelled red, green, and blue in Figure 2.2 as well as Figure 2.3) with pigments of different spectral sensitivity, the so-called L- (long-wave sensitive), M- (medium-wave sensitive) and S-cone (short-wave sensitive) cones. L-, M-, and S-signals for the perception of homogeneous colour fields and for coloured spatial structures (e.g. a red-purple rose with fine colour shadings).

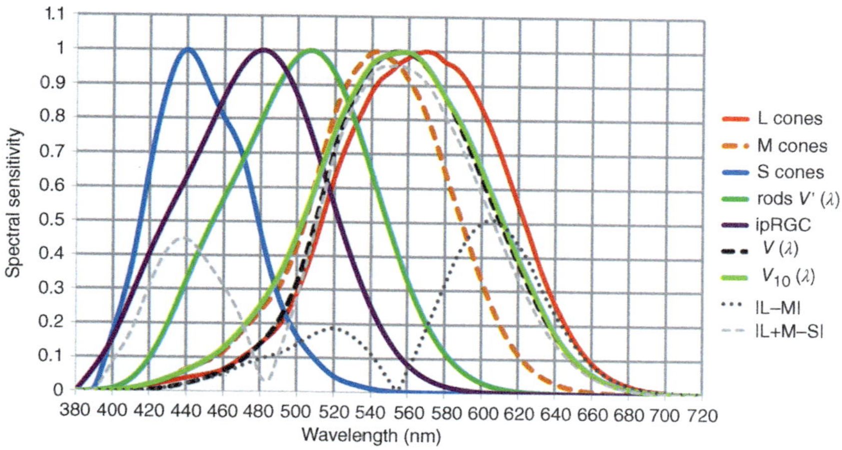

Figure 2.4 Relative spectral sensitivity of the L-, M-, and S-cones (Stockman and Sharpe 2 ° Cone Fundamentals for 2 ° field of view) [6–8] and other visual mechanisms using the LMS cone signals as input. The sensitivity of the ipRGC mechanism is also shown (see Section 2.3). The spectral sensitivity of the rods (dark green curve) is approximated by the so-called $V'(\lambda)$ function. $V(\lambda)$: spectral luminous sensitivity function (see Section 2.2.1) as the basis of photometry, for stimuli of the standard visual angle, about 1–4 °); $V_{10}(\lambda)$: alternative version for stimuli of the larger visual angle (about 10 °). Source: Reproduced with permission from Wiley-VCH [1].

The number of different cone signals (three) has an important psychophysical implication: colour vision is trichromatic, colour spaces have three dimensions (see Section 2.2.3), and there are three independent psychological attributes of colour perception (hue, saturation, and brightness). The relative spectral sensitivities of the L-, M-, and S-cones are shown in Figure 2.4 along with some other important features explained below.

The spectral sensitivities in Figure 2.4 were measured on the cornea of the eye. They thus include the average spectral transmission of the ocular media and macular pigment at a retinal eccentricity of 2 °. These are the so-called Stockman and Sharpe 2 °-Cone Fundamentals [6–8]. As can be seen in Figure 2.4, the spectral bands of the L-, M-, and S-cones result in three receptor signals for further processing in the human visual cortex. From these signals, the retina derives three so-called visual signal processing channels, see Figure 10.6.

Two visual signal processing channels provide so-called *chromatic signals* or *opponent signals:*

1. 'L − M' signal or red–green signal (the mediating signal processing channel is known as the *opponent neural channel*; this channel provides a signal for the visual cortex to compute the perceived red or green component of the visual stimulus); and
2. 'L + M − S' signal (or yellow–blue signal, which is the signal from the second opposing channel that sends a signal to the visual cortex to calculate the perceived yellow or blue component of the visual stimulus).

The signals of the above two opponent channels represent the output variables of the psychophysical models of colour appearance. In such models, the L-, M-, and S-signals are weighted. E.g. with the weighting factors a and b *the* result is $(L + a \cdot M - b \cdot S)$. The foundations of colour appearance modelling are described in Section 2.2.4. In the above formula with the signs '+' and '−', these signs are only symbolic: in physiological reality, the signals are not only weighted but also adapted and compressed, see Section 2.2.4 (Figure 2.10).

The third channel is the so-called *luminance channel* or achromatic channel, which transmits the 'L + M' signal. In this channel, the same output signals of the L-cones and M-cones are not subtracted as in the (first) opposing 'L − M' signal but added. The 'L + M' channel is usually considered the *luminance channel*. The most important role of the luminance channel is that it enables seeing fine image details and the temporal resolution of fast events.

As can be seen from Figure 2.4, the L-, M-, and S-cone sensitivity curves have their maxima at 566, 541, and 441 nm [4], respectively. In- photometry, the spectral response of the L + M channel for stimuli with a visual angle of 1–4° is approximated by the spectral luminous responsivity function $V(\lambda)$ (see Section 2.2.1), which forms the basis of photometry, while for more spatially extended (e.g. 10°) stimuli the so-called $V_{10}(\lambda)$ function (the so-called CIE 10° observer [9]) is used. For spatially extended stimuli covering e.g. a 10° field of view, other LMS spectral sensitivities are to be used; e.g. the so-called Stockman and Sharpe 2000 10°-Cone *Fundamentals* [8].

2.2 Photometric and Colorimetric Quantities

2.2.1 Lighting Technology and Colorimetry

In 1924, the CIE defined the spectral luminous efficiency function known today as the $V(\lambda)$-function for daytime vision (so-called photopic vision) [10] and thus laid a perceptually relevant foundation for the characterisation of light sources and lighting situations. In 1931, the CIE introduced the colorimetric system that is still valid today (see Section 2.2.2) with the so-called 2°- *colour-matching functions*. Its $\bar{y}(\lambda)$-function is equal to the $V(\lambda)$ function (see Figure 2.5) thus colorimetry has a very close relationship to photometry. In colorimetry, which is thus unseparated by the equation $\bar{y}(\lambda) = V(\lambda)$, various photometric and colorimetric parameters are defined, e.g. parameters for the terms colour stimulus, colour temperature, chromatic adaptation, colour space, colour appearance, colour difference, colour difference formula, colour preference, colour rendering, and colour gamut [11, 12], which are explained in Sections 2.2.2 and 2.2.3.

To calculate the so-called *luminance* from the instrumentally measured spectral radiance (as an example of a *photometric* parameter), Eq. (2.1) is used.

$$L_v = 683\ \mathrm{lm\ W^{-1}} \cdot \int V(\lambda) \cdot L_e(\lambda) d\lambda \tag{2.1}$$

with

$V(\lambda)$: the spectral luminous efficiency function for daytime vision.
$L_e(\lambda)$: the spectral radiance of the object or light source.

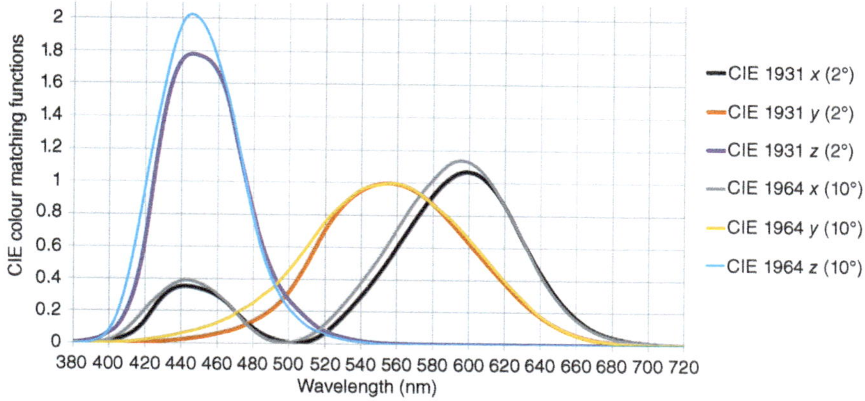

Figure 2.5 Comparison of the standard 2° $\bar{x}(\lambda)$, $\bar{y}(\lambda)$, and $\bar{z}(\lambda)$ colour matching functions and the 10° standard $\overline{x_{10}}(\lambda)$, $\overline{y_{10}}(\lambda)$, and $\overline{z_{10}}(\lambda)$ colour matching functions. Source: TU Darmstadt.

At each wavelength, the spectral radiance of the light source $L_e(\lambda)$ is weighted with the $V(\lambda)$ function in Eq. (2.1) and integrated in the visible wavelength range (i.e. between 380 and 780 nm) to form a total signal (so-called *luminance*), which is used in numerous psychophysical models of *visual performance*, *brightness*, and *colour perception* as a basic numerical parameter – in combination with other signals or parameters. An interpretation of luminance *alone*, according to Eq. (2.1), as a parameter for perceived *brightness is* only a first, rough approximation. For a much better prediction of human brightness perception than with the help of luminance, there are complex models available today, which are presented in detail in Chapters 5 and 7.

2.2.2 Colorimetry: CIE Tristimulus Values and CIE Chromaticity Diagram

The term *colour stimulus* is defined as follows [11]: optical radiation that evokes a colour sensation in the human visual system. A colour stimulus can be characterised by the so-called CIE *tristimulus values* X, Y, and Z. To calculate these tristimulus values, the measured spectral radiance is spectrally weighted and integrated in the visible spectral range (i.e. between 380 and 780 nm) by one of the above-mentioned 2° *colour matching functions* $\bar{x}(\lambda), \bar{y}(\lambda), \bar{z}(\lambda)$ for small colour stimuli (between the visual field sizes of 1–4°). For larger fields of view (>4°, e.g. 10°), the so-called 10°-colour matching functions are used: $\overline{x_{10}}(\lambda), \overline{y_{10}}(\lambda), \overline{z_{10}}(\lambda)$ [12, 13]. These functions are compared with each other in Figure 2.5.

Figure 2.5 shows the differences in the spectral curves of the 2° – colour matching functions and the 10° – colour matching functions. These differences can be attributed to the change in the maximum optical density of the macular pigment and the change in the maximum density of the visual pigments of the three photoreceptor types of the retina (so-called L-, M-, and S-cones) as a function of the visual field size (2° or 10°) [14], see Figure 2.3.

The spectral radiance $L_\lambda(\lambda)$ of the colour stimulus is measured with the help of a spectroradiometer. This is used to calculate the tristimulus values X, Y, and Z according to Eq. (2.2).

$$\begin{aligned} X &= k \cdot \int_{360\text{ nm}}^{830\text{ nm}} L_\lambda(\lambda) \cdot \bar{x}(\lambda) \cdot \mathrm{d}\lambda \\ Y &= k \cdot \int_{360\text{ nm}}^{830\text{ nm}} L_\lambda(\lambda) \cdot \bar{y}(\lambda) \cdot \mathrm{d}\lambda \\ z &= k \cdot \int_{360\text{ nm}}^{830\text{ nm}} L_\lambda(\lambda) \cdot \bar{z}(\lambda) \cdot \mathrm{d}\lambda \end{aligned} \tag{2.2}$$

The value of k can be chosen as 683 lm W^{-1} for self-emitting light sources, making the value of Y equal to the *luminance* of the light source.

The so-called *standard chromaticity coordinates* x, y, and z are defined according to Eq. (2.3).

$$x = \frac{X}{X+Y+Z}, y = \frac{Y}{X+Y+Z}, z = \frac{Z}{X+Y+Z} \tag{2.3}$$

The diagram of the chromaticity coordinates x, y is called the *chromaticity diagram* or CIE 1931 chromaticity diagram, a certain combination of x, y (and $z=1-x-y$) values is called chromaticity coordinates [13]. Figure 2.6 illustrates some

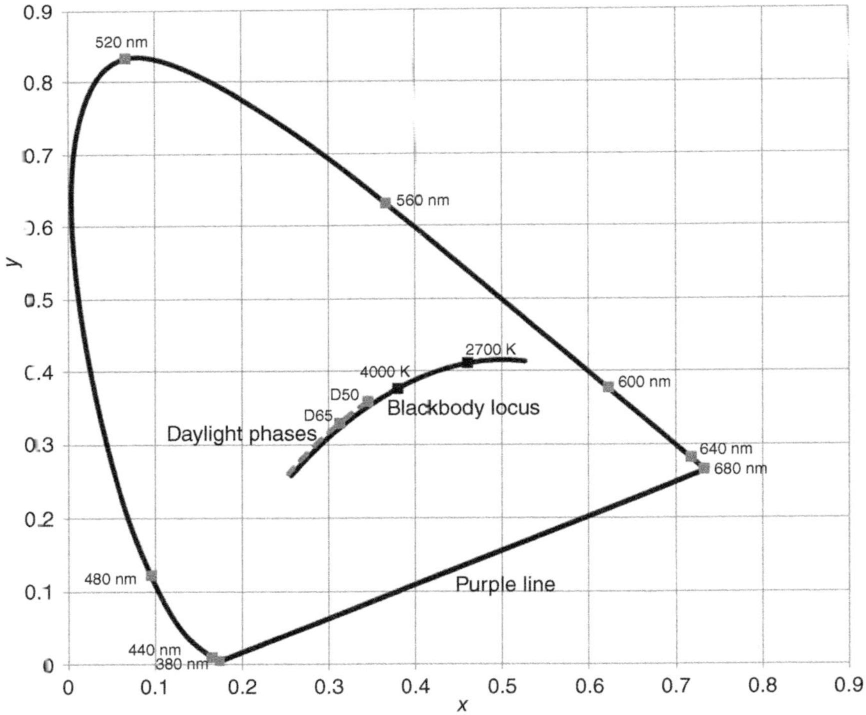

Figure 2.6 Important chromaticities in the x, y chromaticity diagram. Source: TU Darmstadt.

of the chromaticity coordinates in the chromaticity diagram that is important in the practice of colorimetry. These include the spectrally monochromatic chromaticities (that correspond to a single wavelength), the purple line, the blackbody locus, and the locus of daylight phases [15].

The '*spectral*' chromaticities (of '*monochromatic*' radiations with only one wavelength) form the left, upper, and right edges of the shoe-shaped range of all existing chromaticity locations. In Figure 2.6, some '*spectral*' chromaticities (such as 680, 640, 600 nm, etc.) are marked between 380 and 680 nm. The purple line connects the chromaticity locations of the left (380 nm) and right (680 nm) extreme wavelengths. The blackbody locus (or Planckian curve) represents the chromaticity coordinates x and y of the blackbody radiators. Figure 2.6 also illustrates the blackbody radiators with temperatures between 2000 and 20 000 K. The temperatures 2700 and 4000 K are marked. 20 000 K is located at the left end of the blackbody locus.

In connection with the chromaticity diagram, the term *correlated colour temperature* (T_{cp} or *CCT* in *K*) of a light source to be tested (so-called test light source) is defined as follows (this is especially important in the evaluation of *colour rendering*): the temperature of a blackbody radiator whose chromaticity is closest to the chromaticity of the test light source in a uniform chromaticity diagram (in which the distances between two colours, the so-called colour differences, are approximately perceptually represented; the standard x, y chromaticity diagram does not correspond to this criterion) [16]. The so-called UCS chromaticity diagram (CIE 1960) is used as a uniform chromaticity diagram to calculate the correlated colour temperature. Its axes (u and v) are defined according to Eq. (2.4).

$$u = \frac{4x}{-2x + 12y + 3}, v = \frac{6y}{-2x + 12y + 3} \tag{2.4}$$

The chromaticities of the daylight phases with different correlated colour temperatures [15] are located on a curve above the blackbody locus. These are shown in Figure 2.6 between CCT = 5000 and CCT = 20 000 K. In lighting practice, the daylight phases D65 (CCT = 6504 K) and D50 (CCT = 5003 K) are particularly relevant. Figure 2.7 shows the relative spectral radiant power of the blackbody radiators with temperatures 2700 and 4000 K as well as the daylight phases D50 and D65.

Figure 2.7 shows that the black-body radiator with 2700 K (P2700) has a high yellow or red component (therefore its light colour is yellowish, so-called warm white), which becomes smaller and smaller with the types of light P4000, D50, or D65, whereby the blue component in the relative spectral radiant power of these light sources increases in this order. For this reason, the D65 light type has a colder white tone (so-called cold white).

A better approximation of a uniform chromaticity diagram is given by Eq. (2.5), which gives the CIE 1976 UCS colour *diagram* (or u', v' *diagram*). A comparison of Eqs. (2.4) and (2.5) show the following relationship: $u = u'$; $v = (2/3) \cdot v'$.

$$u' = \frac{4x}{-2x + 12y + 3}, v' = \frac{9y}{-2x + 12y + 3} \tag{2.5}$$

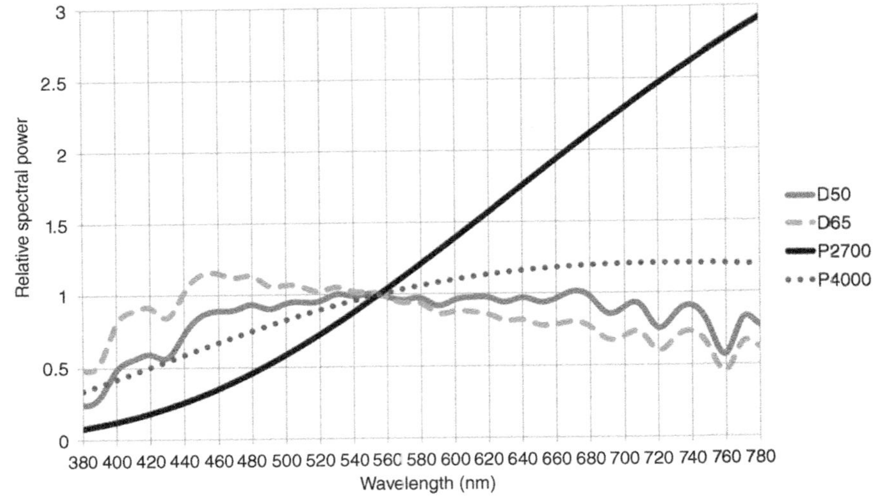

Figure 2.7 Relative spectral radiant power (=1 at 555 nm) of the blackbody radiators with temperatures 2700 K (P2700) and 4000 K (P4000) as well as daylight phases D50 and D65. Source: TU Darmstadt.

2.2.3 Colour Appearance, Colour Matching, Colour Spaces, and Colour Difference Formulas

The tristimulus values X, Y, Z, and standard chromaticity coordinates x and y defined in Section 2.2 are not suitable for *quantifying* colour perception, i.e. to predict the so-called perceived colour attributes. These perceived colour attributes include *hue*, *brightness*, *lightness*, *colourfulness*, *saturation*, and *chroma*, which are defined (according to [11]) as follows.

- *Hue:* the term hue describes between which two basic colours of the set red, yellow, green, and blue, the colour perception of the colour stimulus is located and what percentage it has of the two basic colours. For example, a certain shade of orange is located between red and yellow and is perceived as having 40% red and 60% yellow (hue).
- *Brightness*: how much light the colour stimulus appears to perceptually emit.
- *Lightness*: perceived brightness of the colour stimulus in relation to the brightness of a reference white in the visual field. The reference white can be, for example, a white wall in an interior room.
- *Colourfulness: the* absolute perceived amount of colour in a colour stimulus.
- *Saturation:* the colourfulness of the colour stimulus in relation to the stimulu's own brightness.
- *Chroma:* the colourfulness of the colour stimulus in relation to the brightness of the reference white.

Perceptually, different colour stimuli (e.g. differently coloured printed colour samples) can be arranged manually (with the help of an artist) in a three-dimensional colour atlas in a visually equidistant manner, see Figure 2.8. The resulting colour

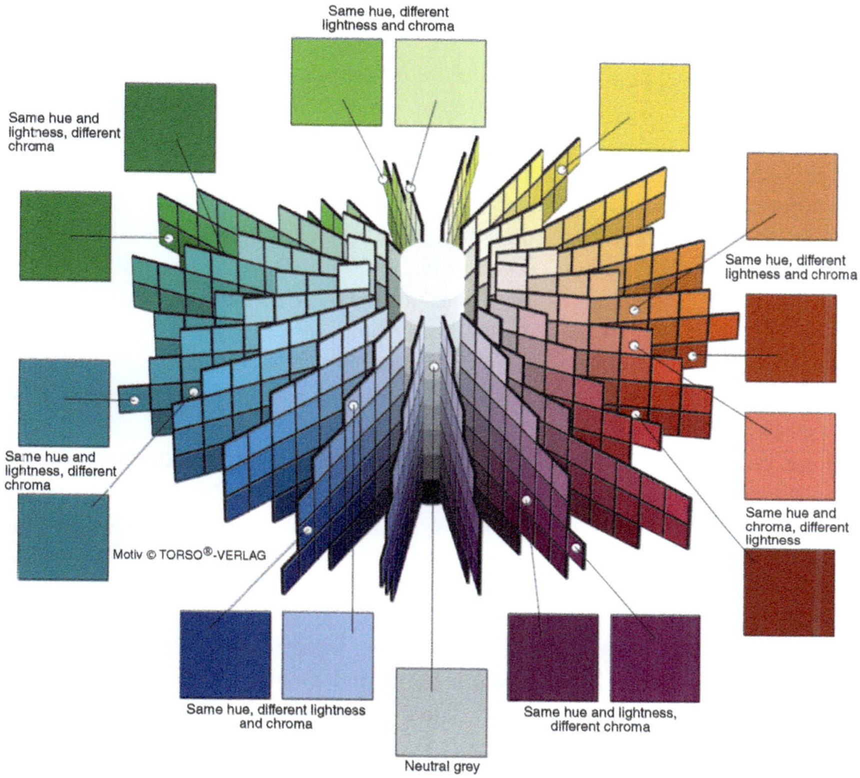

Figure 2.8 Illustration of a three-dimensional colour atlas with the basic properties of the colours. Along the middle axis are the so-called achromatic colours, which have no chromaticity. The lightness increases from the bottom to the top, and the chroma from the inside to the outside. The hue changes continuously around the achromatic axis. Colour spaces represent mathematical models of the colour atlas. Source: https://www.torso.de/en/Tutorials/Tutorials/Torso-ColorTrainings/TORSO-ColorTraining-Basics-1::371.html.

atlas (see Figure 2.8) illustrates the perceived colour characteristics defined above. *Colour spaces* represent mathematical models of the colour atlas, see below.

The colour appearance of a colour stimulus [17, 18], i.e. a spatially limited homogeneously radiating surface (e.g. a homogeneously filled circle with a diameter of 4°) depends, however, not only on the surface itself but also on its surroundings, generally on the entire visual field. Thus, the luminance level and the standard chromaticity coordinates of the white tone prevailing in the visual environment strongly influence the colour perception corresponding to the colour stimulus. This white tone is also called *adapted white point* (or *reference white*), e.g. the white-painted walls of an interior illuminated by an artificial light source.

If the spectral radiant power of the light source changes (e.g. if a *cool white* LED luminaire is switched on instead of a *warm white* compact fluorescent lamp), colour perception changes not only because of the change in the spectral radiant power of the light source, but also because the chromatic adaptation state (or ‘colour mood') of the visual system then changes from 'warm white-adapted' to 'cool white-adapted'.

The effect described above is called *chromatic adaptation* and plays an important role in the evaluation of colour reproduction. With chromatic adaptation – despite changes in spectral radiant power and changes in the tristimulus values of the colour stimulus – the colour perception of the colour stimulus tends to remain constant (this is the so-called colour constancy), provided that the chromaticity of the light source on the CIE standard chromaticity diagram is close to (or immediately on) the blackbody locus or daylight curve marked in Figure 2.6. The perceived colours of such light sources may be called *whites,* although whites of lower colour temperature (such as the white of a gas-filled tungsten filament lamp at 2856 K, which corresponds to the so-called CIE *standard illuminant A*) have a yellow cast. Nevertheless, these are preferred – for certain applications such as living room lighting – by most users in Western culture.

The perceived colour characteristics of a colour stimulus can be mathematically predicted from the measured spectral radiance of the colour stimulus and its surroundings (its background and the reference white). Thus, for each colour attribute, a parameter (or '*correlate*') can be calculated that considers the chromatic adaptation state. For this purpose, the colour stimuli are represented in a so-called *colour space,* the three orthogonal axes of which generally describe the characteristics of lightness, red–green proportion, and yellow–blue proportion in a certain chromatic adaptation state. The value of the *correlate* for the red–green component is positive if the colour perceptually contains red, negative if the colour contains green and zero if the colour has no chromaticity (so-called *achromatic*). The value of the *correlate* for the yellow–blue component is positive if the colour perceptually contains yellow, negative if the colour contains blue and zero if the colour is achromatic. The colour space thus represents a *mathematical model of the colour atlas* (see Figure 2.8).

The colour spaces used in today's colorimetric practice include the so-called CIE 1964 Uniform Colour Space (or U*-V*-W* colour space with the three axes U^*, V^*, and W^* [19], which is already obsolete today), the so-called *CIELAB colour space* (with the three axes L^*, a^*, b^*[13]), and the so-called CAM02-UCS colour space [20] (which is derived from the CIECAM02 colour appearance model [21]). Here, only the CAM02-UCS colour space (and its underlying CIECAM02 colour appearance model) used in modern lighting technology and colour science is discussed, which is important for the formation of the descriptor quantities of *colour quality evaluation* (e.g. colour fidelity index R_f, colour gamut index R_g, see Section 6.7). (A so-called *colour preference index*, CQS Q_a, which is based on the CIELAB colour space, is also presented in Section 6.7).

2.2.4 The CIECAM02 Colour Appearance Model

The difference between CIECAM02 [21] and CAM02-UCS [20] can be explained as follows.

- The CIECAM02 model is a *colour appearance model* whose primary task is to determine reliable *numerical* correlates, i.e. *correlates* that correlate well with human perception, for the various colour *attributes* (e.g. chromaticity, hue, and lightness, see Section 2.2.3) of a colour stimulus from its instrumentally measured spectral radiance (and the photometric parameters of the environment of the colour stimulus) and to take into account the adaptation to the luminance level prevailing in the scene as well as the chromatic adaptation of the human observer. One advantage of the CIECAM02 model is that the scales of the numerical correlates change according to changes in the surrounding, further depending on adaptation luminance level and the current state of chromatic adaptation.

- CAM02-UCS is a *colour space* based on the numerical correlates of the CIECAM02 model whose primary task is to *perceptually* predict *colour differences* between two colour stimuli. Since the numerical scales of the correlates of the CIECAM02 model always depend on the *viewing conditions* of the colour stimulus (such as chromatic adaptation or adaptation luminance level), the CAM02-UCS colour space derived from it also retains this favourable property.
- Although the CIECAM02 model also has a 'built-in' colour space (the so-called *J-a_C-b_C colour space*), it is not presented here because of the superiority of the derived CAM02-UCS colour space.

The mathematical formalism of the CIECAM02 model [17, 18] is not repeated here, instead, only the most important features are explained. The CIECAM02 colour appearance model calculates numerical correlates for the different colour attributes of a colour stimulus. Not only the tristimulus values of the colour stimulus (X, Y, and Z) and the prevailing white tone in the environment (X_n, Y_n, and Z_n) are taken into account, but also so-called *viewing conditions,* such as the lighting parameters of the background and the so-called *surround,* see Figure 2.9.

It can be observed in Figure 2.9 that the colour perception of the colour stimulus depends not only on the colorimetric properties of the colour stimulus itself but also

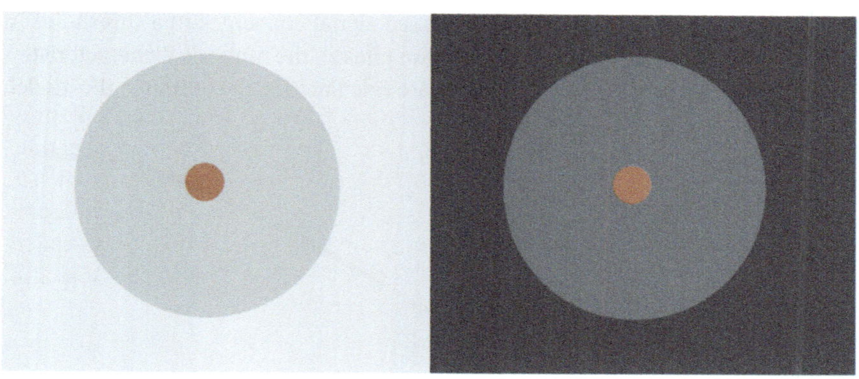

Figure 2.9 Two viewing conditions with the same colour stimulus. Colour stimulus (small circular area in the centre: usually with a diameter of 2 °), background (large circular area: typically with a diameter of 10 °) and surround (the remaining field of view outside the background). Adaptive field = background + surround. Left: Colour stimulus on a bright adaptive field; right: the same colour stimulus on a dark adaptive field. Observe the difference in the two perceived brownish shades. Source: TU Darmstadt.

on the viewing conditions. Accordingly, the CIECAM02 model works with the following parameters for the viewing condition: the mean luminance of the adaptive field or predominant luminance level (L_A), the relative luminance of the background (Y_b), the general degree of adaptation of the visual system (F), the degree of adaptation to the reference white (D), the effect of the environment (c), and the so-called factor for chromatic induction (N_c).

The above parameters for the viewing conditions (L_A, Y_b, F, D, c, and N_c), and the tristimulus values X, Y, Z, X_n, and Y_n Z_n form the input variables of the CIECAM02 model. The output variables of the model are the numerical descriptor quantities (*'correlates'*) for the colour attributes chroma C, lightness J, hue angle h, hue H, brightness Q, saturation s, and colourfulness M. The characteristic H (with values between 0 and 400) predicts between which two basic colours the perceived hue of the colour stimulus is located and what percentage it visually exhibits of these basic colours. For the basic colours red, green, yellow, and blue, respectively, the value of the parameter $H = 0$, 100, 200, and 300, respectively. If the value of the parameter is e.g. $H = 280$, this value means that the colour perception exhibits 80% blue and 20% green. If the value of the parameter is e.g. $H = 220$, this value means that the colour perception exhibits 20% blue and 80% green.

In the first step, the CIECAM02 model calculates an adaptation factor to the prevailing luminance level (the mean luminance of the adaptive field, see Figure 2.9). Then the tristimulus values X, Y, and Z are converted into receptor signals (into the signals of the L-cones, M-cones, and S-cones of the retina, see Figure 2.4). This step is followed by the modelling of chromatic adaptation with a chromatic adaptation algorithm, the so-called CAT02 (chromatic adaptation transformation). The adapted receptor signals are then compressed. Figure 2.10 illustrates this so-called *signal compression*, which is generally important in the human visual system.

The log–log diagram of Figure 2.10 shows the *'S-shaped'* non-linearity of the signal compression. For low receptor signals, noise is present in the receptors, while for high-adapted receptor signals, the compressed signal converges to a threshold. In the CIECAM02 model – after the compression phase – the numerical characteristics for the colour features (C, J, H, Q, s, and M) are calculated as the output of the model.

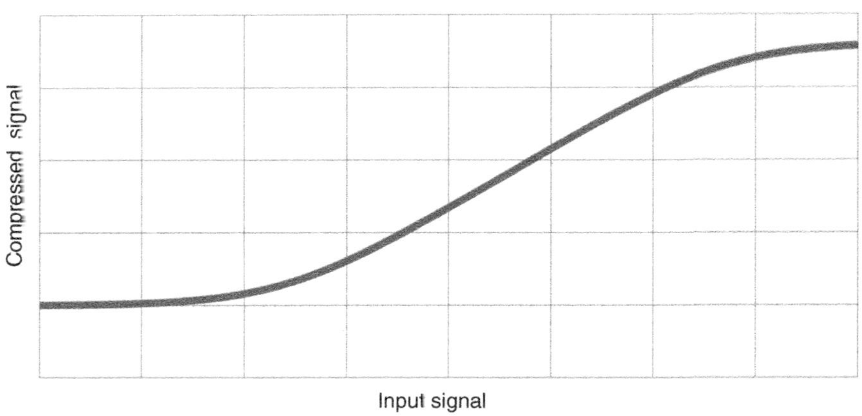

Figure 2.10 Compression of the adapted receptor signals in the CIECAM02 colour appearance model (log–log diagram). Source: TU Darmstadt.

2.2.5 CAM02-UCS Colour Space

The CAM02-UCS colour space is defined on the basis of the CIECAM02 model in two steps as follows: [20].

Step 1: The CIECAM02 parameters lightness (J) and colourfulness (M) are transformed according to Eqs. (2.6) and (2.7). The CIECAM02 chromaticity angle h *is* not transformed.

$$J' = \frac{1.7J}{0.007J + 1} \tag{2.6}$$

$$M' = \frac{1}{0.0228} \ln(1 + 0.0228M) \tag{2.7}$$

Step 2: The variables a' and b' are defined according to Eq. (2.8).

$$a' = M' \cos(h)$$
$$b' = M' \sin(h) \tag{2.8}$$

The variables a' and b' can be interpreted as new correlates of the perceived red–green component and yellow–blue component (compare with Figure 6.29), although their significance is rather that they form two orthogonal axes of a colour space in which the perceived colour differences are quantified by the *Euclidean* distance capable of accurately predicting the human perception of colour differences. Accordingly, to predict the perceived colour difference between *any two*-colour stimuli in the CAM02-UCS colour space, the differences J', a', and b' between the two colours are formed. The so-called CAM02-UCS colour difference $\Delta E'$ is calculated according to Eq. (2.9).

$$\Delta E' = \sqrt{(\Delta J')^2 + (\Delta a')^2 + (\Delta b')^2} \tag{2.9}$$

A visual experiment on colour difference perception with colour differences of all orders of magnitude (colour differences at the perception threshold, as well as small to large colour differences; $0.6 < E^*_{ab} < 50.3$), which occur e.g. in the practice of evaluating the colour fidelity property of common light sources, showed that the CAM02-UCS colour difference metric is capable of predicting all orders of magnitude of colour differences with reasonable accuracy [22]. The latter was also experimentally confirmed and published in the laboratory of the authors of this book [23].

2.3 Basics of the Non-visual Aspects

2.3.1 Melatonin Suppression at Night

For *Human-Centric Lighting*, both the mechanisms of visual perception (perception of light, colour, geometry, movement, and form) and the aspects of non-visual signal processing are important [24]. As mentioned in Section 2.1, the visual and non-visual aspects must always be considered simultaneously. An important result of the contribution of *non-visual* signal processing is the so-called *circadian effect* of light: the effect of electromagnetic radiation on the so-called *circadian clock* of the human organism. This circadian clock organises the temporal coordination of human biological functions in the course of the day, in the so-called *circadian rhythm* [5].

All human organ functions run in this circadian rhythm. Human performance during the course of the day is determined by the various maxima and minima of the individual cell-biologically controlled functions [25]. The reason for the circadian rhythm is that – in a certain time interval in the course of the day, where daylight is available under natural conditions, the secretion of the sleep hormone melatonin is suppressed by light and that of the hormone cortisol is promoted, whereby the performance of the human being increases.

For the practice of *Human-Centric Lighting*, the question now is to what extent the normal circadian rhythm and thus the health, satisfaction, and work performance of the human users of a lighting installation can be promoted by changing the local and spectral radiation distribution of the lighting. In this way, during the day, new lighting concepts come into play if daylight is *lacking*, and the circadian rhythm must be *supported* by artificial light.

Accordingly, during the day, lighting concepts are particularly important that, in addition to supporting the circadian rhythm (e.g. in northern countries in winter), also promote colour quality, visual performance or other biological, or emotional/psychological aspects, see Chapters 3, 9, and 12. In the evening or during the night, however, artificial light has an increased circadian effect, e.g. during shift work. The question is, again, how the circadian rhythm can be promoted with artificial light and how the emission spectrum, the timing, the exposure duration, and the light quantity of the light sources, at any time during the night, can be optimised for optimal circadian effect. Accordingly, Figure 2.11 [5] shows experimental data from literature [26–28] on the effect of the wavelengths of a light source on the *nocturnal* suppression of melatonin levels.

Figure 2.11 shows – when the subject is stimulated with narrowband light (according to Brainard et al. [26] and Thapan et al. [27]) of different wavelengths

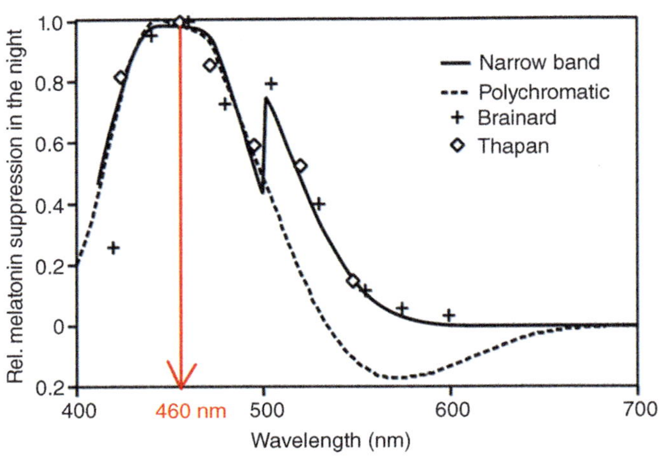

Figure 2.11 Experimental data according to Brainard et al. [26] and Thapan et al. [27] for melatonin suppression during the night. The subjects were stimulated during the night with narrow-band light ([26, 27]), the characteristic wavelength range of which is shown on the abscissa. The ordinate shows melatonin suppression as a response of the subjects. Source: Adapted from Rea and Figuiero [5].

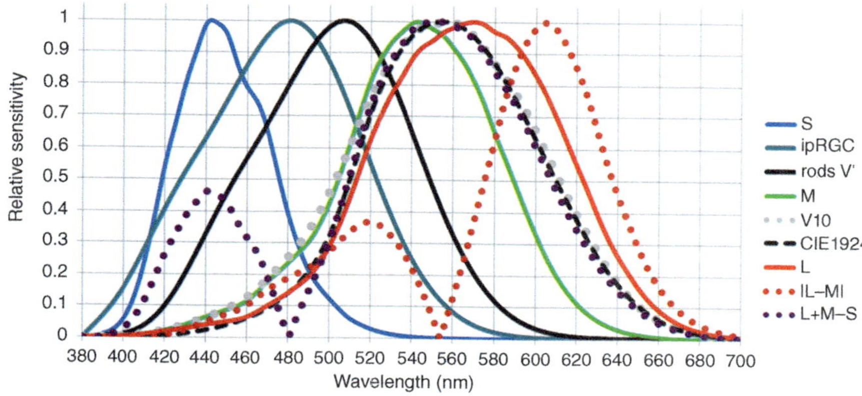

Figure 2.12 Spectral sensitivity of the photoreceptors or retinal mechanisms contributing to the non-visual effect melatonin suppression (modelled by the so-called circadian stimulus CS) in the Rea et al. model [5] (except |L − M|, this signal is shown only for the sake of completeness). Source: TU Darmstadt.

(see Section 9.4.3 and Table 9.7 in this book) – the effect of at least *two* retinal mechanisms, since *two* maxima appear in the visible wavelength range (380–780 nm, in the figure only the range 400–700 nm is shown) [5]. A *local maximum* of approx. 505 nm is also noticeable with narrowband colour stimuli. But this local maximum disappears in the response to a broadband (polychromatic) light stimulus due to a *spectrally opponent* mechanism (see Section 2.1) according to Rea et al. [24] (compare with the red dot curve in Figure 10.7 according to Gall [29]), see the dashed curve in Figure 2.11. The absolute maximum of melatonin suppression is at about 460 nm (see the red arrow in Figure 2.11) and not at 480 nm, where 480 nm corresponds to the maximum spectral sensitivity of *melanopsin*, see the dark green curve in Figure 2.12. As described in Section 2.1, *melanopsin* is the photopigment of ipRGCs, which are themselves photosensitive (see Figure 2.2) and contribute to *circadian clock regulation* [30].

2.3.2 Modelling Melatonin Suppression at Night with the Circadian Stimulus (CS) and the Melanopic Action Factor

In order to model melatonin suppression at night (Figure 2.11) by light, this section describes three parameters that are often used in today's practice: (i) the melanopic action factor; and (ii) the so-called Circadian Stimulus (CS), and (iii) mEDI- metric from CIE. These parameters characterise the effectiveness of a light stimulus for nocturnal melatonin suppression depending on its intensity and spectral composition. The CIE metric mEDI will be described in the Section 2.3.3 and in the Section 10.7 of this book.

The melanopic action factor (of visible radiation) $a_{\mathrm{mel,v}}$ is defined according to DIN SPEC 5031-100 [31] as the 'ratio of the melanopically effective radiation quantity to the photometrically effective radiation quantity evaluated with the sensitivity for day vision'. To calculate a melanopically effective radiant magnitude, the relative radiant power is weighted with the so-called $s_{\mathrm{mel}}(\lambda)$ function [31] according to Lucas et al. [2]. This is a so-called action spectrum, see the bluish-green, continuous curve in Figure 10.7 with a maximum at 490 nm. The action spectrum $s_{mel}(\lambda)$ is based on data from a 32-year reference observer [2, 31]. The DIN standard [31] contains correction factors for other age groups. The numerical method is explained in detail in Section 10.3.2, see also Eq. (10.2).

As already described in Section 2.3.1, the results of Figure 2.11 indicate that not only the above-mentioned $s_{mel}(\lambda)$ function is responsible for melatonin suppression during the night, but a *combination of several retinal mechanisms* (cf. with Figure 2.2), which can be modelled by the so-called *CS* of the Rea and Figuiero *model* [5] in 2016 (so-called CS 2016-version). For this purpose, Figure 2.12 shows in comparison the spectral sensitivity of the receptors of the retinal mechanisms that contribute to melatonin suppression in this model [5].

The following mechanisms can be seen in Figure 2.12:

1. S-cones.
2. intrinsic photosensitive ganglion cells (ipRGCs);
3. rods whose spectral sensitivity is modelled by the $V'(\lambda)$ function.
4. the luminance channel L + M, which, because of the greater extent of a *circadian active* light stimulus (greater than 2°), is modelled by the $V_{10}(\lambda)$ function that applies to the 10° observer; as well as
5. the opponent channels 1–2 (|L − M| or |L + M − S|). The retinal mechanism |L − M| (which forms the difference between the L- and M-cone signals) is shown here only for completeness, this mechanism is not used in the Rea et al. model.

Figure 2.13 illustrates the calculation procedure for the *CS* in the Rea et al. model. The calculation steps are shown in Figure 2.13 are quantified in detail in Section 10.3.2, see Eqs. (10.3) and (10.4).

The input variables of the Rea and Figuiero model [5] are the vertical illuminance at the eye and the relative spectral irradiance of the colour stimulus, from

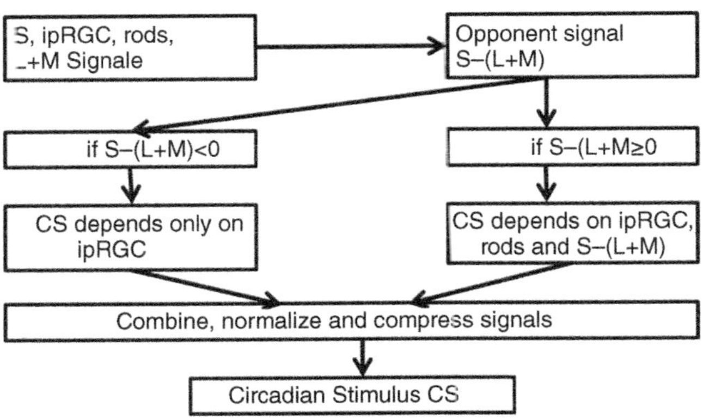

Figure 2.13 Calculation procedure of the Rea and Figuiero model [5]. Input variables of the model: vertical illuminance at the eye and relative spectral radiance of the colour stimulus, from which the S-cone, ipRGC, rod and L + M signals are calculated. Result of the model: Circadian Stimulus (CS), which models the non-visual effect of melatonin suppression of a colour stimulus (see also Section 10.3.2). Source: TU Darmstadt.

which – according to Figure 2.13 – the signals of the S-cones, the ipRGCs, the rods and L + M are calculated in the first step. From these, the spectrally opponent signal S–(L + M) is formed. After that, there is a branching of the calculation procedure according to

- whether the value of S–(L + M) is negative (CS depends only on the ipRGC signal) or
- this value is positive or equal to 0 (CS depends on the ipRGC signal, the rod signal, and S–(L + M)).

After this step, the signals are combined, normalised, and compressed to return the result of the model, the *CS*. The procedure is described quantitatively in Section 10.3.2 (Eqs. (10.3) and (10.4)). In this chapter and in more detail in Section 10.3.2 of Chapter 10, the CS-metric from version 2016 is described. *In the time range between* 2020 *and* 2022, *Rea et al. improved this model based on a number of experimental data on melatonin suppression at the night-time and added the factors for the exposure time (one or three hours) and of the viewing field size into the model.*

To illustrate the meaning of CS (version 2016–2018), Figure 2.14 shows those vertical illuminance values at the observer's eye that is necessary to achieve a defined CS level (0.3; 0.4, or 0.5). These values were calculated using 553 light source spectra of different CCTs that are commonly used today. For a constant value of CS (e.g. CS = 0.4), the higher the colour temperature of the light source is set, the lower the illuminances are required for this.

The CS parameter was originally based on experiments with a fixed exposure time of one hour. A modification of the CS formula that introduces a dependence on the exposure time is shown in Section 9.4.4 of Chapter 9 (Eq. (9.8)).

2.3.3 Spectral Sensitivity Functions According to the CIE

As an alternative to the Rea and Figuiero model [5], the CIE publication [32], which is based on the fundamental research by Lucas et al. [2], works with the (non-combined) signals of the five receptor channels (L, M, S, rods, and ipRGC) calculated using their action spectra so-called '*α-opic spectral weighting functions*' or $s_\alpha(\lambda)$-functions (see Figure 10.7). The CIE method [32] is described and discussed in Section 10.3.2 (in comparison to the Rea et al. model). Two important parameters, which will be used later in this book, are defined here.

The first parameter is the so-called *α-opic irradiance* E_α (in W m^{-2}; also called 'effective photobiological irradiance'), see Eq. (2.10).

$$E_\alpha = \int_{380\ \mathrm{nm}}^{780\ \mathrm{nm}} E_\mathrm{e}(\lambda) \cdot s_\alpha(\lambda) \cdot \mathrm{d}\lambda \tag{2.10}$$

The symbols in Eq. (2.10) have the following meaning:

- E_α: α-opic irradiance.
- $E_\mathrm{e}(\lambda)$: spectral irradiance.

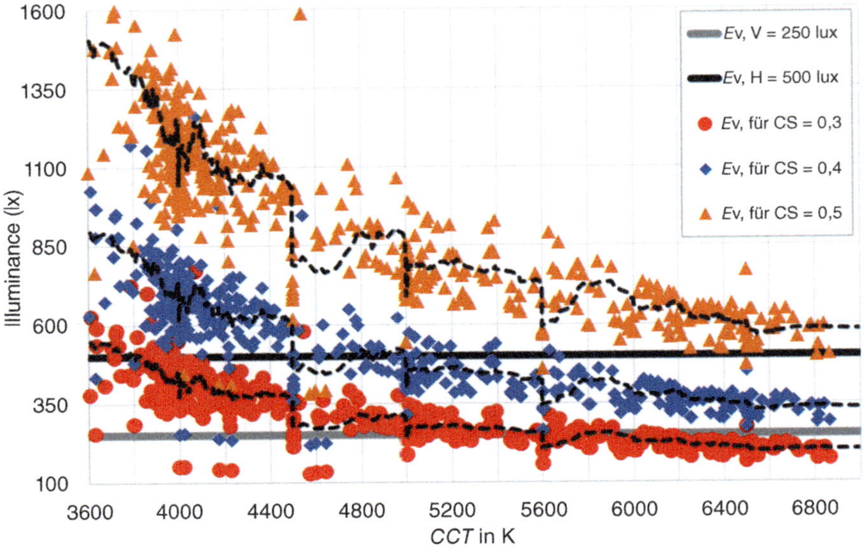

Figure 2.14 Vertical illuminance values (ordinate, in lx) required to achieve a defined CS level (0.3; 0.4 or 0.5), calculated for a set of nowadays common light source spectra of different CCT (abscissa, in *K*). The standard-compliant level of horizontal illuminance (500 lx) and the corresponding vertical illuminance (250 lx) (in a general lighting situation) are marked with horizontal black and grey lines respectively. Source: TU Darmstadt.

- $s_\alpha(\lambda)$: α-opic action spectrum (L, M, S, rods, or ipRGC, see Figure 10.7) as spectral weighting function for the spectral irradiance.

The second parameter is the so-called α-opic equivalent daylight (D65) illuminance $E_{v,\alpha}{}^{D65}$, see Eq. (2.11), which reflects the following definition: '*illuminance, produced by radiation conforming to standard daylight (D65) that provides an equal α-opic irradiance as the test source*'. It is measured in lux (lx) units and is also called '*α-opic EDI*', in case of the ipRGC channel, '*melanopic EDI*' (see Table 2.1) or briefly '*mEDI*'.

$$E_{v,\alpha}{}^{D65} = E_\alpha / K_{\alpha,v}{}^{D65} \tag{2.11}$$

The symbols in Eq. (2.11) have the following meaning:

- E_α: α-opic irradiance;
- $K_{\alpha,v}{}^{D65}$: five different constants for the five α-opic receptor channels (L, M, S, rods or ipRGC, see Table 2.1).

When quantifying the circadian effect of different light source spectra, the parameter $E_{v,mel}{}^{D65}$ (i.e. α = '*mel*' in Eq. (2.11)), which describes the melanopic effect, i.e. the effect of the ipRGC channel, plays an important role in lighting practice. To illustrate the significance of this parameter, Figure 2.15 shows the vertical illuminance values required to achieve a fixed $E_{v,mel}{}^{D65}$ level (240, 500, or 750 lx). These values were calculated for a set of 553 of nowadays common light source spectra at different CCTs.

Table 2.1 Constants for the five α-opic receptor channels in (mW lm^{-1}) units for Eq. (2.11).

Receptor channel	L	M	S	Chopsticks	ipRGC
Name [32]	L-cone-opic	M-cone-opic	S-cone-opic	Rhodopic	Melanopic
$K_{\alpha,v}^{D65}$	0.8173	1.4558	1.6289	1.4497	1.3262

Name: Name of the receptor channel according to the CIE publication [32].
Source: Adapted from CIE [32].

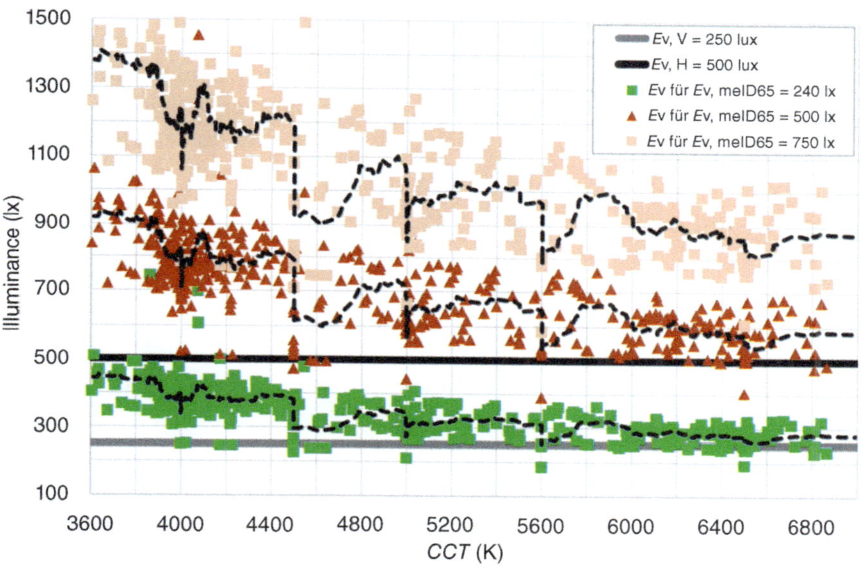

Figure 2.15 Vertical illuminance values (ordinate, in lx) required to achieve a fixed $E_{v,mel}^{D65}$ level (240, 500, or 750 lx) calculated for a set of 553 light source spectra of different CCT commonly used today (abscissa, in *K*). The standard-compliant level of horizontal illuminance (500 lx) and the corresponding vertical illuminance (250 lx) (in a general lighting situation) are marked with horizontal black and grey lines respectively. Source: TU Darmstadt.

2.3.4 Correlations Among Circadian Stimulus CS, Melanopic Illuminance, and D65-Equivalent Illuminance

The correlations among the following circadian effective metrics were computed:

- CS according to M. Rea [5];
- Melanopic illuminance $E_{mel} = a_{mel,v} \cdot E_v$ (see Section 2.3.2)
- Melanopic equivalent daylight (D65) illuminance (mEDI) $E_{v,mel}^{D65}$ (see Eq. (2.11)).

Using the above spectral analysis of 553 light sources (daylight, fluorescent lamps, high-intensity discharge lamps, thermal radiators, and white LEDs), the authors of this book determined mathematical correlations between the D65-equivalent illuminance $E_{v,mel}^{D65}$ to CS and the melanopic illuminance E_{mel} to CS, which are shown in Figure 12.8. The Eqs. (2.12) and (2.13) quantify these correlations with good correlation ($r^2 = 0.82$).

$$E_{\mathrm{mel}} = 1135.1 \cdot \mathrm{CS} - 41.535 \qquad (r^2 = 0.82) \tag{2.12}$$

$$E_{\mathrm{v,mel}}{}^{\mathrm{D65}} = 1022.2 \cdot \mathrm{CS} - 34.102 \qquad (r^2 = 0.82) \tag{2.13}$$

2.3.5 Recommendations of Necessary Melanopic EDI (mEDI) Levels for Optimum Sleep and Daytime Environments and Summary of this Chapter

The quantity melanopic EDI (mEDI, Eq. (2.11), Table 2.1) was used in [33] to derive lighting recommendations with appropriate mEDI ranges for optimum sleep and daytime environments. The findings of this article are summarised in Figure 2.16.

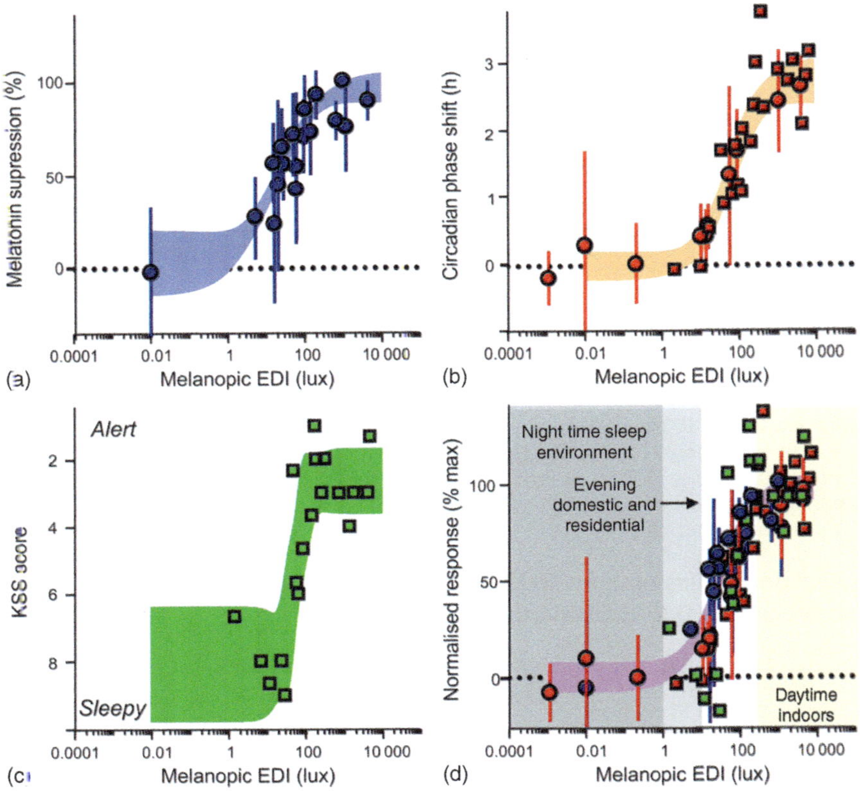

Figure 2.16 Recommendations of necessary melanopic EDI (mEDI) levels for optimum sleep environments and daytime environments [33], see text for explanations. Group data (round symbols) are presented as mean ± SD values; otherwise, data for individual subjects are presented (square symbols). Shading represents the 95% confidence limits of an unconstrained 4-parameter sigmoid fit to the data. Source: Brown et al. [33]/PLOS/CC BY 4.0.

Figure 2.16 shows three important non-visual responses to light, melatonin suppression (a), circadian phase (b), and subjective alerting responses (c) derived from laboratory studies in humans without the use of pupil dilators investigating the impact of long exposures (>2 hours) to primarily broadband light sources, as a function of melanopic EDI (mEDI) [33]. These responses of humans to light exposure characterise circadian behaviour, which plays a very important role in Human Centric (Integrative) Lighting. These three characteristics are described in detail in Chapter 3. Here, only their overall dependence on *mEDI* will be described to emphasize the significance of this quantity.

In Figure 2.16, for comparison across different response types, data sets from (a)–(c) were normalised relative to the range of the curve fit for that response type and presented together (D). Shaded areas in (D) reflect the consensus recommendations of the authors of [33] for sleep, evening, and indoor daytime environments. *These recommendations were intended to provide realistic targets for lighting technology that minimise inappropriate non-visual responses in the sleep environment (mEDI* < 1 lux*) and reduce these so far as is practically possible for pre-sleep (three hours before habitual sleep; mEDI* < 10 lux*), while maximising relevant effects across the intervening daytime hours (mEDI* > 250 lux*)*. The non-shaded region indicates the range of melanopic EDI that should, where possible, be avoided during evening and night-time and is also considered suboptimal for daytime environments. This non-shaded region in Figure 2.16(D) is thus not suitable at all, neither for relaxation nor for activation.

In summary, the objectives of Chapter 2 were to describe the basic features of eye physiology, colorimetry, and non-visual effects, their mathematical correlations, and the metrics or parameters used to quantify them. In Chapter 3 and the following chapters, the non-visual mechanisms, the psychological aspects, the ways of brain signal processing, as well as consequences, and practical hints for *Human-Centric Lighting* are presented.

References

1 Khanh, T.Q., Bodrogi, P., and Vinh, T.Q. (2017). *Color Quality of Semiconductor and Conventional Light Sources*, 1e. Weinheim: Wiley-VCH Verlag GmbH.

2 Lucas, R.J., Peirson, S.N., Berson, D.M. et al. (2014). Measuring and using light in the melanopsin age. *Trends Neurosci.* 37: 1–9.

3 Østerberg, G.A. (1935). Topography of the layer of rods and cones in the human retina. *Acta Ophthalmol.* 6: 1–102.

4 Gegenfurtner, K.R. and Sharpe, L.T. (ed.) (1999). *Color Vision: From Genes to Perception*. Cambridge University Press.

5 Rea, M. and Figuiero, M.G. (2016). Light as a circadian stimulus for architectural lighting. *Light. Res. Technol.* 50 (4): 497–510.

6 Stockman, A., Sharpe, L.T., and Fach, C.C. (1999). The spectral sensitivity of the human short-wavelength cones. *Vis. Res.* 39: 2901–2927.

7 Stockman, A. and Sharpe, L.T. (2000). Spectral sensitivities of the middle- and long-wavelength sensitive cones derived from measurements in observers of known genotype. *Vis. Res.* 40: 1711–1737.

8 *Web Database of the Color & Vision Research Laboratory*. London: University College London, Institute of Ophthalmology www.cvrl.org (8 January 2022).

9 CIE (2005). CIE 10 degree photopic photometric observer. 165: 2005. Publ. CIE.

10 CIE: Principales decisions (6e Session, 1924) (1926). CIE Sixieme Session, Geneve, Juillet 1924, Recueil des Travaux et Compte Rendu de Seances, S. 67–69. Cambridge: University Press.

11 Khanh, T.Q., Bodrogi, P., Vinh, T.Q., and Brückner, S. (2013). *Farbwiedergabe von konventionellen und Halbleiter-Lichtquellen: Theorie, Bewertung, Praxis (Colour Rendering of Conventional and Semiconductor Light Sources: Theory, Evaluation, Practice)*. München: Pflaum Verlag.

12 Schanda, J. (ed.) (2007). *Colorimetry: Understanding the CIE System*. Wiley-Interscience.

13 CIE (2004). Colorimetry, 3. Ed. Publ. CIE 015: 2004.

14 CIE (2006). Fundamental chromaticity diagram with physiological axes – Part I, 170, 171: 2006. Publ. CIE.

15 Judd, D.B., MacAdam, D.L., and Wyszecki, G. (1964). Spectral distribution of typical daylight as a function of correlated color temperature. *J. Opt. Soc. Am.* 54 (8): 1031–1040.

16 CIE (1970). ILV: International Lighting Vocabulary, 3. Ed. CIE –International Commission on Illumination.

17 Fairchild, M.D. (1997). *Color Appearance Models, The Wiley-IS&T Series in Imaging Science and Technology*, 2e. Wiley.

18 Hunt, R.W.G. and Pointer, M.R. (2011). *Measuring Colour, The Wiley-IS&T Series in Imaging Science and Technology*, 4e. Wiley.

19 CIE (1971). CIE Publication No. 15 (E.1.3.1.) 1971: Colorimetry, (updated, see [12]).

20 Luo, M.R., Cui, G., and Li, C. (2006). Uniform color spaces based on CIECAM02 color appearance model. *Color Res. Appl.* 31: 320–330.

21 CIE (2004). A Colour Appearance Model for Colour Management Systems: CIECAM02, 159: 2004. CIE Publ.

22 Wang, H., Cui, G., Luo, M.R., and Xu, H. (2012). Evaluation of colour-difference formulae for different colour-difference magnitudes. *Color Res. Appl.* 37: 316–325.

23 Bodrogi, P., Brückner, S., and Khanh, T.Q. (2011). Ordinal scale based description of colour rendering. *Color Res. Appl.* 36: 272–285.

24 Rea, M.S. and Figueiro, M.G. (2018). Light as a circadian stimulus for architectural lighting. *Light. Res. Technol.* 50 (4): 497–510.

25 Licht.de (ZVEI) (2010). Wirkung des Lichts auf den Menschen (Effect of Light on Human). licht.wissen 19.

26 Brainard, G.C., Hanifin, J.P., Greeson, J.M. et al. (2001). Action spectrum for melatonin regulation in humans: evidence for a novel circadian photoreceptor. *J. Neurosci.* 21: 6405–6412.

27 Thapan, K., Arendt, J., and Skene, D.J. (2001). An action spectrum for melatonin suppression: evidence for a novel non-rod, non-cone photoreceptor system in humans. *J. Physiol.* 535: 261–267.

28 Rea, M.S., Figueiro, M.G., Bullough, J.D., and Bierman, A. (2005). A model of phototransduction by the human circadian system. *Brain Res. Rev.* 50: 213–228.

29 Gall, D. (2004). *Die Messung Circadianer Strahlungsgrößen (Measurement of Circadian Radiation Quantities)*. Ilmenau: TU Ilmenau.

30 Hattar, S., Liao, H.W., Takao, M. et al. (2002). Melanopsin-containing retinal ganglion cells: architecture, projections, and intrinsic photosensitivity. *Science* 295: 1065–1070.

31 DIN SPEC 5031-100 (2015). *Strahlungsphysik im optischen Bereich und Lichttechnik – Teil 100: Über das Auge vermittelte, melanopische Wirkung des Lichts auf den Menschen – Größen, Formelzeichen und Wirkungsspektren (Radiation Physics in the Optical Domain and Lighting Technology - Part 100: Melanopic Effect of Light on Humans Mediated by the Eye - Quantities, Symbols and Action Spectra)*. Beuth Verlag.

32 CIE (2018). CIE system for metrology of optical radiation for ipRGC-influenced responses to light. *S* 026: 2018. https://doi.org/10.25039/S026.2018.

33 Brown, T.M., Brainard, G.C., Cajochen, C. et al. (2022). Recommendations for daytime, evening, and night-time indoor light exposure to best support physiology, sleep, and wakefulness in healthy adults. *PLoS Biol.* 20 (3): e3001571. https://doi.org/10.1371/journal.pbio.3001571.

3

Basic Principles of Human-Centric Lighting and Integrative Lighting

In the previous Chapter 2, the basic knowledge about the photoreceptors on the retina as well as the formation of the photometric, colorimetric, and non-visual parameters were presented. In this chapter, the effects of light and optical radiation are first illustrated in a clear system-technical and bio-cybernetic form. This begins with the input variables as characterising parameters of the light entering the eye in order to characterise the effects of light on humans as a '*system*'.

Subsequently, the optical and signal processing of the ocular apparatus and the brain, the signal networking as well as the significance of these signal communications in the brain regions for the interpretation of the effect of light on human beings are described. This description is based on new literature, which is referred to in the appropriate places.

The output variables – both in subjective and objectively measurable form – embody the various forms of expression in the acute and long-term effects of light on humans and are also described in this chapter together with sleep behaviour. These descriptions of the three system blocks '*input variables – brain as a system – output variables*' then lead at the end of the chapter to the consideration of the concept of Human-Centric Lighting (HCL) as an integrative lighting process. This HCL concept is then broken down into individual, important, and necessary topics in Chapters 4–11 in order to then bundle their important and relevant results together in Chapter 12 with the focus on '*Lighting design as a process*'.

3.1 Basic Questions, General Aspects

Through previous research in the field of biomedicine, neuropsychology, eye physiology, and lighting technology, the relevant processes are known today and are schematically shown in Figure 3.1.

While the pure input variables of physical nature are clearly defined in Figure 3.1, the mechanisms and signal processing in the ocular apparatus and in the body, including the human brain, continues to be the subject of research, with great progress having been made in the last 20 years. While in the age of industrialisation in factories, work performance or error rate under various influences of lighting

Human Centric Integrative Lighting: Technology, Perception, Non-Visual Effects, First Edition.
Tran Quoc Khanh, Peter Bodrogi, and Trinh Quang Vinh.

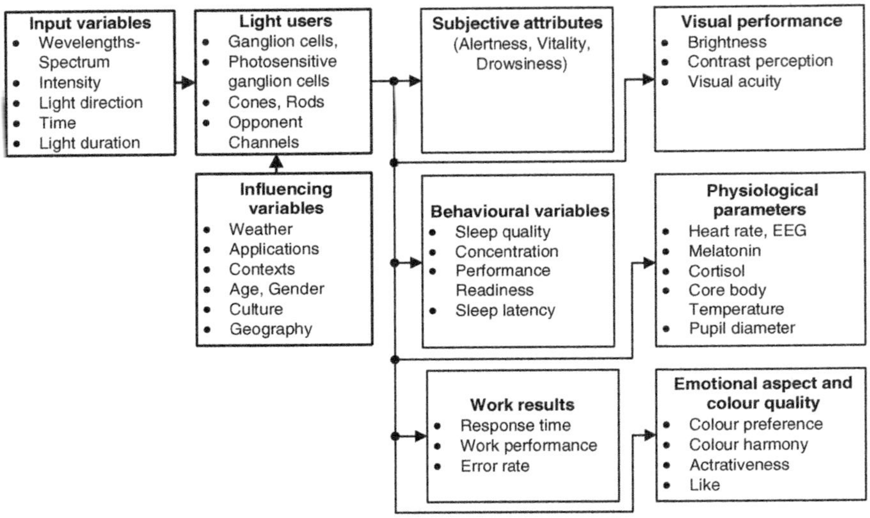

Figure 3.1 Overview of input and influencing variables and output variables in the comprehensive consideration of the effects of light on humans. Source: TU Darmstadt.

effects (light levels, glare, and homogeneity of lighting) were quantified until the 1980s, other output variables such as subjective and emotional parameters have become more important and intensively studied in the information society era in the last 30 years. Physiological parameters such as hormone release (melatonin, cortisol), EEG, pupil diameter, ECG/PEP, and colour quality characteristics such as colour preference and colour naturalness determine today's concepts and research work in the various laboratories worldwide.

There are some important questions in neuroscience, biomedicine, and lighting technology that needs to be answered:

1. How are the physical optical inputs processed in the brain to generate visual and non-visual effects?
2. Do the visual and non-visual processes have factually different processing centres or do they overlap? Also, are there really '*image forming, IF* and *non-image forming, NIF*' processes that run *separately,* as was believed until recently?
3. Do different subgroups of output variables (subgroups of behavioural variables, subjective attributes, visual performance, emotional effects, and subgroups of physiological parameters) have a correlation to each other, i.e. do they behave in the same direction with the same tendency for a certain light intervention or do they partly run counter to each other?

Such questions have practical significance in order to solve the problem, among other things, at which spectrum with which intensity and light duration as well as at which time of day a light stimulus can positively influence both the subjective and the objective states of well-being of light users. Are there also certain light characteristics for which opposing negative effects can be expected?

3.2 Input Variables – A Systematic Approach

Optical signal processing starts with the light source (daylight, sunlight, electric light such as LED, or a fluorescent lamp). The light source (a luminaire, a lamp, or a group of many luminaires) with a specific spectral distribution illuminates the objects in an observer's field of view with their spectral reflectance $R(\lambda)$, which reflects the incoming radiation of the light source towards the observer's eye according to Eq. (3.1), see Figure 3.2.

$$\varphi_e(\lambda) = \varphi_{LQ}(\lambda) \cdot R(\lambda) \tag{3.1}$$

- $\varphi_e(\lambda)$: spectral radiant flux reflected by the object (in W)
- $\varphi_{LQ}(\lambda)$: spectral radiant flux of the light source (in W) arriving at the object
- $R(\lambda)$: spectral reflectance of the specific object

The optical radiation entering the eye (see Figure 3.2) evokes a brightness perception, a colour perception, and non-visual light effects on the retina in the eye and later in further brain processing. For an initial, highly simplified modelling of human brightness perception (Chapter 5 shows advanced models of perceived brightness), the parameter *luminance* according to Eq. (3.2) can be used.

$$L_V = 683\ \mathrm{lm\,W^{-2}} \cdot \int V(\lambda) \cdot L_e(\lambda) \cdot \mathrm{d}\lambda \tag{3.2}$$

- $V(\lambda)$: the spectral luminous efficiency function for day-time vision;
- $L_e(\lambda)$: the spectral radiance of the object or light source.

The colorimetric metrics and attributes of colour perception described in Section 2.2 are available for colour perception analysis. Sections 2.3, 10.3.2, and 10.3.3 also discuss the metrics of non-visual lighting effects in detail. Figure 5.1 will show the interactions and combination of the lighting and colorimetric metrics for the *joint formation of* the visual perception and psychological–emotional evaluation of an indoor scene, see Figure 3.3.

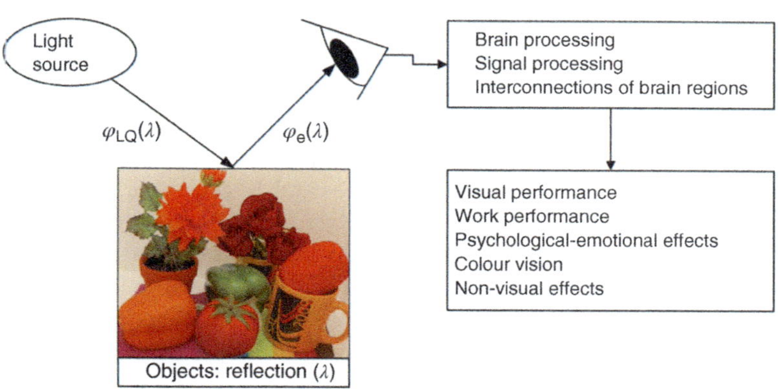

Figure 3.2 Simplified diagram of signal transmission from the light source to the brain. Source: Technische Universität Darmstadt.

Figure 3.3 Example of an illuminated interior scene equipped with objects. Source: Technische Universität Darmstadt.

In the interior scene in Figure 3.3, there are various objects with different colours and colour contrasts in different geometric shapes on the table and on the wall. The table and the walls are illuminated differently by the luminaires with indirect luminous flux components upwards to the ceiling as well as directly downwards to the table and the wall with different brightness levels. All these lighting features are perceived by the observers in the room, with the different zones in the visual field being mapped onto different areas on the retina.

For lighting engineers and occupational physicians working in practice, the photometric, colorimetric, and non-visual parameters are the input variables of the '*human as a system*' process. Thus, many textbooks and research articles also use photometric and colorimetric parameters such as illuminance (in lx) or colour temperature CCT (in Kelvin) as input variables for the consideration of research results. Strictly speaking, the real input variables are the duration of exposure, the time of exposure to light, the location of light incidence on the retina, and the spectral radiance of the radiation entering the eye. The illuminance at the eye, the D65 – equivalent illuminance, the colour temperature or chromaticity, the Circadian Stimulus CS, etc. are already processed with the spectral and spatial sensitivity of the brain apparatus, whereby different eye and brain stages are involved.

In the case of illuminance, the signals of the L- and M-cones are taken into account. In contrast to electrical engineering or physics, the input variables used in lighting technology and human medicine research are already a mixture of the physical parameters (such as a spectral radiance in the visual field) and the system response of the brain (such as the spectral luminous efficiency function $V(\lambda)$). This makes the analysis of research results and the interpretation of practical observations more complicated.

The spectra of the natural and electric light sources whose light people encounter with their ocular apparatus everyday are very diverse and are therefore described

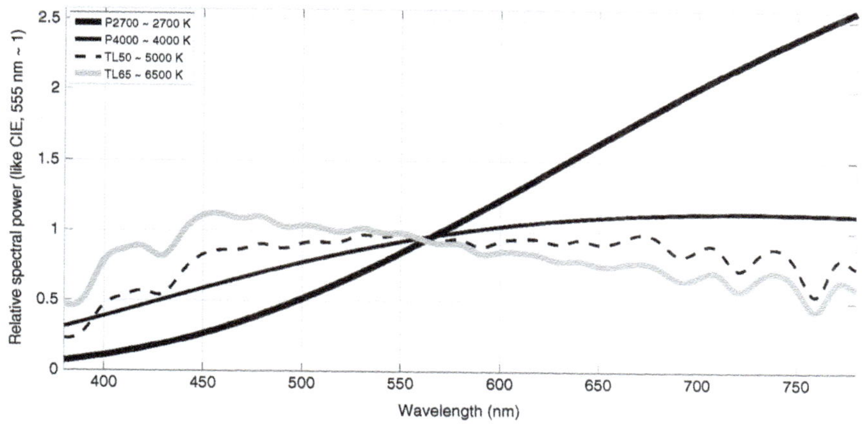

Figure 3.4 Spectra of the daylight phases D50 and D65 as well as the thermal radiators with a colour temperature of 2700 and 4000 K (Planck's radiators P2700 and P4000). Source: TU Darmstadt.

in this book at various points in relation to context. Figure 3.4 shows the spectra of the daylight phases D50 and D65 defined by the International Commission on Illumination (CIE) and the thermal radiators, calculated according to Planck's law, with temperatures of 2700 and 4000 K (Planck's radiators P2700 and P4000).

Figure 3.5 shows the relative spectra of modern semiconductor LEDs at different peak wavelengths.

Such semiconductor LEDs (see Figure 3.5) with electromagnetic radiation in the visible range from about 380 nm (violet LEDs) to 730 nm (deep red LEDs) in

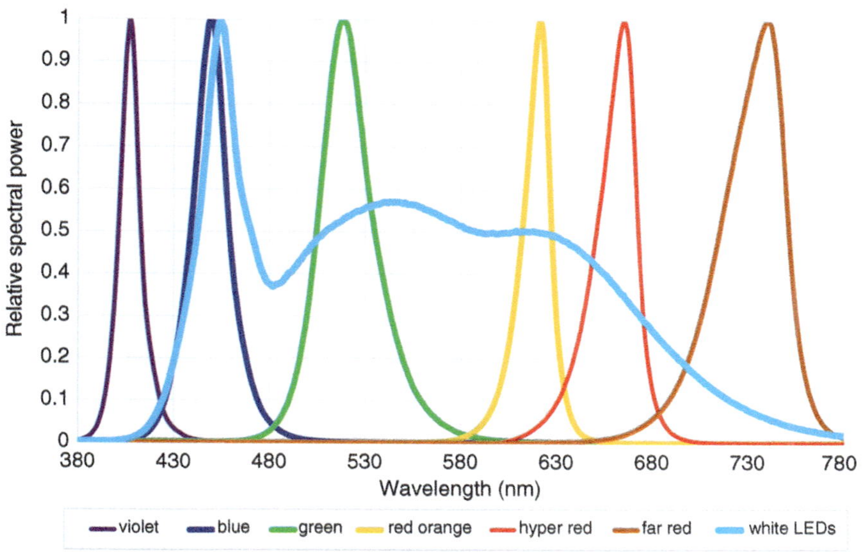

Figure 3.5 Relative spectra of current semiconductor LEDs of different peak wavelengths and a cold white LED type. Source: TU Darmstadt.

very good physical and photometric quality are now available for research and development purposes. The spectra of modern white phosphor-converted LEDs are discussed in detail in Section 11.4.2.2, with Figures 11.18–11.28. The spectra of the daylight phases measured by the authors on a summer day in Darmstadt are shown in Figure 11.29. When looking at all spectra of different light source types, it can be seen that the spectral distributions of these light sources are very different in different wavelength ranges and that these differences in radiation absorption in different receptor types lead to different signal sizes in the five channels for the five receptor types (see Chapter 2).

3.3 Brain Processing for the Formation of Subjective and Objective Behavioural Variables

The spectral radiance, which either comes indirectly from the objects in the observers' field of vision or is emitted directly from the light sources to the eye, is imaged onto the retina through the cornea and the eye lens (see Chapter 2). The spectral absorption processes in the eye lens apparatus and how the pupil diameter changes depending on the age of the observer are described in detail in Section 12.6.1.

According to the illustrations in Section 2.1, the photons of the respective wavelengths of the incident radiation are absorbed in the LMS cones, in the rods, and in the intrinsically photosensitive ganglion cells (ipRGCs) cells and first converted into photochemical signals and later into action impulses. The formation of the signals at the output of the conversion channels for these five receptors has already been described mathematically in Section 2.3.3. These signals are then transmitted through the various nerve tracts to the later brain regions, processed by compression, and by interconnection of the signals of different receptor, types to finally generate photometric, colorimetric, and non-visual parameters. The formation of the photometric quantities and the colorimetric quantities are worked out in detail in Sections 2.2.1 and 2.2.2 mathematically and context-specifically. In Sections 3.3.1 and 3.3.2, signal processing and the networking of different brain regions are described in more detail.

3.3.1 Visual Processing Systems

The process of visual signal processing begins with the absorption processes in the photoreceptors [1]. The cones and the rods absorb certain spectra, which can be determined by psychophysical methods or direct measurement procedures. The action spectrum of a photoreceptor describes the relationship between the irradiation spectrum and the output signal of the photoreceptor. This can be determined after isolation and spectral measurement of the individual cells (suction electrode method). Figure 2.4 shows the relative spectral sensitivities of the rods, the cones, and the ipRGC cells. As in any biological system, these functions differ from one another in their spectral sensitivity among different individuals: the curves shown in Figure 2.4 apply to the so-called '*normal observer*'. However, the most important

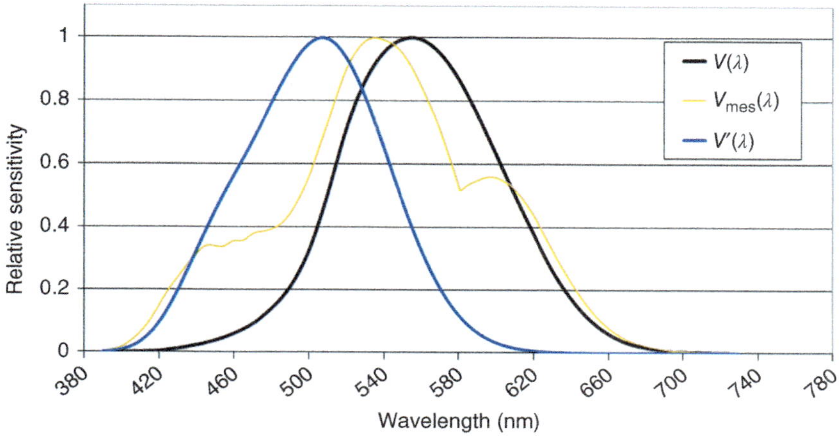

Figure 3.6 Spectral sensitivity curves for day vision, night vision, and one example for mesopic vision. Source: TU Darmstadt.

curve for lighting technology is still the spectral luminous efficiency function, $V(\lambda)$. It is composed of the LMS photoreceptor signals according to Eq. (3.3).

$$V(\lambda) = 0.68 \cdot L(\lambda) + 0.34 \cdot M(\lambda) - 0.02 \cdot S(\lambda) \tag{3.3}$$

The weighting of a spectrum of optical radiation with the luminous efficiency function $V(\lambda)$ is the basis for many photometric evaluation variables. In a first (rough) approximation, it describes the human perception of brightness in the photopic range (daytime vision) and can be seen in Figure 3.6. The other function, $V'(\lambda)$, describes roughly the brightness perception in the scotopic brightness range (night vision). Figure 3.6 also shows an example of a so-called *mesopic* visual sensitivity function (in the transitional range between day vision and night vision, yellow curve). Depending on the luminance level and visual task, very different sensitivity functions apply in the mesopic range, which would go beyond the scope of this book due to its complexity.

3.3.1.1 Horizontal Cells of Bipolar Cell Layer

The horizontal cells of the bipolar layer receive signals from many photoreceptors and send them to the synapses between the receptors and the bipolar cells. Due to their ability of lateral signal inhibition, they are essential for local adaptation and also for the contrast enhancement in the eye.

The bipolar cells (see Figure 3.7) are responsible for the transmission of stimuli from the photoreceptors to the ganglion cells.

At the same time, the bipolar cells convert the constant-voltage signal of the receptors into a frequency-modulated signal. They are essentially divided into two groups: the ON-bipolar cells, which are connected to rods and cones and invert the negative signal of the photoreceptors, and the OFF-bipolar cells, which are non-inverting and only connected to the cones. They are mainly used for sharp vision in the fovea. Furthermore, the OFF cells have no direct connection to the ganglion cells.

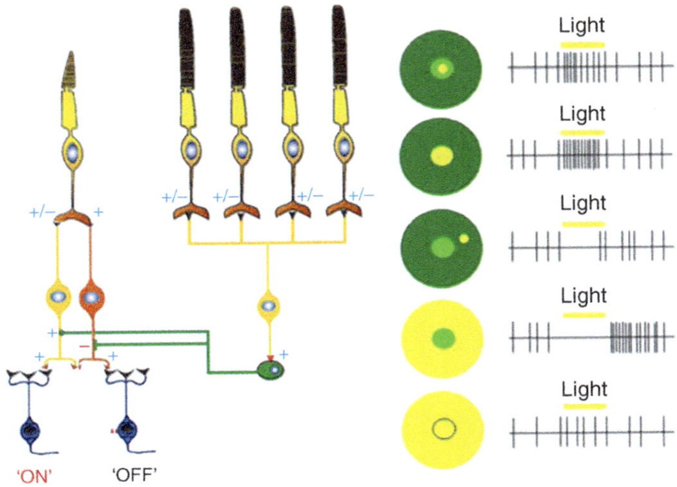

Figure 3.7 Interconnections in the bipolar layer (after [2] Hubel, 1989). Source: Reproduced with permission from Springer Nature Customer Service Center GmbH.

The amacrine cells form another lateral connection between the OFF-bipolar cells and several synapses between the ON-bipolar cells and the ganglion cells. In doing so, the amacrine cell optimises the formation of an ON signal in a double way: It forms synapses with ON-bipolar cells and passes on its excitation to the ON-bipolar cell. Through inhibitory synapses, the amacrine cell additionally suppresses the activity of OFF-bipolar cells. Through this combination of activation of ON-bipolar cells and inhibition of antagonistic OFF-bipolar cells, a useful ON signal can be generated even with a weak excitation of the rod bipolar cell (with little light).

3.3.1.2 Ganglion Layer

The ganglion layer consists of the normal ganglion cells, to which – depending on their location – one (in the fovea) or several photoreceptors (up to 130 in the periphery) are assigned. Due to the lateral interconnection of the retinal cells, each ganglion cell has a relatively large catchment area. The area encompassing all the retinal cells acting on an individual ganglion cell is called the *receptive field*. One can determine the extent of these receptive fields by registering the output signal of a ganglion cell and scanning the retinal cells point by point with a fine beam of light. Each point that triggers a reaction from the ganglion cell when illuminated is part of its receptive field.

In such investigations, receptive fields of variable diameters were found, on average 50 μm. The illumination of the inner area (close to the ganglion cell from which the signal is derived) results in activation of the ganglion cell (a so-called *ON response*). Outside this central ON-field, illumination produces an inhibition of the ganglion cell, because here the lateral inhibition outweighs the inhibition of the ganglion cell by its neighbouring cells.

The knowledge about the wiring of the normal ganglion cells and the ipRGCs to each other and to the receptors as well as the contributions of the cone and rod signals to visual and non-visual processing has grown strongly in recent years due to intensive research work in neurobiology and cell research. Figure 2.2 illustrates the interconnections of the LMS cone and rod signals to each other and to the bipolar cells, which in turn communicate with each other via the amacrine cells and transmit signals to the ipRGC cell. The ipRGC cells are themselves photosensitive and absorb photons in their melanopsin-containing pigments. The total signal generated by an ipRGC cell thus consists of the signals from the cones, rods, and self-generated signals. In contrast, the normal ganglion cells only contain signals from the cones and rods.

One of the most important results of recent neurobiological research was described in [3]. A new type of amacrine cell has been identified that receives excitatory input signals on the input side from ON-bipolar cells that are directly connected to S-cones. On the output side of these amacrine cells, inhibitory signals are transmitted to the ipRGC cell. These results are significant in that they show that the ipRGC cell signals always contain a component of S-cones that have a signal inhibitory effect. An increase in the signals of the S-cones leads to a reduction of the ipRGC-signals and an increase in the pupil's diameter [4, 5].

3.3.1.3 The Visual Pathway

The processing of visual information takes place separately via the different systems of the visual pathway. In the so-called P-system (parvocellular), information about the shape and colour of objects is transported and processed. In the so-called M-system (magnocellular), on the other hand, location and movement information is processed and passed on. The course of the visual pathways is shown in Figure 3.8.

Optic chiasm: The visual fields are sorted in the crossing of the two visual pathways (see Figure 3.8). The right visual fields of both eyes are directed to the right hemisphere of the brain, the left to the left. Above the chiasm, the suprachiasmatic nucleus (SCN) is situated. This nucleus processes a small part of the non-visual eye information, it forms a component of the internal clock and is responsible for circadian rhythms, which will be reported on later in this book. Specifically, the so-called saccades are also controlled here. Saccades are short, irregular movements of the eyes that occur unconsciously and scan the visual field.

Lateral geniculate nucleus (LGN): The LGN is part of the thalamus, which forms the largest part of the diencephalon and is made up of many closely interconnected nuclei. This is where all signals from sensory organs converge and are passed on to the cerebral cortex. The name '*gateway to consciousness*' refers to the thalamus guiding function for sensory information. The thalamus can regulate various signals to give priority to '*more important*' impressions. Visual impressions are filtered in the LGN.

Under the light microscope, six layers can be distinguished in the so-called *corpus geniculatum laterale (CGL)* of primates. Layers 2, 3, and 5 receive fibres from the ipsilateral eye and layers 1, 4, and 6 from the contralateral eye. Within these layers, neighbouring neurons receive signals from neighbouring retinal ganglion cells. Layers 5 and 6 belong to the *magnocellular* layers. They consist of large cells and

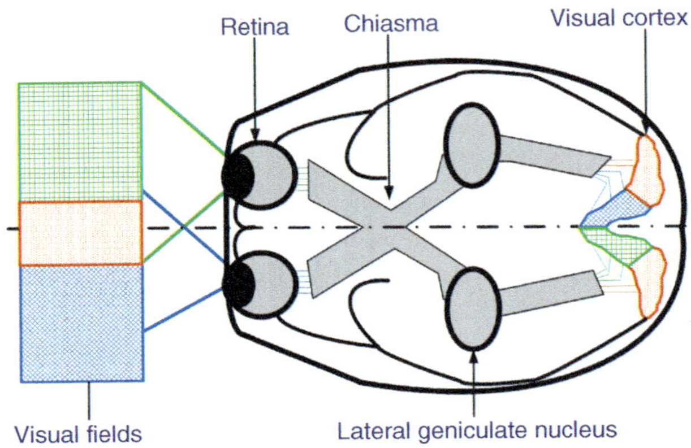

Figure 3.8 Visual pathways in the brain [1]. Source: Adapted from Khanh and Söllner [1].

receive their input signals from the so-called *parasol cells* of the retina. The latter cells have larger dendrite trees, and thus a lower spatial resolution, but a greater temporal resolution. The *parvocellular* layers 1–4 consist of smaller neurons and receive their input signals from so-called *midget cells* of the retina. These have smaller dendrite trees, a high spatial resolution, but a lower temporal resolution.

From the CGL, the signals are transmitted to the primary visual cortex in the occipital lobe via the broadly fanned-out visual signal distribution, partly via the so-called *capsula interna,* another node of large nerve tracts.

The visual cortex: The visual cortex here refers to the so-called *primary* visual cortex. It is divided into two pathways, the so-called *ventral* and the so-called *dorsal* pathway. In brain research, shape and object recognition as well as storage in long-term memory are attributed to the ventral pathway. The dorsal pathway recognises movements and locations; and coordinates the movements of the body. The visual cortex is divided into the visual cortices V1–V5 to which different functions are assigned.

3.3.1.4 Overall Network Structure of the Visual System: An Overview

If the optical radiation enters the human ocular apparatus with a certain intensity and spectrum, the visual processing pathway can be described schematically from today's research perspective as shown in Figure 3.9 [6].

After passing the ocular optics (see Figure 3.9), the optical radiation hits the three cone types (L, M, and S), rods R and the photosensitive ganglion cells *G*. From the signals of the L-, M-, and S-cones, the signals for the two opponent channels (L – M) and S – (L + M) as well as for the luminance channel (L + M) are formed (see also Chapter 2). This formation takes place on the way to the '*normal*' (i.e. non-ipRGC) ganglion cells as well as in the cortex. In the mesopic area (as in the evening with street lighting or at the cinema), the signals of the rods are added [7–10].

Recent studies such as [11] demonstrated the involvement of melanopsin-containing ganglion cells in the formation of brightness perception, which was also confirmed by the modelling of brightness in [12] based on the psychophysical experiments. Thus, brightness and other visual attributes result from all

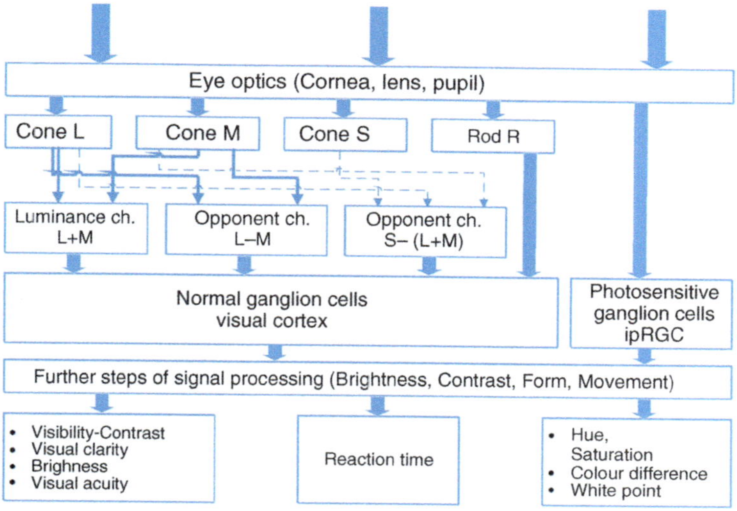

Figure 3.9 Processing circuits of visual perception. Source: TU Darmstadt.

component signals, opponent channels, and melanopsin-containing ganglion cells. Through the development of colour perception models [13, 14] and colour quality models [15, 16], visual perception has already been relatively well understood in lighting and colour science, even if this has not always directly and immediately been taken into account in standards, regulations, and lighting product development. Visual perception and its psychological–emotional forms of expression (such as scene preference, colour harmony, and brightness) are rather acute lighting effects that are formed within a short time (in the range of a few minutes).

Recently, scientists in neurobiology and lighting technology learned that the melanopsin-containing ipRGC ganglion cells were responsible for brightness perception as well as for alertness and hormone regulation. The term IF in contrast to NIF refers to the imaging of objects on the retina and thus to the perception of contrasts (light–dark difference or structural difference). The ipRGC cells do not enable local and temporal contrasts (NIF). In 2019, it was possible [17] to demonstrate the effects and involvement of melanopsin-containing cells at low local frequencies up to 0.8 cpd (cycle per degree, period per degree of visual angle) and up to a temporal frequency up to 0.45 Hz at a Michelson contrast of greater than 14%.

3.3.2 Processing Centres and Transmission Pathways for Non-visual Light Effects

For the non-visual light effects, the ipRGC cells with their intrinsic signal formation in the melanopsin-containing pigments and with the signals coupled from the LMS cones and rods to the ipRGC cells represent the initial steps (see Figure 3.10).

The ipRGC-cell signals are transmitted to the visual cortex V_1 and contribute to visual behaviour (see [18] and Figure 3.9). For the non-visual effects of light

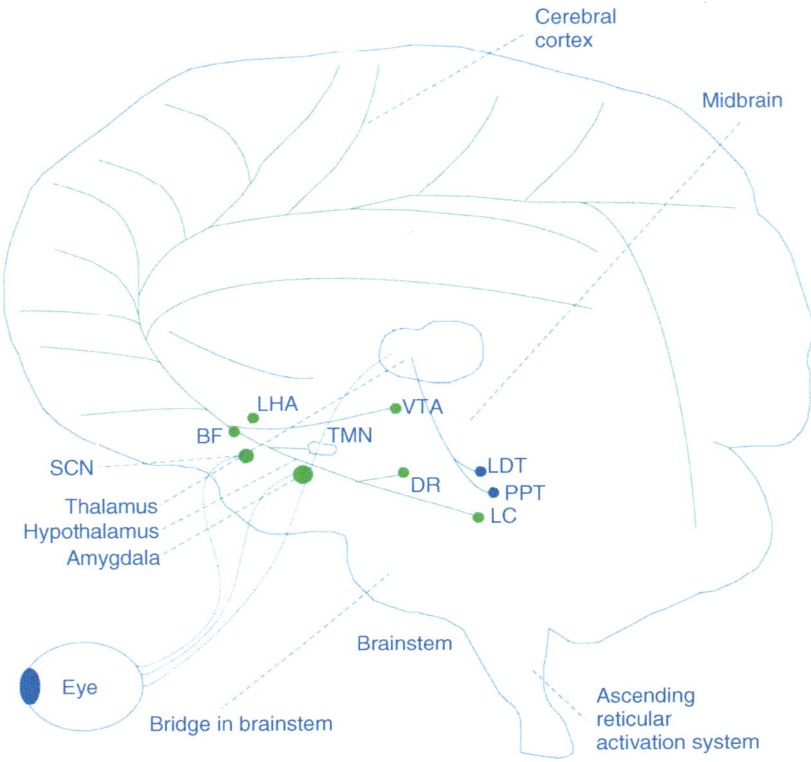

Figure 3.10 Processing centres and transmission paths for non-visual light effects. . Source: TU Darmstadt.

(Figure 3.10), the ipRGC-signals are transmitted to the pacemaker SCN as well as to other processing centres in the brain and are responsible for many physiological and psychological effects (e.g. alertness, emotions, learning, and hormone formation).

The SCN (Figure 3.10) is a structural unit in the hypothalamus and is about 0.25 mm^3 per nucleus [19, 20]. It is the clock for the circadian biorhythm, whose oscillatory capacity is made possible by two interconnected molecular regulatory loops. The timing function and synchronisation with the environment are possible with the SCN on the input side through the incoming ipRGC-signals and through the signals from other body units. In addition – on the output side – there are the signals to the other processing centres in the brain, which contain information about time and phase position (Figure 3.11).

The information from the environment, the so-called *'photic' information*, originates from the ipRGC cells and is transmitted directly to SCN in the retinohypothalamic tract (see Figure 3.11). The signal exchange of SCN with the retina can also take place indirectly via the so-called IGL (*'Intergeniculate Leaflet'*) in the thalamus via the so-called geniculohypothalamic tract (GHT). The signals on the two pathways RHT and GHT, thus contain the information about the light intensity

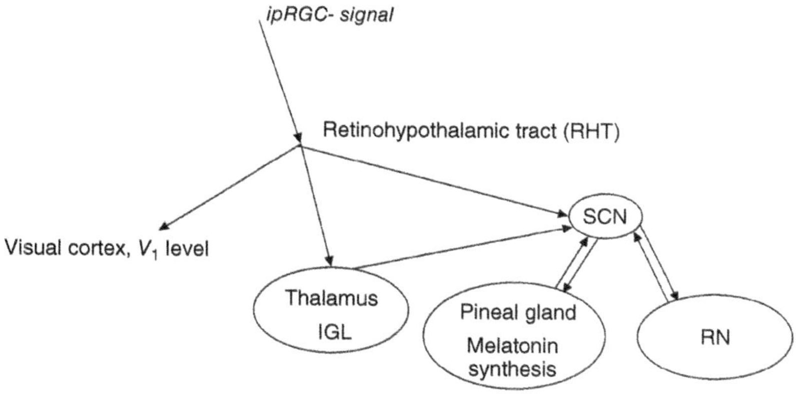

Figure 3.11 Processing centres and transmission paths of the ipRGC signal with the SCN in focus. Source: TU Darmstadt.

and the spectral composition of the incident ambient light. The communication between the SCN and the RN (raphe nucleus), an endogenous, so-called '*non-photic communication*', takes place via the so-called *raphe-hypothalamic tract,* which contains information about the phase of the circadian rhythm and regulates the body's vigilance. The RN itself controls the generation of the hormone *serotonin*, a tissue hormone and neurotransmitter that affects the blood system, the gastrointestinal system, and the cardiovascular system. The main functions of serotonin are to influence perception, sensory function, pain processing, sleep, and temperature regulation.

There is also a communication loop between the SCN and the *pineal gland*. The function of the pineal gland is to generate the hormone melatonin during night-time hours. The hormone docks with the blood vessels in the brain and the cells of the immune system and signals the body when it is dark that it is time to rest. The consequences of this are a reduction in physical energy expenditure, with blood pressure also being lowered. The hormone lowers the body's core temperature and significantly influences the immune system.

By synchronising the SCN pacemaker with the light–dark rhythm of natural light, various physiological body functions are clocked in a circadian periodicity. Figure 3.12 shows the circadian rhythms of the two hormone types melatonin and cortisol and the body's core temperature over two days.

The hormone level of melatonin during the day is low (see Figure 3.12) and is about one-third to one-twelfth of the melatonin level during the night-time hours. The melatonin level starts to rise around 8 p.m. in the evening (local time). Without the effect of light, the melatonin concentration reaches its maximum at 3–4 a.m. The level of cortisol reaches the absolute maximum in the early morning between about 6:30 and 8 a.m. The core body temperature (CBT) passes through its minimum at around 4:30 a.m.

The signalling pathway from ipRGC cells to SCN and via SCN to the brain centres in the *hypothalamus* contributes to hormone control and the control of vegetative

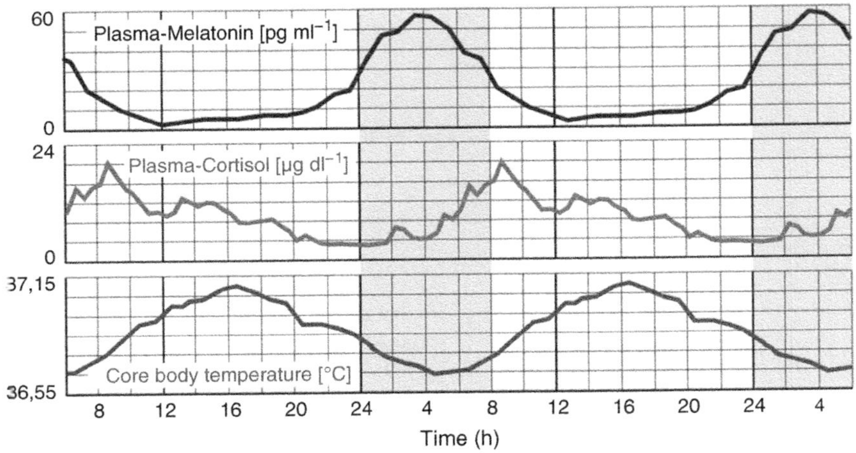

Figure 3.12 Circadian rhythms of the two hormone types melatonin and cortisol and the core body temperature over two days (after [21]). Source: Reproduced with permission of the authors of [21].

functions as described above. The effects of light on these autonomic functions may persist for days to weeks after light exposures have ceased. The direct effects of ipRGC cells to the other processing centres without SCN *as an intermediary* stop quickly after the termination of eye exposure. Other light effects include effects such as learning, cognition, alertness, mood, and emotion, with these five effects influencing each other. The question of how light with its intensity, spectrum, and temporal properties controls these five effects in the individual brain regions and in the network of regions is answered in the following explanations.

3.3.2.1 Light Effects on Mood and Learning

In 2018, in [22] research results were presented, according to which the different subpopulations of ipRGC cells with different proteins or different transcription factors are the originators of the signals that exert effects on learning and mood in the different pathways to the SCN and thalamus, see Figure 3.13.

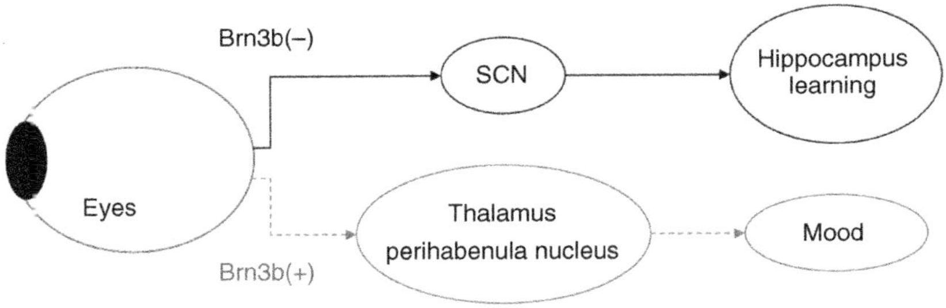

Figure 3.13 Different ipRGC-signalling pathways affect learning and mood (after [22]). Source: Fernandez et al. [22])/with permission of Elsevier.

The ipRGC cells with the transcription factor *Brn3b*(−) send the signals to SCN, which influence learning when they are coupled in the hippocampus (see Figure 3.13). This process does not affect the pacemaker process in SCN. The ipRGC cells with the transcription factor *Brn3b*(+) send the signals to the thalamus, specifically to the peri habenula nuclei (*PHb*), where they influence mood.

3.3.2.2 General Light Effects on Cognition, Emotion, and Alertness

There have been a number of recent neurological studies that have demonstrated that ipRGC cell signals project to different brain regions [23]. These include:

- Ventro-lateral preoptic nucleus (VLPO), an area of hypothalamus that contains sleep-active neurons;
- LHA (Lateral Hypothalamic Area, see Figure 3.10), an area that contains so-called '*orexinergic neurons*' for the regulation of vigilance;
- Olivary pretectal nucleus (OPN) to regulate pupil diameter;
- Amygdala, a centre for emotion regulation.

One of the most modern methods of brain diagnostics in connection with the effect of light is the method of brain imaging ('*neuroimaging*' or tomography), in which – under the effect of white light or coloured lights of different wavelengths and different exposure durations – the regions of the excited brain centres are identified and the strength of the brain activity is recorded. In recent years, the methods '*Positron Emission Tomography*' (PET) [24] for investigating the effects of light at night and '*functional Magnetic Resonance Imaging*' (fMRI) for daytime analysis have been used. In both cases, the illuminated subjects had to perform computational and search tasks that require cognitive load, attention and responsiveness, and require increased auditory and visual sensory responses [25–27]. The brain image observation here took place both during the exposure time and a few minutes after the end of the exposure procedure.

The modulation of brain activity stimulated by light exposure can be observed in the following brain regions:

- In the subcortical structures responsible for alertness, such as the structure in the brainstem [27] in a location similar to LC (locus coeruleus), see Figure 3.10, in the hypothalamus, in a region containing the SCN [24] and in the dorsal and posterior parts of the thalamus [25–27];
- In the region of the hippocampus [27], which is responsible for long-term memory and learning;
- Amygdala, a region responsible for emotions [27].

At the cortical level, brain activity and its modulations can be observed in the regions responsible for a '*top-down*' regulation of attention, such as the dorsolateral prefrontal cortex [27]. The prefrontal cortex, for example, receives sensory signals and is closely linked to the assembly of memory processes and emotional evaluations. The functions and processes of the prefrontal cortex are decisive for action planning and control and for the regulation of emotional processes. The fact that the cortical structures in the brain also show their activity during light effects speaks for a close communication of signals between the retina, the hypothalamic, the thalamic, the limbic (amygdala, habenula) structures, and the subcortical and cortical units in the cerebral cortex.

3.3.2.3 Wavelength Dependence of Brain Activities on Light Exposure

The research work [24–27] aimed, among other things, to investigate the wavelength dependence of brain activity in the individual brain regions during the day [23]. The wavelength dependence of melatonin suppression in the nocturnal hours is already known from the work of Brainard and Thapan (see Chapter 2). In the results in [26, 27], modulations of brain activity by wavelengths of equal photon densities during the performance of tasks of high cognitive load (so-called '*2-back*' tasks according to [28]) by test subjects. Such modulations can be registered in the brainstem in an LC-compatible location, in the thalamus and insula, which is partly responsible for the emotional evaluation of pain and for the perception of the sense of balance. Such modulations were found in the left frontal and parietal cortex [26, 27], which is important for *working memory*.

In the above-mentioned research, it could be demonstrated that blue radiation around 480 nm causes a higher brain activity than the green (550 nm, near the maximum of the $V(\lambda)$ – function or the (L + M)-channel) or than the violet wavelength around 430 nm (near the maximum of the spectral action spectrum of the S-cones). During the initiation of the light pulse (so-called '*light onset*') of the blue wavelength 473 nm, a higher brain activity could be recorded in the thalamus, hippocampus, and amygdala than that caused by light of the green wavelength at 527 nm (near the maximum of the spectral action spectrum of the M-cones). It can thus be argued that a relatively long pulse of blue radiation around 473–480 nm can exert a higher excitation and effect on regions for long-term memory, emotions, and visual perception than green radiation around 527 nm. *The latter speaks much for the importance of the melanopsin-containing ipRGC ganglion cells for cognition, and emotion, as well as for learning behaviour, and alertness.*

3.4 '*Timing System*', Circadian Rhythm, and Sleep Behaviour

3.4.1 Questions

The 'Timing System' in the context of the circadian rhythm exists in all living beings on earth (plants, animal beings, and humans), which are illuminated by sunlight, and for this reason, their biorhythm is closely connected with the light–dark cycle of daylight and the rotation of the earth around its own axis and around the sun. In the analysis of the theme, '*Timing System*', some fundamental questions arise:

1. What is the basic task of the timing system?
2. How is the timing system molecularly constructed in the nervous system?
3. How does one's own biorhythm come into existence and what process sets the pace?

4. How can the rhythm, periodicity, and phasing be stabilised, both in the central nervous system and in the autonomic nervous and functional system?
5. How can the amplitude and the phase of the circadian rhythm be influenced?

3.4.2 Timing System: Entrainment, Timing Role

The timing system coordinates the temporal processes and the temporal allocation of the functions of all mechanisms in the vegetative and physiological systems of humans. It synchronises these processes to the periodic change in environmental conditions. The timers here are light, social events, and appointments (such as meeting dates, working hours, meetings with acquaintances, and meals). Without these timers as an anchor and readjustment, the biorhythm would run free and this would then lead to a de-synchronisation of the vegetative, physiological, and cognitive functions. By means of the timing system, living beings and humans can prepare themselves for the diurnal and seasonal changes in the environment and take precautions. These preparations affect the health, nutritional, social, and even technical aspects of life.

The question '*which timer influences people's circadian rhythms the most?*' was answered in a research paper [29]. This study of the sleep behaviour of 21 600 people in Germany, whose population distribution corresponds to the actual population distribution of the Federal Republic of Germany and whose places of residence are distributed from West to East over 9° (from 6° to 15°) longitude (time zones), pursued the hypothesis that people's sleep behaviour should be independent of time zone *social time* being the most important 'zeitgeber' (i.e. the factor that determines 'physiological time and perceived time'). These residents were further divided into those living in cities with populations up to 300 000, between 300 000 and 500 000, and larger than 500 000. This division pursued the second hypothesis that the sleeping behaviour of the inhabitants of larger cities must be less dependent on the time zone since the way of life in large cities is more indoors with less movement in the open space.

The results [29] show the clear directions that sleep behaviour *does depend on time zone*, i.e. on solar events (sunset, sunrise), although this dependency becomes weaker when larger cities are considered. *Nevertheless, the statistically validated result clearly states that people's timing system is controlled by sunlight.*

Every timing system contains a clock generator, a biological clock, which in biocybernetic terms is a biological system whose input variables receive synchronisation signals from the external environment or from the living being's own internal (endogenous) system. It is far from optimal if one's own biorhythm and the cycles of the environment (such as light–dark cycle, working hours, and shift work) differ greatly from each other. Therefore, a synchronisation between rhythms of the environment and one's own rhythm that is self-sustaining in a natural way is desirable.

The synchronisation between a timer (clock) and one's own organism is called *entrainment*. As described above, natural light (sunlight) is the most important and formative 'zeitgeber'. In mammals and humans, there is a hierarchically structured architecture in the timing system, whose central timer is the SCN unit. The cells within the SCN can synchronise with each other and synchronise externally the rhythms of the individual body organs and body functions of the lower levels. The most important body functions that are subject to circadian periodicity include melatonin secretion, cortisol synthesis throughout the day, and CBT, the daily courses of which are shown in Figure 3.12.

3.4.3 PRC – Function, Phase Shift

After gaining the knowledge that sunlight is the most dominant zeitgeber for the human biorhythm, it now becomes interesting and relevant to ask how the changes in sunlight due to the changing weather influences (rain, snow, storms, and clouds) affect the synchronisation of the biological clock. It is also relevant to ask how the phasing of the circadian periodicity depends on the light intensity and on the time of light exposure. This question leads to the so-called *PRC curve* (*'Phase–Response Curve'*), see Figure 3.14.

This curve, known in entrainment theory (see Figure 3.14) is based on the basic principle that a light stimulus arriving at the retina can shift the biological clock. The amount of phase shift depends on the light intensity and the spectral radiation composition, while the direction of the phase shift depends on the time of light exposure.

The course of the PRC curve in Figure 3.14 shows that the light exposures between 12 o'clock and 8 o'clock cause hardly any or no phase shift. Light exposures between 8 p.m. and 5 a.m. in the early morning on the evening oscillator of the central pacemaker readjust this oscillator (so-called *'delay shift'*). The transition between phase delay (evening oscillator is active) to phase shortening (default, the morning oscillator is active) is very rapid at 5 a.m. The highest efficiency of phase shortening is achieved at 6–7 a.m. A phase shift of about max. three hours can be achieved daily.

In Section 9.4.2 of this book, light effects in the nocturnal hours with *polychromatic white* light, among others the phase shift in Figure 9.31 and in Table 9.5 are described in detail. In [30] a research work was reported that investigated the wavelength dependence of the phase shift of the melatonin rhythm. The wavelengths

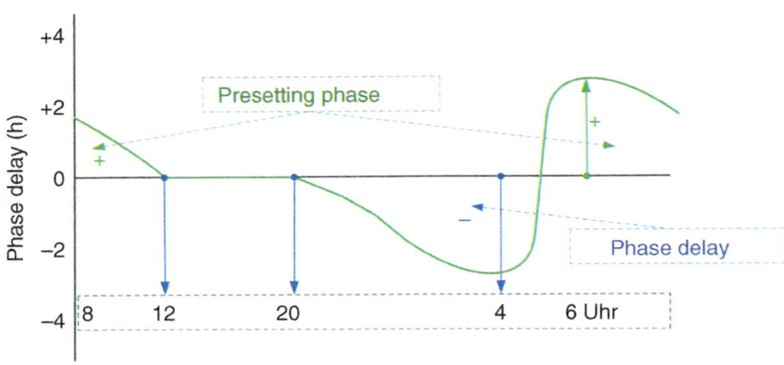

Figure 3.14 PRC (Phase Response Curve). Source: TU Darmstadt.

investigated were 660 nm (red), 595 nm (amber), 525 nm (green), 497 nm (greenish blue), and 470 nm (blue). The light sources were semiconductor LEDs. Fifteen subjects participated in the test with all wavelengths and a reference condition with no light. Each light condition ran for two consecutive nights (night 1 and night 2). Saliva samples for melatonin analysis were always collected for 30 minutes each from 7:00 p.m. to 2:00 a.m. (night 1) and from 7:00 p.m. to 1:00 a.m. (night 2). Light exposure in each test condition was conducted between midnight and 2:00 a.m. for night 1. The timing of the *dim light melatonin onset* (DLMO) in each night was measured and the phase shift was calculated. The *shorter* wavelengths 470 and 497 nm caused the *greater* phase shift between 27 and 36 minutes.

3.4.4 Chronotypes, Sleep Behaviour

The determination of the free-running circadian period duration can be realised by determining the periodicity of the courses of melatonin release, cortisol synthesis, and CBT. Research work on this was carried out on two age groups by Czeisler et al. [31] whose results are shown in Table 3.1.

Table 3.1 shows that the intrinsic circadian period duration does not depend on age and physiological mechanisms (CBT, cortisol, and melatonin). These mechanisms are therefore subject to the central periodic control of the central pacemaker.

The free-running circadian period duration is statistically normally distributed within the population. Those who have a shorter period duration are the so-called *larks*, who get up early in the morning, are tired early in the evening, and therefore want to go to bed early. Those people who have a longer period duration (so-called *owls*) behave the other way around. They get up later and are alert in the evening hours as well as in the early hours of the night. With an average value of the free-running circadian period of about 24 hours and 12 minutes, the biological clock of most people ticks a little too slowly and therefore needs to be exposed to daily morning light.

The wake–sleep cycle contains processes of alertness during the day and the sleep phase at night and, in a first approximation, is composed additively of a circadian rhythm with a circadian component C (see Figure 3.15) and a homeostatic component S of the so-called sleep homeostat. Both components run periodically throughout the day.

The circadian component C of alertness (see Figure 3.15) reaches its minimum at 8 o'clock. Thereafter, it increases steadily until about 8 p.m. During an assumed

Table 3.1 Free-running intrinsic circadian period duration according to [31].

Age group	Middle age (years)	Period duration CBT (hours:minutes)	Period duration Melatonin (hours:minutes)	Period duration Cortisol (hours:minutes)
Young	23.7 ± 2.7	24:10	24:11	24:11
Older	67.4 ± 3.2	24.10	24:13	Not specified

Source: Adapted from Czeisler et al. [31].

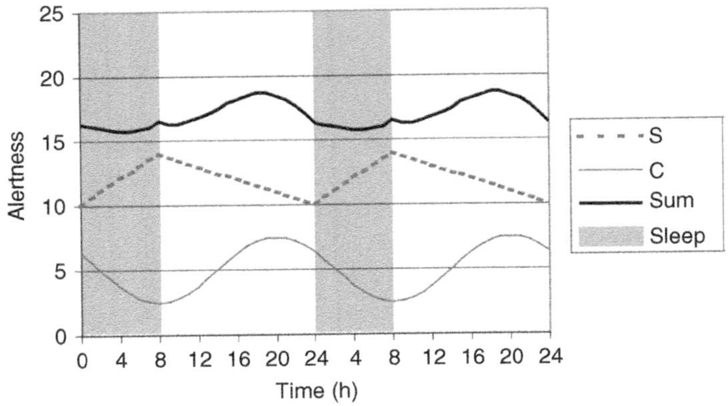

Figure 3.15 Circadian and homeostatic components of alertness over two days (after [21]). Source: TU Darmstadt after [21].

sleep between 0 o'clock and 8 o'clock, this circadian component decreases, while the homeostatic sleep pressure decreases and the homeostatic component S increases. In sum, alertness during night-time sleep decreases until about 5 a.m. in the early morning and increases thereafter. During the day, between about 8 a.m. and 8 p.m., the homeostatic component S steadily decreases again, while the circadian component C increases. As a result, the sum of alertness increases during the day between about 5 a.m. and about 6 p.m., and then it decreases again. We can also see that the two components run in opposite directions between 0 a.m. to 8 p.m. They run in the same direction between 8 p.m. and midnight 0 a.m. and help people prepare for sleep.

The light effects during the night hours are described in Section 9.4.2. The alertness in the daytime hours is described in detail in Section 9.7.2.

3.5 Output Variables of the Visual and Non-visual Brain Processing Apparatus: A Systematics

In Sections 3.2 and 3.3, the input variables of the overall '*brain processing*' system and its visual and non-visual processing networks were presented. In this section, the *output* of this system is described. This output is divided into *physiological-photometric metrics* (visual performance – work performance), *colorimetric-psychological-emotional* attributes, and *non-visual metrics*. However, it is not always possible to separate these three groups so clearly. Emotions, for example, have a good part to do with alertness, and a short reaction time also has its roots in a high-brightness contrast. The separation of the metrics serves the purpose of breaking down the high complexity of signal processing into individual sub-topics and to describe these sub-topics in this book in individual chapters in a result-oriented way. The overall result of the presentations in the individual chapters will finally lead to an *overall lighting concept* in Chapter 12 at the end of this book.

The systematics of the output variables is summarised in Figure 3.16.

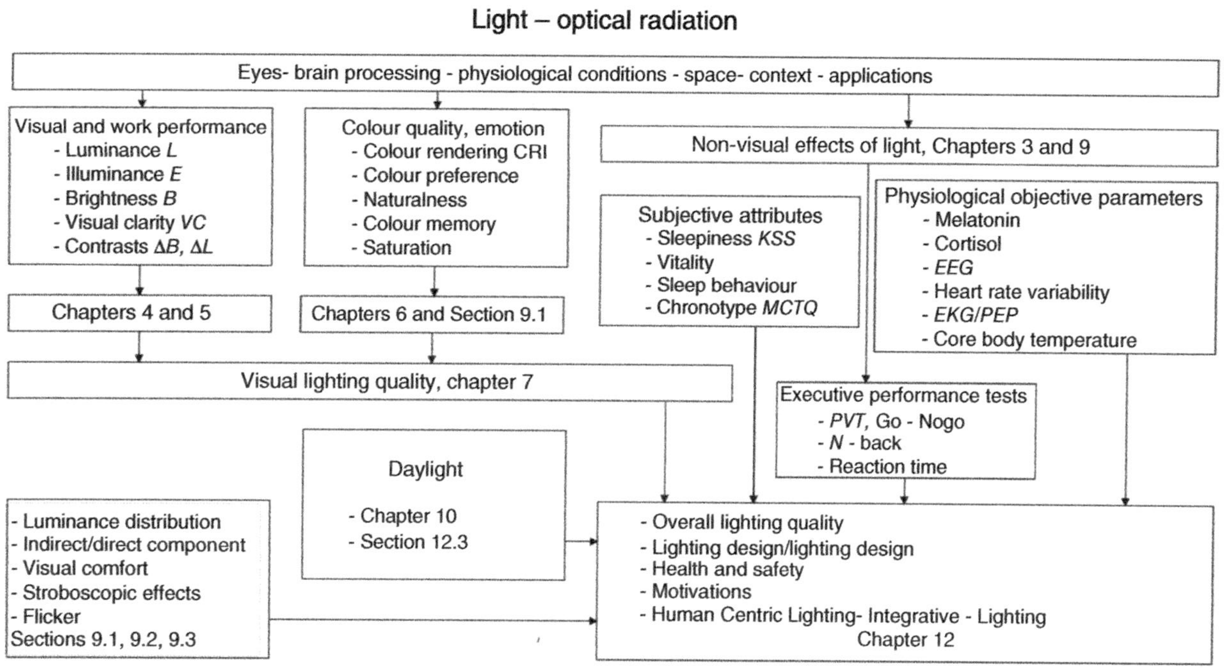

Figure 3.16 Output variables of the visual and non-visual brain processing apparatus: a system. Source: TU Darmstadt.

Figure 3.16 shows the following:

- The photometric parameters are mainly used to characterise visual performance and work performance, as to be presented in Chapter 4;
- The brightness and visual clarity, which include the previous, purely photometric parameters with the signals from the ipRGC cells and the S-cones, are presented in Chapter 5;
- Colour quality has been the focus of light and colour research from around 2002 to the present. Research has made great progress since then and is discussed in Chapter 6;
- The two components of colour quality and work performance/visual performance lead to the new modelling and verification of lighting quality in Chapter 7;
- Visual comfort, luminance distribution of the walls and ceiling, and stroboscopic effects form further aspects of lighting quality in Chapter 9;
- The non-visual aspects are discussed in this Chapter 3 and then also in detail in Chapter 9;
- All the above aspects lead to a concept of lighting design or lighting planning in Chapter 12, which is the result of the strategy and philosophy under the umbrella term 'HLC' or '*Integrative Lighting*'. These two terms have different histories but the same content and the same goals.

3.6 Basic Aspects of Human-Centric Lighting/Integrative Lighting

Even in the first hours of electric lighting technology at the beginning of the twentieth century, lighting research was already working on concepts and characteristic values to improve people's visual performance at work and at home [32]. Humans were and are therefore already at the centre of lighting technology, which has historically passed through different stages of knowledge and technological possibilities. From today's perspective, HLC can be described as follows according to Boyce's formulation (see [33]):

> What people mean when they talk about Human Centric Lighting is lighting that considers both the visual and non-visual effects of exposing humans to light and that widens the range of possible effects from visual performance and comfort to sleep quality, alertness, mood and behaviour with consequences for human health, learning and spending.

In the sense of lighting research, the framework of HCL thus consists of visual and non-visual effects with their components *visual performance*, *emotional aspects* such as mood, and behaviour as well as *non-visual effects* such as sleep quality, and alertness. The term 'HLC' is thus the primary goal of lighting technology and encompasses the activity of developing the appropriate lighting products, lighting design, and the concrete lighting installation and lighting programming on site with the aim of optimally satisfying the lighting needs of users in a concrete context of use.

This includes the follow-up and evaluation of the lighting quality a few weeks or months after the lighting installation. These lighting activities are important and depend on each other to fulfil the HCL goal. A suitable lighting design realised with a great deal of experience and care but without optimal and HCL-promoting lighting products is ultimately limited. Conversely, even the best technical lighting systems cannot produce the intended lighting effect without good lighting design and with little consideration of user needs.

From the perspective of lighting design, characteristic values, and metrics are needed that can describe the complex subject area of 'HCL' in their mutually dependent interactions. From today's perspective, a structure for these HCL metrics or HCL parameters emerges, which is shown in Figure 3.17 is shown.

The parameters in Figure 3.17 obtained from the spectral radiant flux of the light sources (lamps, luminaires) or the spectral distributions on the walls or on the working plane can thus be divided into the three groups of *visual performance, colour quality,* and *non-visual* (melanopic) *effects*. Here, the first two groups produce an immediate effect (recognisable within a few minutes) and impression about the illuminated scenes and the room's environment (e.g. '*The scene is pleasant*', '*The room*

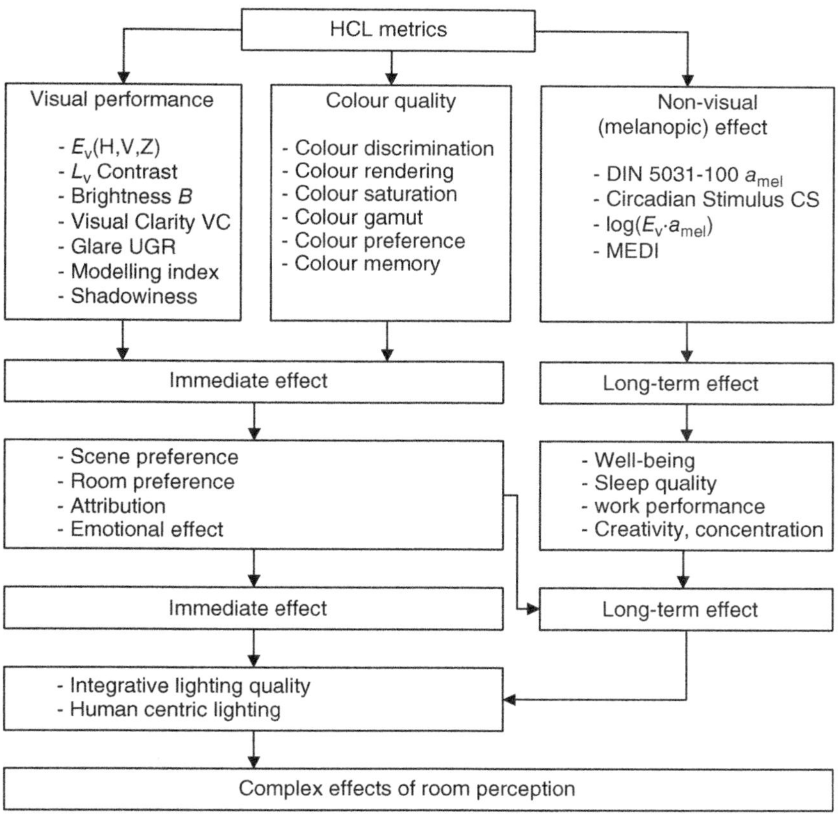

Figure 3.17 Structure and flow diagram of the HCL parameters. Source: TU Darmstadt.

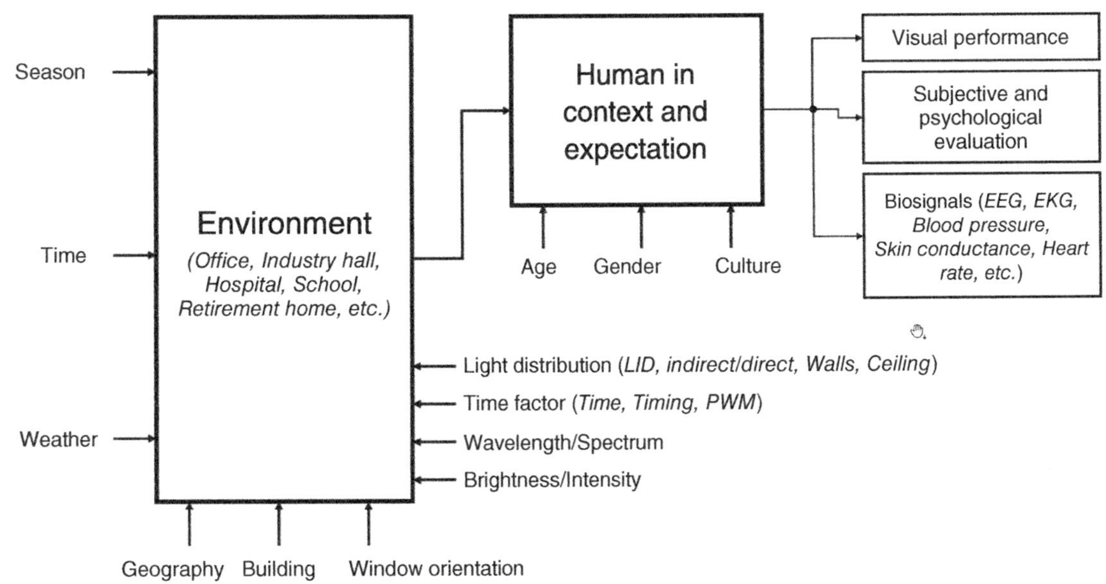

Figure 3.18 System diagram of the HCL complex. Source: TU Darmstadt.

is harmoniously lit', '*The scene is bright and clear*', and '*The room is spacious*'), which can also produce a long-term effect over a longer period of time.

The non-visual effects predominantly relate to long-term sustainability, which only arises and becomes noticeable if the lighting used produces positive effects (such as good sleep quality and concentration) over a longer period of time or negative effects (such as lack of concentration and fatigue) in the bad case. The immediate and long-term effects in the context of HCL (or '*Integrative Lighting Concept*') then lead to a sustainable, positive effect on room perception and a positive effect on health.

In the process of researching the effects of HCL, lighting engineering and science, and sleep research consider the entirety of the above system, see Figure 3.18.

According to Figure 3.18, the HCL system starts with the influencing parameters (time of day, season, weather, geography, building location, and window orientation) and takes into account the applications with their specific contents (e.g. schools, hospitals, industrial halls, and offices). All users have their own life history and individuality (habits, culture, gender, age, and light sensitivity) and their own light perception and visual performance characteristics (e.g. reading time, visual acuity), subjective room assessments (e.g. too bright, dark, pleasant, preferred, or '*not nice*') and their bio-signals (heart rate, EEG, blood pressure, melatonin level, and stress hormone level).

Figure 3.18 also outlines the control and regulation parameters with which luminaire developers, programmers of cloud software, and building technicians can influence the quality of lighting and its effect on room perception and assessment with the aid of lighting technology. The technical aspects of Smart Lighting in connection with HLC are presented in detail in Chapter 11.

3.7 Tools and Methods for Determining the Subjectively and Objectively Measurable Lighting Effects

Psychophysical and occupational science methods and tools have been developed and applied for determining the necessary visual performance or work performance since the beginning of the twentieth century until today. These methods are discussed in detail in Chapter 4. For the determination of the colour quality characteristics of a lighting scene, some questionnaires are presented in Chapter 6 in direct context with the colour experiments described there. Section 3.7 summarises some methods and tools for determining non-visual lighting effects and lighting quality. Readers who wish to obtain more detailed information are given literature references to specific articles at the appropriate places.

3.7.1 Questionnaires for Comprehensive Subjective Determination of Indoor Lighting Quality

Questionnaires are often used to characterise the quality of indoor lighting in order to qualitatively determine the user's assessments. Such quality features include

Extremely awake	1 -	☐
-	2 -	☐
Awake	3 -	☐
-	4 -	☐
Neither awake nor sleepy	5 -	☐
-	6 -	☐
Sleepy (but not fighting sleep)	7 -	☐
-	8 -	☐
Extremely sleepy (fighting sleepiness)	9 -	☐

Figure 3.19 KSS – Questionnaire (Karolinska Sleepiness Scale) according to [36] and [35]. Source: Adapted from Akerstedt and Gillberg [35].

brightness level, luminance distribution of walls, daylight, glare, colour and colour rendering, reflections, and ease of use. Such a questionnaire has been developed within the German Lighting Society [34] and can be used both for research purposes in laboratories and in field studies. All questions are linked to a bipolar answer scale and are presented in Appendix 3.A of this book.

This questionnaire [34] comprises 25 questions that take about 10 minutes to answer. In laboratory experiments, a 7-point scale is used and the field '*not specified*' should be removed. In field trials, the 5-point scale and the '*not specified*' field are used. In addition, personal data of the observers such as age (in years or year of birth), gender, and the presence of glasses are requested.

3.7.2 Questionnaires on Sleep Behaviour, Sleepiness, and Alertness: The Subjective Basis

The sleepiness of the test subjects is measured with the subjective questionnaire *Karolinska Sleepiness Scale* (KSS) according to Akerstedt and Gillberg [35]. The KSS questionnaire is a 9-step scale, where 1 indicates the lowest sleepiness and 9 the highest sleepiness, see Figure 3.19.

The observer is asked (see Figure 3.19) about their sleepiness at a given time and they are asked to assess their sleepiness using the predefined scale. In laboratory or field studies, this questionnaire is often used to assess sleepiness on an hourly basis.

The questionnaire MCTQ ('*Munich Chrono Type Questionnaire*') pursues the goal of determining sleep behaviour and chronotype on the basis of the subjective assessment of observers [37]. The questionnaire is divided into what happens on workdays and on days off as well as on weekend days. The workload (number of working days per week), sleep behaviour at the beginning of sleep (time at which the observer goes to bed, time of sleep onset, and sleep latency time) as well as after waking up (time of waking up, duration of time to get up after waking up), the duration of time spent outdoors under daylight, as well as drinking habits, etc. are queried.

One method of determining sleep quality in laboratory and field studies on a subjective basis is the so-called '*Pittsburgh Sleep Quality Index*' (PSQI), which was created by Buysse et al. [38] and validated by Backhaus et al. [39]. The PSQI questionnaire uses 19 self-report questions covering sleep behaviour over the past four weeks. The total score categorises the observer according to a *PSQI score* in a range of 0–21. A higher score means worse sleep behaviour. If the PSQI score is greater than 5, the sleep behaviour can be considered poor.

The 19 questions used to assess one's own sleep are divided into 7 components that take into account aspects such as subjective sleep quality, sleep latency, sleep duration, habitual sleep efficiency, sleep disturbances, sleep medication use, and daytime sleepiness. Such aspects are often consulted in clinical interviews of patients with sleep deficits.

3.7.3 Objective Methods and Tools

Objective measurement parameters for determining the non-visual effects of light are the pupil diameter, which assumes a certain time-dependent size (diameter) under the influence of light with a certain spectrum and with a certain luminance. At the beginning of the light exposure, the L- and the M-cone signals control the pupil diameter and only after about 300 s the control is taken over by the ipRGC signals [40].

An important representative for the physiological assessment of alertness is the analysis of the EEG-signals (electroencephalography), which represent the activity of the central nervous system. Here, fluctuations in electrical voltages of the order of microvolts are measured on the surface of the scalp. After [41], the observer wears an EEG cap on the head with 16 electrodes. The 16 electrodes are placed at certain defined points on the surface of the skull. To ensure good electrical conductivity between the scalp and the electrodes, a so-called electrode conductive gel is injected at each electrode. During recording, the electrode signals are referred to as a reference electrode (so-called *Cz electrode*) and digitised at a sampling rate of usually 250–500 Hz with a high-resolution AD converter (14 or 16 bit). The EEG signals are evaluated with Fast Fourier Transformations and the theta, alpha, and beta waves obtained are analysed, the power density of which provides information about the state of alertness [42]. Theta waves and alpha waves are signals in the frequency range between 4–8 and 8–13 Hz, respectively. Section 9.4.2 describes the effects of light during night-time hours with polychromatic white light, whereby the EEG activities are shown as a function of the illuminance at the eye.

In a basic study on alertness, Kaida et al. [43] carried out tests on 16 healthy test subjects (age 33–43 years, mean 38.1 ± 2.68) with parameters such as KSS and EEG signals. The KSS values correlated well with the alpha and theta waves in the open-eye state. Alertness in the context of the subjective evaluation according to KSS means a lower performance in the two frequency bands mentioned.

The best marker or evidence for a 24-hour wake–sleep cycle, coupled to the rotation of the earth and the associated light–dark rhythm, is the hormone melatonin, which is generated by the pineal gland in the brain at the onset of the dark hours (about 8 p.m.) and reaches its maximum at 4 a.m. (see Figure 3.12). Melatonin is also known as the *sleep hormone* and its frequent withdrawal during the night hours objectively shows the level of melatonin production and allows an assessment of possible sleep disturbances at different night phases. Melatonin can be found in serum, whole blood, urine, and saliva and can be taken from there. Of the possibilities mentioned, a saliva sample is the easiest to handle. Saliva contains only free *biologically active* hormones whose concentration corresponds most closely to the body's actual hormone activity (see [44]).

Major glucocorticoid hormones, of which cortisol is the most important hormone in humans, are produced in the adrenal cortex [45, 46]. Its production follows a circadian rhythm (peak in the morning 30 to 60 minutes after waking and decline through the night, see Figure 3.12). In contrast to melatonin, the daily (circadian) course of cortisol is more complex, because cortisol levels rise independently of the circadian rhythm in response to stress. Cortisol can also be taken relatively conveniently with a saliva sample like melatonin. Salivary cortisol correlates reliably with serum levels [44].

The so-called '*n-back test*' is a software-supported visual or auditory psychological test. This cognitive endurance test is used for the analysis of working memory. In methods such as PET and fMRI of brain tomography, the brain activities of the observers are stimulated during the performance of so-called *n-back tasks* (see Section 3.3.2). This *n*-back test was introduced by Kirchner in [47], in which the subjects were shown a sequence of stimuli either auditorily or visually. The subjects had to signal by pressing a button when the current stimulus matched the stimulus that was present in the series *n* steps earlier. The variation in the number of *n* speaks for the level of difficulty of the task. As a rule, 1- , 2-, or 3-back tasks are the most commonly used. The effects of light on concentration, alertness, and working memory can be investigated by having the observer sit in a laboratory room and perform such tasks while systematically and gradually changing the lighting conditions (illuminance on the worktable in front of them or on their eye or the spectrum or colour temperature of room lighting).

References

1 Khanh, T.Q. and Söllner, S. (2016). *Vorlesungsskripte für Grundlagen der Lichttechnik LT1 (Lecture Notes for "Foundations of Lighting Engineering LT1"), Lehrstuhl für Lichttechnik (Laboratory of Lighting Technology)*. TU Darmstadt.

2 Hubel, D.H. (1989). *Auge und Gehirn. Neurobiologie des Sehens. (Eye and Brain. Neurobiology of Vision) Spektrum der Wissenschaft*. ISBN: 978-3922508922.

3 Patterson, S., Kuchenbecker, J.A., Anderson, J.R. et al. (2020). A color vision circuit for non-image-forming vision in the primate retina. *Curr. Biol.* 30: 1269–1274.

4 Spitschan, M., Jain, S., Brainard, D.H. et al. (2014). Opponent melanopsin and S-cone signals in the human pupillary light response. *Proc. Natl. Acad. Sci. U.S.A* 111: 15568–15572.

5 Woelders, T., Leenheers, T., Gordijn, M.C.M. et al. (2018). Melanopsin- and L-cone-induced pupil constriction is inhibited by S- and M-cones in humans. *Proc. Natl. Acad. Sci. U.S.A* 115: 792–797.

6 Vinh, Q.T. and Khanh, T.Q. (2019). Visuelle und nichtvisuelle Systeme – Eine ganzheitliche Betrachtung (Visual and Nonvisual Sys). *Z. Licht* 7: 68–75.

7 Roufs, J.A.J. (1978). Light as a true visual quantity: principles of measurement. In: CIE publication; Vol. 41, Bureau Central de la CIE, Paris). CIE: Commission Internationale de l'Éclairage.

8 Lennie, P., Pokorny, J., and Smith, V.C. (1993). Luminance. *J. Opt. Soc. Am. A* 10: 1283–1293.

9 Lee, B.B., Martin, P.R., and Valberg, A. (1988). The physiological basis of heterochromatic flicker photometry demonstrated inthe ganglion cells of the macaque retina. *J. Physiol. (Oxford, UK)* 404: 323–347.

10 Kaiser, P.K. (1971). Luminance and brightness. *Appl. Opt.* 10: 2768–2770.

11 Zele, A.J., Adhikari, P., Feigl, B., and Cao, D. (2018). Cone and melanopsin contributions to human brightness estimation. *Josa A* 35 (4): B19–B25.

12 Bodrogi, P., Guo, X., und Khanh, T.Q. (2018). Helligkeit und Sehklarheit statt Leuchtdichte in der Bewertung der Beleuchtung von Innenräumen (Brightness and visual clarity instead of luminance in indoor lighting assessment). Tagung LICHT (Conference LIGHT) 2018. Davos, Schweiz.

13 Khanh, T.Q., Bodrogi, P., and Vinh, Q.T. (2016). *Color Quality of Semiconductor and Conventional Light Sources*. Wiley-VCH Verlag GmbH ISBN-13: 978-3527341665.

14 Moroney, N., Fairchild, M.D., Li, C. et al. (2002). The CIECAM02 Color Appearance Model. In: *Proceedings of IS&T/SID 10th IS&T/SID Color Imaging Conference*.

15 CIE (2017). *Color Fidelity Index for Accurate Scientific Use*, 224. CIE Publication.

16 Davis, W. and Ohno, Y. (2010). Color quality scale. *Opt. Eng.* 49: 033602.

17 Allen, A.E., Martial, F.P., and Lucas, R.J. (2019). Form vision from melanopsin in humans. *Nat. Commun.* 10: 2274.

18 Spitschan, M., Bock, A.S., Ryan, J. et al. (2017). The human visual cortex response to melanopsindirected stimulation is accompanied by a distinct perceptual experience. *Proc. Natl. Acad. Sci. U.S.A* 114: 12291–12296.

19 Blume, C., Garbazza, C., and Spitschan, M. (2019). Effects of light on human circadian rhythms, sleep and mood. *Somnologie* 23: 147–156. https://doi.org/10.1007/s11818-019-00215-x.

20 Ibuka, N. and Kawamura, H. (1975). Loss of circadian rhythm in sleep-wakefulness cycle in the rat by suprachiasmatic nucleuslesions. *Brain Res.* 96: 76–81.

21 Vandahl, C., Bieske, K., Neuhäuser, S., and Schierz, C. (2009). *Optimale Beleuchtung bei Schichtarbeit, Literaturstudie*. TU Ilmenau.

22 Fernandez, D.C., Fogerson, P.M., Lazzerini Ospri, L. et al. (2018). Light affects mood and learning through distinct retina-brain pathways. *Cell* 175 (71–84): e18.

23 Vandewalle, G., Maquet, P., and Dijk, D.J. (2009). Light as a modulator of cognitive brain function. *Trends Cogn. Sci.* 13 (10): 429–438.

24 Perrin, F., Peigneux, P., Fuchs, S. et al. (2004). Nonvisual responses to light exposure in the human brain during the circadian night. *Curr. Biol.* 14: 1842–1846.

25 Vandewalle, G., Balteau, E., Phillips, C. et al. (2006). Daytime light exposure dynamically enhances brain responses. *Curr. Biol.* 16: 1616–1621.

26 Vandewalle, G., Gais, S., Schabus, M. et al. (2007). Wavelength-dependent modulation of brain responses to a working memory task by daytime light exposure. *Cereb. Cortex* 17: 2788–2795.

27 Vandewalle, G., Schmidt, C., Albouy, G. et al. (2007). Brain responses to violet, blue, and green monochromatic light exposures in humans: prominent role of blue light and the brainstem. *PLoS One* 2: e1247.

28 Cohen, J.D., Perlstein, W.M., Braver, T.S. et al. (1997). Temporal dynamics of brain activation during a working memory task. *Nature* 386: 604–608.

29 Roenneberg, T., Kumar, J., and Merrow, M. (2007). The human circadian clock entrains to sun time. *Curr. Biol.* 17 (2): R44–R45. https://doi.org/10.1016/j.cub.2006.12.011.

30 Wright, H.R. and Lack, L.C. (2001). Effect of light wavelength on suppression and phase delay of the melatonin rhythm. *Chronobiol. Int.* 18: 801–808.

31 Czeisler, C.A., Duffy, J.F., Shanahan, T.L. et al. (1999). Stability, precision, and near-24-hour period of the human circadian pacemaker. *Science* 284: 2177–2181.

32 Khanh, T.Q., Bodrogi, P., and Trinh, Q.V. (2018). Intelligente Beleuchtung und HCL-Beleuchtung Grundsätze und Zusammenhänge (Intelligent lighting and HCL lighting principles and contexts). *Z. Licht* 9: 72–79.

33 Boyce, P. (2016). Editorial: Exploring human-centric lighting. *Light. Res. Technol.* 48: 101.

34 Vandahl, C., Moosmann, C., und Funke, C. (2016). Einheitliche Bewertung von Lichtsituationen mit dem LiTG-Fragebogen – erste Erfahrungen (Uniform assessment of lighting situations with the LiTG Questionnaire – first experiences). LICHT 2016, Karlsruhe, 25–28 September, 27.

35 Akerstedt, T. and Gillberg, M. (1990). Subjective and objective sleepiness in the active individual. *Int. J. Neurosci.* 52: 29–37.

36 Truong, W. and Khanh, T.Q. (2021). Einfluss der Lichtintensität auf die Schläfrigkeit? Feldstudie zum Einfluss der Lichtintensität auf ältere Arbeiter in der Frühschicht (Influence of light intensity on sleepiness? Field study on the influence of light intensity on older workers on the early shift). *Z. Licht* 1 (1-2021): 64–70. ISSN: 0024-2861.

37 Juda, M., Vetter, C., and Roenneberg, T. (2013). The Munich ChronoType Questionnaire for shift-workers (MCTQShift). *J. Biol. Rhythms* 28 (2): 130–140. The adaptation of the MCTQ for use with shift workers & instructions for its evaluation. https://www.thewep.org/documentations/mctq (Zugriff am 13 January 2022).

38 Buysse, D.J., Reynolds, C.F., Monk, T.H. et al. (1989). The Pittsburgh sleep quality index: a new instrument for psychiatric practice and research. *Psych. Res.* 28 (2): 193–213. https://doi.org/10.1016/0165-1781(89)90047-4.

39 Backhaus, J., Junghanns, K., Broocks, A. et al. (2002). Test-retest reliability and validity of the Pittsburgh sleep quality index in primary insomnia. *J. Psychosom. Res.* 53 (3): 737–740. ISSN: 00223999. https://doi.org/10.1016/S0022-3999(02)00330-6.

40 Zandi, B., Klabes, J., and Khanh, T.Q. (2020). Prediction accuracy of L-and M-cone based human pupil light models. *Sci. Rep.* 10 (1): 1–14.

41 Brain Products GmbH (2014). https://pressrelease.brainproducts.com/brain-products-acticap-xpress-much-more-than-just-a-dry-electrode-system/ (accessed on 13 January 2022).

42 Santamaria, J. and Chiappa, K.H. (1987). The EEG of drowsiness in normal adults. *J. Clin. Neurophysiol.* 4 (4): 327–382.

43 Kaida, K., Takahashi, M., Akerstedt, T. et al. (2006). Validation of the Karolinska sleepiness scale against performance and EEG variables. *Clin. Neurophysiol.* 117 (7): 1574–1581. https://doi.org/10.1016/j.clinph.2006.03.011.

44 Saliva Lab Trier (2013). https://daacro.de/de/saliva-lab-trier (Zugriff am 13 January 2022).

45 Nicolson, N.A. (2008). Measurement of cortisol. In: *Handbook of Physiological Research Methods in Health PhysiologyPt. II: Physiological Sysems and Assessments: Hormonal* (ed. L.J. Luecken and L.C. Gallo), 37–74. Sage Publications.

46 Halberg, F., Barnum, C.P., Silber, R.H., and Bittner, J.J. (1958). 24-hour rhythms at several levels of integration in mice on different lighting regimens. *Proc. Soc. Exp. Biol. Med.icine* 97 (4): 897–900.

47 Kirchner, W.K. (1958). Age differences in short-term retention of rapidly changing information. *J. Exp. Psychol.* 55 (4): 352–358.

3.A Appendix A

Uniform assessment of lighting situations with the LiTG (German Lighting Society) questionnaire

Field: 5-point scale, ‘not specified’ / Laboratory: 7-point scale, ‘not specified’ not applicable

1. How are you feeling at the moment?

								Not specified
Uncomfortable ☐	☐	☐	☐	☐	☐	☐ Well		☐

2. How do you feel about the space in which you work?

							Not specified
Ugly ☐	☐	☐	☐	☐	☐	☐ Beautiful	☐
Cramped ☐	☐	☐	☐	☐	☐	☐ Spacious	☐
Boring ☐	☐	☐	☐	☐	☐	☐ Interesting	☐

3. How satisfied are you with the lighting with daylight?

							Not specified
Very dissatisfied ☐	☐	☐	☐	☐	☐	☐ Very satisfied	☐

4. How satisfied are you with the lighting with artificial light?

							Not specified
Very dissatisfied ☐	☐	☐	☐	☐	☐	☐ Very satisfied	☐

Brightness

5. How do you feel about the brightness at your workplace at the moment?

								Not specified
Very dark	☐	☐	☐	☐	☐	☐	☐ Very light	☐
I'd rather have... Clearly darker	☐	☐	☐	☐	☐	☐	☐ Clearly brighter	☐

6. How do you feel about the brightness on the walls at the moment?

								Not specified
Very dark	☐	☐	☐	☐	☐	☐	☐ Very light	☐
I'd rather have... Clearly darker	☐	☐	☐	☐	☐	☐	☐ Clearly brighter	☐

7. How do you feel about the overall brightness in the room at the moment?

								Not specified
Very dark	☐	☐	☐	☐	☐	☐	☐ Very light	☐
I'd rather have... Clearly darker	☐	☐	☐	☐	☐	☐	☐ Clearly brighter	☐

8. How do you feel about the light distribution in your room?

								Not specified
Very uneven	☐	☐	☐	☐	☐	☐	☐ Very even	☐
I'd rather have... Clearly more uneven	☐	☐	☐	☐	☐	☐	☐ Clearly more even	☐

Colour

9. How do you feel about the light colour?

								Not specified
Very cold	☐	☐	☐	☐	☐	☐	☐ Very warm	☐
I'd rather have... Distinctly colder	☐	☐	☐	☐	☐	☐	☐ Significantly warmer	☐

10. How do you feel about the colour of the objects in the room?

								Not specified
Very unnatural	☐	☐	☐	☐	☐	☐	☐ Very natural	☐

Operability, control

11. How satisfied are you with the ease of use of your sunshade?

								Not specified
Very dissatisfied	☐	☐	☐	☐	☐	☐	☐ Very satisfied	☐

12. If your room is equipped with an automatically controlled sunshade: How satisfied are you with the automatic control of your sunshade?

								Not specified
Very dissatisfied	☐	☐	☐	☐	☐	☐	☐ Very satisfied	☐

13. How satisfied are you with the **operability** of the artificial lighting?

Very dissatisfied ☐ ☐ ☐ ☐ ☐ ☐ ☐ Very satisfied — Not specified ☐

14. If the artificial light in your room is controlled automatically: How satisfied are you with the **automatic control** of the lighting?

Very dissatisfied ☐ ☐ ☐ ☐ ☐ ☐ ☐ Very satisfied — Not specified ☐

Reflections (mirroring)

15. Do you notice **reflections on** the table or on working materials at the moment?

Clearly perceptible ☐ ☐ ☐ ☐ ☐ ☐ ☐ Not perceptible — Not specified ☐

I would rather have...
More reflections ☐ ☐ ☐ ☐ ☐ ☐ ☐ Less reflections ☐

16. Do you notice **reflections** on the screen at the moment?

Clearly perceptible ☐ ☐ ☐ ☐ ☐ ☐ ☐ Not perceptible — Not specified ☐

Glare

17. **In general,** how often are you blinded by **daylight** in this room?

Actually always ☐ ☐ ☐ ☐ ☐ ☐ ☐ Never really — Not specified ☐

18. Do you perceive **glare** from **daylight** at the moment?

Intolerable ☐ ☐ ☐ ☐ ☐ ☐ ☐ Imperceptible — Not specified ☐

19. Do you **directly** perceive **glare** from **luminaires** at the moment?

Intolerable ☐ ☐ ☐ ☐ ☐ ☐ ☐ Imperceptible — Not specified ☐

In conclusion

20. How do you feel about the **view from your window** at this time of year?

Uncomfortable ☐ ☐ ☐ ☐ ☐ ☐ ☐ Pleasant — Not specified ☐

21. How satisfied are you with the illumination of your workplace by **daylight** at this time of year when the **sunshade is open**?

Very dissatisfied ☐ ☐ ☐ ☐ ☐ ☐ ☐ Very satisfied — Not specified ☐

Reason: ______________________________

22. How satisfied are you with the illumination of your workplace by **daylight** when the **sunshade is closed** as far as you consider necessary at this time of year?

Very dissatisfied ☐ ☐ ☐ ☐ ☐ ☐ ☐ Very satisfied — Not specified ☐

Reason: ______________________________

23. Do you notice a **flickering** of the lighting?

Strong flickering ☐ ☐ ☐ ☐ ☐ ☐ ☐ No flicker — Not specified ☐

24. How satisfied are you with the **artificial lighting** in your workplace at this time of year?

Very dissatisfied ☐ ☐ ☐ ☐ ☐ ☐ ☐ Very satisfied — Not specified ☐

Reason: ______________________________

25. After answering all the questions, how satisfied are you with the **general lighting situation** in your room?

Very dissatisfied ☐ ☐ ☐ ☐ ☐ ☐ ☐ Very satisfied — Not specified ☐

Source: Vandahl et al. [34]. https://www.litg.de/Service/Links-und-Downloads.html.

4

Visual Performance–Work Performance

4.1 Status of Standardisation for Interior Lighting Using the Example of DIN EN 12464

The electrification at the end of the nineteenth century and the beginning of the twentieth century in the major industrial nations of North America, Europe, and East Asia led to value-added production being shifted indoors and the industrial process being organised over a 24-hour period every day. With this electrification and with electric light, lighting science also began to investigate the relationship between visual performance, work productivity, the number of occupational accidents, and photometric parameters such as lighting levels, luminance distributions, homogeneity, and glare. Such investigations were carried out intensively in industrial environments up to the 1960s and later in offices with different forms of organisation. The accompanying circumstances of such investigations are, on the one hand, the energy crises such as the international oil crisis in 1973, which forced energy conservation and a questioning of the significance of lighting energy consumption in this particular context, and, on the other hand, the increasing applications of computer monitors and other visual devices in offices in the 1990s.

The investigations in industrial production are basically concerned with the question of how high the level of illuminance must be in order to reduce occupational accidents to a minimum order of magnitude and to increase work productivity. In addition to pure visual performance, factors such as motor ability, attention, and identification with the work and work organisation as well as fatigue also tend to enter work productivity, whereby the *long-term effect of lighting* plays an important role. Likewise, in typical office work, attention, glare, or homogeneity must be taken into account in addition to the pure visual performance for recognising *writing* with different font sizes and paper qualities.

All these investigations, definitions of terms, and international agreements find their results in today's lighting technology standards. In the context of this book, the authors concentrate on the European standard DIN EN 12464 [1] for interior lighting. In general, the standard contains the photometric, i.e. $V(\lambda)$-based, parameters for planning and evaluating indoor lighting, which is shown in Figure 4.1.

In this chapter, the two most important parameters, illuminance and luminance distribution, and their relationship to visual performance and work performance

Human Centric Integrative Lighting: Technology, Perception, Non-Visual Effects, First Edition.
Tran Quoc Khanh, Peter Bodrogi, and Trinh Quang Vinh.

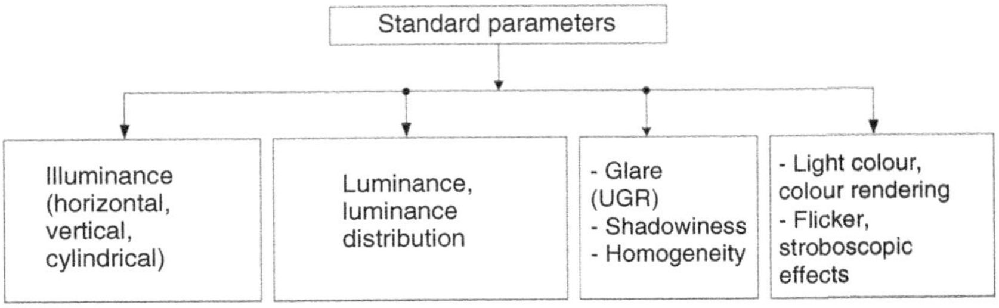

Figure 4.1 Photometric parameters in the international standards for indoor lighting. Source: TU Darmstadt.

Table 4.1 Industrial and craft activities – electrical and electronic industry.

No.	Type of indoor space (area), visual task area or activity area	$\bar{E}_m$ lx	UGR_L –	U_o –	R_a –	Specific conditions
1	Cable and wire production	300	25	0.60	80	—
2	Winding:	300	25	0.60	80	—
	– large bobbins	500	22	0.60	80	
	– medium bobbins	750	19	0.70	80	
	– fine bobbins					
3	Coil impregnation	300	25	0.60	80	—
4	Electroplating	300	25	0.60	80	—
5	Assembly work:	300	25	0.60	80	—
	– coarse, e.g. large transformers	500	22	0.60	80	
	– medium-fine, e.g. switchboards	750	19	0.70	80	
	– fine, e.g. telephones, radios, and IT products (computers)	1000	16	0.70	80	
	– very fine, e.g. measuring instruments, circuit boards					
6	Electronics workshops, testing, and adjusting	1500	16	0.70	80	—

Source: Adapted from DIN EN 12464:2011-08.

are presented. In the context of this book, the analyses of the human-centric lighting (HCL) aspects assume that certain *perceptual thresholds* exist for the parameters glare, shadowiness, and homogeneity (see Figure 4.1). The discussion of the preferred chromaticities of white light (via correlated colour temperature, CCT) as well as colour rendering, and its semantic scaling will be shown in detail in Chapters 6 and 9.

Tables 4.1–4.3 show extracts from DIN EN 12464-1:2011-08.

The standard DIN EN 12464-01 states the importance of illuminance levels and the related requirements:

> Illuminance levels and their distribution in the visual task area and in the surrounding area have a great influence on how quickly, how safely and how easily a person grasps and performs the visual task. All illuminance values specified in this European Standard are maintenance illuminance values and take into account visual comfort and performance requirements.

Table 4.2 Offices.

No.	Type of indoor space (area), visual task area/activity area	$\bar{E}_m$ lx	UGR_L -	U_o -	R_a -	Specific conditions
1	Filing, copying, etc.	300	19	0.40	80	—
2	Writing, typing, reading, and data processing	500	19	0.60	80	Screen work
3	Technical drawing	750	16	0.70	80	—
4	CAD workstations	500	19	0.60	80	Screen work
5	Conference and meeting rooms	500	19	0.60	80	Lighting should be adjustable
6	Reception desk	300	22	0.60	80	—
7	Archives	200	25	0.40	80	—

Source: Adapted from DIN EN 12464:2011-08.

Table 4.3 Health care facilities – Examination rooms (general).

No.	Type of indoor space (area), visual task area or activity area	$\bar{E}_m$ lx	UGR_L -	U_o -	R_a -	Specific conditions
1	General lighting	500	19	0.60	90	$4000\,K \leq CCT \leq 5000\,K$
2	Examinations and treatments	1000	19	0.70	90	—

Source: Adapted from DIN EN 12464:2011-08.

The principles for designing different lighting levels are based on the following aspects in accordance with DIN 12464-01 (DIN 12464-01, pages 10–11):

> The values given in Section 5 (see Tables 4.1–4.3) are maintenance values of illuminance in the visual task area on the assessment surface, which maybe horizontal, vertical, or inclined. Regardless of the age and condition of the lighting installation, the average illuminance for the task in question must not fall below the value given in Section 5.

The values apply to normal viewing conditions and take the following factors into account:

- Psychophysical and physiological aspects such as visual comfort and well-being.
- Requirements for visual tasks.
- Visual ergonomics.
- Practical experience.
- Operational safety.
- Economic efficiency.

The illuminance value maybe adjusted by at least one level on the illuminance scale (see Section 4.3.2) if the visual conditions deviate from the usual assumptions. The required maintenance value of illuminance should be increased if:

- The visual task is critical to the workflow.
- The correction of errors is cost intensive.

- Accuracy, higher productivity, or increased concentration are of great importance.
- Task details are unusually small or have unusually low contrast.
- The task must be carried out for an unusually long time.
- The worker's eyesight is below normal vision.

It is thus clear that the maintenance values of illuminance levels in the currently applicable international and national standards for indoor lighting are predominantly dimensioned on the principles of visual performance (i.e. establishing '*fast*' and '*safe*' visual tasks, types of task details, and visual ergonomics) and work performance (accuracy, productivity) as well as long-term effects (well-being). With such goals and attributes, lighting science attempted to address the effect of light on attention, reaction time, and accuracy in its early stages. The non-visual research following the discovery of the ipRGC cells of the human retina at the beginning of the twenty-first century can therefore be considered the second phase of lighting research in this context. In other words, there has already been 'HCL' lighting research from the very beginning, with different accents.

According to Kokoschka [2], the design of the required illuminance levels is based on the approach of achieving a relative visual performance (RVP) of at least 95%, depending on the difficulty of the visual task. A RVP of 100% (as a reference value) is usually defined – for a specific visual task – at a very high illuminance (e.g. 10 000 lx).

With regard to HCL technology in the coming years with modern and intelligent lighting systems, it is very important to answer the following questions:

1. Are today's standard values correctly dimensioned to specify relative performance at the 95% level according to visual and work performance criteria for the most typical visual tasks and work tasks? Can this dimensioning be substantiated by the best visual performance models?
2. Is it always necessary to raise the lighting level to achieve a RVP of 95% or can this also be achieved by varying the size and spatial structure of the visual objects?
3. Is it possible to prove the dimensioning of the standard values on the basis of *work performance* in real test and production conditions over a longer working period?
4. Based on measurements of work productivity, what do we know today about how this work productivity can be influenced by *lighting levels* in different work tasks?
5. What role do the psychological–emotional aspects play in determining work productivity in terms of user acceptance over time? Do psychological–emotional criteria and visual performance criteria lead to fundamentally different illuminance levels?

The first four questions are answered in detail in this chapter by analysing and contrasting modern visual performance models and the results on work performance from the scientific literature (see Figure 4.2). Question 5 will be partly answered in Chapters 6 and 7 and finally – in connection with the non-visual criteria – also in Chapter 9.

It is known from the literature that the necessary lighting level can be determined according to *ocular physiological* and psychological criteria. According to Völker [3], the procedures for determining lighting levels can be subdivided into physiological–optical, work-physiological, and psychological aspects. The overview is shown in Figure 4.2.

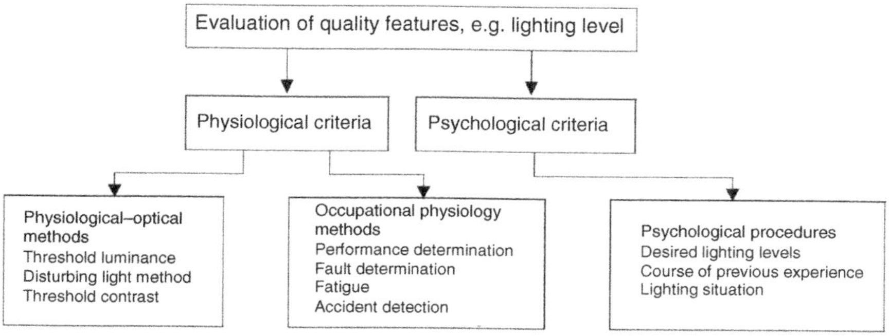

Figure 4.2 Evaluation procedure to determine the necessary lighting levels. Source: TU Darmstadt, taken from the dissertation by S. Völker [3].

By the early 1980s, visual performance models using physiological–optical methods such as the threshold contrast method led to the CIE publication 19/2.1-1981 '*An analytical model for describing the influence of lighting parameters upon visual performance*' [4]. Such methods nevertheless only determined the *threshold* illumination level or the *threshold* contrast in case of a specific visual task, and it is difficult to determine and use so-called *practice factors* or *field test factors* to move from a visual condition at *threshold* contrast or *threshold* luminance to a visual condition that corresponds to a relative performance of 95%. The two visual performance models presented in the next Section 4.2 have been developed since the mid-1980s [5] and in the year 2000 [2], and are more akin to work with physiological methods under conditioned laboratory conditions for reading and concentration tasks (e.g. in offices). These models are frequently used in the scientific lighting literature and are based on numerous reliable laboratory data.

4.2 Visual Performance

4.2.1 Definition and Influencing Factors

Visual performance has been interpreted in the lighting literature for decades. Völker [3] formulated: '*Visual performance is the ability of the visual organ to grasp and perceive brightness, colour and form structures of the outside world more or less quickly*'.

Rea [5] elaborates on its definition as follows:

> Visual performance ... is defined in terms of how quickly and accurately people can process visual stimuli that are defined in terms of adaptation luminance, target contrast, and target size. One goal of lighting and vision research has been to develop a general technique for assessing visual performance in actual environments such as offices, schools, and roadways. Such a technique would permit the evaluation of real environments by criteria relevant to the visual system, rather than by intermediary criteria such as illumination levels.

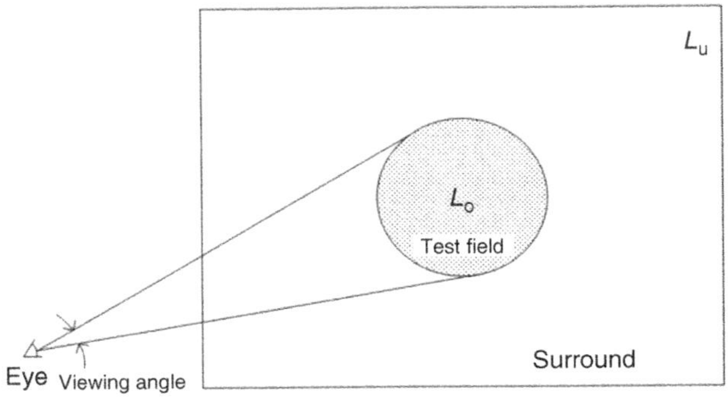

Figure 4.3 Illustration of the concept of contrast in Eq. (4.1). Source: TU Darmstadt.

In contrast to the term '*visual performance*', work performance in *work science* is represented by the time required to complete a work task and the errors made during this process. In the classical lighting experiments of Rea [6, 7], Weston [8], and Boyce [9], the search time and the reaction time as well as the error frequency were measured as a function of the lighting level, whereby the measurement result (time value) consists of a purely visual (lighting-dependent) and a motor-related and therefore lighting-independent time component. The illumination-independent time is, for example, the time for pressing a button or for crossing a visual mark on the paper with the pencil and is largely constant during the experiment. Kokoschka [2] wrote about this: '*In this sense, visual performance is understood to mean only the purely visually conditioned performance component*'.

Visual performance depends not only on the level of illumination but also on *visual detail* such as the contrast of the object related to its background, spatial structures, and contrasts inside the vial object (e.g. a letter) and the size of the viewing field. Figure 4.3. illustrates the concept of *contrast*, which is defined by Eq. (4.1).

$$\text{Contrast: } K = C = \frac{L_o - L_u}{L_u} = \frac{\Delta L}{L_u} \tag{4.1}$$

Detail size or field of view size is most often illustrated in vision science with the Landolt ring. At a certain contrast between the ring and the background (e.g. paper) as well as at a certain size, the recognisability of the ring by the test person can be demonstrated by their answer of where the ring gap is oriented (see Figure 4.4, left). The visual angle of the gap can be calculated by the size of the gap and the distance

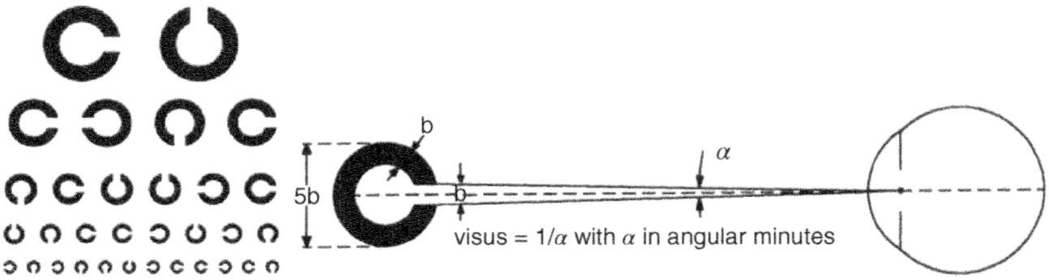

Figure 4.4 Landolt rings with different sizes and orientations of ring gaps. Source: TU Darmstadt.

between the ring and the eye position (in angular minutes) and the reciprocal value of this is defined as *visual acuity* (or visus see Figure 4.4, right). A normally sighted young person with a visual acuity of 2 is able to resolve a visual object (e.g. a letter or a Landolt ring) with a visual angle of 0.5 angular minutes. To illustrate the above considerations, Figure 4.5 shows rows of letters with different contrasts and letter sizes (detail sizes).

There have been numerous studies in recent decades on the actual contrasts in a typical office with paper documents. In Table 4.4, contrasts of typical reading materials are shown (after Van Bergem-Janssen [10]). In a field study by R.F. Dillon et al. [11] it was found in the tested American offices that contrasts with $K > 0.7$ occur very frequently.

Table 4.5 shows the luminance levels of paper materials with different degrees of reflectance at different illuminance levels on the worktable according to DIN EN 12464-01.

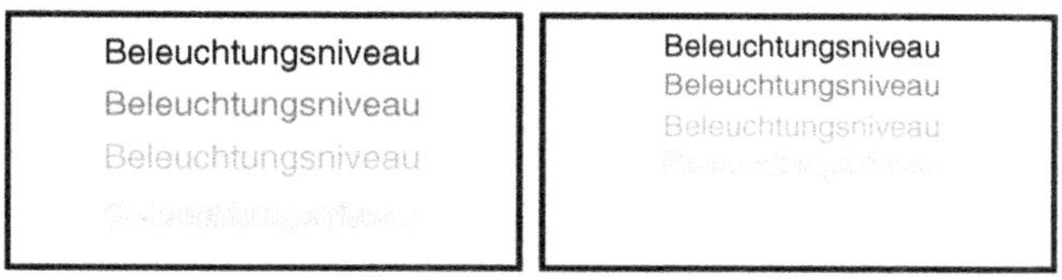

Figure 4.5 Series of fonts with different contrasts and font sizes. 'Beleuchtungsniveau' means 'illuminance level'. Source: TU Darmstadt.

Table 4.4 Contrasts of typical reading materials.

Reading material	**Contrast**
Hard pencil on white paper	0.3
Soft pencil on white paper	0.7
Print letters in newspaper	0.8
Black ink on white paper	0.9

Source: Adapted from Van Bergem [10].

Table 4.5 Luminance values of paper materials with different reflectance values at different illuminance levels.

	Paper $\rho = 0.7$	**Paper $\rho = 0.9$**
Illuminance (lx)	**Luminance in cd m^{-2}**	**Luminance in cd m^{-2}**
300	66.9	86
500	111	143.3
750	166.5	215
1000	222	286.6

Source: TU Darmstadt, compiled according to DIN EN 12464-01.

Table 4.6 Field of view sizes in angular minutes for fluent reading with different font sizes at a viewing distance of 500 mm.

Reading task/font size (*p*)	Font height (Didot point, mm)	Critical detail (mm)	Field of view size in angular minutes at a viewing distance of 500 mm
10	3.75	0.75	5.1′
11	4.125	0.825	5.7′
12	4.5	0.9	6.3′

Source: Adapted from [12] CODE-Knacker https://www.code-knacker.de/typografisch.htm/last (accessed 20 January 2023).

Table 4.7 Visual object contrasts and critical detail dimensions to characterise differently difficult visual tasks.

Visual task	Contrast *K*	Critical detail size α
Easy	0.9	3′
Moderately difficult	0.9	1.5′
Difficult	0.3	1.5′

Source: Adapted from Kokoschka [2].

The typical tasks in an office include reading documents with different fonts. In the case of alphanumeric characters, experience has shown that the size of critical visual detail corresponds to about 1/5 of the font height (Kokoschka [2]). According to [12], for *fluent reading*, the font sizes in Table 4.6 should be used.

Table 4.7 lists visual object contrasts and critical detail dimensions to characterise differently difficult visual tasks.

4.2.2 Rea's RVP Model, 1991

After the procedures with threshold luminance and threshold contrasts had been analysed and had proved to be methodologically difficult and problematic in practice, a research team led by Rea [5–7] looked at visual and work performance *above threshold* in two different visual tasks in both 1986 and 1988 and combined the results of the two experiments into a RVP model in 1991 [5]. In this section, these two experiments and their important results are brought together and analysed for interpretation in terms of HCL illumination.

4.2.2.1 Experiments and Results from 1986

These experiments were conducted to evaluate the effects of adaptation luminance and object contrast on speed and accuracy in processing visual information while performing a simulated, near-realistic task [6]. Subjects were 22 years old on average and were given the task of comparing two lists of 20 pairs of 5-digit '*number words*'. The measured time for this task was related relative to the time under a reference condition with a high luminance. The time of *marking* the non-applicable digits (a *motoric* procedure) was subtracted from the time measured in case of each comparison procedure.

The visual condition was varied by changing the adaptation luminance L_A between 12 and 169 cd m^{-2} and the contrast K between 0.092 and 0.894. The *RVP values* are related to a visual output at $L_A = 169$ cd m^{-2} and can assume a value greater than 1.0 if the adaptation luminance is greater than 169 cd m^{-2}, which corresponds approximately to an illuminance of 750 lx at a paper reflectance of 0.7 according to the representation in Table 4.4. The mathematical formulation is given in [6] and in [5] in detail.

If one applies the RVP model to reading tasks and *writing recognition* tasks, one obtains the two Figures 4.6 and 4.7. Figure 4.6 shows the RVP as a function of contrast (from 0.092 to 0.894) with *adaptation luminance* as a parameter. The curves increase very steeply up to a contrast of about 0.20–0.25 and then slowly go into saturation, regardless of the adaptation luminance. A RVP of 0.95 = 95% can already be achieved at $L_A = 169$ cd m^{-2} with $K = 0.2$ or at 22 cd m^{-2} (i.e. at about 100 lx and with a paper reflectance of 0.7) and $K = 0.4$. It can be assumed that the RVP from $K = 0.7$ depends only on the adaptation luminance.

From Figure 4.7, which illustrates the RVP as a function of adaptation luminance with *contrast* as a parameter, it is clear that RVP increases strongly with contrast. But this increase slows down from the contrast of about $K = 0.2$.

The model and the contents of Figures 4.6 and 4.7 ultimately show that for typical contrast ratios in today's offices of $K > 0.7$, an adaptation luminance of higher than 22 cd m^{-2} is sufficient to achieve a RVP of 95% (compared to 169 cd m^{-2}). An adaptation luminance of 22 cd m^{-2} corresponds to an illuminance of about 100 lx

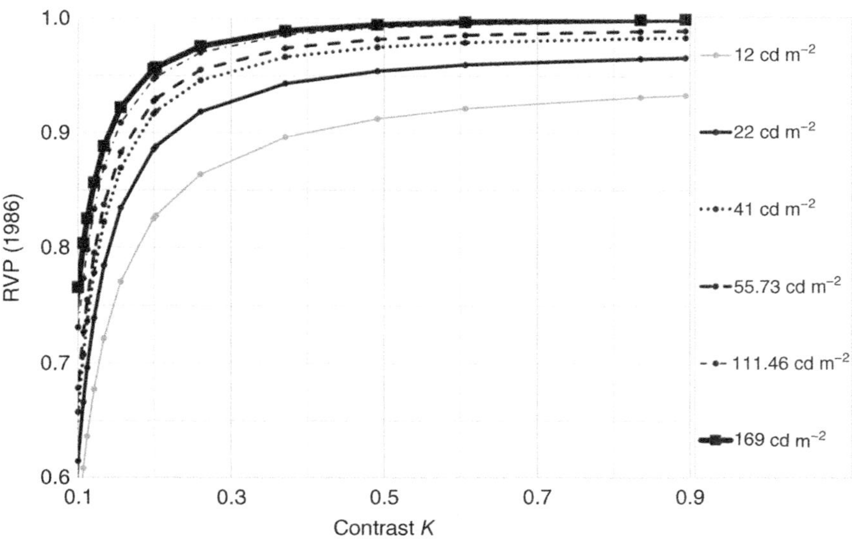

Figure 4.6 Relative visual performance (RVP) as a function of contrast with the adaptation luminance as parameters according to [5, 6]. Source: TU Darmstadt.

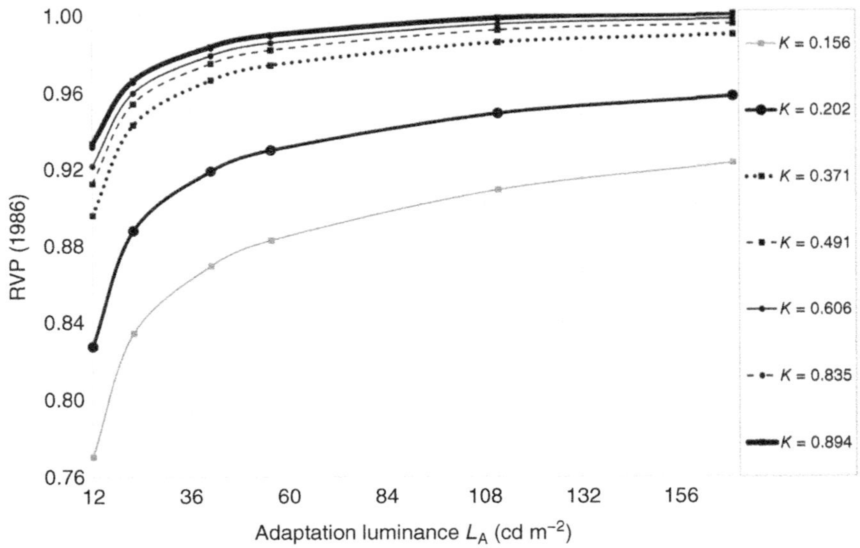

Figure 4.7 Relative visual power RVP as a function of adaptation luminance with contrasts as parameters according to [5, 6]. Source: TU Darmstadt.

with a paper reflectance of 0.7 (see Table 4.4). However, this statement only applies to young people (aged about 22).

4.2.2.2 Experiments and Results from 1988 and Modelling from 1991

In 1988, Rea and Ouelette conducted a contrast test in which young subjects with a median age of 22 years viewed a monitor with variable luminance and a constant artificial eye pupil of 2 mm, yielding retinal illuminances between 0.53 troland and 801 troland [5, 7]. Square objects with different positive and negative contrasts and a field of view between 2×10^{-6} sr and 2.8×10^{-3} sr were imaged on the monitor, and *threshold contrast* and *reaction time* were measured. In this way, reaction times were measured from the threshold contrast condition to saturation at a reference condition. The reference condition was set to a high luminance of the monitor (corresponding to 801 Troland), a high contrast and with a large field of view of $\omega > 1.3 \times 10^{-4}$ sr. This modelling was then transformed into the above 1986 model in 1991. The final model is shown in Figure 4.8. In this new modelling, the aspects of visual performance at different ages of observers from 20 to 65 years and pupil diameters were also considered.

Figure 4.9 shows – as an application of the Rea and Oulette model of 1991 – the RVP values at a contrast of $K = 0.57$, which is quite typical in office rooms, at different adaptation luminance for three age groups (20-year-olds, 50-year-olds, and 65-year-olds), where the field of view equals 4.8×10^{-6} sr, which is equivalent to a detailed size of 7.7′ angular minutes. While a RVP of 95% is achieved at a luminance of 15 cd m^{-2} for a 20-year-old, it is achieved at 30 cd m^{-2} for a 50-year-old and 80 cd m^{-2} for a 65-year-old. A luminance of 80 cd m^{-2} is achieved with a paper reflectance of 0.7 at an illuminance on a paper of 360 lx. With a nominal illuminance

The RVP model of M. Rea and Ouelette from 1991 with 7 basic equations (4.5)

(1) Retinal Illuminance - $I_{R'} = P \cdot L_A \cdot \pi r^2$
- Age factor $P = 1 - 0.017(a - 20)$, a = age between 20 and 65
- Pupil radius in mm $r = 2.3859 - [1.2204 \cdot \tanh(0.3 \cdot \lg(L_A)]$
- Pupil diameter $D = 2 \cdot r = 4.77 - [2.44 \cdot \tanh(0.3 \cdot \lg 10(L_A))]$

(2) Retinal contrast ε and threshold contrast $C'_{t,d}$
- $\varepsilon = 1 + [(\frac{0.113}{45}) \cdot (a - 20)]$
- $C'_{t,d} = \varepsilon \cdot 10^{(-1.36 - 0.179 \cdot A - 0.813 \cdot L + 0.226 \cdot A^2 - 0.0772 \cdot L^2 + 0.169 \cdot A \cdot L)}$
  - $A = \lg(\tanh(20\,000 \cdot \omega)$
  - $L = \lg(\lg(10 \cdot I_{R'}/\pi)$
  - $I_{R'} = P \cdot L_A \cdot \pi r^2$ = Retinal Illuminance (0.53 – 801 Troland)
  - ω = Solid angle of critical viewing object details $\approx \alpha^2$; from $2.0 \cdot 10^{-6}$ to $2.8 \cdot 10^{-6}$ in steradian (Sr)

(3) Half - saturation constant K
- $\lg(K) = -1.76 - 0.175 \cdot A^* - 0.0310 \cdot L^* + 0.112 \cdot A^{*2} + 0.171 \cdot L^{*2} + 0.0622 \cdot A^* \cdot L^*$
  - $A^* = \lg(\tanh(5000 \cdot \omega))$
  - $L^* = \lg(\tanh(0.04 \cdot I_{R'}/\pi)$

(4) Maximal visual performance $R_{max} = 0.000196 \cdot \lg(IR') + 0.00270$

(5) Search performance R and its reciprocal value RT
- $R = \frac{\Delta C_d^{0.97}}{\Delta C_d^{0.97} + K^{0.97}} \cdot R_{max}$
- $RT = 1/R$
- Contrast difference $\Delta C_d = C_v - C_{t,d}$; $\Delta C_d > 0$, $C_v = \frac{|L_o - L_u|}{L_u}$ like $|K|$ in Eq. (4.1)

(6) Search time ΔT_{vis}
- $\Delta T_{vis} = RT_{ref} - RT$
  - RT_{ref} is the random reaction time associated with the following stimulus conditions: $\omega > 13 \times 10^{-5}$ Sr, $I_{R'} = 101$ troland and $C \to \infty$
  - $RT_{ref} = 305.38324$ ms

(7) Visual performance
- $\text{RVP}_{RT} = \text{RVP}' \cdot \left(\frac{\Delta T_{vis} - \Delta T_{vis,r}}{\Delta T'_{vis} - \Delta T_{vis,r}}\right) \cdot 100$
- $\Delta T'_{vis} = -22.992718$ ms
- $\text{RVP}' = 0.998$
- $\Delta T_{vis,r} = -800$ ms

Figure 4.8 The RVP model of Rea and Ouelette from 1991 in mathematical form [5]. Source: TU Darmstadt.

of about 500 lx according to DIN EN 12464-01, this model thus achieves a RVP of 95% even for 65-year-old subjects. Nevertheless, this statement must be strongly relativised because the detail size of 7.7′ is relatively large.

4.2.3 The Model of Kokoschka on the Data Basis of Weston

Weston [8] published experimental results on visual performance in 1945 [2, 8]. The test subjects had the task of reading black Landolt rings printed on white paper *line by line* and marking those Landolt rings with ring gaps that deviated from a predefined standard orientation. The search time required to find these special ring gaps was determined, from which the time required to mark the rings (between 0.625 and 0.7 seconds) was subtracted. The test conditions are shown in Table 4.8.

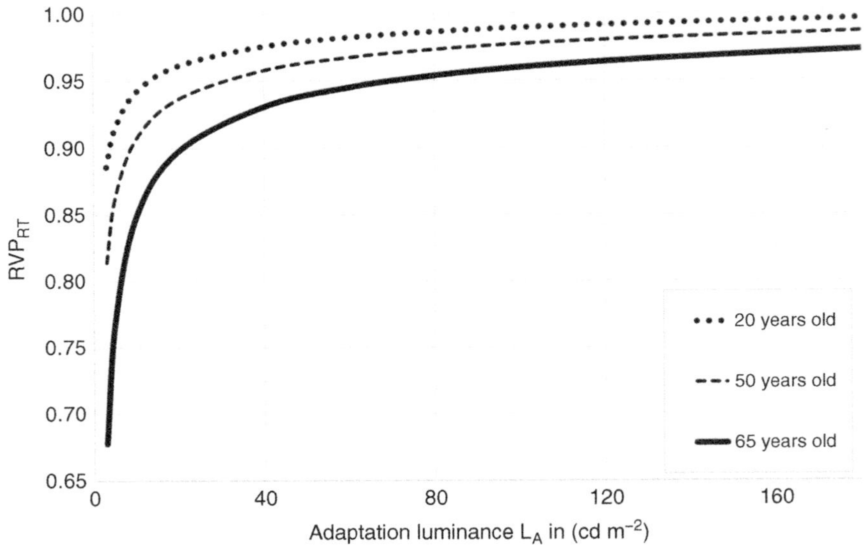

Figure 4.9 Relative visual performance of three different age groups at a constant contrast of 0.57 and a detailed size of 7.7 angular minutes. Source: TU Darmstadt.

Table 4.8 Test conditions of the visual performance tests by Weston (1945).

Number of test subjects	**12**
Test number	78 tests per set, with 3 detail sizes at 4 contrast levels plus a '*reserve contrast*' at one detail size, at 6 lighting levels. The '*reserve contrast*' occurred with bright Landolt rings printed on dark paper and this experiment was conducted at 3′
Scope	One session in the morning and one in the afternoon per subject, to be completed with one set of tests per session per subject
Test duration	Effectively 78 minutes plus adaptation time and lighting level changeover, 120 minutes in total
Contrast levels	0.97–0.56–0.39–0.28
Paper property	Reflectance 0.9; the four contrasts above were achieved by varying the reflectance of inks
Illuminance in lx	$E = 5.4–21.5–86.1–344.3–1377–5509$
Luminance on paper in cd m^{-2}	$L_A = 1.54–6.16–24.67–98.67–394.6–1578.9$
Visual angle	1.5′, 3′, and 4.5′

Source: Adapted from Weston [8].

The advantage of using the data from Weston [8] is that these lighting conditions range from as low as $E_v = 5.4$ lx to as high as $E_v = 5509$ lx or $L_A = 1578.9$ cd m^{-2}. This covers the lighting conditions of almost all industrial and office lighting applications with and without daylight. The performance measure is defined as the product of the *reciprocal of search time per search character* and a so-called *correctness factor*, which is calculated as the ratio of correctly marked characters to all displayed characters.

These data from Weston were used by Kokoschka in 2000 [2] for the construction of a mathematical model. This model is described in Eqs. (4.2)–(4.4).

$$P = 100 \cdot \frac{1}{1 + \left(\frac{L_u}{L_o}\right)^{-0.6}} \cdot \frac{1}{1 + \left(\frac{z}{1.12}\right)^{-2.5}} \tag{4.2}$$

$$L_o = 5.7 \cdot (\alpha \cdot K)^{-3} \tag{4.3}$$

$$z = \alpha \cdot K^{0.5} \tag{4.4}$$

- P: Visual performance in the range between 0 and 100
- L_u: Adaptation-relevant ambient luminance [$= (\rho \cdot E)/\pi$] in cd m^{-2}
- α: Critical detail size of the Landolt rings in arcminutes
- K: Luminance contrast of the visual objects ($= \Delta L/L_u$)
- ρ: Reflectance of the visual objects
- E: Illuminance on the subject's working object in lux

The model in Eqs. (4.2)–(4.4) contains the expressions L_u (adaptation luminance) and L_o, as well as functions of the critical detail size of the Landolt, rings (α) and the contrast K. At a high luminance (approximately greater than 1500 cd m^{-2}), the maximum visual performance approaches the value in Eq. (4.5).

$$P_{max} = 100 \cdot \frac{1}{1 + \left(\frac{z}{1.12}\right)^{-2.5}} \tag{4.5}$$

If, for a given visual object, the measured visual performance is related to a maximum value achievable at a very high illumination level, this results in a RVP denoted by p here, see Eq. (4.6).

$$p = 100 \cdot \frac{P}{P_{max}} = 100 \cdot \frac{1}{1 + \left(\frac{L_u}{L_o}\right)^{-0.6}} \tag{4.6}$$

This RVP p thus depends on the adaptation luminance (illuminance on the object), on the critical object detail size and on the contrast.

From Figure 4.10 in which the absolute visual performance P *is* illustrated as a function of illuminance, contrast, and critical detail size, the following aspects can be recognised:

- As illuminance increases, absolute visual performance P increases.
- With low contrasts such as $K = 0.3$ and a critical detail size of 1.5′, absolute visual performance is not higher than 30, even with a very high illuminance of 10 000 lx. If contrast is increased to a high value of 0.9, absolute visual performance is still only 65 at $E_v = 10\,000$ lx with a critical detail size of 1.5′.
- Consequently, one should increase the critical detail size from 1.5′ to 3′, if possible, to increase the visual performance. With a contrast of $K = 0.9$, detail size of 3′, one achieves an absolute visual performance of 90 at $E_v = 1000$ lx.

With the above model of Kokoschka [2] based on Weston's data for Landolt ring recognition, which closely resembles the reading task in an office environment, the magnitudes of illuminance required to achieve a RVP of 95% with different typical office paper materials and ink–paper contrasts should be calculated for a character

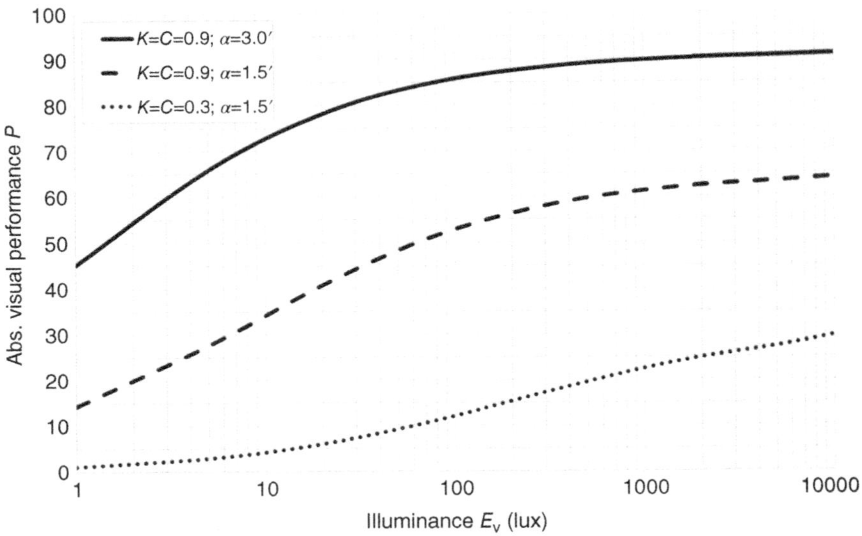

Figure 4.10 Absolute visual performance P as a function of illuminance, contrast, and critical detail size according to [2]. Source: TU Darmstadt.

Table 4.9 Required illuminance levels for a relative visual performance $p = 95\%$ with different typical office paper materials and ink–paper-contrasts and with a critical visual detail size of 2.2′ according to the model [2].

Visual task	Paper reflectance $\rho = 0.75$	Paper reflectance $\rho = 0.9$
Contrast $K = 0.7$; $\alpha = 2.2'$	880 lx	740 lx
Contrast $K = 0.56$; $\alpha = 2.2'$	1725 lx	1440 lx

Source: TU Darmstadt.

height of 1.54 mm (small letters on A5 format paper), which corresponds to a critical detail size of 2.2′ at a reading distance of 480 mm.

A further analysis (see Table 4.9) makes it clear at this critical level of visual detail size that the change in paper reflectance plays a relatively minor role in achieving visual performance. Object contrast K is much more significant and can be improved by using different inks, ink types, and pencil types. For typical office paper reflectance, contrast, and A5 book size, the illuminance required to achieve 95% RVP is between 740 and 1725 lx.

4.3 Work Performance

4.3.1 Assignment of Work Performance Aspects

In a work process in an office, a nursing home, or a factory floor, work performance depends not only on the performance in solving the immediate visual task but also on the work organisation, the motivation of workers and the interaction of people

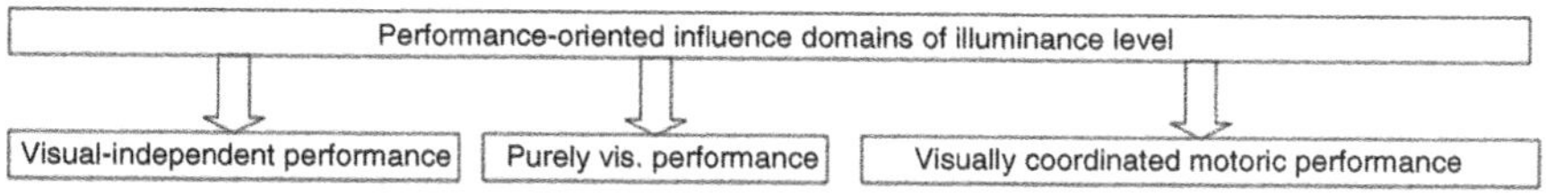

Figure 4.11 Performance-oriented spheres of influence of the lighting level according to [13]. Source: TU Darmstadt.

with each other, as well as also on *visual performance independent* and *visually coordinated* motor performance. The performance and degrees of fulfilment of the work organisation, the process design as well as the interaction of workers with each other are strongly influenced by the psychological, emotional, and health aspects of the workers involved, which are discussed in Chapter 9.

The human being in *work processes* and in *social life processes* perceives a certain amount of information from his environment per unit of time, 80–90% of which is *of optical nature*. The visual information density is characterised by the variety of geometry, shape, movement, dynamics, colours, and intensity of the representation of the environment. The clearer the information structure, the clearer people's reactions, and associated actions. In order to analyse the relationship between work performance and the level of lighting, Lindner [13] presented the scheme in Figure 4.11.

Lindner [13] formulated the following:

> Sight-independent performances include activities that take place completely without the involvement of the visual organ or in which the visual organ plays only a subordinate role, for example, solving an arithmetic problem. The purely visual performances characterise activities in which the visual organ alone or predominantly alone determines the performance, for example, the recognition of a very small number. Finally, visually coordinated motor performances are to be understood as activities in which the visual organ assumes a coordinating function between different motor and visual performances, for example, leading a thread into a needle [13].

In this way, work performance in an office environment tends to consist of *pure* visual performance and *vision-independent* performance. In an industrial environment in the classical sense with a moderate degree of automation, work performance is mainly formed by pure visual performance and visually coordinated motor performance. The work performance obtained in a work process depends to a decisive degree on the degree of difficulty of the task (among other things, visual tasks, combination of work steps one after the other, cognitive and physical demands, duration of time, and work pressure), so that the correlation between work performance and the lighting level is to be determined over a longer period of time and for a variety of the work tasks. In Section 4.3.2, the relationship between work performance, motivation, and stress is discussed and consequences for the dimensioning of the necessary lighting levels are derived. The content of this section is based on the publication by Krüger and Schierz [14] and is supplemented by the further interpretation of the authors of this book.

4.3.2 Model for Stress Regulation Under Poor Lighting

In a publication from 1950, the relationship between work performance and motivation was characterised as follows [14, 15]: *'The real work performance in industry can be regarded as the product of ability and motivation'* [16]. Ability includes intelligence, experience, and knowledge of the specific work process. Motivation is defined by Gendolla and Richter in 2010 as *'the process that determines the direction and energization of behaviour'* [17]. According to Krüger [14] *'motivation has a directional component, which indicates the direction of energy mobilisation, and an energetic aspect. The energetic aspect refers to the mobilisation of effort and can be explained by the construct 'effort''*. The directional component includes the goals and expectations for the work as well as the conviction and identification with the content of the work, see Figure 4.12.

The generation of performance depends on the degree of difficulty of the task as well as on the so-called *disturbance variables* and is maintained according to the so-called Hockey model [15] as follows: *'The energetical-control framework argues that the maintenance of performance stability under demanding conditions is an active process under the control of the individual, requiring the management of cognitive resources through the mobilisation of a mental effort'* ([15], p. 78). However, if the difficulty of the tasks, the duration of the work, and the disturbance variables (e.g. poor lighting conditions) increase, performance and stress *compensation* occurs. Krüger [14] formulated the following:

According to Hockey, the following four compensation strategies can be explained with the model (Hockey 1997):

- Power reduction in secondary tasks
- Acceptance of higher stress (power compensation)
- Strategic adjustments of work performance or quality of work (stress compensation)
- Consequences of fatigue

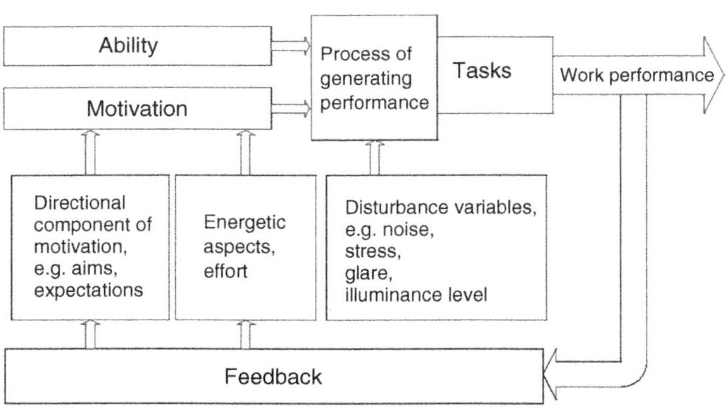

Figure 4.12 Process and components for work performance generation and control according to [14]. Source: TU Darmstadt.

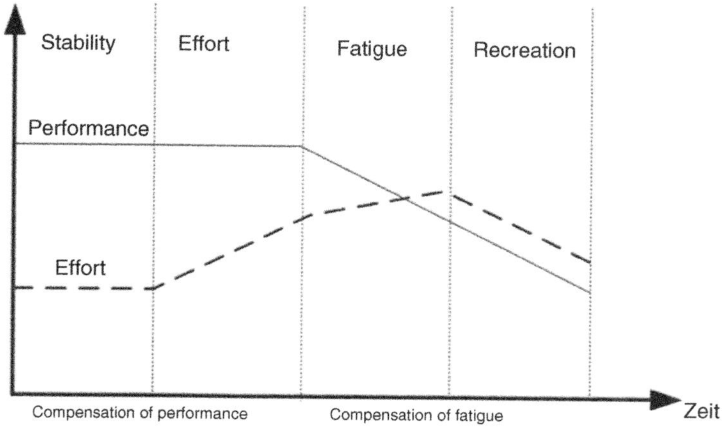

Figure 4.13 Compensation phases according to Haider and Rohmert (1978) after [18] (cited by [14]). Source: TU Darmstadt.

The four points are performance patterns under stress conditions in which compensation processes come into play that are intended to ensure the stability of the system.

These four phases of compensation are shown in Figure 4.13.

The *performance compensation* phase runs through the first two phases of Figure 4.13. In the first phase, both performance and effort are constant. The work content is filled with routine and low-threshold disturbances. In the second phase (effort increases), the disturbance variables (environmental influences, lighting conditions such as glare or lower illuminance levels) increase, although the performance can still be kept constant if a greater effort is made. The last two phases form the area of *stress compensation.* In the case of stress compensation, work objectives, speed, quantity, and/or quality of task performance are reduced.

In the following two Sections 4.3.3. and 4.3.4, the correlations between lighting level and work performance, accident records and work productivity in the office branch and at industrial workplaces are analysed on the basis of literature evaluations. Optimal illuminance levels from the point of view of work performance will be compared with the nominal values of today's lighting technology standards.

4.3.3 Influence of Lighting Level on Mental Work

In the office environment, the most common visual and work task is the reading task, both with paper documents and with monitors. For reading tasks with paper documents, the horizontal illuminance on the document or on the worktable is the relevant photometric quantity. When capturing visual information on a display, the monitor luminance and the luminance in the field of vision (e.g. the luminance of the wall opposite the workstation) are important. Optimal wall luminance is described in both Section 6.3 and Chapter 9 of this book. In a meeting room, cylindrical (vertical) illuminance at the participants' faces is crucial for visual communication. For this reason, there has been a great deal of fundamental research in lighting science for several decades to investigate the relationship between office work performance and lighting conditions. In this section, the fundamental work of Boyce [19] and Lindner [13] are analysed.

4.3.3.1 The Experiments of Boyce

After the experiments of Weston [8] in the 1940s and before Rea's experiments in the 1980s and 1990s on visual and work performance [5–7], Boyce conducted a series of basic visual experiments in the 1970s, the results of which are still valid today [9, 19]. The advantages of his experiments [19] are the test methodology, the statistical evaluation and the large, and relevant variation range of the illuminances used, e.g. between 160 or 220 lx and 1550 or 1600 lx (see Table 4.10). The test duration of about one hour is moderate.

As with Weston's experiments, this study also used [19] *Landolt rings* as visual objects, choosing one of the following two test modes:

- *So-called 'self-paced' Landolt ring procedure*, in which the test person could determine the viewing duration of the visual sign. The subject's head was fixed at a distance of 410 mm from the test panel, which contained 100 Landolt rings in a matrix of 10 × 10 rings. The subject's task was to read the matrix and cross as quickly and accurately as possible those rings that had no ring gap or a gap in a particular orientation. The time taken to complete the tests and the errors made were recorded and converted to a measure of *work performance*.
- *So-called 'paced' Landolt ring procedure*, in which the viewing duration of the visual sign was specified as 0.65 seconds by the *test administrator*. Here the subject's head was also fixed at a distance of 410 mm from the Landolt ring. The 46 Landolt rings themselves were located on a roller and were presented in two roller rotations at the time of 0.65 s per visual sign. A total of 92 Landolt rings were thus randomly presented per test.

The photometric properties of the Landolt rings are shown in Table 4.11.

The results of Boyce [19] were analysed by Krüger [14]. Table 4.10 (extract from [14]) contains the relevant data. In the diagrams contained in Table 4.10, the subjective effort data (black continuous curves) and the objective performance data (black dashed curves) collected in the questionnaires were standardised by a z-transformation and plotted as a function of illuminance. In this way, these data could be presented together in the same diagram with a uniform y-axis scaling.

Analysing these three diagrams in Table 4.10, the following results can be obtained for the dimensioning of illuminance levels for reading tasks in the office branch:

- In the case of ring-type A with a lower contrast of 0.4 and consequently, with a higher degree of difficulty, the effort decreases monotonically with increasing illuminance. This is not the case with ring type B with a higher contrast of 0.69. The effort shows a minimum at about 650 lx and increases again with higher and lower illumination.

Table 4.10 Test procedures and results of the experiments by Boyce [19] (cited by Krüger and Schierz [14]). Subjective effort data (Tasks variable, continuous curves) and objective performance data (Performance variable, dashed curves).

Lighting	Performance parameters	Duration of experiment, number of subjects N	Diagram
Illuminance levels: 220, 440, 700, 1050, and 1600 lx	*'Self-Paced'* Landolt task (the *test subject* can freely select the viewing time of the visual sign, ring-type A, see Table 4.11)	Age: 31–45 yr, $N = 150$, duration = 1 h	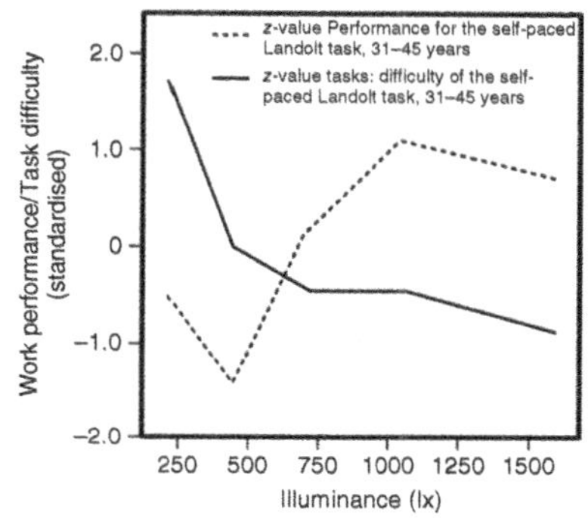
Illuminance levels: 160, 400, 650, 1000, and 1550 lx	*'Paced'* Landolt task (viewing time of the visual object set by the *experimenter* = 0.65 s; ring-type A, see Table 4.11)	Age: 31–45 yr, $N = 150$, duration = 1 h	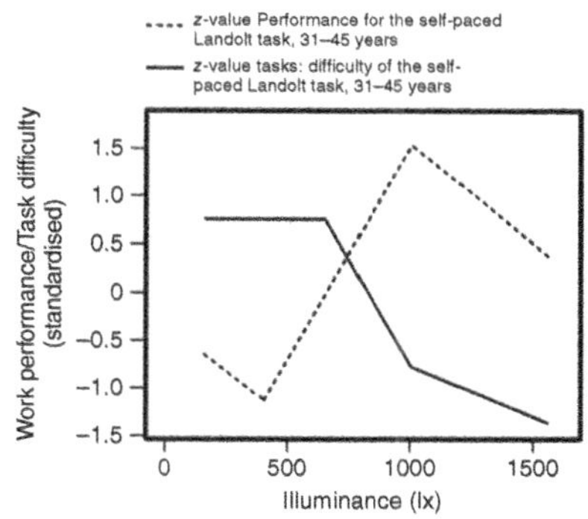
Illuminance levels: 160, 400, 650, 1000, and 1550 lx	*'Paced'* Landolt task (viewing time of the visual sign = 0.65 s, ring-type B, see Table 4.11)	Age group 31–45 yr, $N = 150$, duration = 1 h	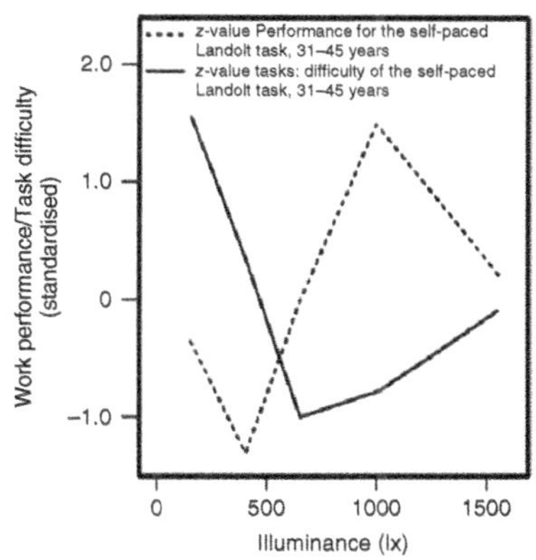

Source: Boyce [19]; Krüger and Schierz [14].

Table 4.11 Photometric properties of the Landolt ring (types A and B) in the experiments of Boyce [19].

Ring type	Visual object size	Paper reflectance	Ink reflectance	Contrast
A	0.45 mrad	0.65	0.39	0.40
B	0.45 mrad	0.65	0.20	0.69

Source: TU Darmstadt.

- The objective performance first increases with illuminance up to a maximum of 1000 lx (horizontal illuminance) and then it decreases again. This means that the performance decreases again at the higher illuminance values of 1550 or 1600 lx.

The significance of a horizontal illuminance of 1000 lx comes into play again as soon as we discuss the experiments of Lindner [13].

4.3.3.2 The Experiments of Lindner

With a large number of 96 test subjects, divided into two age groups with an average age of 30.1 and 56.0 years, respectively (see Table 4.12), Lindner pursued the goal of investigating the dependence of concentration on illuminance using the so-called *Bourdon-Wiersma test chart* [13]. This test chart consists of clusters of 3, 4, and 5 dots of which the '*4-dot-cluster*' should always be selected by the subject. On the test sheet, 1250 clusters of dots were provided among which 462 '*4-dot-clusters*' were included. In general, the contrasts and the size of the dots were relatively large, so that concentration, as well as vigilance over the entire test duration of 2.5 hours, was the focus of these tests and not necessarily visibility.

In a 1 m × 1.5 m observation booth with fluorescent lamps with white light, the illuminances on the test panel were systematically varied between 60 and 3000 lx (60, 250, 1000, and 3000 lx). The viewing distance was 40 cm. In order to remove the influence of practice, a pre-test at 250 lx was carried out before the main test, which was not included in the evaluation. In total, the following three performance parameters were determined [13]:

- Total time t (s) for completing the test sheet. From this, the time performance $1/t$ (in s^{-1}) was calculated.
- Total error F, which means the number of omitted or incorrectly ticked 'dot clusters' (i.e. with 3 and 5 points).
- The following value W, which takes into account both the errors and the time: $W = t + (1.3\ \text{seconds}) \times \text{F}$

Table 4.12 Test subjects in the experiments of Lindner [13].

Test group	Age group/years	Total	Number of men	Number of women	Average(Age/years)
1	20–40	48	35	13	30.1
2	50–60	48	37	11	56.0

Source: Adapted from Lindner [13].

Table 4.13 Result of the experiments by Lindner [13] for the age group 20–40 years.

Illuminance	60 lx	250 lx	701 lx[a)]	1000 lx	3000 lx	10 000 lx[a)]
Time performance ($s^{-1}10^{-4}$)	16.76	17.20	—	17.51	17.83	—
Error	9.3	8.1	—	7.3	8.2	—
Total performance ($s^{-1}10^{-4}$)	16.43	16.89	**17.13**	17.23	17.50	**17.85**
Relative performance	0.91	0.94	**0.95**	0.96	0.97	**0.99**

a) Extrapolated and interpolated values.
Source: Adapted from Lindner [13].

Table 4.14 Result of the experiments by Lindner [13] for the age group 50–60 years. *Extrapolated and interpolated values.

Illuminance	60 lx	250 lx	1000 lx	3000 lx	5129 lx[a)]	10 000 lx[a)]
Time performance ($s^{-1}10^{-4}$)	14.13	14.72	14.99	15.20	—	—
Error	14.7	13.1	11.7	13.1	—	—
Total performance ($s^{-1}10^{-4}$)	13.90	14.45	14.65	14.81	**14.95**	**15.58**
Relative performance	0.88	0.92	0.93	0.94	**0.95**	**0.99**

a) Extrapolated and interpolated values.
Source: Adapted from Lindner [13].

From the above value W, the total performance was determined as $(1/W)$ in s^{-1} units.

The test results are shown in Table 4.13 for the age group 20–40 years and in Table 4.14 for the age group 50–60 years, see also Figure 4.14.

For this book, the authors took the data from Lindner [13] for the total performance and extrapolated it up to a reference illuminance of 10 000 lx. Then the total performance at this illuminance was set to 99%. The total performance at lower illuminance levels was converted according to the relative performance and is shown in the two Tables 4.13 and 4.14.

From the analysis of the Tables 4.13 and 4.14 as well as Figure 4.14, we can see the following findings:

- The time performance and the total performance increase with higher illuminance levels.
- The error is smallest at $E = 1000$ lx. At lower and higher illuminances, the error increases.
- At 1000 lx, the 20–40-year age group has a 17.6% higher total performance and 38% fewer errors than the 50–60 age group.
- While for the young age group, the total performance at 1000 lx differs relatively by 2% and 1.5% from the total performance at 250 and 3000 lx, respectively, the error at 1000 lx differs by about 10% from the error at 250 and 3000 lx, respectively. If error rate and error correction are important for the work process, then an illuminance of 1000 lx will thus be of particular interest.

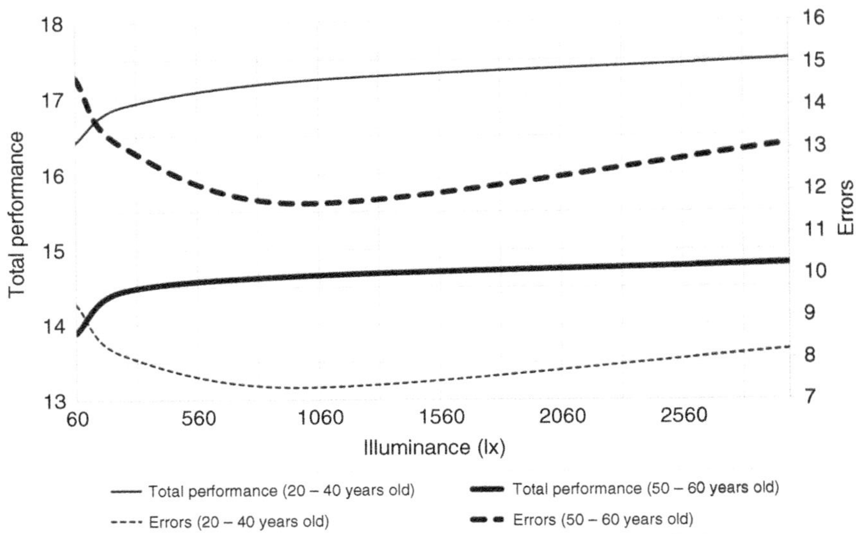

Figure 4.14 Performance and error as a function of illuminance in the experiments by Lindner [13]. The total performance on the ordinate is given in ($s^{-1} \times 10^{-4}$) units. Source: Adapted from Linder [13].

If we combine the results of the Boyce and Lindner experiments described above, we will arrive at the result that an illuminance on the work surface of 1000 lx is a possible optimum from a work-physiological point of view. However, if we look at the statements in Table 4.9 for Kokoschka's visual performance model, which is based on Weston's data for Landolt rings, for a RVP of $p = 95\%$ for different typical office paper materials and ink–paper contrasts and for a critical visual object detail size of 2.2′, one will arrive at a range between 740 and 1725 lx. These intermediate results apply more to reading and concentration tasks in an office environment. The next important question is whether an illumination level of this magnitude in the core area around 1000 lx is also highly relevant for industrial workplaces. This question will be answered in the next Section 4.3.4, on the basis of further short-term and long-term studies.

4.3.4 Influence of Lighting Levels on Work Performance in Industrial Workplaces

4.3.4.1 Literature Review Until 1971

As early as the beginning of the twentieth century, lighting science was concerned with investigating the relationship between lighting levels and work productivity as well as accidents in industrial environments. The studies conducted up to the beginning of the 1970s were systematically listed in the work of Lindner [20] in 1975. They were subdivided into laboratory experiments and factory trials. However, most of the older studies lacked a detailed description of the test conditions (lighting levels, number and structure of test subjects, and their statistics), so from today's point of

Table 4.15 Occupational–physiological assessments in laboratory tests of industrial character.

Year	Author	Type of work	No. of test subj.	Detail size	Performance time	Work durat.	Increase of lighting from–to	Performance increase
1926	Stoer	Measuring calibre with calliper	60	Not specif.	Any	20 min	50 to 600 lx	−55% Error
1930	Golstern	Recognition of threads	15	1.8′	0.5 s	1 h	100 to 1000 lx	−11% Error
1956	Studies about light	Stringing beads	15	13′–15′	Any	6 h	100 to 1000 lx	+5.5% Power
1971	Summer	Detection of defects on steel sheets	9	Not specif.	0.5 s	5 h	20 to 400 lx	8% resp. 11% for light resp. dark stripes

Source: Adapted from Lindner [20].

Table 4.16 Occupational–physiological evaluations in industrial operating tests.

Year	Author	Type of work	No. of test subj.	Trial duration	Increase in lighting from–to	Performance increase
1962	Stenzel	Punch leather	12	2 years	350 to 1000 lx	7.6%
1962	Stenzel	Camera mounting	13	2 years	370 to 1000 lx	7.4%
1969	Honestly	Mechanical manufacturing	24	2 years	270 to 760 lx	7.5% and 28% fewer accidents

Source: Adapted from Lindner [20].

view these lists of literature were reduced to a manageable number and integrated by the authors of this book into Tables 4.15 and 4.16. The criteria for inclusion in these tables are a considerable number of test subjects and experiment performance of an industrial character. Experiments with Landolt rings are thus excluded. It can be seen from Table 4.15 that illuminances in the range of up to only 100 lx cause more errors.

Despite the large number of occupational–physiological experiments since 1925, Table 4.16 can only list three experiments up to 1969 that gave an indication of the number of test subjects and had a test subject number greater than 11. These three operational trials lasted two years and documented a high data relevance for industrial work typical in precision engineering and consumer goods industries (camera assembly, leather stamping). To investigate performance increase, the authors [20] started with the range between 270 and 370 lx and increased the illuminance levels to a range between 760 and 1000 lx. In all three cases, the increase in performance was about 7.5%. It is of great relevance to note that an increase from 270 to 760 lx

(Table 4.16) reduced accidents by 28% in a mechanical manufacturing environment, in addition to the effect of the 7.5% increase in performance.

4.3.4.2 Lindner's Experiments in 1976

Lindner's experiments in [21] pursued the intention of determining the influence of illuminance on a defined practical visual–motor activity. The activity of screwing in cheese head screws with a screwdriver was selected for this purpose. The 60 screws with different slot widths of 0.15, 0.3, 0.6, and 1.0 mm were placed on four horizontal test plates and had to be rotated 90° to the left and to the right one after the other. The rotation was indicated to the test leader by a lamp control. The number of test runs and the cycle time were determined by a counter or an electric timer.

In [21], Lindner applied a metric for *visual detail structure* X according to Hentschel [22, 23] according to Eq. (4.7).

$$X = \frac{1}{\alpha} \cdot \sqrt{\frac{1}{K}} \tag{4.7}$$

- α: Object detail size in arc minutes (projected slot width)
- K: Contrast between the luminance of the screw slot and the screw heads or screw plate.

The geometry and photometric properties of the test items are shown in Table 4.17. Twenty-four normally sighted test subjects between the ages of 20 and 30 took part in the experiments. The illuminance levels varied between 22, 190, 750, and 2700 lx. To prevent exercise and fatigue influences, a pre-test was conducted at 190 lx (as a training session). Since the duration of the test was longer (4.5 hours), including a 30-minute break, the work performances determined also included components such as fatigue, concentration, visual ability, and endurance. The performance parameter considered is the turnaround time for the 4 × 60 screws, whereby possible errors due to incomplete turns or omitted screws were registered electronically and reworked by additional turning operations. This reworking time is therefore also included in the total test time. The performance was quantified using the reciprocal of the turnaround time according to Eq. (4.8):

$$L_D = \frac{1}{t_p} \tag{4.8}$$

Table 4.17 Geometry and photometric properties of the test items according to [21], see Eq. (4.7). The symbol X denotes a metric for 'visual detail structure' according to Hentschel [22, 23].

Slot width in mm	α (object size in arc minutes)	K (contrast)	X
0.15	1.25	0.66	0.98
0.3	2.49	0.64	0.50
0.6	4.98	0.41	0.31
1.0	8.3	0.31	0.21

Source: Hentschel [22]; Hentschel [23]; Lindner [21].

Table 4.18 Illuminances calculated by the authors of this book for a 95% power in the experiments of Lindner [21].

Visual detail structure X	Threshold luminance in cd m^{-2}	Increase factor from 50% perf. to 95% perf.	Luminance at 95%-perf. in cd m^{-2}	Illuminance in lx assuming a reflectance of the screw plate of $\rho = 0.7$
0.21	0.0314	550	17.3	77.6
0.31	0.062	550	34.1	153.1
0.50	0.144	550	79.2	355.6
0.98	0.4729	550	260	1167

Source: Adapted from Lindner [21].

From the results, Lindner determined the threshold luminance L_{50} for 50%-performance as a function of the 'visual detail structure' X. The 99% performance was defined at an illuminance of 10 000 lx and determined by extrapolating the performance curve over the four measuring points at 22, 190, 750, and 2700 lx. To get from a 50%-performance to a 95%-performance, an increase factor was calculated. The data in the diagrams in the work by Lindner [21] result – by the calculation of the authors of this book – in the estimated values in Table 4.18.

Combining the results of Tables 4.17 and 4.18, the necessary illuminance on the screw plate is 1167 lx, if a 95% performance is to be achieved for young people between 20 and 30 years of age when the cheese-head screws of slot width 0.15 mm are to be screwed in by the subjects.

4.3.5 Summary of the Significance of the Visual Performance and Work Performance Results – Preliminary Consequences for Indoor Lighting

In this chapter, the correlations between visual performance and work performance on the one hand and illuminance levels on the other were worked out using visual performance models according to Kokoschka and Rea, with the experiments by Lindner and Boyce for reading and concentration tasks for office-like tasks, and with the experiments in industrial operations up to 1969 and in 1976 by Lindner with an assembly activity. These results can be summarised as follows:

- Table 4.9 lists the illuminance levels required for a RVP of $p = 95\%$ with different typical office paper materials and ink–paper contrasts and at a critical visual detail size of 2.2′. These illuminances range from 740 to 1725 lx.
- The experiments by Boyce (Table 4.10) showed the highest performance at 1000 lx on the test panel with Landolt rings. Likewise, the results of the concentration experiments with office and school characters by Lindner (Table 4.13 and Table 4.14) show that an illuminance of 1000 lx allows the smallest error.
- In the experiments by Lindner (Table 4.18) with assembly activities (screwdriving), an illuminance on the screw plate of 1167 lx is required if a 95% performance is to be achieved for young people between 20 and 30 years of age when screwing in cheese-head screws with a slot width of 0.15 mm.

Table 4.19 Nominal illuminance and direction of light according to [24] (also cited in [25]) for workplaces with increased requirements.

Activity group	Diffuse lighting only		Directional and diffuse lighting		Light incidence angle with additional directional ltn	
	E_h in lx	E_z/E_h	E_h in lx	E_z/E_h	Vertical component β in °	Horizontal component ε in °
Dental laboratory						
Wax modelling	2900	0.4–0.45	3500	0.4–0.45	50–70	175–180
Ceramic processing	3000	0.4–0.5	6200	0.3	90	—
					79	185
Elaboration	3300	0.45	6800	0.3	75–90	170–180
Goldsmith's workshop						
Easy assembly	—	—	1700	0.5	65–80	Approx. 60
Fine mech. machining	—	—	3100	0.4	65–80	Approx. 60
Melting silver	—	—	2900	0.4	70–80	Approx. 60
Electrotechnical company						
Mechanical assembly	700	0.45	850	0.5	40–50	130–180
Electromech. assembly	900	0.4–0.5	1250	0.5–0.55	40–60	125–180
Platinum soldering	900	0.4–0.45	1000	0.6–0.65	40–60	125–190
Test station	1000	0.55	1100	0.55	50–60	60–65

Source: Schmidt-Clausen and Finsterer [24]; Vandahl [25].

From the analysis of visual and work performance for realistic tasks in an office and industrial environment, a range of illuminance levels between 740 and 1725 lx can be determined for 95% work performance, with a focus approximately between 1000 and 1200 lx. This analysis is supported if one also analyses the recommendation of Schmidt-Clausen and Finsterer in [24]. There, a study is described for assessing the lighting of a workplace with increased requirements in the field of electronics and precision mechanics. For these workplaces studied, the nominal illuminance levels according to current standards are not sufficient. In addition, not only the horizontal but also the cylindrical (vertical) illuminance levels are of great importance there. These requirements for high vertical (cylindrical) illuminance levels also enable a high circadian-effective effect and should be favoured by higher diffuse and additional in-directional lighting (see Table 4.19).

References

1 Deutsches Institut für Normung (2019). DIN EN 12464-1 Light and lighting - Lighting of work places - Part 1: Indoor work places. Available online at: https://www.din.de/en/getting-involved/standards-committees/fnl/drafts/wdc-beuth:din21:302583817?destinationLanguage=&sourceLanguage= (accessed 21 April 2023).

2 Kokoschka, S. (2000). Modelle der visuellen Leistungsfähigkeit für die Innenbeleuchtung (Models of visual performance for indoor lighting), pp. 569–580. Proceedings of the Conference *Licht 2000*, Goslar.
3 Völker, S. (1998). Eignung von Methoden zur Ermittlung eines notwendigen Beleuchtungsniveaus (Suitability of methods for determining a necessary lighting level). Dissertation, TU Ilmenau.
4 CIE (1981). An analytical model for describing the influence of lighting parameters upon visual performance. CIE-Publication 19/2.1-1981.
5 Rea, M. and Ouelette, J. (1991). Realative visual performance: a basis for application. *Light. Res. Technol.* 23 (3): 135–144.
6 Rea, M.S. (1986). Toward a model of visual performance: foundations and data. *J. Illum. Eng. Soc.* 15 (2): 41–57.
7 Rea, M.S. and Ouellette, M.J. (1988). Visual performance using reaction times. *Light. Res. Technol.* 20 (4): 139–153.
8 Weston, H.C. (1945). *The Relation Between Illumination and Visual Efficiency: The Effect of Brighness Contrast, Report No. 87*, 1–35. London: Industrial Health Research Board, Great Britain Medical Research Council.
9 Boyce, P.R. (2014). *Human Factors in Lighting*, 3e. CRC Press, Taylor& Francis Group.
10 Van Bergem-Janssen, P.M. (1990). Sehleistung oder Sehkomfort als Kriterium für das Lichtniveau am Arbeitsplatz (Visual Performance or Visual Comfort as a Criterion for Lighting Levels in the Workplace), Proceedings of the Conference Licht, Rotterdam, 94–107.
11 Dillon, R.F., Pasini, I.C., und Rea, M.S. (1987). Survey of visual contrast in office forms: 21st Session of the Commission Internationale de l'Eclairage, Venice, Italy (17–25 June). Paris: CIE.
12 CODE-Knacker: Lexikon der Codes – Symbole – Kurzzeichen, English: "Lexicon of codes - symbols – tokens". https://www.code-knacker.de/typografisch.htm (Reached on 22 April 2023).
13 Lindner, H. (1973). Der Einfluss des Beleuchtungsniveaus auf geistige Leistungen (The influence of lighting level on mental performance). *Z. Ges. Hyg.* 19 (2): 116–119.
14 Krüger, J. und Schierz, C. (2018). Modell zur Beanspruchungsregulation bei schlechter Beleuchtung (Model of stress regulation for insufficient lighting). Tagung LICHT 2018, Davos 9.–12.9.2018.
15 Hockey, G.R. (1997). Compensatory control in the regulation of human performance under stress and high workload; a cognitive-energetical framework. *Biol. Psychol.* 45: 73–93.
16 Brozek, J. and Monke, J.V. (1950). Miniature work situations as a research tool. *Arch. Ind. Hyg. Occup. Med.* 2: 63–97.
17 Gendolla, G.H.E. and Richter, M. (2010). Effort mobilization when the self is involved: some lessons from the cardiovascular system. *Rev. Gen. Psychol.* 14: 212–226.
18 Haider, E. and Rohmert, W. (1978). Performance- and strain-compensation with informatory work. *Eur. J. Appl. Physiol.* 39: 263–276.

19 Boyce, P.R. (1973). Age, illuminance, visual performance and preference. *Light. Res. Technol.* 5: 125–144.

20 Lindner, H. (1975). Beleuchtung und Arbeitsleistung – Systematik experimenteller Grundlagen (Lighting and work performance – systematics of experimental foundations). *Z. Ges. Hyg.* 21 (2): 101–106.

21 Lindner, H. (1976). Einfluß der Beleuchtungsniveaus auf visuell-motorische Leistungen (Influence of lighting levels on visual-motor performance). *Lichttechnik* 28 (7): 299–305.

22 Hentschel, H.-J. (1968). Beleuchtungsniveaus in betriebswirtschaftlicher Sicht(Lighting levels from an economic point of view). *Lichttechnik* 20 (7): 75A–80A.

23 Hentschel, H.-J. (1971). Zur Frage der leistungsorientierten Bewertung der Beleuchtung (On the question of performance-based assessment of lighting). *Lichttechnik* 23 (5): 304–307.

24 Schmidt-Clausen, H.-J. and Finsterer, H. (1989). Beleuchtung eines Arbeitsplatzes mit erhöhten Anforderungen im Bereich der Elektronik und Feinmechanik (Lighting of a workplace with increased requirements in the field of electronics and precision mechanics). In: *Schriftenreihe der Bundesanstalt für Arbeitsschutz, Fb 593*. Dortmund: Wirtschaftsverl NW.

25 Vandahl, C. (1998). Zur Festlegung von Gütemerkmalen der Einzelplatzbeleuchtung (On the determination of quality characteristics of single-user lighting). Dissertation, TU Ilmenau.

5

Modern Aspects of Brightness and Visual Clarity in the Context of Light Quality and Visual Performance

5.1 Introduction

The development of lighting technology started at the beginning of the twentieth century in the era of industrialisation and electrification and was intended to provide indoor workplaces with light around the clock, thereby minimising accidents at work. The research focus of lighting technology was initially on industrial lighting and later, in the 1940s, also on office lighting, with the aim of determining lighting levels on the basis of the criteria of contrast perception, object size, and adaptive luminance. In the course of this, the minimum illuminance at the working level was defined in the range of 350–500 lx, which is anchored in today's lighting technology standards [1].

In the 1960s, and then increasingly in the 1990s, the psychological component of lighting technology was intensively investigated with the aspects of homogeneity, colour distribution, colour rendering, colour temperature, and light direction with indirect and direct light components in the room [2–5]. For this integration of psychological components into lighting technology, the chromatic aspects of the radiation components reaching the eye were also considered. Nevertheless, the focus of the discussion was still predominantly on the photometric parameters, such as illuminance and luminance.

Since the beginning of the twenty-first century, there have been three new focal points in lighting technology. The first focus – following the discovery of light-sensitive ganglion cells [6, 7] – is the substantiation, detection, and quantification of the effect of optical radiation of different wavelengths and radiation components on sleep quality, work productivity, alertness, and well-being. A few years ago, it was recognised that, in this context, it is not possible to clearly separate the non-visual and visual effects of light, because the two signal streams, non-visual and visual, are partly processed in the same brain centres. The second focus is the necessity to describe the colour quality of the illumination of an object not only by the colour rendering index (CRI): for the assessment of colour quality, further parameters of other properties such as colour gamut, colour memory, and colour saturation shall be added [8, 9].

The third focus is the connection of lighting technology with architecture. In addition to a visual effect, lighting in the context of a building's architecture also has an

Human Centric Integrative Lighting: Technology, Perception, Non-Visual Effects, First Edition.
Tran Quoc Khanh, Peter Bodrogi, and Trinh Quang Vinh.

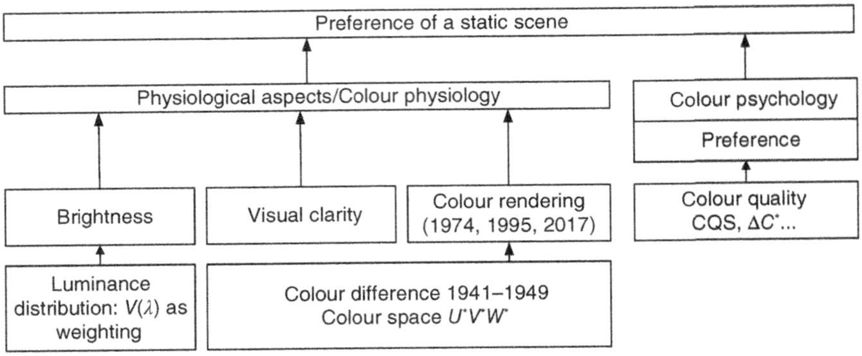

Figure 5.1 Stages in the development of lighting parameters to describe the preference of a static scene in interior lighting [10]. Source: TU Darmstadt.

emotional effect that depends on the room context and can be influenced by the different lighting parameters (together with daylight integration) and by emphasising certain objects in the room. In addition to brightness or brightness distributions in the room, other factors shall come into play: the so-called visual clarity, colour shadings on object surfaces, room preference, and overall visual efficiency, which are currently being investigated in lighting research. The stages in the development of lighting parameters to describe the user preference for a static scene in interior lighting are illustrated in Figure 5.1 [10].

The investigations into *spatial brightness* and *visual clarity*, which form the subject of this chapter, have been carried out experimentally and theoretically in lighting science and especially in lighting architecture since the late 1960s [11–13] and continued until 2012 with an overall evaluation of the experimental methodology and the knowledge gained up to that time. There is a definition for *spatial brightness* formulated by IESNA (Illuminating Engineering Society of North America) and reproduced in [14]:

> Spatial brightness describes the visual perception of the strength of the lighting in an environment, e.g., a room or a lit street. In general, ambient lighting creates atmosphere and facilitates visual tasks such as safe traffic and visual communication. This brightness perception includes overall perception based on the response of a large portion of the visual field extending beyond the fovea. Spatial brightness is perceived while the observer is immersed in a room, or when a room is observed from a distance but fills a large part of the visual field. Spatial brightness does not necessarily refer to the brightness of individual (small) objects or surfaces in the environment but can be influenced by the brightness of these individual objects.

The following formulations or views exist for the definition of *visual clarity*:

(a) Thornton and Chen stated in 1978 [12] that clarity of vision means the same as distinctness of detail and is directly related to brightness.

(b) In a fundamental article, Hashimoto and Nayatani [13] formulated as follows: 'Visual clarity is caused by the *feeling of contrast* between coloured objects under illumination'. This formulation was also supported by Fotios and Atli [14].

(c) The authors of this book describe visual clarity as follows (see Figures 5.2 and 5.5): 'Visual clarity is the clear visibility of continuous colour transitions, fine colour shadings on the object surfaces, and clearly visible contrasts (large colour differences and/or large luminance differences) between the different coloured objects'.

These above explanations have in common that the distinguishability of object details is in the foreground of the observation, in which both the luminance contrast (achromatic contrast) and the contrast between coloured objects (difference in hue and saturation) play a role [15]. This aspect is illustrated in Figure 5.2 (left and right: photo of the same objects).

Of great importance for lighting architecture and practical lighting technology are the findings of the study by Flynn and Spencer in 1977 [16]. These authors

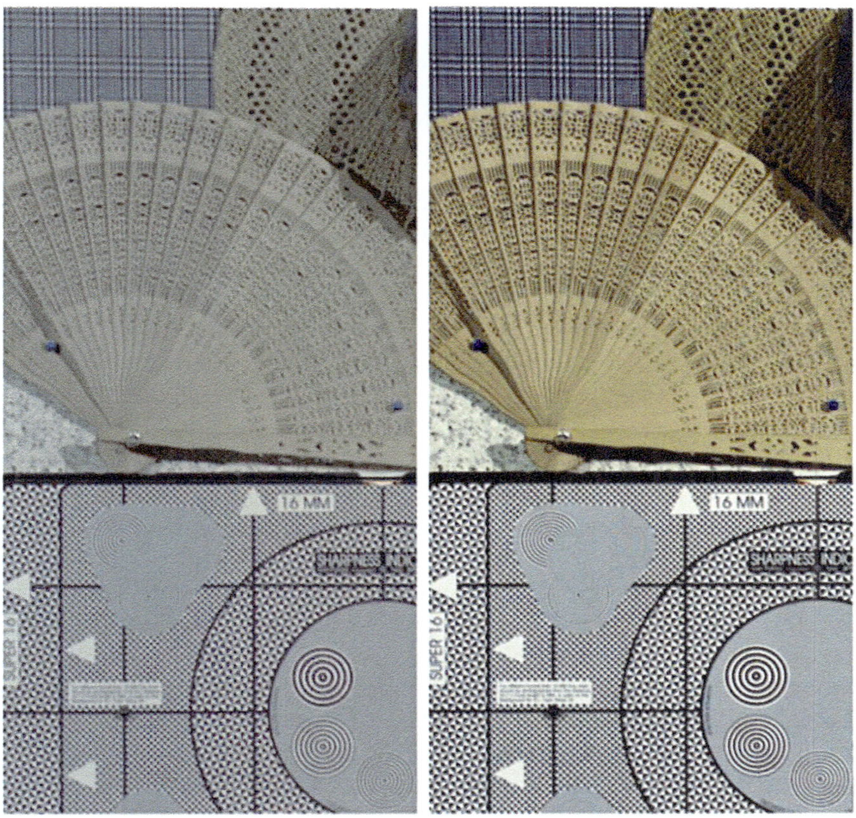

Figure 5.2 Visual clarity with achromatic and chromatic contrasts. Source: Reproduced with permission of ARRI AG. Left: low visual clarity; right: good visual clarity.

experimentally investigated the effect of light source spectra on the basis of a series of adjectives with a seven-step scaling, among others for visual clarity with 'dull…clear' and for brightness with 'dark…bright'. The *factor analysis* of the results obtained showed a good correlation between visual clarity and brightness perception by the test subjects.

This good correlation was also confirmed by Fotios and Cheal [17], Viénot et al. [18], and Fotios and Atli [14]. In reviewing all previous literature on visual clarity [14], it became known that visual clarity increases steadily with illuminance up to about 800–1000 lx, with the *correlated colour temperature* (CCT) of illumination also significantly affecting visual clarity.

In addition to the experimental description of the characteristics of visual clarity and brightness, however, the following questions are also important for analytical lighting science, vision science as well as for luminaire development and light source development:

- Is the parameter illuminance or luminance *alone* decisive for the perceived brightness and visual clarity of neutral (colour tone-free) and coloured objects with *local spatial structures* (e.g. wool, letters, and paintings with conspicuous brush strokes)?

  If the answer to the above question is 'No', then:

- How can new numerical parameters for the mathematical modelling of perceived brightness and visual clarity be determined?

These questions are also important because numerous publications – when describing the test conditions and interpreting the test results – predominantly used only *$V(\lambda)$-weighted* parameters, e.g. the *illuminance* at the object level, the *luminance* of objects, and the light emission properties of the surrounding walls of a room. Important for today's vision science and lighting technology, on the other hand, is also the knowledge of whether and to what extent – apart from $V(\lambda)$ – there are other signals or signal combinations of the various photoreceptors of the retina, i.e. the cones (L, M, and S), the rods and the light-sensitive ganglion cells (ipRGC) that should contribute to the perception of *visual clarity* and *brightness*, and thus to the perception of the overall atmosphere of an illuminated room and also, which wavelength ranges constitute the main part of electromagnetic radiation for *perceived* brightness and *perceived* visual clarity.

From the visual experience of lighting engineers and lighting designers, it is known that *neutral white* or *cool white* light at a certain constant luminance or illuminance level evokes a higher brightness perception than *warm white* light. The reason for this is that not only the $V(\lambda)$ signals (L + M), but also the signals of the other retinal mechanisms mentioned above (S, rods, and ipRGC) contribute to *brightness perception* [19]. The rods responsible for night vision are irrelevant for indoor lighting (in the subject area of the present book) because they are not active at the higher illuminance levels (e.g. 700 lx) that prevail in the ideal case. As described in Chapter 2, electrical impulses are formed in the photoreceptors – after photon absorption – and transmitted to the various brain processing stages. Within

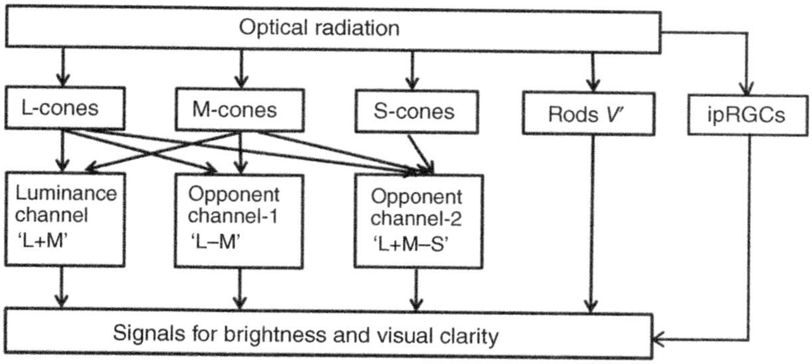

Figure 5.3 Signal components of the human visual system to form a numerical parameter for brightness perception [10]. The rod signals are irrelevant for most interior lighting applications with higher illuminance levels; they only make a contribution in dark scenes (e.g. in street lighting, mentioned here only for the sake of completeness). Source: TU Darmstadt.

these stages, the signals of the individual receptors or receptor groups are combined with each other (additively or subtractively).

As described in Chapter 2, the sum of the L- and M-cone signals forms the so-called *luminance signal*. The latter is modelled in today's practice with the help of the quantities of conventional photometry, i.e. with the help of the $V(\lambda)$-function, e.g. luminance, illuminance. The difference between the L and M-cone signals forms the so-called opponent Channel-1. The difference between the (L + M) and the S-signals form the so-called opponent Channel 2. From the signals (L + M), (L − M), (L + M − S), and the ipRGC, the signals for the attributes *brightness* and *visual clarity* arise in the brain, see Figure 5.3 [10]. It should be emphasised that – according to Figure 5.3 – not only the $V(\lambda)$-function, but – on the short-wave side of the visible spectrum – also the wavelength ranges around 440 nm (S-cones) and 480 nm (ipRGCs) [20] contribute to the formation of the signals for the attributes *brightness* and *visual clarity* (at least in the photopic range, where, as mentioned above, the rod signals are missing), see Figure 5.4.

Accordingly, *perceived brightness* has been modelled in literature with various so-called *brightness metrics*. One such model, the so-called *equivalent luminance* (L_{eq}) according to Fotios and Levermore [21] divides the signal of the S-cones (S) by the signal of the $V(\lambda)$-function and then calculates with the 0.24 power, see Eq. (5.1).

$$L_{eq} = L_v\,(S/V)0.24 \tag{5.1}$$

In Eq. (5.1), the exponent of (conventional) luminance (L_v) remains equal to 1. To calculate the signals S or V, the relative spectral radiant flux of the light source must be multiplied by the spectral sensitivity function of the S-cones or by the $V(\lambda)$-function, respectively, and this product must be integrated in the visible wavelength range. Visual experiments [19] showed that the S-cone signal is also

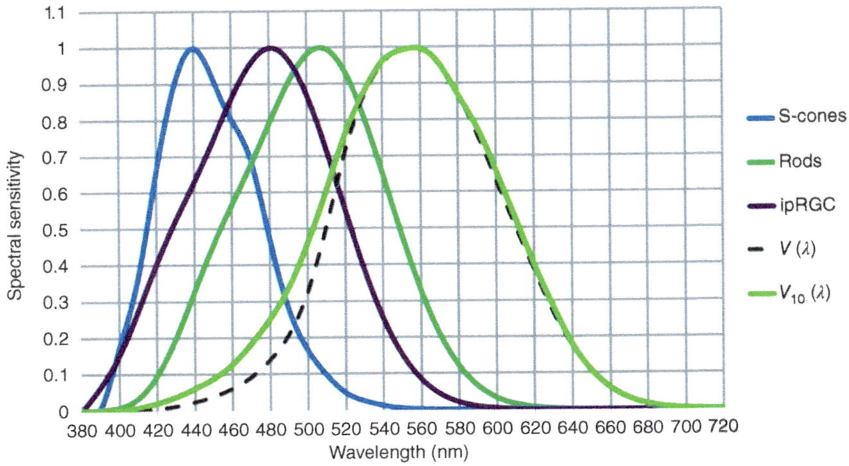

Figure 5.4 Spectral sensitivity of rods, S-cones and photosensitive ganglion cells (ipRGCs) compared to the $V(\lambda)$-function and the $V_{10}(\lambda)$-function [20]. Source: TU Darmstadt.

suitable for describing *visual clarity* in the photopic range. Similar to Eq. (5.1), the *equivalent illuminance* can be defined according to Eq. (5.2).

$$E_{eq} = E_v\ (S/V)^{0.24} \tag{5.2}$$

In the above-mentioned studies [19, 21], however, either – instead of a practice-relevant real room – an observation box or a room model was used, or *only* visual clarity *alone* or *only* brightness *alone* was investigated. In this chapter, the two attributes brightness and visual clarity are modelled with the S-cone signals and the ipRGC signals – based on the results from a subjective study conducted in a real room [22]. The motivation is that – with the help of this model (see Section 5.3), which *also* works with the S-cone and ipRGC signals instead of *only* luminance or *only* illuminance – brightness and visual clarity can be predicted accurately.

5.2 Experimental Method of the Subjective Study

The subjective study [22] took place in a real, office-like room without daylight, with homogeneous white and matt-painted walls and neutral flooring. The test subjects sat on chairs and looked at a homogeneously illuminated table with a white tablecloth (see Figure 5.5). The subjects first adapted to this white tablecloth for about two to four minutes. The test subjects scaled their impression of the brightness and visual clarity of a scene with various coloured and achromatic objects (a doll with long hair, a jumper with colourful patterns, an artificial water lily, and a black and white test pattern with fine line structure), which were illuminated in this room with two RGBW LED lights.

The horizontal illuminance on the table was 45, 90, 470, 1000, and 2000 lx respectively. The CCT – at each illuminance level – was also varied: 2700, 3100, 4100, 5000,

Figure 5.5 Experimental set up for the subjective assessment of brightness and visual clarity. Source: Reproduced with permission from Khanh et al. [22], SAGE PUBLICATIONS, INC.

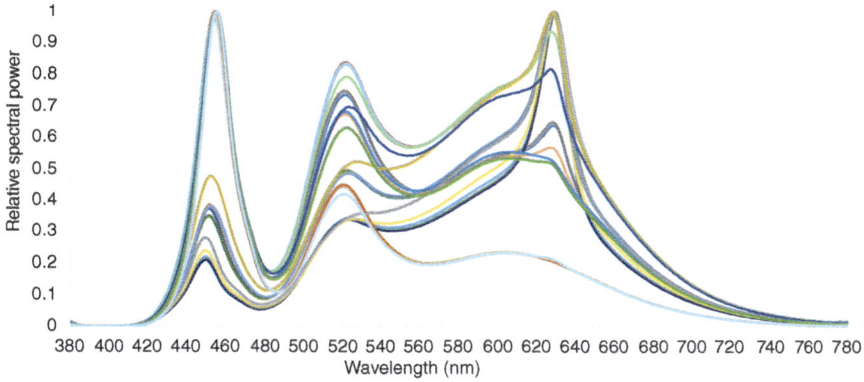

Figure 5.6 Spectra in the 25 lighting situations examined in the subjective study for brightness and visual clarity. Source: Reproduced with permission from Lighting Research and Technology [22].

and 10 000 K. Thus, the subjects were shown 25 different ($E_v \times$ CCT) combinations, waiting about one minute for re-adaptation after adjusting each light combination until the subjective evaluation was made. The illuminating RGBW LED light sources were optimised to ensure the same, high CRI level ($92 \leq R_a \leq 96$), for each spectrum ($5 \times 5 = 25$). The illumination spectra are shown in Figure 5.6.

Twenty-eight subjects with normal or corrected visual acuity and without colour vision deficiencies scaled their impression of *brightness* and *visual clarity* on two scales, a continuous *brightness scale* between 0 and 100 and a *visual clarity scale* between 0 and 100 *with categories*, see Figure 7.2. The subjects wrote an appropriate cross on the scales. The categories were only used for visual clarity and corresponded to the ratings 'excellent', 'very good', 'good', 'moderate', 'low', 'poor,' and 'very poor', whereby the test person was also allowed to rate their visual clarity impression between two neighbouring categories.

5.3 Modelling Brightness and Visual Clarity

The subject's mean visual scale scores were modelled – for both *brightness* and *visual clarity* – using the quantity M [10, 23] defined in Eq. (5.3).

$$M = a\ \left[E_{\mathrm{v}}^{\ \gamma}\ (\alpha\ \mathrm{S}^{\delta} + \beta\ \mathrm{G}^{\delta})\right] + b \tag{5.3}$$

The symbols in Eq. (5.3) have the following meaning. The parameter γ is an exponent for the compression of the *illuminance scale* and the parameter δ is an exponent for the compression of the S and the G signals. Parameters a and b are the parameters of a linear transformation to fit the calculated M values to the mean visual scale values W (either for brightness or for visual clarity) of the test subjects for the 25 light source spectra ($i = 1$–25, see Figure 5.6) with the smallest error F. In practice, the light source spectra with low visual rating ($W_i \leq 60$) do not play a role, therefore the modelling quantity M was not adjusted to the latter data ($W_i \leq 60$). The error size F has defined accordingly in Eq. (5.4).

$$F = \left[\Sigma_i (M_i - W_i)^2\right]^{(1/2)} \qquad (W_i > 60) \tag{5.4}$$

The parameters of Eq. (5.3) are weighting parameters of the relative signals of the mechanisms S (S-cone) and G (ipRGCs: photosensitive ganglion cells). The relative signals were calculated from the 25 relative spectra (see Figure 5.6). To calculate these signals (S and G), the relative spectral radiant flux of the light source must be multiplied by the spectral sensitivity function of the S-cones or the ipRGCs and this product shall be integrated in the visible wavelength range. Afterwards, these signals are divided by the so-called V-signal. To calculate the V-signal, the relative spectral radiant flux of the light source shall be multiplied by the $V(\lambda)$-function and this product shall be integrated in the visible wavelength range.

Table 5.1 shows the optimal parameter values of Eq. (5.3) and the error F of Eq. (5.4) for the attribute *brightness*. The parameters a, b, γ, δ of Eq. (5.3) were optimised and are shown in rows No. 1–5 of Table 5.1, with different combinations of the constant values (α, β), in such a way that the error F of Eq. (5.4) was minimised. In optimisation No. 6, only a and b were changed, while the four other parameters remained constant. Latter parameters were adjusted to reflect the *equivalent illuminance* formula according to Fotios and Levermore [21] as per Eq. (5.2).

Line No. 7 in Table 5.1 shows the optimal parameter values a, b, γ, and the error F for Eq. (5.5), which disregards the S and G signals and calculates only with *illuminance*, as an input parameter.

$$M = a\ E_{\mathrm{v}}^{\ \gamma} + b \tag{5.5}$$

Row No. 8 in Table 5.1 shows the optimal parameter values (a, b) of Eq. (5.5) with $\gamma = 1$ constant, i.e. without compression of the illuminance scale in Eq. (5.5) as well as the error F for Eq. (5.5).

Table 5.1 shows that the smallest error ($F = 10.32$) results in line no. 2, where the weighting of the S-signal is equal to 1 and the weighting of the ipRGC-(G-) signal is equal to 0. If the parameters (α, β) are being optimised, then the same result

Table 5.1 Parameter values of Eq. (5.3) and the corresponding values of the error F of Eq. (5.4) for brightness [10, 23]. Weight of S-cones: α; the weight of ipRGC: β; exponent for illuminance: γ; exponent for S and G signals: δ; a, b: parameters of the linear transformation to approximate the mean visual scale values of the subjective study of Section 5.2.

No.	α (S)	β (G)	a	b	γ (E_v)	δ (S and G)	F
1	0.000	1.000	1.902	39.57	0.444	0.171	10.44
2	1.000	0.000	2.566	37.04	0.417	0.116	10.32
3	0.500	0.500	2.169	31.54	0.367	0.120	10.38
4	0.333	0.667	0.940	40.92	0.461	0.150	10.39
5	0.667	0.333	1.485	36.31	0.410	0.132	10.37
6[a)]	1.000	0.000	0.020	60.51	1.000	0.240	11.42
7[b)]	—	—	3.744	31.81	0.362	—	14.69
8[c)]	—	—	0.015	60.68	1.000	—	15.37

a) Optimisation no. 6: these parameter values represent Eq. (5.2) according to Fotios and Levermore [21]; in no. 6: α, β, γ, and δ were constant and only a and b were optimised;
b) No. 7: optimal parameters and error F of Eq. (5.5) with γ = variable;
c) No. 8: optimal parameters and error F of Eq. (5.5) with γ = const. = 1.
Source: TU Darmstadt.

is obtained: i.e. there is a global optimum at $\alpha = 1$; $\beta = 0$. For other fixed (α, β) parameter values (lines nos. 1, 3, 4, and 5 in Table 5.1) there are only slightly larger errors (between $F = 10.37$ and $F = 10.44$) and only slightly different optimum (γ, δ) values. The latter indicates that, on the basis of the present experimental data set, it is not possible to decide, which signal (S or G = ipRGC) is decisive for brightness perception.

For modelling perceived brightness, row no. 2 of Table 5.1 with the smallest error F is proposed here, which is also reproduced as Eq. (5.6). Equation (5.6) is scaled so that the model values M reflect the same scale used by the subjects (see Figure 7.2, left). The numerical values of Eq. (5.6) are intentionally given with seven decimal places, this should not be a problem in the age of fast computers.

$$M(\text{Brightness}) = 2.566\,\left[E_\text{v}^{\,0.4167494} \cdot \text{S}^{0.1153348}\right] + 37.04 \tag{5.6}$$

As mentioned above, the parameters α, β, γ, and δ in line no. 6 of Table 5.1 remained constant and only the parameters a and b were optimised, which linearly fit the scale of the Fotios–Levermore model [21] to the brightness scale of the present experiment (Section 5.2). In this case, a significantly larger error results, $F = 11.42$. The reason for this is that the Fotios–Levermore model [21] does not compress the illuminance scale (E_v), i.e. works with exponent $\gamma = 1.000$.

Equation (5.5), which ignores the shorter wavelength sensitive (or 'blue sensitive') signals S and G = ipRGC (see Figure 5.4), leads to larger errors (14.69 and 15.37 in rows no. 7 and 8 of Table 5.1, respectively). This indicates that it is very important to include a 'blue-sensitive' signal in the brightness model. Equation (5.6) represents such a brightness model by the term $\text{S}^{0.1153348}$. Figure 5.7 visualises the mean, visually scaled brightness values of all observers from the experiment in Section 5.2

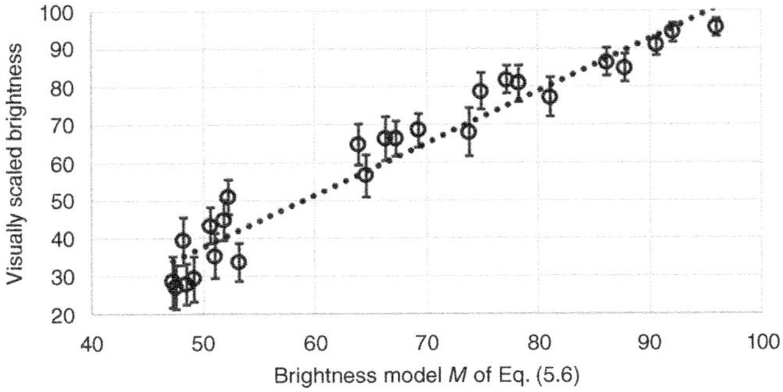

Figure 5.7 Mean, visually scaled brightness values of all observers from the experiment in Section 5.2 and their 95% confidence intervals as a function of the parameter M (brightness) of Eq. (5.6), with $r^2 = 0.94$. Source: TU Darmstadt.

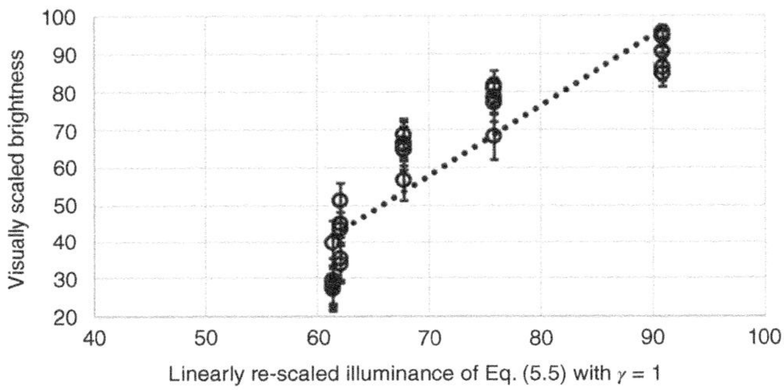

Figure 5.8 Mean, visually scaled brightness values of all observers from the experiment in Section 5.2 and their 95% confidence intervals as a function of Eq. (5.5) with $\gamma = 1$, i.e. the linearly rescaled illuminance E_v, with $r^2 = 0.81$. Source: TU Darmstadt.

and their 95% confidence intervals as a function of the parameter M (brightness) of Eq. (5.6), with $r^2 = 0.94$. In comparison, Figure 5.8 shows the same brightness values as a function of Eq. (5.5) with $\gamma = 1$, i.e. the linearly rescaled illuminance E_v, with $r^2 = 0.81$. From the comparison of Figures 5.7 and 5.8, it is evident how important it is, on the one hand, to compress the scale of the illuminance with a suitable exponent and, on the other hand, to refine the model with the help of a 'blue-sensitive' signal.

Table 5.2 shows the optimal parameter values of Eq. (5.3) and the error F of Eq. (5.4) in case of the other attribute, *visual clarity*. The parameters a, b, γ, δ of Eq. (5.3) were optimised in rows No. 1–5 of Table 5.2 in case of different combinations of the constant values (α, β) in such a way that the error F of Eq. (5.4) was minimised. In optimisation no. 6, only a and b were changed, while the four parameters α, β, γ, and δ remained constant. The latter parameters were adjusted to reflect the equivalent illuminance formula of *Fotios and Levermore* [21] according to Eq. (5.2). Although the latter formula is intended for modelling *brightness* [21], it is also used here for *visual clarity* because, as mentioned in Section 5.1, the study

Table 5.2 Parameter values of Eq. (5.3) and the corresponding values of the error F of Eq. (5.4) for visual clarity [10, 23]. α: weight of S-cones; β: weight of ipRGC; γ: exponent for illuminance; δ: exponent for S and G signals; a, b: parameters of the linear transformation to approximate the mean visual scale values of the subjective study in Eq. (5.2).

No.	α (S)	β (G)	a	b	γ (E_v)	δ (S and G)	F
1	0.000	1.000	0.010	27.47	0.167	0.113	10.97
2	1.000	0.000	20.27	11.96	0.195	0.093	10.75
3	0.500	0.500	0.010	14.96	0.167	0.094	10.84
4	0.333	0.667	0.010	15.03	0.167	0.095	10.85
5	0.667	0.333	0.010	15.04	0.167	0.094	10.84
6[a)]	1.000	0.000	0.017	67.20	1.000	0.240	13.78
7[b)]	—	—	2.045	47.95	0.406	—	17.89
8[c)]	—	—	0.013	67.65	1.000	—	18.26

a) Optimisation No. 6: these parameter values reproduce Eq. (5.2) according to Fotios and Levermore [21]; in No. 6, α, β, γ, and δ remained constant and only a and b were optimised;
b) No. 7: Optimal parameters and error F of Eq. (5.5) with γ = variable;
c) No. 8: Optimal parameters and error F of Eq. (5.5) with γ = const. = 1.
Source: TU Darmstadt.

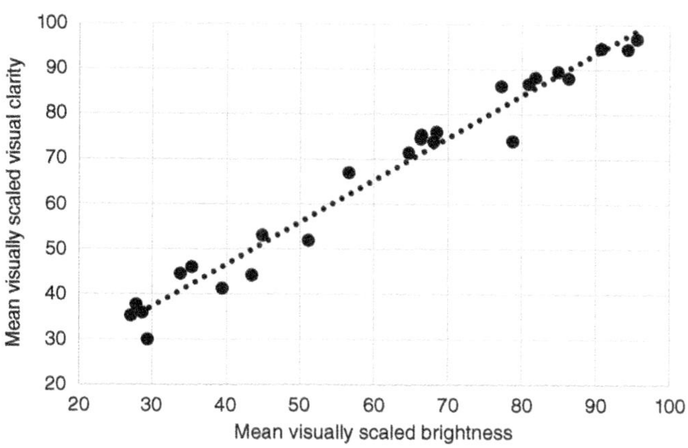

Figure 5.9 Correlation ($r^2 = 0.97$) between the mean, visually scaled visual clarity and the mean, visually scaled brightness of the test subjects in the present study (Section 5.2). Source: TU Darmstadt.

by Flynn and Spencer [16] found a good correlation between the subjects' *visual clarity* and *brightness* perceptions. This correlation was also very good ($r^2 = 0.97$) in the present study, see Figure 5.9.

Line No. 7 in Table 5.2 shows the optimal parameter values a, b, γ, as well as the error F for Eq. (5.5), which *disregards* the S and G-signals and calculates only with the illuminance as an input parameter. Line No. 8 in Table 5.2 shows the optimal parameter values (a, b) of Eq. (5.5) with $\gamma = 1 =$ const. i.e. without compression of the illuminance scale in Eq. (5.5) as well as the error F for Eq. (5.5).

Table 5.2 shows that the smallest error ($F = 10.75$) (compare with Table 5.1) results in line no. 2 again, where the weighting of the S-signal is equal to 1 and the weighting of the ipRGC-(G-) signal is equal to 0. If the parameters (α, β) are not being optimised then, again, the same result is obtained: a global optimum at $\alpha = 1$; $\beta = 0$. For other fixed (α, β)-parameter values (lines nos. 1, 3, 4, and 5 in Table 5.2) there are only insignificantly larger errors (between $F = 10.84$ and $F = 10.97$) and equal or only slightly different optimal (γ, δ) values. The latter indicates that it is not possible to decide, which signal (S or G = ipRGC) is decisive for *visual clarity* on the basis of the present experimental data set (Section 5.2).

For modelling *visual clarity*, row no. 2 of Table 5.2 with the smallest error F is proposed here, which is also reproduced as Eq. (5.7). Equation (5.7) is scaled so that the model values M reflect the same scale, which was anchored with categories in the subjective study ('excellent', 'very good', ... to 'very poor', see Figure 7.2, right). Thus, for a given light source, the lighting designer has the possibility to calculate the illuminance levels associated with the criteria 'good' or 'very good' visual clarity.

Equation (5.7), however, only applies to the *high* CRI levels ($R_a \geq 92$) of the present subjective study (see Section 5.2). This is because *visual clarity* also depends on the colour rendering or, more precisely, the colour saturation property of the light source. For the general case, i.e. for a light source with any colour saturation property, Eq. (7.3) is therefore proposed, which is not only valid for high CRI levels ($R_a \geq 92$).

$$M\,(\text{visual clarity}) = \left[20.27\, E_v^{\,0.1951482} \cdot S^{0.0931085}\right] + 11.96 \tag{5.7}$$

The numerical values of Eq. (5.7) are intentionally given with seven decimal places, this should not be a problem in the age of fast computers.

As mentioned above, the parameters $(\alpha, \beta, \gamma, \delta)$ remained constant at line no. 6 of Table 5.2 and only the parameters a and b were optimised, which linearly transform the scale of the Fotios–Levermore model [21] to the visual clarity scale of the present experiment (Section 5.2). In this case, a significantly larger error, $F = 13.78$ results. The reason for this is again that the Fotios–Levermore model [21] does not compress the illuminance scale (E_v) but works with the exponent $\gamma = 1.000$.

Equation (5.5), which ignores the shorter wavelength sensitive (or 'blue sensitive') signals S and G = ipRGC (see Figure 5.4), leads to larger errors (17.89 and 18.26, respectively, in rows Nos. 7 and 8 of Table 5.2). This indicates that it is very important to include a 'blue-sensitive' signal in the *visual clarity* model. Equation (5.7) represents such a visual clarity model that includes the term $S^{0.0931085}$.

Figure 5.10 visualises the mean, visually scaled *visual clarity* values of all observers from the experiment in Section 5.2 and their 95% confidence intervals as a function of the parameter M(visual clarity) of Eq. (5.7), with $r^2 = 0.95$. In comparison, Figure 5.11 shows the same visual clarity values as a function of Eq. (5.5) with $\gamma = 1$, i.e. the linearly rescaled illuminance E_v, with $r^2 = 0.77$. From the comparison of Figures 5.10 and 5.11 it is again evident how important it is to include the exponents ($\neq 1$) and the 'blue-sensitive' signal.

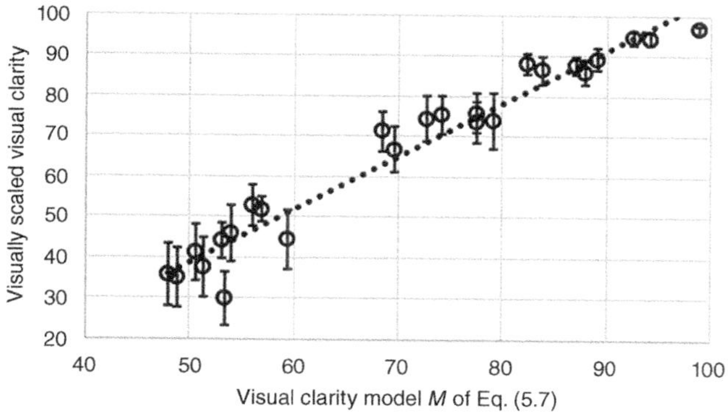

Figure 5.10 Mean visually scaled visual clarity values of all observers from the experiment in Section 5.2 and their 95% confidence intervals as a function of the parameter *M* (visual clarity) of Eq. (5.7), with $r^2 = 0.95$. Source: TU Darmstadt.

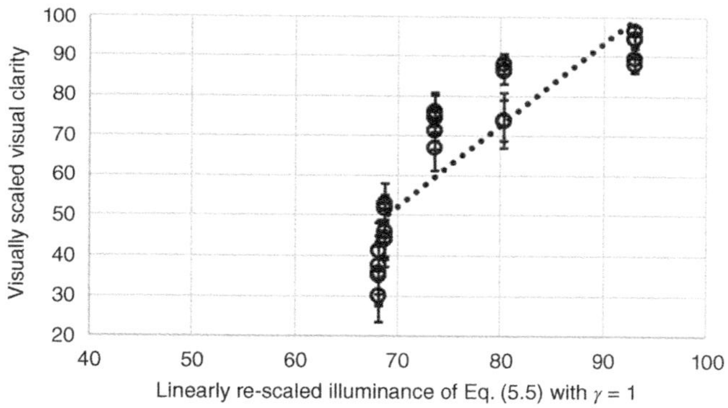

Figure 5.11 Mean visually scaled visual clarity values of all observers from the experiment in Section 5.2 and their 95% confidence intervals as a function of Eq. (5.5) with $\gamma = 1$, i.e. the linearly rescaled illuminance E_v, with $r^2 = 0.77$. Source: TU Darmstadt.

5.4 Summary

The visual perception of the users of indoor space with regard to the visual characteristics *brightness* and *visual clarity* cannot be *fully* described using illuminance or luminance *alone* as in Eq. (5.5) because these parameters only use the $V(\lambda)$-function, which disregards the signals from the S-cones (S) and/or the intrinsically light-sensitive ganglion cells (G = ipRGC).

However, if these signals (or one signal of them, e.g. the S-signal) are taken into account as in Eq. (5.6) and in Eq. (5.7), one gets a better prediction of the mean, visually scaled results of brightness and visual clarity. According to the results of the present study, either the ipRGC signal (which is also called G-signal here), the S-signal, or both signals play an important role. Whether the S or the G signal is decisive could not be concluded from the results of the present study.

In any case, it is very important to include a 'blue-sensitive' signal in the model. Including a compression of the scales of the input variables (E_V, S, and G) by suitable exponents (<1), as γ and δ in Eq. (5.3), is also relevant because signal compression is generally a significant property of the human perceptual system.

Paying attention to a 'blue-sensitive' signal (S or G = ipRGC) means for the lighting industry and lighting designers that the range around 420–520 nm (see the blue curve and the purple curve in Figure 5.4) plays an important role in the design of light spectra and the brightness and visual clarity perception of a room.

References

1 DIN EN 12464-1 (2011). Beleuchtung von Arbeitsstätten – Teil 1: Arbeitsstätten in Innenräumen (Lighting of workplaces – Part 1: Indoor Workplaces). German version EN 12464-1:2021). Beuth, Berlin.

2 Flynn, J.E., Spencer, T.J., Martyniuk, O., and Hendrick, C. (1973). Interim study of procedures for investigating the effect of light on impression and behaviour. *J. Illum. Eng. Soc.* 3 (1): 87–94.

3 Flynn, J.E., AIA/FIES, Hendrick, C. et al. (1979). A guide to methodology procedures for measuring subjective impressions in lighting. *J. Illum. Eng. Soc.* 8: 95–110.

4 Rowlands, E., Loe, D.L., McIntosh, R.M. et al. (1985). Lighting adequacy and quality in office interiors by consideration of subjective assessment and physical measurement. *CIE J.* 4: 23–37.

5 Veitch, J.A. and Newsham, G.R. (1998). Lighting quality and energy-efficiency effects on task performance, mood, health, satisfaction, and comfort. *J. Illum. Eng. Soc.* 27: 107–129.

6 Thapan, K., Arendt, J., and Skene, D.J. (2001). An action spectrum for melatonin suppression: evidence for a novel non-rod, non-cone photoreceptor system in humans. *J. Physiol.* 535 (1): 261–267.

7 Brainard, G.C., Hanifin, J.P., Greeson, J.M. et al. (2001). Action spectrum for melatonin regulation in humans: evidence for a novel circadian photoreceptor. *J. Neurosci.* 21 (16): 6405–6412.

8 Yaguchi, H., A. David, T. Fuchida et al. (2017). CIE 2017 color fidelity index for accurate scientific use. CIE 224: 2017.

9 Davis, W. and Ohno, Y. (2010). Color quality scale. *Opt. Eng.* 49: 033602.

10 Bodrogi, P. and Khanh, T.Q. (2019). Beschreibung von Helligkeit und Sehklarheit für die Innenraumbeleuchtung (Describing brightness and visual clarity for interior lighting). *Z. Licht* 1/2019 (2): 80–84.

11 Aston, S.M. and Bellchambers, H.E. (1969). Illumination, colour rendering and visual clarity. *Light. Res. Technol.* 1 (4): 259–261.

12 Thornton, W.A. and Chen, E. (1978). What is visual clarity? *J. Illum. Eng. Soc.* 7 (2): 85–94.

13 Hashimoto, K. and Nayatani, Y. (1994). Visual clarity and feeling of contrast. *Color Res. Appl.* 19 (3): 171–185.

14 Fotios, S. and Atli, D. (2012). Comparing judgments of visual clarity and spatial, brightness through an analysis of studies using the category rating procedure. *Leukos* 8 (4): 261–281.

15 Khanh, T.Q. and Bodrogi, P. (2016). Sehklarheit und räumliche Helligkeit in der modernen Beleuchtungstechnik – Experimente und Modellierung (Visual clarity and spatial brightness in modern lighting technology – Experiments and modelling). *Z. Licht* 7, 8/2016: 2–7.

16 Flynn, J.E. and Spencer, T.J. (1977). The effects of light source colour on user impression and satisfaction. *J. Illum. Eng. Soc.* 6 (3): 167–179.

17 Fotios, S.A. and Cheal, C. (2007). Lighting for subsidiary streets: investigation of lamps of different SPD. Part 2 – Brightness. *Light. Res. Technol.* 39 (3): 233–252.

18 Vienot, F., Durand, M.-L., and Mahler, E. (2009). Kruithof's rule revisited using LED illumination. *J. Mod. Opt.* 56 (13): 1433–1446.

19 Bodrogi, P. and Khanh, T.Q. (2017). Visual clarity and brightness in indoor and outdoor lighting: Experiments and modelling In: *Proceedings of the CIE Midterm Meeting 2017*. CIE. Jeju, Republic of Korea, ISBN: 978-3-901906-95-4, DOI: https://doi.org/10.25039/x044.2017, pp. 45.

20 Lucas, R.J., Peirson, S.N., Berson, D.M. et al. (2014). Measuring and using light in the melanopsin age. *Trends Neurosci.* 37: 1–9.

21 Fotios, S.A. and Levermore, G.J. (1998). Chromatic effect on apparent brightness in interior spaces, II: SWS lumens model. *Light. Res. Technol.* 30 (3): 103–106.

22 Khanh, T.Q., Bodrogi, P., Guo, X., and Anh, P.Q. (2018). Towards a user preference model for interior lighting Part 1: concept of the user preference model and experimental method. *Light. Res. Technol.* 51 (7): 1014–1029.

23 Bodrogi, P., Guo, X., and Khanh, T.Q. (2018). Brightness and visual clarity rather than luminance in indoor lighting assessment. Light 2018, Creative Lighting Solutions, 23rd European Lighting Congress. Davos, Switzerland.

6

Colour Quality and Psychophysical–Emotional Aspects, Laboratory Experiments

6.1 Introduction

This chapter discusses colour quality as well as the psychophysical–emotional aspects of light sources and lighting installations, which is particularly relevant to Human-Centric Lighting (HCL) (see Chapter 3) because of the associated psychological and emotional aspects. Colour quality can generally be defined as '*the observer's judgement of the pleasant, harmonious, or emotionally positive perception of illuminated object colours under a particular illumination, in relation to the light levels prevailing in the visual environment or the white tone of the current light source*'. Accordingly, the following basic parameters of lighting, which in this way strongly influence the perceived colour quality in an illuminated environment and its emotional aspects, will be discussed as follows:

1. The illuminance level (Section 6.2).
2. The level of the prevailing luminance of the room walls (Section 6.3).
3. The correlated colour temperature of the lighting (Section 6.4).
4. The preference of illuminance *and* the colour temperature in the so-called modified *Kruithof diagram* and their dependence on age, gender, or activation/relaxation (Section 6.5); as well as
5. The distance of the white point of the light source from the curve of the blackbody radiators or the daylight phases, i.e. the effect of any perceived hue in the white tone (e.g. a slightly bluish or yellowish-white tone, Section 6.6).

The way in which the above parameters (1–5) interact with colour quality is described and modelled in detail in Chapter 7 ('Light quality models'). Further development of the conventional *Kruithof diagram* is also given in Chapter 7 (see Section 7.5). Visual attributes (e.g. colour preference, vividness, memory colours) and some further perceptual aspects (e.g. Helmholtz–Kohlrausch effect, Hunt effect) of colour quality are discussed in Section 6.7, while Section 6.8 deals in detail with the effect of the lighting spectrum on the colour quality of *skin tones,* which are particularly important in lighting practice. Finally, Section 6.9 deals with the so-called *semantic interpretation of* the colour-rendering properties of light sources, and how the various R_a values can be interpreted for lighting practice with the help

Human Centric Integrative Lighting: Technology, Perception, Non-Visual Effects, First Edition.
Tran Quoc Khanh, Peter Bodrogi, and Trinh Quang Vinh.

of categories. At the end (Section 6.10) there is a summary with the consequences for indoor lighting.

All these results and findings were obtained under laboratory conditions or under constant, conditioned room conditions. Further psychological–emotional aspects under field test conditions are described in Chapter 9.

6.2 Preferred Horizontal Illuminance Levels

While some lighting research in the past was concerned with determining the threshold values of illuminance levels and luminance levels for a specific work or visual performance, some research groups also addressed the psychophysically preferred optimum of horizontal illuminance [1]. These studies were often conducted in the laboratory under a defined static visual condition with reading tasks. Table 6.1 shows the preferred illuminance levels of some research groups in the 1960s and 1970s for workplaces without VDUs. It is noticeable that these illuminances are relatively high in the value range between 1300 and 2250 lx. Therefore, further field studies with VDU work are desirable, see Section 6.3.

In numerous studies on the determination of preferred illuminance levels, a large individual variation has been registered. This variation may depend on the application, the type of work, and the age and gender of the user. Most of the studies were conducted in rooms with office character. From 1994 to 1996, a working group led by Begemann et al. [7] carried out a study in offices with windows, in which 96 test persons took part who had chosen an illuminance for artificial light of 800 lx on average in addition to daylight.

The interpretation of these data by the authors of this book is that, according to the study by Seidl [8], a daylight quotient of 0.9% and, according to [9], a global illuminance of about 6000 lx can be assumed in the winter months in Germany and of 70 000 lx in the summer months (except for the peak values). This means that the additional daylight component can be calculated to be between 54 and 630 lx, so that the total illuminance value can lie in the preferred range between 854 and 1430 lx. Begemann et al. did not register any age influence or season variation in their experiments. Tops et al. [10] re-analysed the data of this research group a few years later and gave a mean preferred illuminance around 1400 lx.

Probably the most comprehensive field study to date was conducted by Moosmann [11] in nine office buildings in Germany. The results of the evaluation of almost 1000 questionnaires are summarised below:

- 67% of the respondents stated that they would not like any change in the lighting level at the mean value of the horizontal illuminance at the workplace of 966 lx.
- The lighting situations where the light users did not express any wishes to change the lighting level are subject to the influence of age. The horizontal illuminance at the workplace where light users did not make any change requests was 1168 lx for people under 30 years of age, 1006 lx between 30 and 39 years of age, 883 lx between 40 and 49 years of age, and 834 lx above 50 years of age. These results were statistically significantly different.

- The desired horizontal illuminance depends on the season and is higher in spring (1046 lx) than in summer (927 lx) and autumn (875 lx) and winter (787 lx).

These results by Moosmann from 2014 are thus smaller than the preferred illuminance levels in Table 6.1. The results of Begemann and Moosmann were obtained in offices (university buildings, bank buildings, and research institute buildings) in Central Europe. There has been very little research on lighting in industrial workplaces in countries with a long winter. An exception to this is Juslen's PhD work in 2007 in six industrial plants in Holland, Finland, and Germany [11]. In an industrial plant in Finland, a field study was conducted over two years with 23 assembly workers (3 men, 20 women with an average age around 35) and with fluorescent luminaires of 4000 K colour temperature. The desired illuminance levels were set with an IR remote control. The set illuminance levels were between 1340 and 1370 lx and thus far higher than the minimum values of the standard DIN EN 12464-1. The results of Begemann, Tops, Moosmann, and Juslen are summarised in Table 6.2.

For the setting and dimensioning of indoor lighting solutions, an important range of horizontal illuminance levels for possible satisfaction of light users is thus conceivable between 850 and 1430 lx, which is to be considered in the context of HCL lighting and the promotion of people's health. Office areas as well as industrial halls

Table 6.1 Preferred illuminance levels from studies in the 1960s and 1970s (cited by [2]).

Authors	Preferred illuminance levels (lx)
Muck and Bodmann [3]	Work 1: 1300 (with high contrast) Work 2: 1800 (with moderate contrast)
Söllner [4]	1750
Westhoff and Horemann [5]	2250
Bodmann et al. [6]	1600
Boyce [2]	1550

Source: TU Darmstadt.

Table 6.2 Preferred horizontal illuminance levels in three field studies.

Study	Result
Begemann et al. [7]	854 and 1430 lx (according to the interpretation of the authors of this book)
Tops et al. [10]	1400 lx
Mossmann [12]	966 lx: depending on age (834 lx for ages over 50 and 1168 lx for people under 30) and depending on the time of year
Juslen [11]	1340–1370 lx

Source: TU Darmstadt (based on the studies).

have thus been examined in terms of lighting technology. The question of whether the preferred illuminance levels, e.g. in salesroom lighting, combined with a necessary colour quality and colour saturation for a certain brilliance of the goods and objects, deviate from the above values in Table 6.2, is answered in Chapter 7 (Light quality models).

6.3 Preferred Luminance Levels on the Wall for a Computer Monitor

6.3.1 Introduction

The analysis and definition of horizontal illuminance levels at a given working level are quite historical and justified by the fact that current international lighting standards still mostly specify *illuminance levels*. Communication between lighting engineers, lighting designers, and professional lighting users is still being determined by illuminance levels [1]. In a new research project in 2017 [13] to investigate the Perceived Adequacy of Illumination (PAI) and the spatial brightness of a small office space, Pearson's correlation coefficient between PAI and horizontal illuminance equalled only $r^2 = 0.19$. The perception of spatial brightness also only had a low correlation coefficient of $r^2 = 0.14$ to horizontal illuminance. The horizontal illuminance, therefore, does not adequately describe spatial brightness.

The room appearance characteristics brightness, homogeneity, or inhomogeneity were most often described with luminance and luminance distribution. As early as the mid-1990s, the research group around Loe and Rowlands found out through experiments that the relevant part of the visual field for assessing the illumination of a room is about 40° (±20°) visual angle wide and centred on the eye level of the seated person (about 1.2–1.3 m) [14]. If we consider the room height of about 2.8–3.0 m in a typical office room today, the illumination of the wall areas around the centre of the room height is crucial for adaptation and room assessment. If we now also consider the new knowledge about non-visual effects, that the light-sensitive ganglion cells (ipRGCs) are more abundant in the lower half of the retinal surface than on the upper half, it is necessary to illuminate the upper half of the visual field sufficiently strongly. For a one-person or two-person office room with a distance of about 2–3 m between the sitting position and the wall, this knowledge means that the upper half of the vertical wall must be illuminated sufficiently strongly. Table 6.3 contains the preferred luminance ratios and wall luminances based on the literature.

In [17], the two lighting researchers De Boer and Fischer stated in 1978 that at a horizontal illuminance of 1000 lx (which roughly corresponds to the preferred illuminance according to the above-mentioned study by Moosmann [12]), a luminance for the ceiling of 200 cd m^{-2} and for the walls of 100 cd m^{-2} are preferred. At a horizontal illuminance of 500 lx (as defined as a minimum by DIN 12464-1), the above values can be estimated as about 210 cd m^{-2} (ceiling) and 70 cd m^{-2} (walls). For a possible control of the next generation of intelligent luminaires, this means that ceiling luminance should remain almost constant at 200–210 cd m^{-2}. Another

Table 6.3 Preferred luminance ratios and luminance levels of the wall.

Authors	Luminance of the wall	Wall luminance ratio task	Ceiling-to-wall luminance ratio
Ooyen et al. [15]	20–45 $cd\,m^{-2}$ for monitor work 30–60 $cd\,m^{-2}$ for other reading tasks	3.3–1	Not specified
Tregenza et al. [16]	Not specified	2–1	1.6 : 1

Source: TU Darmstadt (based on literature data).

specification of the preferred luminance at 75 $cd\,m^{-2}$ on the wall at a horizontal illuminance of $E_v = 500$ lx was found in Balder's publication in 1957 [18].

These preferred wall luminances of around 70–100 $cd\,m^{-2}$ were determined before or in the time of the old CRT monitor technology with a maximum monitor luminance around 75–85 $cd\,m^{-2}$, i.e. of the same order of magnitude. With today's OLED display technology or LCD technology with LED backlighting, the monitor luminance in the working range is around 175–250 $cd\,m^{-2}$, so it is possible that the optimum luminance of the wall behind the monitor must be designed differently. Accordingly, in another study [19] from 2018, the most comfortable wall luminance for two monitor workstations was $L_v = 238\,cd\,m^{-2}$ (desktop monitor) and $L_v = 176\,cd\,m^{-2}$ (laptop display), respectively. The method and results of this study [19] are summarised in this Section 6.3.

6.3.2 Experimental Method

The study [19] was conducted in a real room (W: L: H = 3.02 m: 3.06 m: 2.54 m). The ceiling and the walls were painted white. A white light-proof curtain was placed in front of the large window. A worktable with a white matt tablecloth and two chairs were placed in the room, see Figure 6.1. Above the table and just below the ceiling of the room, two commercial diffused LED luminaires with RGB LEDs and warm white LEDs were mounted on a stable support. The intensities of the LED channels of these two LED luminaires were optimised to achieve different correlated colour temperatures (CCTs) between 2700 and 6500 K, always with very high colour rendering. The colour mixing and the control of these two luminaires were done with a separate computer in order to illuminate the opposite wall homogeneously and diffusely.

In the study [19], two different displays with different luminances and chromaticities of the monitor white point were used: a laptop display (34.5 cm × 19.6 cm) and a large desktop monitor (59.5 cm × 33.8 cm). Each subject had to perform two rounds of testing on two different days, once on the laptop and once on the monitor. In order to avoid interference between the two test subjects, who looked at the monitor and the laptop in parallel, a noticeboard was placed in the middle. The distance between the vertical wall and the monitor or the laptop display was 199cm and the distance between the screen and the eyes was about 75 cm, which, at the present

Figure 6.1 Test set-up with table, two displays in front of the test subjects (laptop and desktop monitor), two test subjects, and the two RGBW LED luminaires (top), which illuminated the opposite wall homogeneously and diffusely. A black curtain hangs on the wall behind the subjects to reduce reflections on the displays. The other laptop (left) was used to control the two LED lights. The two test subjects shown here have consented to the publication of this picture. Source: Technische Universität Darmstadt.

luminance levels of the monitors and the wall, ensured almost optimal adaptation conditions for the eyes. The size of the wall surface opposite the test subjects was 254 cm (H) × 302 cm (W). From a distance of 199 + 75 cm = 274 cm, this corresponds to a field of view of 50° (H) × 58° (W) from the subject's point of view. The luminance of the wall was measured in the middle of the wall. The deviations of the luminances on the whole wall from this measured value in the middle were max. 20%.

Each subject performed an entire test procedure on only one particular display per day. The tests on the other display were carried out on a different day. A total of 23 young subjects with normal colour vision took part in the study, 12 of them women and 11 men (age between 20 and 47; mean value: 26.4). Each complete test procedure consisted of two test rounds. Before the test, each subject filled a questionnaire about their current state of well-being (MSLT-30 state of well-being, in the so-called Multiple Sleep Latency Test [20]). This questionnaire is used to assess daytime sleepiness and to diagnose sleep disorders and is suitable for these experiments. The test persons are asked to indicate how they are feeling at the moment by making a cross on the following five scales '*in the place that corresponds to their current state*': relaxed–tense, wide awake–sleepy, concentrated–not concentrated, listless–interested, anxious–calm.

6.3.2.1 Test Series 1: Determining the Most Comfortable Display Brightness at a Constant, Typical Wall Luminance

The constant wall luminance with the white LED light with a CCT of CCT = 5013 K ($x = 0.3451$; $y = 0.3551$ according to the 2° colour matching function, CIE 1931) was set according to the following consideration. According to the study by Moosmann in [12] in nine office buildings, the horizontal illuminance at which office users were satisfied with the brightness averaged 1168 lx for users under 30 and 834 lx for users over 50 years of age. On average, therefore, it equals roughly 1000 lx. Experience shows that the vertical illuminance E_V in the room is about half the horizontal illuminance, i.e. about 500 lx.

If one uses the formula for the wall luminance L

$$L = \frac{\rho \cdot E_V}{\pi} \tag{6.1}$$

with a measured reflectance ρ of the vertical wall of 0.7, this results in a typical wall luminance of about 111 cd m^{-2}. Therefore, the practical luminaire setting in the experiment was such that a wall luminance of $L = 107$ cd m^{-2} was established.

With this constant wall luminance ($L = 107$ cd m^{-2}), the most pleasant luminance on the respective display was then experimentally determined. To do this, 20 white or grey PowerPoint slides with different background luminances (between 9.2 and 213.1 cd m^{-2} on the laptop display and 18.0 and 262.0 cd m^{-2} on the desktop monitor) were created with the well-known first verse from the poem 'Lied von der Loreley' by Heinrich Heine depicted on them in black letters. In order to achieve the same white points for all 20 slides from dark grey, light grey to white, the RGB control values of the images of the individual slides on the monitor and the laptop were corrected by *monitor calibration* before the test in order to obtain the same chromaticity coordinates as from slide 1 as a reference (with $r = g = b = 255$). A mean CCT = 7357 ± 195 K was measured on the laptop and CCT = 6943 ± 61 K on the desktop monitor for these 20 slides. Figure 6.2 shows the display of slide #1.

Each subject was given the following instruction:

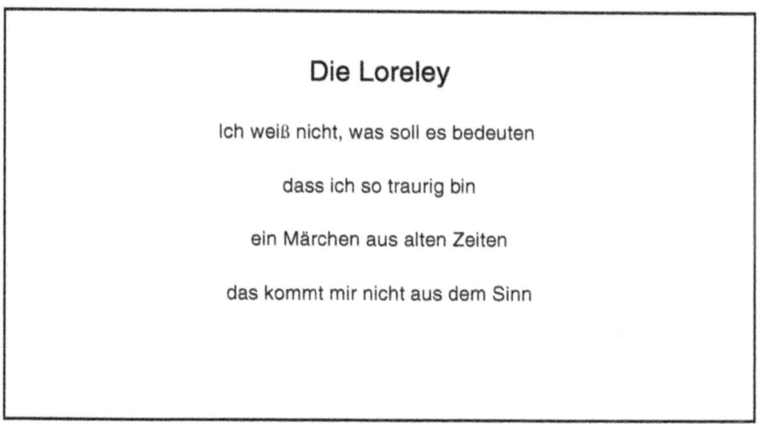

Figure 6.2 Display with slide #1 (with $r = g = b = 255$). The size of the laptop display was 34.5 cm × 19.6 cm (field of view size from the observation distance of 75 cm: 26° × 15°; deviations of the luminance on the total area: max. 14%). The size of the desktop monitor was 59.5 cm × 33.8 cm (field of view size from 75 cm: 43° × 25°; deviations of the luminance on the total area: max. 16%). The title 'Die Loreley' was written in 'Calibri Light 44 pt', the text in 'Calibri 28 pt'. Source: TU Darmstadt.

You must first find out your most comfortable display setting (i.e., the most comfortable brightness of the background for long-term office work) using the pptx slides. You are to read the poem while repeatedly looking at the wall behind the screen for 5 seconds. While reading, please pay attention to the comfortable balance between the brightness of the wall and the brightness of the screen background and choose the best slide accordingly. Keep considering both lighter and darker slides and then again slightly lighter and slightly darker slides before making your final decision. At the end of this procedure, please write the last (i.e. the most pleasant) slide number on the questionnaire.

The 20 slides were available to the test subjects in a randomised order according to luminance. As a result, an average display luminance of $L = 238 \pm 13$ cd m^{-2} was determined on the desktop monitor and $L = 176 \pm 13$ cd m^{-2} on average on the laptop display. As a partial summary for the first round of tests, it can be stated that – with a wall luminance of 107 cd m^{-2} kept constant – the luminance of the desktop monitor and the laptop for a pleasant brightness of the rather young test persons is 238 and 176 cd m^{-2}, respectively. This is – in absolute terms – far higher than the results of Veitch and Newsham of 32.7 cd m^{-2} for the monitor with a visual field luminance of 39.2 cd m^{-2} (see [21]).

Apparently, the lower luminance of the (CRT) monitor technology of the time about two decades ago meant that the luminance of the background also had to be kept low in absolute terms. With standard-compliant lighting according to DIN EN 12464-1 with the frequently used minimum value of 500 lx for horizontal illuminance, the wall luminance is more like 55 cd m^{-2} according to the above Eq. (6.1), so a research question should be posed as to how high the wall luminance should be by keeping the monitor or laptop luminance constant at the value set individually by each test subject and not changed thereafter (238 ± 13 cd m^{-2} in the laptop test and 176 ± 13 cd m^{-2}). This led to the second series of tests.

6.3.2.2 Test Series 2: Determining the Most Pleasant Luminance and Colour Temperature on the Wall with Constant Display Luminance

In this test series, the luminance of the display (as mentioned above) was kept constant for each test subject – corresponding to their own setting in test series #1. A total of 32 light settings of the LED lights were designed, with four colour temperatures (3100, 4100, 5000, and 6500 K) and eight different luminance levels per colour temperature on the facing white wall (15, 30, 45, 75, 100, 200, 500, and 1000 cd m^{-2}). At each colour temperature, there was always a different randomised order of presentation of these eight luminance levels. At each new luminance setting, the subject was asked to adapt for two minutes before evaluating the lighting situation. There was also an adaptation period of two minutes before each colour temperature change. The 32 spectra were optimised for maximum colour rendering ($R_a > 93$) according to CIE 1995. Figure 6.3 shows the four mean (averaged over the eight luminance levels) relative spectra of the light in the centre on the wall at

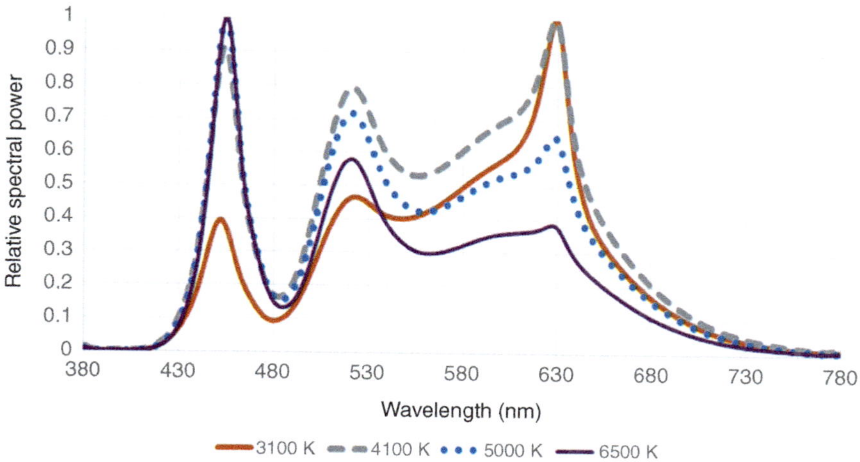

Figure 6.3 *Mean relative spectra (averaged over the* eight *luminance levels) of the light on the opposite wall* (see *Figure* 6.1) at the four colour temperatures 3100 K with $R_a = 95 \pm 0.3$; 4100 K with $R_a = 96 \pm 0.7$; 5000 K with $R_a = 95 \pm 0.1$; and 6500 K with $R_a = 95 \pm 0.8$. Source: TU Darmstadt.

the four colour temperatures investigated (the four relative spectra had only minor deviations over the eight luminance levels).

The test persons had to rate all lighting situations with a cross on each of four scales:

1. Brightness of the wall.
2. Hue of the wall.
3. Hue of the screen.
4. *General pleasantness* of the lighting situation for office work at the screen.

The present analysis deals with the evaluation of the fourth question *'General pleasantness of the lighting situation'*. For each of the eight lighting situations per colour temperature on the wall, the test person rated the pleasantness after the adaptation time using the continuous scale in Figure 6.4 with attributes between 'very poor' and 'excellent'. These ratings were read and stored as continuous numerical values between 0 and 100. In order to give the test persons, the opportunity to form their inner personal attitude to each lighting in a visual way, a training phase was scheduled for each test person before the main test with 32 test light sources, in which the test leader showed four different brightness settings of the wall, once the darkest as the lower 'anchor' and also once the brightest as the upper 'anchor'. Again, there was an adaptation time of two minutes for each test situation before the internal evaluation took place.

6.3.3 Evaluation of the Results

Figure 6.5 shows four diagrams with the mean ratings of the 23 test subjects for the desktop monitor and the laptop display for the four CCTs. In each of the four

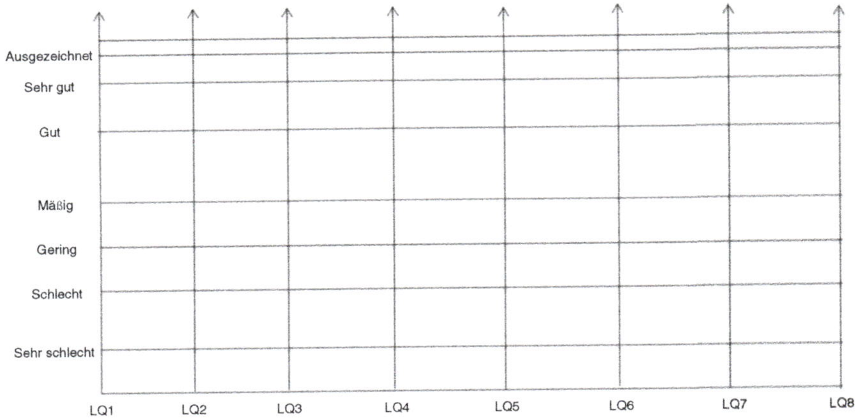

Figure 6.4 Subjects' evaluation sheet for question no. 4 ('general pleasantness of the lighting situation for office work at the computer screen') with the categories between 'very poor' ('sehr schlecht') and 'excellent' ('ausgezeichnet') for the eight luminance levels (LQ1–LQ8; randomised). One such questionnaire was completed for each colour temperature (3100, 4100, 5000, and 6500 K). The way of labelling this continuous rating scale with the categories is as follows: excellent (97.9); very good ('sehr gut', 91.6); good ('gut', 79.6); moderate ('mäßig', 52.9); low ('gering', 41.2); poor ('schlecht', 26.5); and very poor ('sehr schlecht', 12.8). The lower line corresponds to the value 0 and the upper line to the value 100. Source: TU Darmstadt.

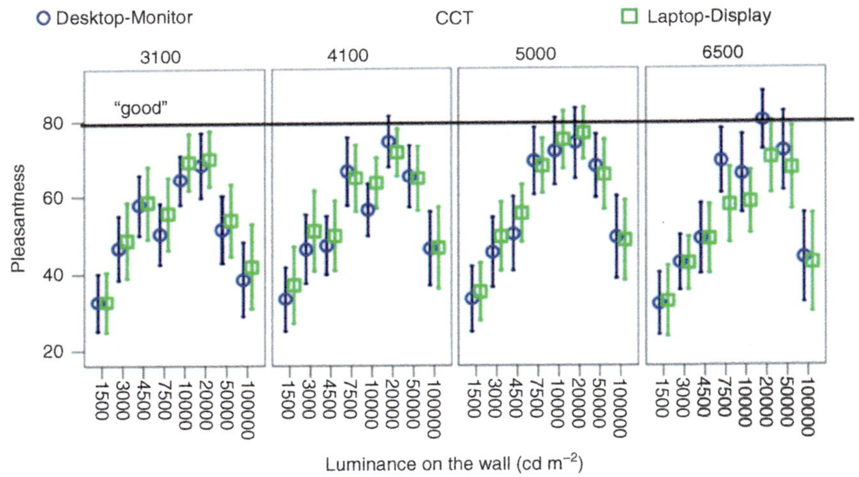

Figure 6.5 Mean rating of the 23 test subjects for the general pleasantness of the lighting situation with 95% confidence intervals for the desktop monitor (blue circles) and the laptop display (green squares), for the four correlated colour temperatures (CCTs) examined, at the different luminance levels on the wall. Source: TU Darmstadt.

diagrams, the luminance L on the opposite wall is on the abscissa and the rating on the scale between 0 and 100 (see Figure 6.4) is plotted on the ordinate. The scale value of 79.6 corresponds to the category 'good'. If all four diagrams in Figure 6.5 are considered, the following can be determined:

- The luminances around the value $L_v = 200\,\text{cd m}^{-2}$ on the wall show – independent of the colour temperature – the highest rating of the test persons. This means that $200\,\text{cd m}^{-2}$ provides the best pleasantness of the lighting situation at the monitor's luminance of $238\,\text{cd m}^{-2}$ and the laptop's luminance of $176\,\text{cd m}^{-2}$.
- The rating of the large desktop monitor tends to be highest (slightly better than 'good' on average) of all four colour temperatures at the CCT on the wall of CCT = 6500 K. Since this desktop monitor has a colour temperature of 6943 K, it is closer to assume that this does not disturb the chromatic adaptation between the light colour (i.e. white hue) of the wall and the emission colour (i.e. white point) of the monitor as much as in the case of the warmer wall light colours at CCT = 3100 or 4100 K.

6.3.4 Summary

If the results of all four colour temperatures are combined, the diagram in Figure 6.6 emerges: across all four colour temperatures, wall luminance around $200\,\text{cd m}^{-2}$ provides the most pleasant lighting in an office with VDU workstations with modern screens with a luminance also around $200\,\text{cd m}^{-2}$. This luminance is thus considerably higher than Ooyen's et al. [15] preferred luminance in Table 6.3. This luminance is thus much higher than the preferred luminance of Ooyen et al. [15] in Table 6.3.

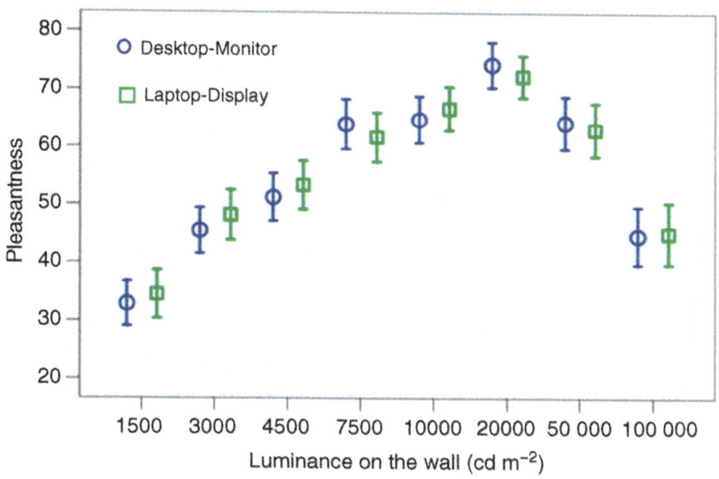

Figure 6.6 Mean rating of the 23 test subjects for the general pleasantness of the lighting situation with 95% confidence intervals for the desktop monitor (blue circles) and the laptop display (green squares) as mean value for all four correlated colour temperatures examined. Source: TU Darmstadt.

However, the *vertical* wall luminance of 200 cd m^{-2} is only achieved with a *horizontal* illuminance in the room of around 1800 lx (see Eq. (6.1)), which is far higher than the minimum value specified in DIN EN 12464-1. These minimum values in this standard as well as in other international standards for lighting technology can only be considered in the context of modern HCL design if user preference (and not only user acceptance) and the demographic development of the ageing society are taken into account.

6.4 Preferred Colour Temperatures

6.4.1 Introduction

The lighting in an interior with neutral and coloured objects (e.g. furnishings, floor, textiles, carpets, and skin tone of the occupants) is initially determined by the spectrum of the light sources [22]. The chromaticities or colour temperatures resulting from the light spectrum determine the balance of visual perception, not least because the white walls and ceilings contain the main part of the interior. The chromatic adaptation of the visual apparatus with its receptors is primarily oriented towards these white walls and other white or grey objects. Coloured objects of different more or less saturated colours can be perceived and assessed in their hues and perceived saturations within this state of chromatic adaptation. One can also formulate that the white walls and other white or grey objects in the room together with the chromaticities of the light sources represent a stable *anchor point* for the chromatic adaptation of the human visual system in the inhomogeneous visual environment of a room that contains several different coloured objects.

Figure 6.7 shows the spectra of two typical daylight phases (D50 and D65) and of two Planckian (blackbody) radiators (temperatures 2700 and 3200 K). While the

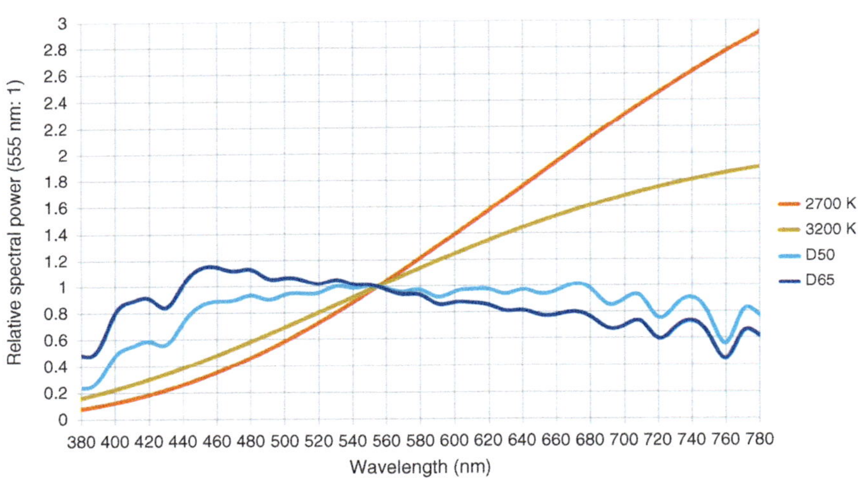

Figure 6.7 Spectra of two typical daylight phases (D50 and D65) and of two Planckian (blackbody) radiators (2700 K; 3200 K). Source: TU Darmstadt.

two Planckian emitters have very few radiation components in the blue range up to 470 nm, the two daylight phases are characterised by spectral balance over the visible wavelength range from 380 to 780 nm. This also means that the hues of objects from violet (400–420 nm) to deep red (650–680 nm) can be reproduced with them.

From an ocular physiological, cognitive, and evolutionary point of view, humans become accustomed over time to daylight from 5000 K (D50) as soon as they spend time outdoors. There, during the day, from about 7 o'clock in the morning to about 5–6 o'clock in the late afternoon, there is an illuminance of 10 000 to 100 000 lx, so that objects appear brilliantly coloured and rich in colour contrast. In interior lighting in an office, private home, hotel, or department store, illuminance levels not only drop to around 300–2000 lx. The presence of objects in the interior (such as yellow floor parquet, furniture tending towards warmer shades, curtains and drapes, and table surfaces) causes the colour temperature to drop to around 4000–4800 K even on a bright summer day. This situation has shaped generations of people for many decades. In our time with modern LED luminaires, where the light can be dynamically varied in luminous intensity, direction (direct/indirect) and chromaticity or colour temperature, the following two questions arise:

1. Which chromaticity ranges or colour temperature ranges are preferred?
2. Does the preferred colour temperature change when the lighting level in the room changes?

A study [22] was conducted on these issues, the methodology and results of which are summarised in this section.

6.4.2 Experimental Method

A room with white diffuse reflecting walls was built for the study [22]. The window was made light-proof with a white diffuse curtain. The different light sources were installed on the ceiling with a selective matt diffuser disc and configured to be switched on and off and dimmable by means of different electric circuits and a controller. Figure 6.8 shows the spectra of the light sources examined.

For the measurement of these spectra, the respective light source type was switched on, and, after more than half an hour, the spectral radiance of the white standard on the table in the middle of the room was measured. Thus, the measured spectrum already includes the multiple selective reflections of the objects in the room. It can be seen from Figure 6.8 that the two LED light sources for the warm white (2748 K, like a previous incandescent lamp) as well as for the neutral light (4020 K) exhibit a full spectrum (i.e. without deep local minima in the visible wavelength range). The spectra of the cool white LEDs with a colour temperature of 5021 K as well as 5725 K also have an emission maximum at 625 nm. The halogen incandescent lamp with a diffusing lens delivers a colour temperature of 2320 K and very little radiation in the 380–500 nm range. The neutral white fluorescent lamp with 4117 K has many radiation gaps in its spectrum in the range 500–530 nm (green range), 560–580 nm (yellow-green range), and 630 nm (red range). Table 6.4

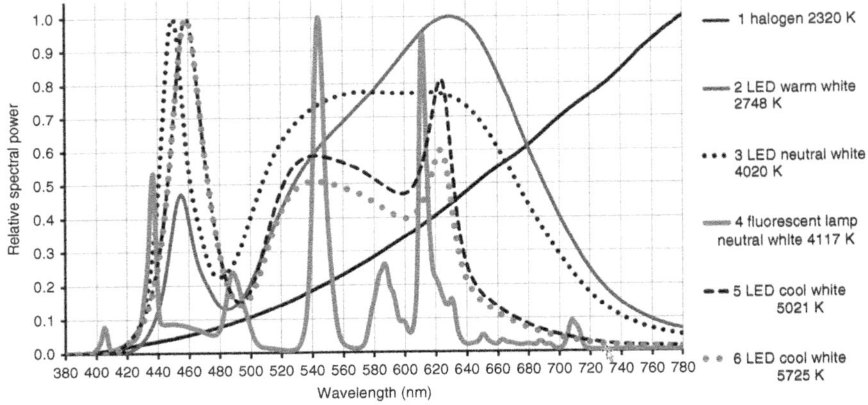

Figure 6.8 Spectra of the investigated test light sources in the study [22]. Source: TU Darmstadt.

Table 6.4 Correlated colour temperatures (CCT) as well as the general colour-rendering indices CIE CRI R_a in the study [22].

Type	Name	CCT (K)	R_a
1	Halogen bulb	2320	99
2	LED warm white	2748	89
3	LED neutral white	4020	88
4	Fluorescent lamp neutral white	4117	81
5	LED cold white	5021	88
6	LED cold white	5725	87

Source: TU Darmstadt.

lists the CCTs as well as the general colour-rendering indices R_a (for the eight desaturated CIE test colours).

Thus, the experimenters had a problem at the beginning of test design. Since the light sources have different colour-rendering properties for the CIE test colours and the objective of the tests was to find out the most preferred colour temperatures, it has therefore been advisable to select the objects in the room from white and neutral colours so that the colour-rendering indices of these light sources for white objects are almost identical and mostly at a high level ($R_a > 90$). This ensured that, during the tests, the test subjects focused their assessment mainly on the white tone of the light and not on the colour rendering of the coloured objects. The illumination of the test room with white objects on the table with white tablecloths at the warm white light type (CCT = 2320 K) as well as at the cool white light colour (CCT = 5725 K) is illustrated in Figure 6.9, as an example.

The white or neutral objects must be spectrally aselective in their reflective properties and typical of everyday professional and private life. For this reason, white paper

Figure 6.9 Illumination of the test room with white objects on the table with white tablecloths in the warm white lighting mode (at 500 lx) and in the cool white lighting mode (at 1200 lx). Source: Technische Universität Darmstadt.

flowers, plates, bowls, cups, white cloths as well as rice were selected for this test purpose and measured spectrally with a spectroradiometer in comparison to a white standard. The spectral reflectance values of these objects are shown in Figure 6.10.

These spectral reflectances are predominantly constant over the wavelength range of 420–700 nm according to Figure 6.10. This information was used to calculate the object-specific colour-rendering indices of the test light sources (see Figure 6.11) for these white and neutral objects. The object-specific colour-rendering indices determined are illustrated in Figure 6.11 for the white objects (Figure 6.9) and the six test light sources (Figure 6.8). It can be seen from Figure 6.11 that most of the object-specific colour-rendering indices are greater than 95.

To experimentally answer the question *'Does the preferred colour temperature change when the lighting level in the room changes?'*, two light levels, 500 and 1200 lx, were chosen. The illuminance of 500 lx represents the minimum indoor

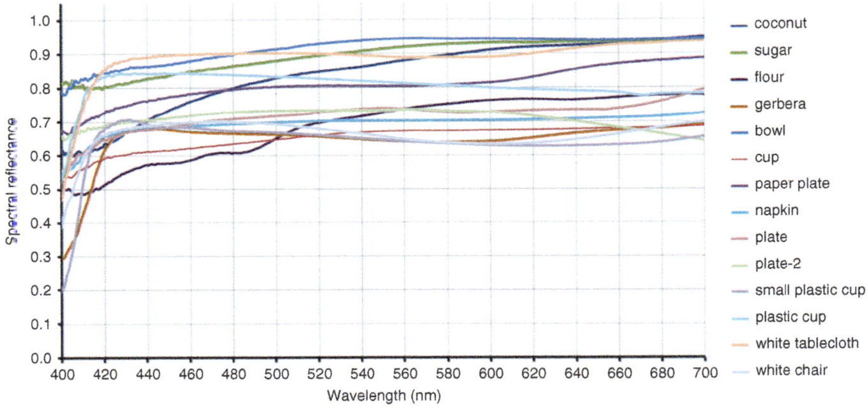

Figure 6.10 Spectral reflectance values of the white and grey (neutral) objects in the test room (see also Figure 6.9). Source: TU Darmstadt.

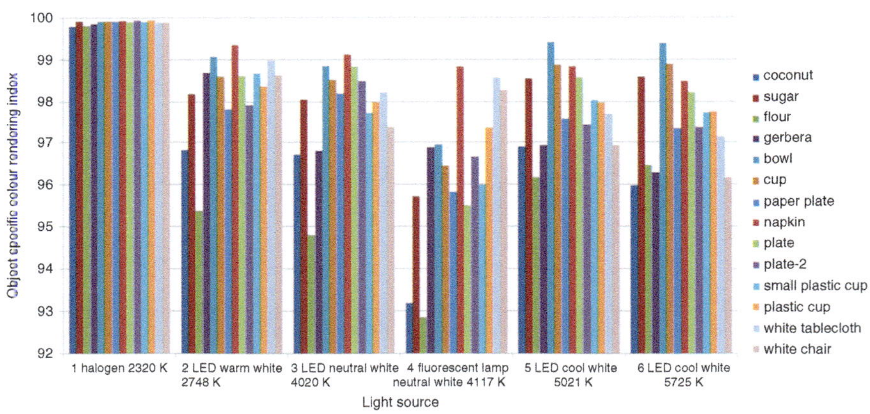

Figure 6.11 Object-specific colour-rendering indices for the white or neutral objects (see Figure 6.9) and the six test light sources (see Figure 6.8). Source: TU Darmstadt.

illuminance value according to DIN EN 12464 for most work tasks. The value of 1200 lx is a feasible value for dynamic lighting control (dynamic lighting) at the higher level without calling energy use into question too much. For the higher illuminance, more light sources of the same type were added without changing the spectra and the good homogeneity of the light distribution in the room.

In the experiment, 20 European test subjects were tested (17 men, 3 women). The average age was 25.8 years (minimum: 22 years, maximum: 37 years). All subjects were tested for colour vision deficiency. All subjects wore a white shirt or grey gown to avoid colour objects in the room. Before the test, the test persons were given a training phase to get used to the scenes as well as the tasks and were able to adapt to the brightness levels. They were given the following instruction:

White interiors are very popular today, they are there for a celebratory mood. This room is an example of that. Imagine: You are attending a feast, you are invited guests, similar to the 'white dressed' feast ('Dîner en Blanc') in the Herrngarten (a park in the City of Darmstadt in Germany, next to the University), which you probably have seen. You can sit down at the table, talk to each other, but please never talk about the furnishings and the lighting situation in the room. You can also stand up and walk around the table, looking at the individual white objects.

Your task is to rate the scene of objects according to your preference on an interval scale between 0 and 100. 100 corresponds to your maximum like (your maximum preference) when you look at all the objects including furniture as you would like the interior to be. We illuminate the objects with different light sources, to which the numbers 1–6 correspond. Rate your preference on the corresponding scale. The test leader names the number of the light source, and you first rate the object scene with a pencil with a cross along the scale. Each lighting situation always lasts 1.5 minutes (90 seconds) and is always repeated three times.

You have to finalise your answer about your preference on the third pass with the biros. There are two test series each with six lighting situations (once darker, then brighter). Please never look at the ceiling where the light sources are. Please wait half a minute after switching and only then start evaluating the new situation. This is a game of patience and an easy experiment. Please look patiently at the objects, and the furnishings, think, and evaluate – remembering that you are attending a 'white' banquet as 'guests'.

In accordance with the instructions described above, the questionnaire for the two light levels and six test light sources is shown in Figure 6.12.

6.4.3 Results and Discussion

Figure 6.13 shows the mean values and their 95% confidence intervals of the visually assessed preference (0–100, see Figure 6.12) for each colour temperature and per colour temperature for the two light levels 500 and 1200 lx.

Figure 6.13 shows the following:

1. The test persons favour the colour temperature range around 4000–4100 K and partly 5000 K.
2. The colour temperature range 2320–2748 K as well as 5725 K are not preferred.
3. In general, the ratings at 1200 lx are better than at 500 lx, regardless of colour temperature.
4. The test instructions with the context 'the guests take part in a feast' as well as the objects such as white plates, flowers, cups, and bowls *ensure the context* of the experiment both in the private and in the social sphere.

The results of a comparison of the mean visual preference values (see Figure 6.13) between the two luminance levels at each CCTare shown in Table 6.5.

Table 6.5 shows the following:

(a) Although the mean values of the visual ratings of preference are different at 500 and 1200 lx, there is no significance (=0.05) from 4020 K inclusive. This means

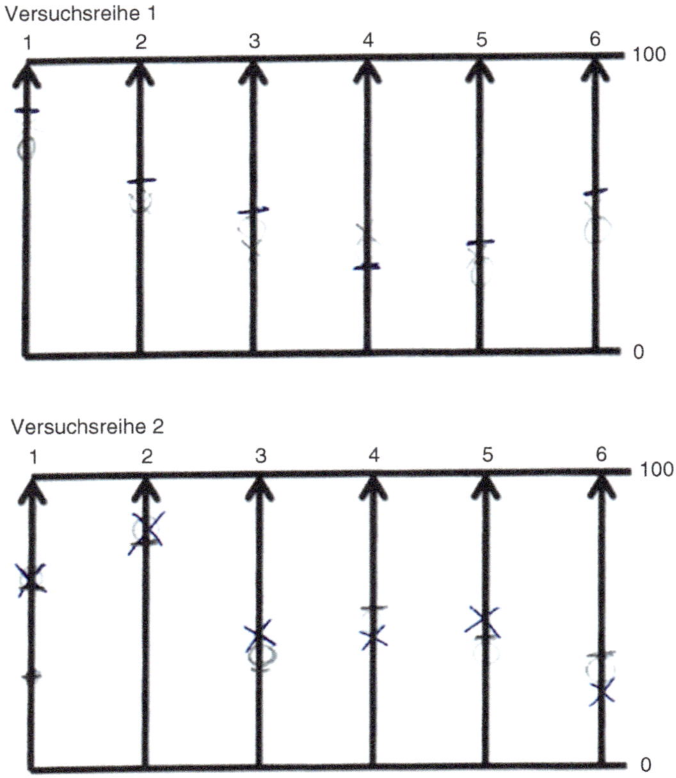

Figure 6.12 Questionnaire for the test subjects at the two tested light levels in the two series of experiments ('Versuchsreihe' 1 and 2) for the six test light sources. Source: TU Darmstadt.

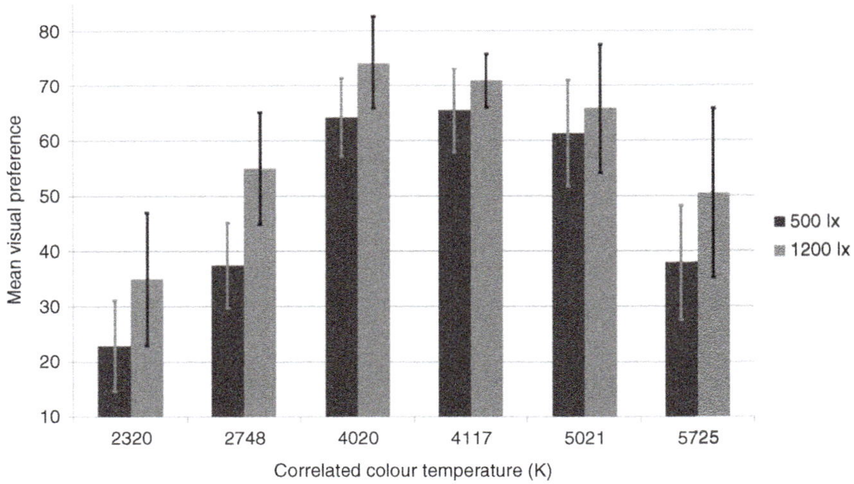

Figure 6.13 Mean visual preference values and their 95% confidence intervals for each colour temperature at the two light levels 500 and 1200 lx. Source: TU Darmstadt.

Table 6.5 Comparison of mean visual preference values (see Figure 6.13) between the two luminance levels at each correlated colour temperature.

CCT (K)	Mean value of visual preference at 500 lx	Mean value of visual preference at 1200 lx	Significance p
2320	23[a)]	35[a)]	0.046
2748	38[a)]	55[a)]	0.000
4020	64	74	0.056
4117	65	71	0.084
5021	61	66	0.317
5725	38	50	0.148

a) Difference significant at the 5% level.
Source: TU Darmstadt.

that there is no significant difference between 500 and 1200 lx in the colour temperature range between 4020 and 5725 K. In this test, the results of Kruithof [23] (see Section 6.5), according to which lower colour temperatures are preferred at low lighting levels and higher colour temperatures are preferred at higher lighting levels, could not be confirmed.

(b) The differences in preference evaluation between 500 and 1200 lx in the colour temperature range 2320–2748 K are significant, i.e. the evaluation at 1200 lx is clearly better than at 500 lx. The thesis of Kruithof [23] is hereby disproved for this colour temperature range and these light levels. The relationship between colour temperature and illuminance – from the point of view of user preference – is discussed in detail in Section 6.5.

6.5 Preferred Ranges of Colour Temperatures and Illuminances

6.5.1 The Nature of Illuminance and Colour Temperature

The concept of illuminance E_v is defined according to Eq. (6.2) [24, 25].

$$E_v = K_m \cdot \int E_e(\lambda) \cdot V(\lambda)\, d\lambda \tag{6.2}$$

In Eq. (6.2), $E_e(\lambda)$ is the spectral irradiance on a working plane and $V(\lambda)$ is the spectral luminous efficiency function for a 2° field of view for day vision. The $V(\lambda)$ function only considers the components of the L- and M-cone signals, i.e. only the achromatic components. The chromatic components (red-to-green and blue-to-yellow) of the incident optical radiation, i.e. (L − M) and S − (L + M), are not considered. In this $V(\lambda)$ function, the signal components of the blue cones $S(\lambda)$ are not present at all or only slightly [26, 27]. This means that the components of the chromatic components, the blue cones, and the light-sensitive ganglion cells (ipRGC) must be added to the parameter *illuminance* (or luminance) to describe the complex perception of 'brightness' (see Chapter 5).

Table 6.6 Correlation coefficients of the correlated colour temperature (CCT) to the retinal receptor signals and the chromatic channel red–green according to [28].

Designation	Meaning	CCT (K)
L_{rel}	Relative signal of the L-cones	−0.940
M_{rel}	Relative signal of the M-cones	0.962
S_{rel}	Relative signal of the S-cones	0.977
R_{rel}	Relative signal of the rods	0.914
G_{rel}	Relative signal of the ipRGC cells	0.924
$\|L-M\|_{rel}$	Relative signal of the chromatic channel 'red–green'	−0.957

Source: TU Darmstadt.

Physiologically and colorimetrically, the CCT is closely related to the chromaticity coordinate of the radiant source to be analysed, which should be on or near the Planckian locus or the locus of daylight phases of the same CCT. Naturally, this light source tends to have (roughly speaking) a similar spectral distribution as the associated reference light source. For physiological reasons, the CCT of a light source correlates with the signals of the M-cones, S-cones, rods, and photosensitive ganglion cells (ipRGC), as calculated in an analysis [28] with 294-light spectra, see Table 6.6.

Table 6.6 clearly shows the tendency that the CCT (in K) is substantially shaped by the signals of the blue- and green-containing receptors of the retina and not by the red–yellow-containing channels. In the end, the colour temperature correlates very strongly with the brightness B, which can be modelled, e.g. with the formula of Fotios and Levermore [29] in Eq. (6.3) (the quantity S is the signal of the S-cones), see Figure 6.14.

$$B = L_V\,(S/V)^{0.24} \tag{6.3}$$

The combination of illuminance E_V or luminance L_V with colour temperature thus determines the perception of brightness when viewing a scene or an illuminated room.

6.5.2 Illuminance and Colour Temperature in the Literature

The first study on the joint effect of illuminance and colour temperature was carried out by Kruithof in 1941 [30], which is still considered one of the most discussed works in the history of lighting technology. Up to the present day, very little information has survived about Kruithof's test condition. There was little description of the test method, how the light stimuli were presented as a combination between colour temperature and illuminance, the test apparatus, and the number of test subjects. Incandescent lamps were used in the CCT <2850 K range, and daylight and fluorescent lamps similar to daylight were used in the higher colour temperature range [31, 32]. Despite this major shortcoming, the so-called Kruithof diagram is nevertheless often cited in the lighting technology literature.

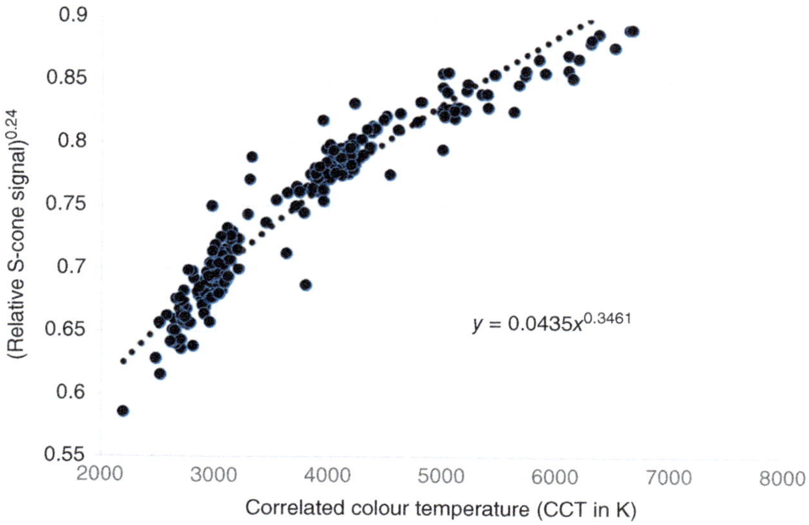

Figure 6.14 Correlation between the quantity (relative S-cone signal)$^{0.24}$ according to Fotios, Eq. (6.3) and the CCT, calculated with 294-light spectra [28]. Source: TU Darmstadt.

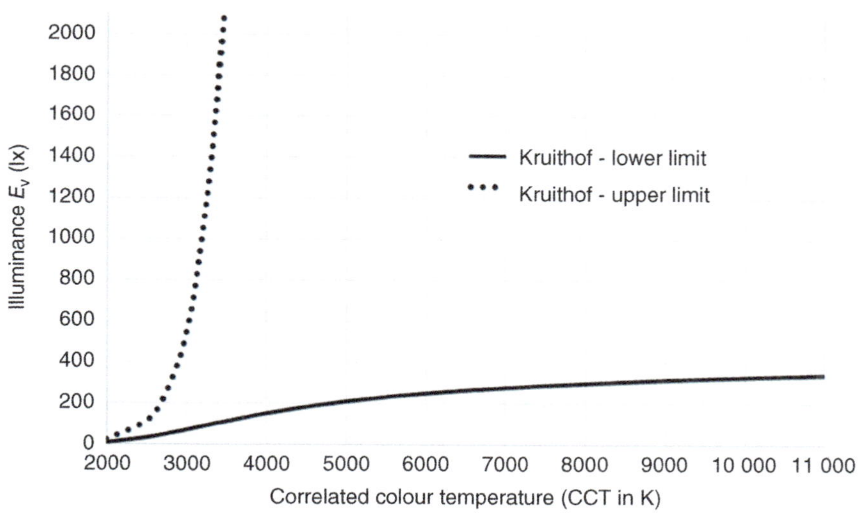

Figure 6.15 The Kruithof diagram, reconstructed by the authors of this book from [32]. Source: TU Darmstadt.

This diagram (see Figure 6.15) was reconstructed by the authors of this book from [32] and shows two ranges of colour temperature and illuminance (below the lower limit and above the upper limit in Figure 6.15) in which the lighting situation does *not* appear *pleasant*. The area with 'pleasant' is in the middle of the diagram and means that a lighting situation appears pleasant if the user selects a suitably low illuminance at a low colour temperature or a suitably high illuminance at a high colour temperature between 6000 and 8000 K.

For many decades, numerous international research groups have dealt with the topic in order to investigate the validity of this diagram (Figure 6.15). A total of 29 publications are known. These were analysed by Fotios in 2017 [31]. Based on a CIE publication (CIE Publ. 212:2014 [33]) with recommendations on how such experiments should be designed (randomised light stimulus presentations, careful statistical data evaluation, and detailed method description), only nine publications are still considered serious after analysis in [31]. Surprisingly, the majority of the 20 non-relevant publications were published in the more recent period since 2003–2015.

This detailed analysis [31] of the nine significant papers showed the following results:

- The majority of publications contradict the statements of the Kruithof diagram.
- Fotios [31] considers 'pleasant' equivalent to 'preferred' or to 'natural in chromaticity'.
- The variation of colour temperature between 2500 and 6500 K has a negligible influence on the rating of 'pleasant and pleasing'. These results contradict the experience of the present authors who therefore conducted another experiment.
- Illuminance levels below 300 lx cannot produce the rating 'pleasant'. Illuminance levels of 500 lx are sufficient to provide this positive rating (i.e. 'pleasant'); and illuminance levels higher than 500 lx provide only a slight improvement, cf. with Table 6.1.

The analysis of the 29 publications by Fotios [31] described above and the closer examination of the nine relevant publications [31] mentioned above prompted the authors of this book to carry out a series of new experiments to investigate the combination of *colour temperature and illuminance* with multichannel LEDs (see Section 6.5.3), which were specified as follows:

- The illuminance should reach the level of about 1800 to 2000 lx.
- To also investigate the effect of colour rendering, the LED light spectra were designed for different colour-rendering indices (from about $R_a = 25$ to $R_a = 92$–96).
- It is not sufficient to find out only the 'pleasant' range of colour temperature–illuminance combinations. The authors used a series of categories (so-called semantic levels), e.g. 'moderate', 'good', and 'very good', to be able to categorise the different combinations of colour temperature and illuminance. These categories should also serve as the basis of recommendations in lighting practice.

6.5.3 Visual Experiments on the Combined Effect of Colour Temperature and Illuminance

Two series of experiments were conducted at a time interval of four months. In the first experiment, the spectra of two LED multichannel luminaires (RGBW) were optimised to vary the colour temperature and illuminance over large ranges with very high, almost constant colour-rendering indices (R_a = 92–96) and the corresponding low colour oversaturation of the objects (approx. $0 < \Delta C^* < 1$, compared to the reference situation). In the second experiment, four months later, the same luminaires were optimised so that light spectra with very different colour temperatures, illuminances, and colour-rendering indices were set.

The first experiment [34, 35] was conducted in winter 2017 in a real room with white and matt painted walls and ceiling and neutral grey flooring. During the test sessions, the window was closed light-tight, and the room was supplied with clean air. Two diffuse luminaires with four LED channels (red, green, blue, and warm white, so-called RGBW) homogeneously illuminated a table with coloured objects and a white tablecloth, see Figure 6.16. Four chairs were placed around the table for the four test subjects. The coloured objects were partly freshly bought food in different hues and saturations, which represented a true, three-dimensional object form, so that the object arrangement appeared realistic and identification-awakening.

The 25 light spectra in Figure 5.6 were used to illuminate the objects in Figure 6.16. The horizontal illuminance on the table was 45, 90, 470, 1000, and 2000 lx. The CCT – at each illuminance level – was also varied: the authors used 2700, 3100, 4100,

Figure 6.16 Table with coloured objects and a white tablecloth in the first experiment. Source: Reproduced with permission from Khanh et al. [34], SAGE PUBLICATIONS, INC.

5000, and 10 000 K. The colour temperature was also varied. Thus, the subjects were shown 25 different E_v – CCT combinations (cf. with Figure 6.15) in a *randomised manner*, always waiting about 1.5 minutes for readaptation after adjusting each light combination until the subjective evaluation was made. The RGBW LEDs were optimised to ensure the same, high colour-rendering index level ($92 \leq R_a \leq 96$) for each spectrum ($5 \times 5 = 25$).

A total of 21 test subjects took part in the first experiment. These 7 men and 14 women, aged between 20 and 47 (with a mean age of 25.2), had normal colour vision. Before the main experiment took place, a training phase was introduced with eight-light spectra for very dark or very bright scenes with very low and high colour temperatures as anchor stimuli (according to the CIE recommendation [33]). Then, the main experiment began, with the subjects rating their impression of the respective scene on a scale between 0 and 100 with categories. The subjects wrote a corresponding cross on the scales. The categories corresponded to the ratings 'excellent', 'very good', 'good', 'moderate', 'low', 'bad', and 'very bad', whereby the test persons were also allowed to rate their impression between two neighbouring categories, see Figure 7.2 (right). The scene preference '*I prefer this scene or not*' is equivalent to the attributes 'pleasing' or 'pleasant'.

The second experiment is presented in detail in Section 7.3. In this section, the scene preference results of Section 7.3 are analysed with a different method, i.e. from the point of view of the Kruithoff diagram (see Figure 6.15). Other coloured test objects were arranged, see Figure 7.1 (Scene preference, bottom right). The coloured objects emphasised the context of a familiar environment for the test subjects with real objects such as fruits, flowers, and books as well as a highly placed landscape painting, which aroused the perception of the three-dimensionality of the object arrangement. Figure 7.3 shows the 36 light spectra used. Table 7.1 shows their photometric and colorimetric characteristics. A total of 30 test subjects took part in the second experiment. These 14 men and 16 women, aged between 19 and 32 (with a mean age of 25.8), had normal colour vision. The test method was the same as in the first experiment.

6.5.4 Results: Combined Effect of Colour Temperature and Illuminance

Figure 6.17 shows the category (according to Figure 7.2, right) of the mean scene preference rating of all participating subjects in experiments 1 and 2, for each lighting situation as a combination of illuminance and colour temperature. Each lighting situation studied corresponds to a coloured symbol in Figure 6.17.

The following trends can be seen from the results in Figure 6.17:

- The lighting situations with illuminances lower than 200 lx were evaluated with the categories 'low' or 'poor' independent of the colour temperature. Since all spectra examined here have a general colour-rendering index of better than 87, it can also be formulated that a lighting scene with a relatively high colour rendering but with an illuminance lower than 200 lx is rated as 'low' or 'poor'.
- The lighting situations around 500 lx were rated 'moderate' or 'moderate–good', also independent of colour temperature.

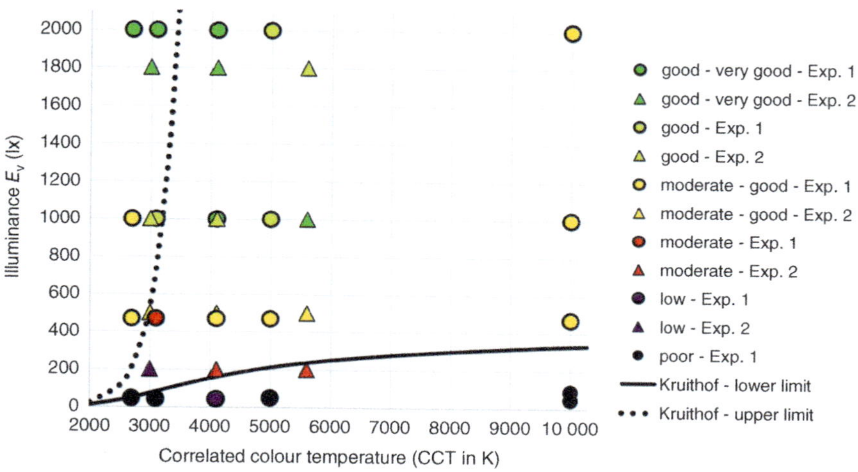

Figure 6.17 Category (as per Figure 7.2, right) of the mean scene preference rating of all participating subjects in Experiments 1 and 2, for each lighting situation as a combination of illuminance and colour temperature. A coloured symbol corresponds to each lighting situation examined. In this section, only the lighting situations of the two experiments were shown that had a general colour-rendering index with CIE CRI $R_a > 87$. Source: Darmstadt University of Technology.

- The lighting situations with CCT = 10 000 K were also rated as only 'moderately good', also independent of illuminance. Experiments with very high colour temperatures such as 10000 K produce a high circadian stimulus but are psychologically and emotionally not useful.
- 'Good' to 'very good' ratings were found in lighting situations at CCT > 3000 K and with a relatively high illuminance between 1000 and 2000 lx.

In the upper left corner of Figure 6.17, outside the upper limit of Kruithof (i.e. 'not pleasant' according to Kruithof), some light situations with 1800–2000lx and with colour temperature around 3000–4000 K were classified as 'good–very good' (i.e. green symbols). Thus, the original Kruithof results [30] cannot be confirmed in part. A further development of the conventional Kruithof diagram is given in Section 7.5, see Figure 7.16–7.18.

6.5.5 Dependence of Preferred Colour Temperature and Illuminance on Age and Gender for Activation and Relaxation

The preference for a particular colour temperature or illuminance level is also usually dependent on the current purpose of the lighting, e.g. to improve productivity during activation or to aid recovery during relaxation [25], which has a major impact on well-being and health [25] and also depends on the age and gender of the user [25]. In this section, the results of a study [25] that investigated subjective preference regarding light settings for activation and relaxation are summarised. In this study [25], a surface of an observation box reflected different light source spectra, and the subject could adjust the illuminance (between 350 and 5500 lx) and colour temperature (between 2500 and 7000 K). In the study [25], 80 subjects – belonging to four groups that differed in gender and age – were asked to imagine *activating* and *relaxing* situations and to reproduce them on the surface of the observation box – by adjusting the illuminance or the colour temperature – so that the resulting lighting situation corresponds to lighting that is suitable and pleasant for the purpose (activation *or* relaxation). Table 6.7 summarises the mean tendency of these results.

The results (see Table 6.7) showed significant differences for the preferred lighting conditions in the two situations (activating, relaxing). Combined gender- and age-specific differences were also revealed. These results are summarised as follows:

- There is a clear difference between the activating and the relaxing light situation. Young subjects distinguish more strongly between activation and relaxation than older subjects.
- All subjects (i.e. men and women, young, and old) generally tend to prefer bluish light (with higher *CCT*) for activation and prefer yellowish light (with lower CCT) for relaxation.
- The subjects prefer a higher illuminance for activation and a lower illuminance for relaxation.
- Gender and age are relevant for the preferred settings of illuminance and CCT: young males prefer significantly more illuminance for activation than all others. Younger women prefer lower illuminance for activation than older women.
- In terms of relaxation, illuminance preferences are similar for all men regardless of age. However, younger women prefer lower illuminance levels than older women.
- Young men prefer a much higher CCT for activation than all others, while young women prefer a lower CCT for activation than older women.
- For relaxation, women in all age groups prefer about the same CCT. However, young men prefer a lower CCT than older men.
- For activation (for example for work in an office, at an industrial workplace, or in a nursing home), the test persons prefer an illuminance higher than 862 lx. For relaxation, illuminance levels higher than 494 lx are desired.

Table 6.7 Dependence of preferred colour temperature (CCT in K) and illuminance (E_v in lx) on age and gender (M: male, W: female) for activation and relaxation. Values according to the predictions of the linear models from reference [25]. The authors of this book read off the values in Figure 6.9 [25] with 2% accuracy, resulting in Table 6.7.

Age	CCT (K)				E_v (lx)			
	Activation		Relaxation		Activation		Relaxation	
	M	W	M	W	M	W	M	W
18	4142	3630	3100	3217	1356	866	585	494
48	3946	3743	3304	3189	1109	953	569	628
78	3750	3856	3507	3160	862	1040	552	761

Source: TU Darmstadt.

6.6 Preferred White Chromaticities

6.6.1 Introduction

The white tone (white chromaticity or white point) of the light spectrum – with a specified, fixed CCT – can have different hues (e.g. more or less bluish) depending on the distance from the locus of the blackbody radiators or the locus of the daylight phases [36]. Observers' preference for these tints (hues) in the chromaticity of lighting (also known as *whiteness preference*) is an important aspect of light quality in a general sense, which also influences perceived colour quality. Although in most practical lighting situations the light source (lamp, luminaire) itself is rarely directly visible, there are achromatic (white, grey) objects (e.g. windowsills, walls, tablecloths, and grey furniture) in the user's field of vision in the illuminated environment, which should generally appear achromatic after reflection of light from the light source. The actual colour quality, i.e. the observer's judgement of the colour appearance of *coloured* objects (e.g. office tools, books, fruits, vegetables, flowers, and clothing) illuminated by the same light source, is visually assessed in relation to these white objects. A shade of white that includes a faint but visible odd hue (e.g. greenish or violet) is not preferred.

In today's lighting practice, the white points of the various manufacturers are usually located in a certain range near the *Planckian curve* (or in other words curve of the *blackbody radiators*) for CCTs below 5000 K in the chromaticity diagram. The white points are located in an area near the curve of the *daylight phases* for CCTs equal to or above 5000 K. These ranges form the most widely used (so-called '*industrial*') white points for spectral optimisation of light sources in *industrial practice.* This convention corresponds to the concept of CIE Publ. 13.3 on colour rendering, in which 'the reference light for light sources with a CCT below 5000 K is a Planckian radiator and above 5000 K a daylight phase' and the reference illuminant should be chosen so that the difference *DC* is less than 0.0054. DC denotes the Euclidean distance between two points in the CIE u, v chromaticity diagram.

This concept defines the above two ranges following the Planckian locus (<5000 K) and the locus of the daylight phases (≥5000 K), see the u'-v' diagram in Figure 6.18, which also shows 896 white points of today's practical indoor light sources (warm white, neutral white, and cool white; LEDs and fluorescent light sources).

As can be seen from Figure 6.18, the white points (circles) of most (95%) of the 896 selected practical light sources are in the range of the 'industrial' white points (in the sense defined above; see the cloud of small dots), 75% of the white points are even in the range of $\Delta u'v' \leq 0.001$. The symbol $\Delta u'v'$ denotes the distance in the u'-v'-chromaticity diagram, measured from the two loci (Planckian and daylight). However, Rea and Freyssinier [37] found in an experiment that the 'industrial' white dots (cloud of small dots in Figure 6.18) perceptually contained purple, violet, green, and yellow, while the location of perceptually *pure* white (without perceived chromaticity in the white) was *below* the Planckian curve at CCT < 4100 K, see Figure 6.20. A similar result was found by Liu et al. [38]. Perz et al. [39] found that

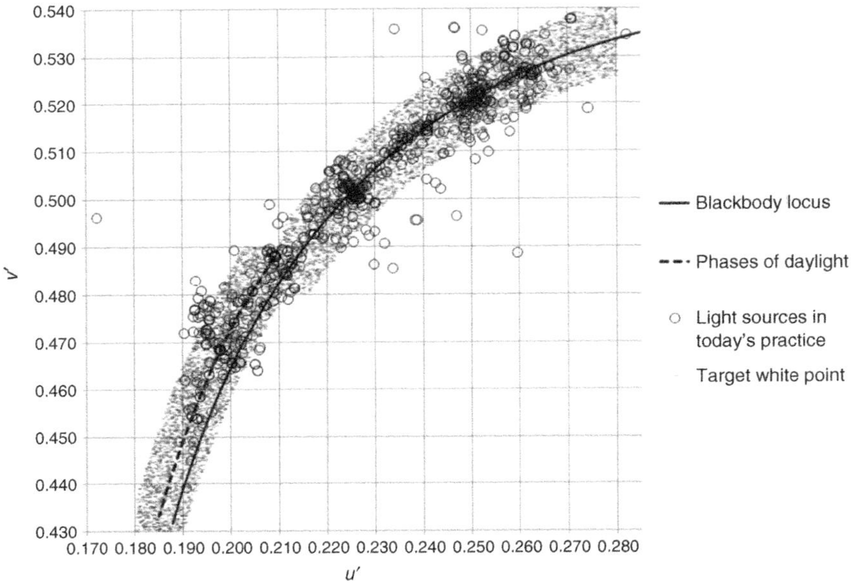

Figure 6.18 Cloud of small dots: range of so-called 'industrial' white points ('target white points' according to CIE Publ. No. 13.3: along the blackbody locus, up to CCT = 5000 K; and along the locus of the phases of daylight, above 5000 K). Circles: White points of a representative set of 896 practical indoor light source spectra ('light sources in today's practice') taken from today's lighting practice (warm white, neutral white, and cool white; LEDs and fluorescent light sources). The locus of blackbody radiators is shown from 2000 to 12 000 K, and the locus of daylight phases from 5000 to 12 000 K. Source: Reproduced with permission from Color Research and Application [36].

white points with the highest perceived whiteness level were *below* the Planckian curve at about $\Delta u'v' = 0.005–0.012$ (depending on CCT; see Figure 6.20).

When considering the problem of white tone preference, we must distinguish between the *absence of a perceived hue* in the white tone (a physiological perceptual problem) and the *preference of* the light source user for the white tone (a subjective, cognitive problem) [40]. For example, warm white tones with a distinct perceptible yellow tone are preferred for evening lighting in homes or in restaurants (at least in Western culture). The question is how strong this yellowish hue should be for the best preference and whether the point of maximum perceived whiteness (i.e. the lowest yellowish hue) in the warm white range corresponds to the most preferred white hue.

Experiments showed that '*lights with colour loci below the Planckian curve provide preferential lighting for interiors*' [41]. This hypothesis was tested in an experiment in which subjects had to judge the naturalness of a scene with several coloured objects (furniture, fruits, and vegetables). On average, a *Duv* value (distance from the Planckian curve in the *u-v* chromaticity diagram) of about –0.015 (the negative sign expresses a white point *below* the Planckian curve) was considered the most natural, lying outside the 'industrial' white point range mentioned above [41].

Figure 6.19 Chambers of an observation box with coloured objects used in Experiment 1 (left; illuminated in the neutral-cold white range) and Experiment 2 (right; illuminated in the warm-neutral white range). Source: Reproduced with permission from Bodrogi et al. [36], John Wiley & Sons.

In this section, the experimental methodology and results of a selected visual experiment [36] are described in detail, where observers rated their white tone preference as well as the percentage of a perceived hue in the white tone.

6.6.2 Experimental Method

In this experiment [36], the observers sat in a dark room at a distance of 60 cm in front of an observation box (Figure 6.19) with a chamber containing small, coloured objects such as textiles, fruits, vegetables, and flowers. The chamber was diffusely illuminated from above at an illuminance of 1000 lx (horizontal; bottom of the chamber) with different spectra with high CRI. In experiment 1, a neutral-cold white group of spectra was investigated using 54 white points with CCTs between 4600 and 9700 K, while in experiment 2, a warm white group and a neutral white group were investigated using a total of 38 white points with CCTs between 3000 and 4500 K. In the warm white group, there were 19 white points at around 3200 K and in the neutral white group another 19 white points at 4200 K.

The observers rated their whiteness preference on an interval scale with rating categories (e.g. ‘moderate’ or ‘good’, see Figure 7.2, right) and the percentage of a visible hue in the whiteness on a continuous rating scale between 0% and 10%.

6.6.3 Results

The correlation between the mean white preference and the mean amount of perceived hue in the white was strongly negative, confirming the hypothesis that those white points containing the minimum hue are generally most preferred [36]. Regarding the resulting mean preference values of the examined white points within and outside the range of the ‘industrial’ white points (see Figure 6.18), the hypothesis was confirmed that the most preferred white shades (in the sense of ‘good’ to ‘very good’ ratings) are within the range of the ‘industrial’ white points. The present whiteness preference data [36] were compared (in case of ‘moderately good’, ‘good’, and ‘good – very good’) with three selected other whiteness preference data sets, the Rea

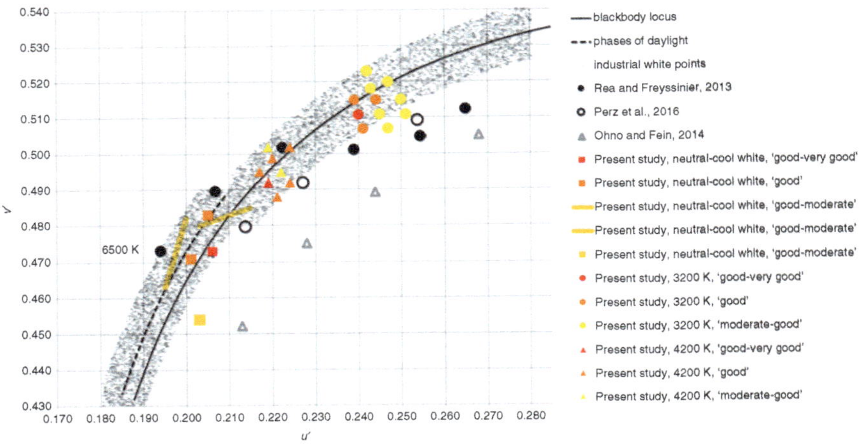

Figure 6.20 Comparison of different experimental data for white preference. Data from the study analyzed in detail in Section 6.6 [36]: 'moderately good' and 'good–moderate' with yellow symbols and broad yellow lines, respectively; 'good' with orange symbols and 'good–very good' with red symbols. Cloud of small dots: Region of 'industrial' white dots (see Section 6.6.1). Source: Reproduced with permission from Color Research and Application [36].

and Freyssinier data [37], the Perz et al. data [39], and the Ohno and Fein data [41], see Figure 6.20.

As can be seen from Figure 6.20, the preferred whites of the various studies differ. The results of Perz et al. [39] and Ohno and Fein [41] are also below the locus of the blackbody radiators. However, the results of the study [36] analysed in detail in Section 6.6 are closer to this locus. At 3000 K, the data of Rea and Freyssinier [37] and of Perz et al. [39] correspond to the 'moderate' to 'moderately good' values (yellow circles) of the study [36]. At 4100 K, the data of Rea and Freyssinier [37] correspond to the 'good' value (orange triangle) of the study [36]. At about 4200 K, the data of Perz et al. [39] correspond to the 'good' value (orange triangle) of the study [36]. The Ohno and Fein data [41], however, do not correspond to the preferred white points of the study [36]. Reasons for the partly different results in the compared studies are differences in the adaptation time of the test subjects to the predominant white tone (between 10 and 90 seconds) and the experimental set-up (empty boxes, large coloured objects in a real room, or small coloured objects in a chamber).

In summary, the following can be stated. The most important result of Section 6.6 is that the *most preferred white points* (with '*good – very good*') of the study [36] presented in detail in this section are in the range of 'industrial' white points (see the small dots in Figure 6.20), all *below* the curve of the blackbody radiators. This result indicates that the whites for indoor lighting practice should be in this range (red symbols in Figure 6.20).

6.7 Colour Quality

The term '*colour quality*' has already been defined in general terms in Section 6.1. This section describes the most important components (or perceptual aspects) of

Table 6.8 Most important perceptual aspects of colour quality. The observer's subjective judgement of each aspect changes as the illumination spectrum changes. The individual aspects and their numerical parameters are explained further in the corresponding sections.

Aspect	Explanation	Section
Naturalness, colour fidelity (colour-rendering)	How natural do the object colours appear?	6.7.1.1
Vividness	How saturated ('vivid') do the object colours appear?	6.7.1.2
Chromatic lightness	Saturated object colours appear bright; this is increased by increasing the luminance level	6.7.1.3
Colour preference	The object colours are subjectively more or less preferred	6.7.1.4
Memory colours	Memory colours (such as 'typical green grass') are preferred	6.7.1.5

Source: TU Darmstadt.

colour quality, their experimental methodology, and the parameters of mathematical modelling.

6.7.1 Perceptual Aspects of Colour Quality

Table 6.8 summarises the most important perceptual aspects of colour quality. Further aspects as well as a comprehensive consideration can be found in the paper of the German Society for Lighting on Colour Quality [42].

The observer's subjective judgement of each aspect in Table 6.8 changes as the illumination spectrum changes. The subjective judgement of an imaginary 'mean' or 'characteristic' observer about a particular aspect of colour quality is often described in practice by a numerical descriptor parameter (usually internationally agreed) determined from the spectrum of illumination (and more rarely the prevailing illuminance level) using a mathematical model of the human visual system.

6.7.1.1 Naturalness, Colour Fidelity (Colour Rendering)

The so-called general colour-rendering index (CIE CRI R_a), for example, models the judgement of a characteristic observer about 'naturalness' (or 'colour fidelity'). If, for example, instead of a halogen incandescent lamp ($R_a = 99$, left in Figure 6.21) a highly oversaturated RGB LED spectrum ($R_a = 10$, right in Figure 6.21) is used to illuminate a combination of coloured objects (so-called 'still life'), then the still life on the right will appear unnatural.

The colour fidelity or *colour rendering of* a light source is defined – according to Figure 6.21 – as the effect of a certain type of light source on the colour impression of the objects illuminated by it, in conscious or unconscious comparison with the colour impression of the same objects under a reference illuminant [43]. The method of calculating the general colour-rendering index (CIE CRI R_a) is described in DIN 6169 [44] based on CIE publication 13.2 from 1974. Updated recommendations of the

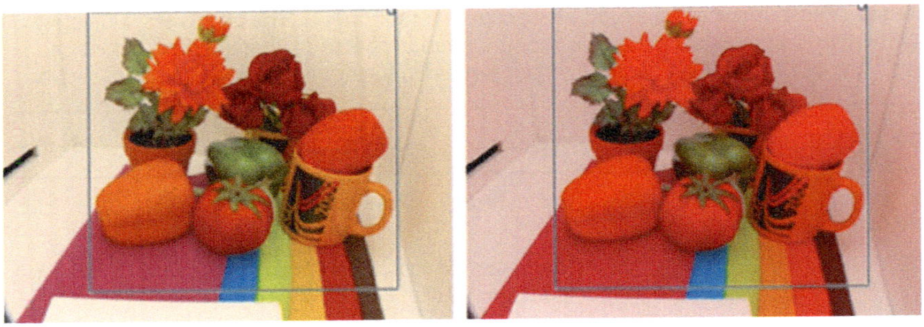

Figure 6.21 Illustration of the naturalness (colour fidelity or colour-rendering property) of light sources. The same test objects under a halogen incandescent lamp (left: high general colour-rendering index, $R_a = 99$) and a highly oversaturating RGB LED light source (right: low general colour-rendering index, $R_a = 10$). When observers form a judgement, they compare the current colour appearance with an (imaginary) reference illuminant (daylight or blackbody radiator, see text). Source: Technische Universität Darmstadt.

CIE are available in CIE 13.3 of 1995 [45]. In the following, only the most important features of this so-called CIE CRI method are summarised. The colour-rendering index is a test colour-based index: from the differences in colour perception (ΔE_i) between test illumination and reference, a special colour-rendering index R_i is calculated for each test colour ($i = 1$–14):

$$R_i = 100 - 4.6 \cdot \Delta E_i \tag{6.4}$$

The arithmetic mean of the special indices of the first eight CIE test colours is used to form the general colour-rendering index R_a:

$$R_a = \frac{1}{8} \cdot \sum_{i=1}^{8} R_i \tag{6.5}$$

As a reference source or reference illuminant, a type of light is chosen that has the same or almost the same chromaticity as the light source to be characterised. The colour-rendering index in this form only applies to the so-called photopic range of vision (for luminance levels greater than 10 cd m^{-2}). Light sources can reproduce the perceived colour appearance of coloured objects better at higher luminance levels (e.g. 200 cd m^{-2}) than at lower luminance levels (e.g. 10 cd m^{-2}).

Visual experiments on the colour-rendering properties of light sources showed that the general colour-rendering index CIE CRI R_a does not correctly describe the perceived colour-rendering properties of light sources. Accordingly, a new colour-rendering index was defined that correlates better with perception, the so-called IES TM-30-15 Color *Fidelity* Index [46, 47] (R_f). Based on this new colour fidelity index, the CIE published the so-called 'CIE 2017 Colour Fidelity Index' [48] in 2017 as a new method for *scientific application,* which is also designated R_f.

In this new definition, the so-called CAM02-UCS colour space is used. 99 colours are used as test colours. In the definition of R_f, a reference light source of the same CCT (as in the CIE R_a method) is always used. Up to 4500 K, the reference light source is a blackbody radiator and from 5500 K it is a daylight phase. Between 4500 and 5500 K, a linear combination of the two types of light is used. For the 99 colours illuminated by the test light source and reference light source, a mean colour difference (ΔE) between their colour appearance under the test and reference light sources in the CAM02-UCS colour space is calculated. The 10° standard colour matching functions are used for this. The value of R_f is thus obtained according to Eq. (6.6). The exact definition of the method can be found in the publication [48].

$$R_f = 100 - 7.54\,\Delta E \tag{6.6}$$

6.7.1.2 Vividness

Vividness means how vivid or saturated the object colours appear under a given illumination spectrum. This property is often modelled using the so-called colour gamut: the area or volume enclosed by the colours of the test colour samples in a chromaticity diagram or colour space. The larger the colour gamut, the higher the perceived vividness of the object colours. The IES TM-30-15 method [47] also defines the so-called IES TM-30-15 Gamut Index (R_g; so-called *colour gamut index*), which is often used today to describe vividness in practice. The colour gamut index R_g is calculated as follows. The a'-b′ plane of the CAM02-UCS colour space is divided into 16 hue groups. The red hues that are particularly important in lighting practice are, for example, in hue group 1 (orange–red to red) and in hue group 16 (purple–red to red). Within each colour group, the mean values of the CAM02-UCS a' and b' values of the test colours are calculated under the reference illuminant and also under the test illuminant. If the resulting 16 points (16 points under the reference illuminant, see the black circle in Figure 6.29; and then 16 points under the test illuminant, see the red curve in Figure 6.29) are connected, this creates two areas (A_{test} and A_{ref}) in the a'-b′ diagram and the ratio of the two areas is then formed according to Eq. (6.7).

$$R_g = 100 - (A_{test}/A_{ref}) \tag{6.7}$$

6.7.1.3 Chromatic Lightness (Brilliance)

The chromatic lightness or brilliance is the bright, brilliant appearance of saturated object colours. Not only the achromatic but also the chromatic channels of colour vision (see Chapter 2) contribute to the perception of the brightness and also to the lightness (the perceived lightness of an object is its perceived brightness in a perceptual relation to the brightness of a reference white) of a coloured object, this is the so-called Helmholtz–Kohlrausch effect. This makes saturated coloured objects of the same luminance appear brighter than unsaturated objects. The brilliance of object colours is also increased by increasing the generally prevailing luminance level, this is the so-called Hunt effect: the perceived colourfulness, i.e. the absolute perceived amount of colour in a colour stimulus, increases when the general luminance level (e.g. from 10 to 1000 cd m^{-2}) of a coloured object increases. The brilliance or chromatic lightness can be modelled, for example, with the help of the parameter L^{**} of Fairchild and Pirrotta [49], which is defined as follows:

$$L^{**} = L^* + [f(L^*) \cdot g(h) \cdot C^*] \tag{6.8}$$

$$f(L^*) = 2.5 - 0.025L^* \tag{6.9}$$

$$g(h) = 0.116 \mid \sin((h-90)/2) \mid +0.085 \tag{6.10}$$

The symbols in Eqs. (6.8)–(6.10) have the following meaning:

- $f(L^*)$: Correction function based on the parameter CIELAB L^*
- $g(h)$: Correction function based on the parameter CIELAB h (in degrees). The minimum of this correction function $g(h)$ is at the yellow hue (h =90°) where the chromatic contribution is lowest.
- C: *CIELAB chromaticity: with increasing correction factor $[f(L^*) \cdot g(h) \cdot C^*]$, the brilliance increases linearly as the chromaticity of a coloured object (C^*) increases.
- h: CIELAB hue angle in degrees.

6.7.1.4 Colour Preference

Colour preference represents the most important perceptual aspect of colour quality from the perspective of HCL because of the associated emotional effect. When the observer looks at the illuminated object colours under a certain lighting spectrum at a certain illuminance level, this appearance evokes a certain aesthetic impression and consequently an emotion. If the colour appearance under the current illumination is pleasing, then this emotion is positive – a positive colour preference judgement is then formed. Colour preference is a subjective judgement and depends, among other things, on age, gender, and cultural background as well as the current mood of the observer and the combination of the currently viewed coloured objects (e.g. fashion, flowers, food, and make-up).

In the literature, there are several numerical descriptor parameters for predicting colour preference from the illumination spectrum and/or the illuminance level [42]. One important parameter is the so-called CQS *metric* (Colour Quality Scale) [50] developed by NIST (National Institute for Standards and Technology, USA). In its version 9 from 2011, there are three variants of the CQS:

- CQS Q_a: Colour preference index that favours a *moderate* increase in the chromaticity of the object colours under the current light source – in contrast to the colour-rendering indices. If the object colours are heavily oversaturated, however, the value of the index is lowered, see Figure 6.21 (right);
- CQS Q_f: Colour fidelity index or fidelity index that evaluates test light sources similar to the CIE CRI R_a rating;
- CQS Q_g: Gamut index which, similar to the parameter R_g (Eq. (6.7)), evaluates the size of the test colour area in the colour space (relative to a reference).

Instead of the 14 CIE test colours (Figure 6.22), 15 colours with high saturation are used in the CQS method, see Figure 6.23. The details of the calculation of the indices Q_a, Q_f, and Q_g can be found in the publication [50].

6.7.1.5 Memory Colours

If the observer is able to identify a coloured object (e.g. the red rose in Figure 6.21), the cognitively determined name of the object (here: 'red rose') evokes a memory of

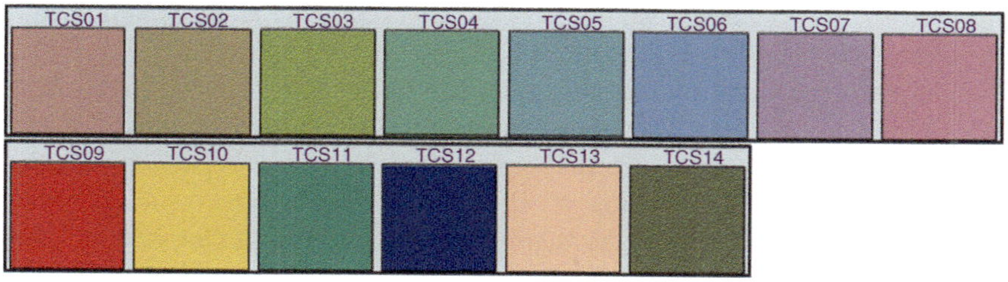

Figure 6.22 The 14 CIE test colour samples (TCS01-14). Only the first eight desaturated colours are used to calculate the current general CIE colour-rendering index (R_a). Source: Davis and Ohno [50]/SPIE/CC BY 4.0.

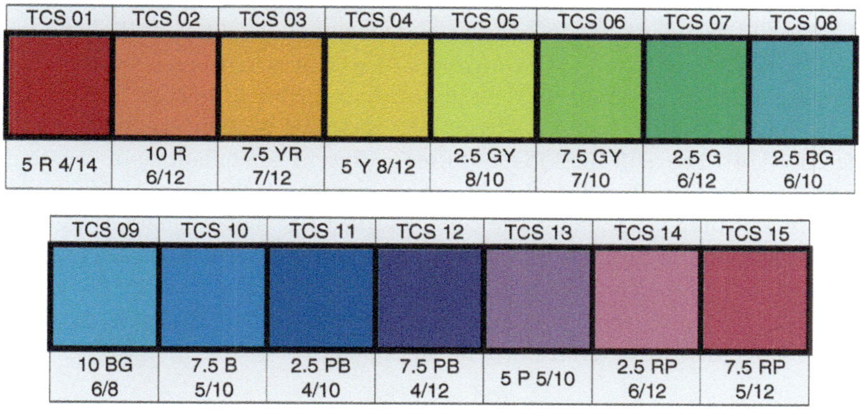

Figure 6.23 The 15 CQS test colours. Source: Davis and Ohno [50]/SPIE/CC BY 4.0.

a *typical object colour* of such an object because the observer often saw this object in the past. This typical object colour (*long-term memory colour*) is compared with the colour appearance of the object under the current illumination spectrum. If this current colour appearance is *similar* to the long-term memory colour, then the colour appearance will be *preferred*.

Several numerical descriptor parameters exist that quantify this similarity to the long-term memory colours of frequently seen objects (e.g. blue sky, green grass, various skin tones, red rose, red strawberry, banana, orange, etc.). In this section, two parameters (i) according to Smet et al. [51] and (ii) according to Babilon and Khanh [52] are mentioned. These parameters express the similarity of the current colour appearance of test objects to the corresponding long-term memory colours using two-dimensional Gaussian functions.

Thus, the index according to Smet et al., the so-called S_a [51] works with the following test objects: apple (green), banana (yellow), orange, lavender, smurf (blue), strawberry yoghurt, peeled cucumber, cauliflower, skin colour (Caucasian), and Munsell grey (N4). Babilon and Khanh [52] conducted a visual experiment with real objects in a real room and determined the long-term memory colours of 12 familiar objects: banana, green lettuce, broccoli, blue jeans, blueberry, red cabbage, red rose, carrot, butternut squash, flower pot made of concrete, Caucasian skin, and Asian skin.

The observer data were modelled – similar to Smet et al. [51] in the CIECAM02 colour space using bivariate Gaussian functions whose centroids defined the corresponding long-term memory colours for each test object. A comparison with the results of Smet et al. [51] showed no significant differences in the centroids of the long-term memory colour centres, but significant deviations in the shape of the fitted Gaussian functions.

6.7.2 Modelling Colour Preference, Naturalness, and Vividness

Colour preference, naturalness, and vividness are important aspects of colour quality (see Table 6.8). In this section, two visual experiments at two different illuminance levels (750 lx [53] and 2000 lx [54]) with different 'still life' arrangements of coloured objects are presented. At 750 lx [53] all three colour quality aspects (colour preference, naturalness, and vividness) were subjectively scaled, at 2000 lx only the colour preference [54]. The 'still lifes' [53, 54] used represent combinations of coloured objects illuminated with different light source spectra, on the basis of which the observers rated the colour quality. The same, so-called 'CQ' model equation ('CQ' is a numeric descriptor parameter of colour quality) is used to model both data sets [53, 54]. The results of the two experiments are compared. Further, all-encompassing light-quality models that also model colour preference as a partial aspect (and their experimental basis) are described in detail in Chapter 7.

6.7.2.1 Modelling of Colour Preference, Naturalness, and Vividness at 750 lx

Subjective colour preference, naturalness, and vividness ratings of two different colourful *still-life* arrangements at a vertical illuminance of 750 lx (see Figure 6.24) were analysed and modelled using the above-mentioned CQ model equation [53]. The quantity CQ (Eq. (6.11)) combines the R_f *colour fidelity metric* (Eq. (6.6)) with the chromaticity difference metric ΔC^*. The latter metric is defined as the mean

Figure 6.24 Still-life arrangements of the coloured objects. The vertical illuminance was 750 lx at the objects on the wall. Source: Reproduced with permission from Khanh et al. [53], SAGE PUBLICATIONS, INC.

of the chromaticity differences of the 15 CQS test colours [50] (Figure 6.23) under the test and reference light sources. The greater the value of ΔC^* for a particular light source spectrum, the more saturated the object colours appear under that light source spectrum.

The coloured *still lifes* of Figure 6.24 were illuminated in an experimental room with a four-channel LED luminaire with diffuse light and its 36 different LED combination spectra at four CCTs (3100, 4100, 5000, and 5600 K) and nine ΔC^* *levels* (between 'not oversaturated' and 'highly oversaturated', compared to the CIE CRI reference spectrum). The observers rated the colour preference, naturalness, and vividness of the *still lifes* on an interval scale labelled with categories (e.g. 'very good', 'good', and 'moderate') (see Figure 7.2).

The mean subjective ratings of the observers implied a significant dependence of the subjective judgements on the CCT and ΔC^* *value*. An important tendency was that the CCT values at 4100, 5000, and 5600 K resulted in higher mean subjectively assessed colour preference, naturalness, and vividness than the warm white CCTs (3100 K). Mean subjective preference and naturalness showed maxima at moderate object saturation levels (ΔC^* values).

However, if the still lifes were oversaturated by light source spectra with too high ΔC^*-values then much lower preference and naturalness judgements were obtained. Vividness showed a monotonically increasing trend as a function of the ΔC^* value at each CCT. Figure 6.25 shows an example of these tendencies for CCT = 3100 K. For the mathematical modelling of these trends, the model equation with the CQ parameter according to Eq. (6.11) was developed [53, 54]. Table 6.9 shows the parameter values of Eq. (6.11) for the different CCTs.

$$\mathrm{CQ} = a + b\,\Delta C^{*2} + c\,\Delta C^* + d\,R_\mathrm{f} \tag{6.11}$$

The CQ parameter of Eq. (6.11) describes the absolute magnitudes of the mean visual scales, which is particularly important for lighting practice because of the prediction of the emotional effect in the HCL context (since semantic categories are assigned to the visual scales according to Figure 7.2). Thus, the value of 79.6 can be used as an *acceptance criterion* for a 'good' colour preference, naturalness, and vividness rating. In order to apply the CQ parameter according to Eq. (6.11) in lighting practice (here at least for illuminance levels of 750 lx), the *target* of colour quality optimisation should first be defined, see below.

Preference should be used in the most general case, and for shorter stays in the space to be lit with more creative contexts (e.g. gallery, museum, fashion show, sales room, and art class), *naturalness* for longer stays and low-emotion applications (e.g. office space, meeting room, and library), and vividness for special applications where the goal is to achieve a very strong emotional effect. Once the target has been established, the CCT to be used should be determined and the optimum parameter values of Table 6.9 should be replaced according to the next possible CCT (3100, 4100, 5000, or 5600 K) or interpolated in between. The value of CQ should be maximised by an algorithm, e.g. by changing the relative intensities of the individual LED channels and keeping the chromaticity coordinates of the white tone of the spectrum in the visually optimal range (see Section 6.6).

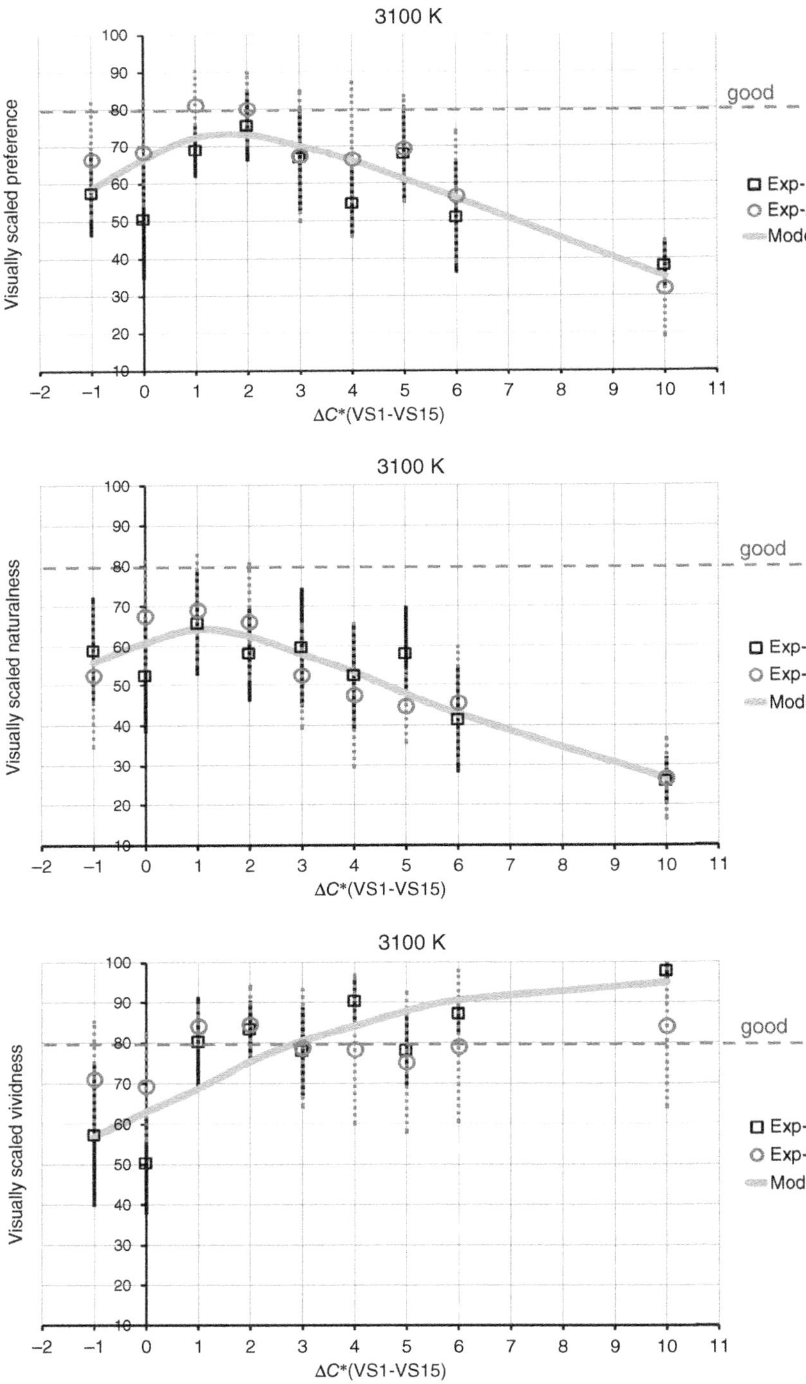

Figure 6.25 Mean visual scale values of colour preference, naturalness, and vividness ratings and their 95% confidence intervals as a function of ΔC^* (metric of mean object saturation) for 3100 K compared to the parameter CQ of Eq. (6.11). Source: Reproduced with permission from Color Research and Application [53].

Table 6.9 Parameter values of Eq. (6.11) for the different CCTs at 750 lx [53]. Only the values of the so-called cluster-1 are shown.

Preference I CCT (K)	3100	4100	5000	5600
a (const)	−28.491	69.688	−129.929	68.255
$b\,(\Delta C^{*2})$	−0.200	−1.014	−0.568	−0.993
$c\,(\Delta C^{*})$	2.069	6.884	10.707	6.076
$d\,(R_f)$	1.067	0.000	2.227	0.000
Naturalness I CCT(K)	3100	4100	5000	5600
a (const)	−27.208	71.292	−186.467	28.904
$b\,(\Delta C^{*2})$	−0.0439	−0.773	−0.185	−0.677
$c\,(\Delta C^{*})$	0.000	3.290	7.988	4.483
$d\,(R_f)$	0.977	0.000	2.911	0.498
Vividness I CCT(K)	3100	4100	5000	5600
a (const)	63.782	48.459	53.752	50.692
$b\,(\Delta C^{*2})$	−0.324	−0.358	−0.377	−0.002
$c\,(\Delta C^{*})$	6.355	7.565	7.313	4.633
$d\,(R_f)$	0.000	0.000	0.000	0.000

Source: Reproduced with permission from Color Research and Application [53].

Table 6.10 Parameter values of Eq. (6.11) for the different CCTs at 2000 lx [54]. At 2000 lx, the naturalness and vividness were not assessed in the experiment [54].

Preference I CCT (K)	3200	4200	5000	5600
a (const)	−12.505	−51.220	120.662	167.554
$b\,(\Delta C^{*2})$	0.223	−0.397	−1.379	−0.761
$c\,(\Delta C^{*})$	−3.703	4.821	10.322	2.408
$d\,(R_f)$	0.767	1.270	−0.727	−1.089

Source: Reproduced with permission from Color Research and Application [54].

6.7.2.2 Modelling Colour Preference at 2000 lx. Comparison of Colour Preference Between 750 and 2000 lx

Subjective colour preference ratings of a colourful (in this case horizontal) *still life* (see Figure 6.26) at a *horizontal* illuminance of 2000 lx were analysed and modelled with the same model equation (CQ, Eq. (6.11)) as in Section 6.7.2.1 [54]. At 2000 lx, naturalness and vividness were not assessed in the test [54]. The *still life* in Figure 6.26 was illuminated with the same LED luminaire with similar spectra as in Section 6.7.2.1. Compared to Section 6.7.2.1, 3200 K was used instead of 3100 and 4200 K instead of 4100 K; the other two CCTs remained the same.

The model curves of Tables 6.9 (750 lx) and 6.10 (2000 lx) are shown in Figure 6.27. Table 6.11 summarises the results shown in Figure 6.27.

As can be seen from Figure 6.27 and Table 6.11, the maximum CQ values (i.e. the metric colour preference descriptor values) according to Eq. (6.11) at 2000 lx

Figure 6.26 Still life in the experiment [54] for the subjective evaluation of colour preference. The coloured objects No. 1–24 are the colour patches of the MacBeth ColourChecker® chart. Source: Reproduced with permission from Khanh et al. [54].

Table 6.11 Summary of colour preference results of Figure 6.27, according to Eq. (6.11). First number: results at 750 lx. Results at 2000 lx are given in brackets thereafter.

E_v (lx)	750 (2000)	750 (2000)	750 (2000)	750 (2000)
CCT (K)	3200 (3100)	4200 (4100)	5000 (5000)	5600 (5600)
$\Delta C^*_{\text{VS1-VS15}}$ at max. pref.	0.1 (1.7)	2.3 (3–4)	3.5–5.5 (2.1)	1.0–5.0 (2.0–4.0)
Max. CQ	57 (73)	75 (82)	77 (83)	73 (77)
Max. category reached	Moderate (good–moderate)	Good (good)	Good (good)	Good–moderate (good)

Source: Reproduced with permission from Color Research and Application [54].

are greater than the maximum CQ values at 750 lx. This result implies that colour preference depends not only on ΔC^* or CCT, but also on illuminance. This effect is discussed in detail in Chapter 7 (Light quality models).

6.7.3 Consideration of Red Object Colours in the Colour Preference Model

In certain applications, such as in the film and television industry, in medicine, shop and museum lighting, often special objects (e.g. artist's paints, special foods, or textiles) must be illuminated whose spectral reflectance values may differ from the test colours specified in standards and/or CIE publications

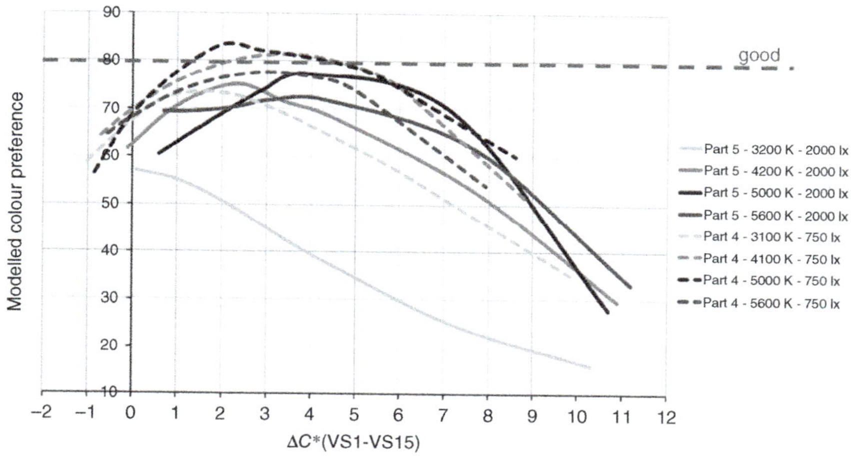

Figure 6.27 Model curves of colour preference with the quantity CQ (Eq. (6.11)) and the parameter values from Tables 6.9 (for 750 lx) and Table 6.10 (for 2000 lx).
Source: Reproduced with permission from Color Research and Application [54].

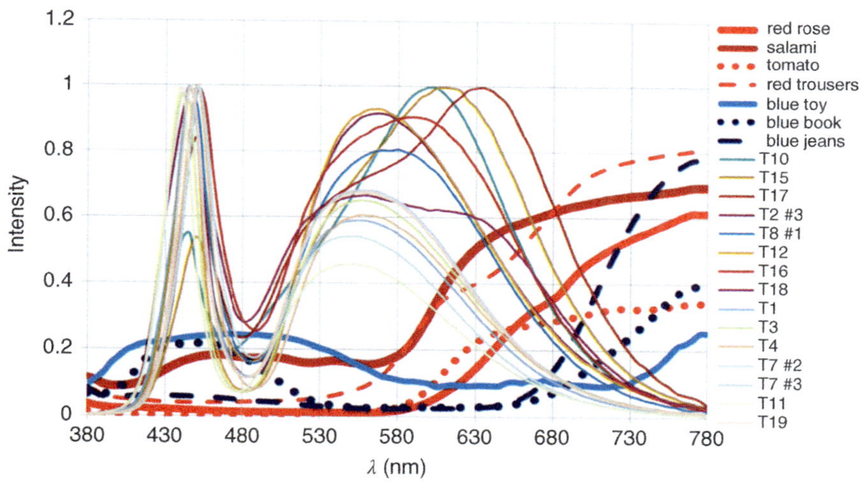

Figure 6.28 Spectral reflectance of red and blue objects and the relative emission spectra of some white LEDs [55]. Abscissa: wavelength λ in nm; ordinate: relative intensity, either spectral reflectance or relative spectral radiant flux of the emission of the white LED light sources (T10–T19). Source: TU Darmstadt.

(Figures 6.22 and 6.23 as well as the 99 test colours of the R_f method). In this respect, skin tones play the most important role, so that Section 6.8 is dedicated to this topic alone. From the point of view of the emotional effect, the reddish and the blue object colours are also important: with multichannel LED systems, for example, either the red or the blue object colours can be emphasised, which strongly influences the *colour mood* of the illuminated scene. Figure 6.28 shows the spectral reflectance of reddish (red salami sausage, red tomato, rose, and red trousers) and bluish objects (jeans, trousers, books, and toys) in connection with some selected fluorescent converted LED spectra [55].

The reflection spectra of the red colours in Figure 6.28 show the fact that the white LED light sources must have a high radiation component from 600 up to 680 nm in order to display the red object colours – more or less saturated, as desired. However, this can only be achieved with LED combinations that have LED channels with red maxima at least from 600 nm, e.g. a 'deep red' LED at 660 nm and a 'red' LED at 630 nm. The same applies to the blue range; blue LED light sources with maxima between 420 and 480 nm, e.g. a 'deep blue' LED at 440 nm and a 'cyan' LED at 480 nm, should be used for this. This allows one to adjust the perceived saturation strength of blue objects as needed.

The perceptual effect of the saturation strength of *red* objects on subjective colour preference (Section 6.7.1.4) is so strong that it has been taken into account in colour preference models. In the following, the Royer et al. model [56] is described as an example. For the development of this colour preference model, a visual experiment was conducted. 28 test subjects evaluated the illumination with 26 different light sources in a large room filled with colourful objects. The values of the IES TM-30-15 colour fidelity index (R_f) ranged from 64 to 93, the values of the IES TM-30 colour gamut index (R_g) ranged from 79 to 117, and the values of the percentage red chromaticity shift in chromaticity group 16 (i.e. the parameter $R_{cs,h16}$, see Section 6.7.1.2 and Figure 6.29) ranged from −19% to 26% (see Table 1 in [56]). The parameter $R_{cs,h16}$ quantifies the change in chromaticity of the purple-reddish hues when the

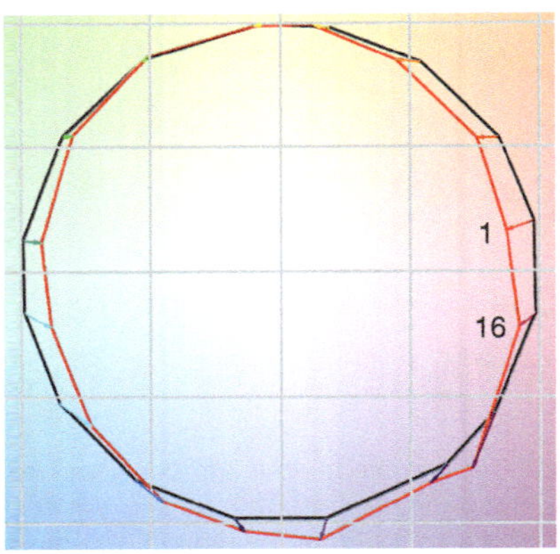

Figure 6.29 Example for the 'IES TM-30 Color Vector Graphics' diagram [47]. Abscissa: normalised CAM02-UCS a' values, ordinate: normalised CAM02-UCS b' values. The normalisation is done in such a way that the a'–b' mean values in the 16 hue groups under the reference illuminant (black polygon course) result in a circle. Red polygon: a'–b' mean values in the 16 hue groups for the test light source. (1) hue group 1 (orange-red to red); (2) hue group 16 (purple-red to red). The parameter $R_{cs,h16}$ corresponds to the radial component of the relative length of the purple-red arrow ('displacement of chromaticity group 16, between test and reference') at marker '16'. The parameter $R_{cs,h16}$ is measured in percentage of the distance of the corner points of the black polygon course from its centre. Source: Adapted from Illuminating Engineering Society of North America [47].

purple-reddish test colours of hue group 16 are illuminated with the current test light source instead of the reference light type. Positive percentage values of the parameter $R_{cs,h16}$ express an oversaturation, and negative percentage values a desaturation of the purple-reddish object colours by the current light source.

The study by Royer et al. [56] found a significant influence of the parameter $R_{cs,h16}$, i.e. the percentage red chromaticity shift on colour preference, which was modelled in [56] by Eq. (6.12).

$$\text{Preference (Royer et al.[57])} = 7.446 - 0.041R_f - 9.99R_{cs,h16} - 0.90(R_{cs,h16})^2 + 106.6(R_{cs,h16})^3 \quad (6.12)$$

The model Eq. (6.12) shows that the *shape* of the colour gamut under the current test light source (see the red polygon in Figure 6.29) is more important for modelling colour preference than the average gamut area (the latter is expressed by the colour gamut index R_g according to Eq. (6.7)). When modelling the shape of the colour gamut, the reddish hues play a very important role according to the experimental results [56]. Equation (6.12) also shows that the colour fidelity indices IES TM-30 R_f or CIE CRI R_a *alone* are *not* able to describe *colour preference*. In the experiment [56], 9 out of the 12 best-rated light sources had a CIE CRI R_a value of 73 or less, indicating that an exclusion criterion of less than CIE R_a = 80 would possibly exclude many preferred red test colours; hence, we would then exclude numerous 'perceptually preferred', i.e. *moderately* red-oversaturating light sources.

6.8 Colour Preference for Skin Tone Lighting

6.8.1 Introduction

In the context of lighting technology with people as light users at the centre HCL, the aesthetic aspects of lighting are of great importance [57]. In the social life of people of different professions, origins, ages, and genders, the importance of visual appearance, gestures, body language as well as facial expressions is increasing strongly. In this context, skin tone plays a central role in visual communication, in living together, in art as well as in film and television. In the meantime, skin creams, lipsticks, and make-up accessories have become permanent in the home. Natural skin tone colours depend on the type of pigmentation, on genetics, on the degree of tanning, and to a considerable extent on the geographical location of the people's habitat. Nevertheless, even in the case of the use of daylight, many people favour the use of make-up colours to change their skin tone according to a certain affection and their understanding of beauty.

This development of cosmetic products as well as the mental adjustment to different skin tones took place in the last decades without much consideration of spectra and colour temperature of lighting systems, as the spectral variation of conventional light sources (discharge lamps, halogen tungsten lamps) was very limited and as the effect of light source spectra and spectral reflectance of skin tones was not in the focus of light and colour science. This has been changing very quickly since 2010 until today with the dynamic development of LED technology in terms of integrating LED light sources into indoor spaces (offices, galleries, shops, beauty salons, and hospitals), with more and more LED light sources with high colour quality at different CCTs available on the market.

Colour science began to explore the interaction between light source spectra and skin tone colour preference in the context of 'Human-Centric Lighting' [58, 59]. In the process, the variety of LED spectra as well as the colorimetric parameters used were still limited and posed the following questions:

- Can LED light spectra with a chromaticity on the Planckian locus or on the locus of daylight phases and with the same colour temperature but with different saturation levels (ΔC^*) elicit different evaluations and acceptance levels when illuminating skin tones?
- Which colorimetric parameters correspond to the subject's assessment of the colour preference of the skin tone illumination?
- How far does the CCT influence the optimal saturation level when illuminating skin tones with the best colour preference?

An experiment was conducted on the above questions [57], which is described below.

6.8.2 Method of the Colour Preference Experiment for Skin Tone Illumination

First, the spectral reflectance values of a set of skin tones were measured on different test subjects [57]. This was followed by the illumination of these skin tones at different colour temperatures with different saturation levels and the subjective evaluation of the colour preference of skin tones under the different illumination spectra.

6.8.2.1 Spectral Measurement of Skin Tones

For the spectral measurement of the skin tones, a certain test person was placed on a chair behind a black cloth. The opening of the black cloth served as the measurement opening, to which the test person held his or her cheek for the measurement (see Figure 6.30). The cloth and the cheek of the test person were homogeneously irradiated with a power-stabilised halogen–tungsten lamp from 380 to 780 nm. First, a white standard was measured at the opening with the spectroradiometer (CS 2000 from Konica-Minolta). Then the spectral reflectance property of the cheek of the respective test person was recorded. The spectral radiance factor of the respective skin tone was calculated from the ratio of the data of the two measurements.

The spectral radiance factors of skin tones of the 15 test subjects are shown in Figure 6.31. This shows that the curves differ in absolute values. There are lighter and comparatively darker skin tones. Around about 420 and 540 nm as well as shortly before 580 nm there are local minima of the spectral reflectance properties. This means that the skin absorbs more at these wavelength ranges. From 580 nm onwards, the curve of the spectral radiance factors rises to different heights in the

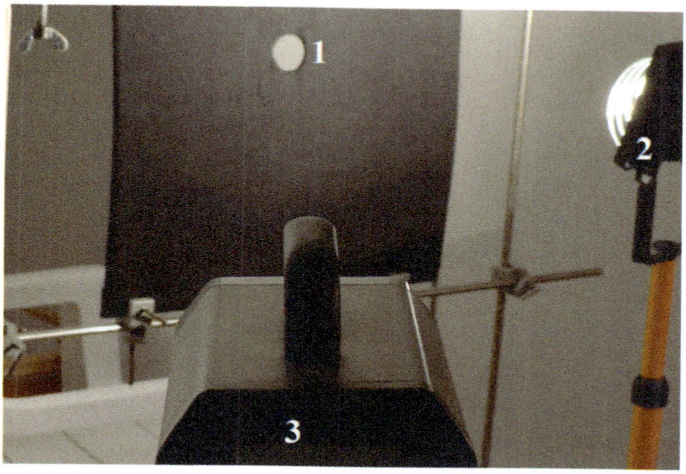

Figure 6.30 Measurement set-up for measuring the spectral radiance factor of skin tones. (1) face position; (2) halogen–tungsten lamp; (3) imaging spectroradiometer. Source: Technische Universität Darmstadt.

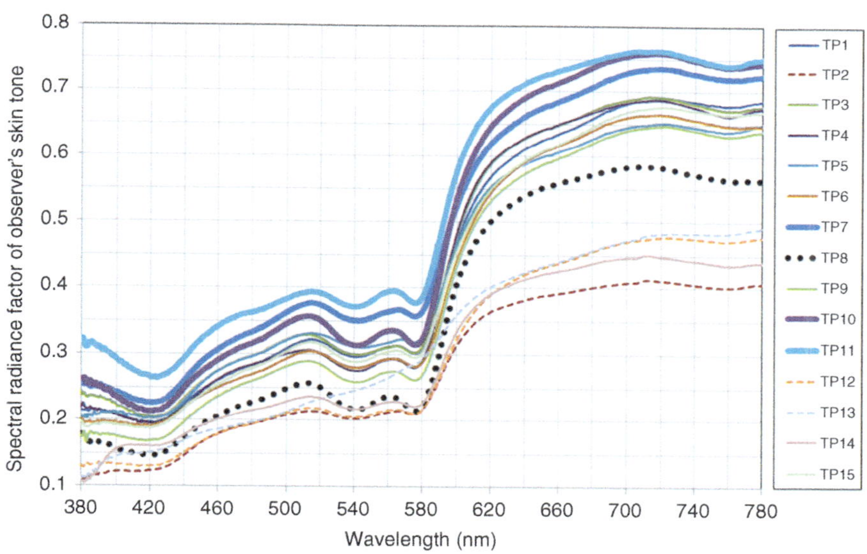

Figure 6.31 Spectral radiance factors of the skin tones of the 15 test persons (TP1–TP15). Source: TU Darmstadt.

range between 620 and 780 nm. Between 580 and 620 nm, a strong increase is visible for all skin types.

Earlier (in 2012), the skin tones of 32 other test persons at the TU Darmstadt were measured spectrally in the same way [55]. The majority of these curves in Figure 6.31 are similar to the curves in Figure 6.32 except for five curves with lower reflection of test persons of African origin (see Figure 6.32). There, not only are the absolute

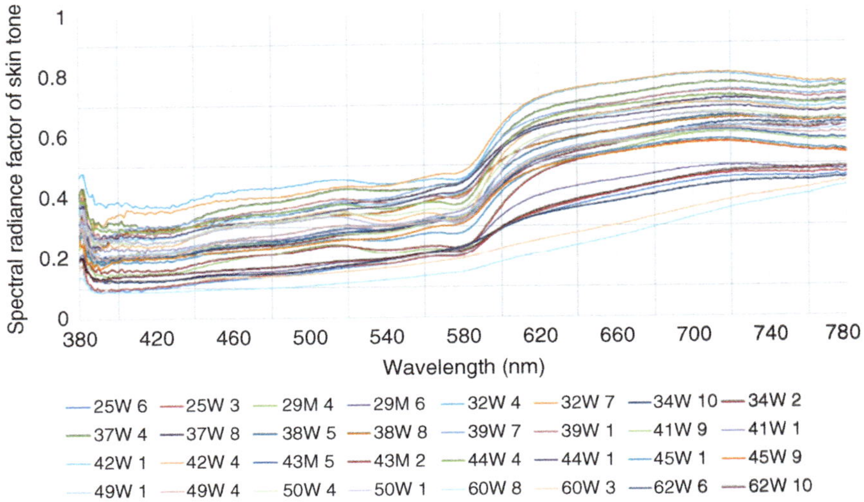

Figure 6.32 Spectral radiance factors of the skin tones of the 32 test subjects from 2012 [55]. Source: Technische Universität Darmstadt.

values of the spectral radiance factors smaller, but the increase from 580 to 620 nm is also not as pronounced.

6.8.2.2 Characterisation of the Light Sources Used

For the psychophysical experiments, a high-quality commercial LED luminaire with max. 20 000 lm and a diffuser lens as a cover lens for diffuse lighting were used. Two CCTs of 3200 K (preferred for residential lighting as in the film industry) and 5600 K (for the film television industry and in office or hospital lighting) were provided for the experiments. At each colour temperature, eleven different saturation levels were spectrally optimised.

The test person sat on a chair and was illuminated from above (see Figure 6.33) by the diffuse luminaire homogeneously and with soft radiation without glare. A large standing mirror was positioned in front of the test person. The test person looked at his or her own face in this mirror for two minutes per light setting and evaluated the colour preference of his or her own face using the semantic scaling from '*very poor*' to '*excellent*', see Figure 6.4. The subjective scaling level 79.6 is defined as 'good'. Below the subjective evaluation level of 52.9, the lighting is evaluated as worse than '*moderate*' and is no longer considered for practical applications.

Since the test person did not look at the face illuminated by the LED luminaire directly but via a mirror, the mirror was also spectrally characterised and taken into account in the spectral optimisation of the LED luminaire. Figure 6.34 shows the LED spectra of eleven different saturation levels (ΔC^*) at the CCT of 3200 K together with the corresponding Planckian radiator.

According to Figure 6.34, the spectrum at each saturation level consists of a blue LED channel with a peak wavelength around 462 nm, a green LED channel at 525 nm, a red LED channel at 635 nm, and a channel with warm white

Figure 6.33 Arrangement for the visual assessment of skin tone illumination. (1) Multichannel LED luminaire; (2) mirror; (3) light sensor; 4: observer's seat. Source: TU Darmstadt.

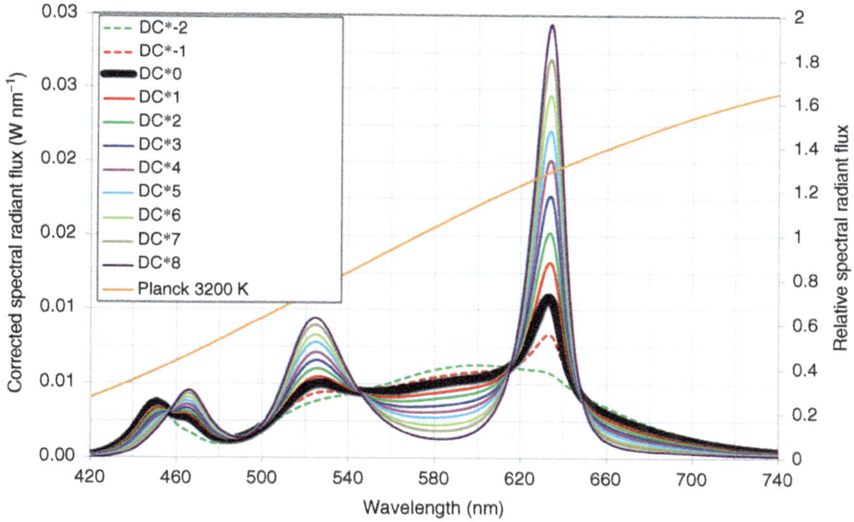

Figure 6.34 LED spectra at eleven different colour saturation levels (ΔC^*) and a blackbody radiator at the CCT of 3200 K ('Planck 3200 K'). Source: TU Darmstadt.

phosphor-converted LEDs. With different intensity levels of the individual LED channels, different saturation levels (ΔC^* between −2 and +8) were optimised with respect to a Planckian radiator at 3200 K as a reference light source, always setting the same chromaticity on the Planckian locus and always setting a constant illuminance on the test person's face at 2400 lx. Table 6.12 lists the photometric and colorimetric characteristics of the eleven spectra for skin-tone illumination at 3200 K.

The 11 saturation levels in Table 6.12 contain two desaturations ($\Delta C^* = -2$ and -1) as well as increased saturations ($\Delta C^* = +1$ to $+8$). The saturation level $\Delta C^* = 0$ also corresponds to the spectrum with the highest colour rendering (in terms of R_a or R_f). The parameter Q_p for the colour preference and the parameter R_g for the colour gamut are also included in Table 6.12. It can be seen from Table 6.12 that the higher the saturation level ΔC^* is selected, the greater colour gamut R_g is obtained.

Figure 6.35 shows the 10 LED spectra with different saturation levels and the corresponding daylight spectrum at the colour temperature of 5600 K. In both Figure 6.34 for 3200 K and Figure 6.35 for 5600 K, the highlighted black curves illustrate the respective spectrum with the highest colour rendering. The spectra with the higher saturation levels are characterised by the fact that the peaks

Table 6.12 Colorimetric parameters of the eleven LED spectra in skin tone illumination at 3200 K. CCT values are given in K, E_v values are in lx units.

ΔC^*	−2	−1	0	1	2	3	4	5	6	7	8
CCT	3200	3214	3205	3190	3214	3205	3182	3228	3211	3205	3192
E_v	2429	2425	2406	2388	2386	2390	2407	2428	2412	2441	2420
R_a	82	91	96	89	82	74	66	60	51	44	36
Q_p	72	83	92	98	99	98	94	90	86	81	77
R_f	82	88	92	91	87	82	77	73	67	62	57
R_g	97	101	103	106	107	109	110	110	111	111	112

Source: TU Darmstadt.

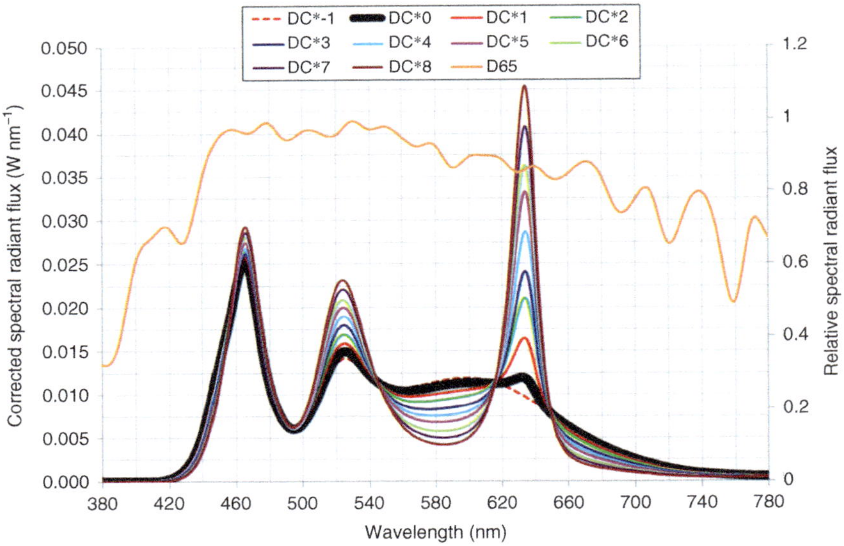

Figure 6.35 Spectra at 10 colour saturation levels at the CCT of 5600 K and the daylight phase at 5600 K ('D65'). Source: TU Darmstadt.

Table 6.13 Colorimetric parameters of the 10 LED spectra in skin-tone illumination at 5600 K. CCT values are given in K, E_v values are in lx units.

ΔC^*	−1	0	1	2	3	4	5	6	7	8
CCT	5526	5626	5562	5508	5567	5525	5481	5632	5582	5552
E_v	2400	2400	2400	2400	2400	2400	2400	2400	2400	2400
R_a	93	93	90	84	80	74	68	62	56	50
Q_p	83	86	93	98	98	97	94	91	87	84
R_f	85	86	85	83	81	77	74	70	66	63
R_g	95	96	99	101	103	105	106	107	109	110

Source: TU Darmstadt.

in the spectrum are very pronounced around 525 nm for the green LEDs and around 635 nm for the red LEDs. Table 6.13 lists the photometric and colorimetric characteristics of the 11 spectra in skin-tone illumination at 5600 K.

6.8.3 Results of Subjective Scaling of Colour Preference for Skin Tone. Optimal Saturation Levels

The resulting mean subjective colour preference values for skin tone of 15 young normally sighted subjects and their 95% confidence intervals are shown in Figure 6.36.

Figure 6.36 shows that the maximum subjective colour preference for skin tone at 3200 K can be achieved with the spectrum with the maximum colour rendering.

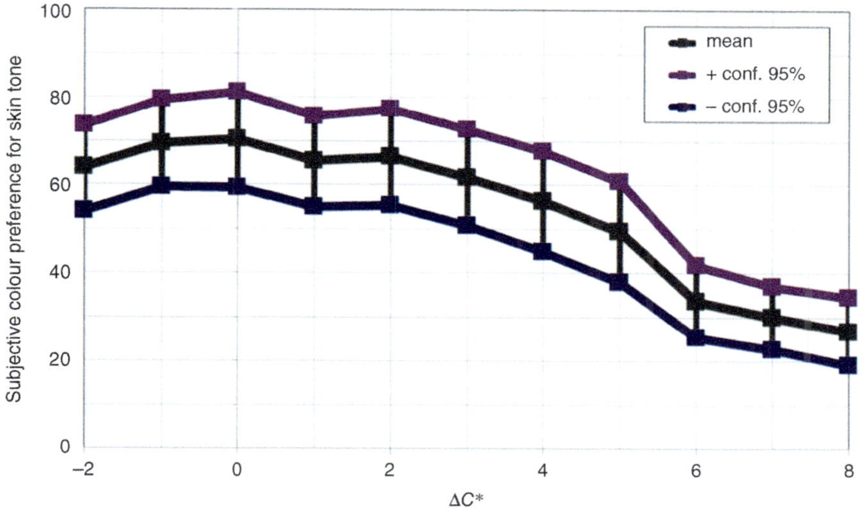

Figure 6.36 Mean subjective colour preference values for skin tone obtained from 15 young normally sighted test subjects (mean black curve) and their 95% confidence intervals (dark blue and purple curves) at 3200 K. Source: TU Darmstadt.

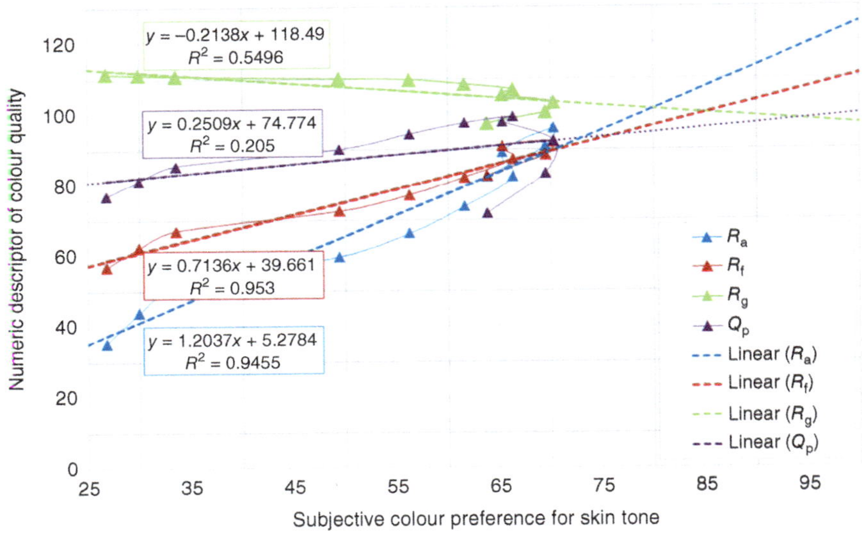

Figure 6.37 Correlation between the subjective colour preference rating for skin tone (abscissa) and the parameters R_a, R_f, Q_p, and R_g (numeric descriptors of colour quality, ordinate) at 3200 K. Source: TU Darmstadt.

This also means that no increase in saturation is necessary, as apparently the red component in this spectrum is sufficient for the skin-tone rating. The maximum rating reached the mean value of 70 and with a saturation increase or desaturation, the rating is worse.

Figure 6.37 illustrates the correlations between the evaluation of the colour preference for skin tone at 3200 K and the numeric descriptor parameters R_a, R_f, Q_p, and R_g. It can be seen that neither the colour gamut index R_g nor the colour preference index CQS Q_p correlates with the subjective colour preference values. These two parameters are therefore not suitable for skin tone quality assessment. In contrast, the correlation between the two indices R_a and R_f and the subjective ratings is very high ($r^2 = 0.95$).

In the case of LED illumination at 5600 K, the best subjective colour preference rating for skin tone is obtained at a saturation level of $\Delta C^* = +2$, see Figure 6.38. This means that some increase in saturation is necessary for illumination at 5600 K, and the light source with the *best* colour rendering ($R_a = 93.2$) does *not* provide the best preference. Compared to the illumination at 3200 K (see Figure 6.36), the maximum rating this time is not at 70 (3200 K) but at 78 (5600 K), so that it can be claimed that the skin-tone rating for the daylight-like spectrum at 5600 K with a saturation level of +2 is better than in the 3200 K case.

It can be seen from Figure 6.39 that – even at 5600 K – the visual colour preference for skin tone correlates better with R_a and R_f than with Q_p and R_g. One can thus summarise the results as follows:

- The colour-rendering indices describe the visual assessment of colour preference for skin-tone lighting very well.

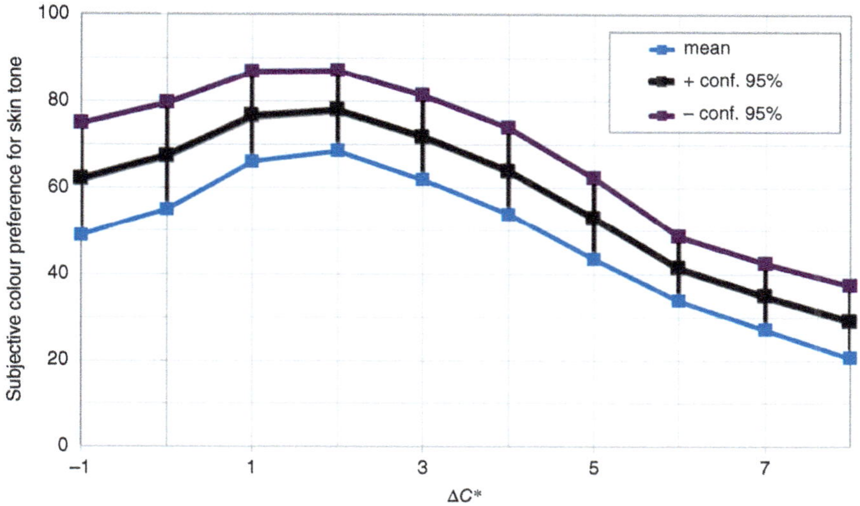

Figure 6.38 Mean subjective colour preference values for skin tone of 15 young normally sighted test subjects (mean black curve) and their 95% confidence intervals (dark blue and purple curves) at 5600 K. Source: TU Darmstadt.

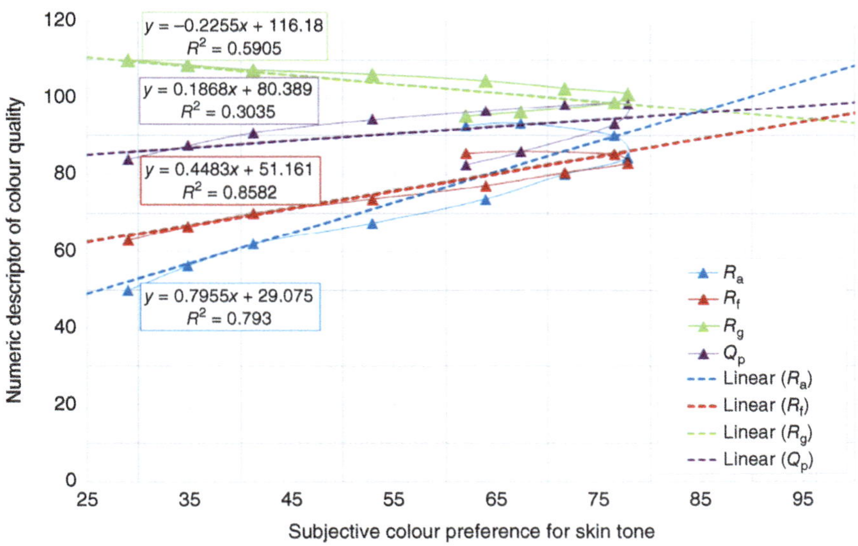

Figure 6.39 Correlation between the subjective colour preference rating for skin tone and the parameters R_a, R_f, Q_p, and R_g (numeric descriptors of colour quality, ordinate) at 5600 K. Source: TU Darmstadt.

- At the colour temperature of 3200 K, no saturation increase is necessary, at 5600 K a saturation level of $\Delta C^* = +2$ is necessary to get the best colour preference rating for skin tone.
- A conventional halogen incandescent lamp at 3200 K or LED lighting at 3200 K with a very good colour rendering $R_a = 95–96$ is the right choice in the warm hue range. At 5600 K, daylight alone or LED lighting similar to daylight with very high colour rendering is no guarantee of the best skin-tone lighting. For this, the saturation should be increased by $\Delta C^* = +2$.

6.9 Colour-Rendering Indices and Their Semantic Interpretation

6.9.1 Introduction

The colour-rendering index was defined in Section 6.7.1.1 as a numerical descriptor parameter of the *colour quality aspect* 'colour naturalness' (or 'colour fidelity') [60]. This section deals with the question of how the different R_a values or R_f values can be assigned a semantic interpretation such as 'very good', 'good', 'moderate', or 'poor' – similar to the parameter *CQ* (Eq. (6.11)) according to the scheme of Figure 7.2 (right). Does, e.g. $R_a = 90$ correspond to a 'moderate', 'good', or 'very good' colour-rendering property? To answer this question, an experiment and its results are described in this section [60].

6.9.2 Methodology of the Experiment on the Semantic Interpretation of the Colour-Rendering Indices

In the experiment, the continuous rating scale already mentioned in this chapter was used with the categories 'excellent' = 97.9; 'very good' = 91.6; 'good' = 79.6; 'moderate' = 52.9; 'low' = 41.2; 'poor' = 26.5; 'very poor' = 12.8 (Figure 7.2, right) [60]. The experiments took place in a windowless room. The interior walls and ceiling were painted white and diffused, see Figure 6.40.

In the centre of the room, there was a table with a white tablecloth and four white chairs for a group of four test subjects (see Figure 6.40). A large LED luminaire with built-in RGB-WW LED arrays and a plexiglass diffuser panel at high mounting height on the ceiling provided bright and diffuse illumination of the objects on the table as well as the subject's faces without glare. At all light settings, the luminance of the white tablecloth equalled 242 cd m^{-2}. This value corresponds approximately to the luminance of modern LCD monitors at the white point and to the optimal luminance of the walls (see Section 6.3). With this value of luminance, the *photopic* vision was guaranteed.

The light source contained four LED channels with red, green, blue, and warm white LED light sources and was set to four different CCTs (3100, 4100, 5000, and 5600 K). For each colour temperature, nine different colour mixtures were programmed in for different colour-rendering indices ranging from 29.2 to 96.4. Table 6.14 shows the colorimetric properties of these 36 spectra.

Figure 6.40 Arrangement of the experiment on the semantic interpretation of the colour-rendering property of light sources. Source: Khanh and Bodrogi [60], Richard Pflaum Verlag GmbH & Co. KG; TU Darmstadt.

Table 6.14 Colorimetric data of the 36 spectra in the experiment to determine the semantic interpretation of the colour-rendering index. The four reference spectra with maximum colour-rendering index at each correlated colour temperature are emphasized with bold numbers [60].

No.	CCT (K)	$\Delta u'v'$	CIE R_a	CIE R_f	No.	CCT (K)	$\Delta u'v'$	CIE R_a	CIE R_f
1	3100	−0.002	80.6	88.1	19	5000	−0.001	66.5	77.0
2	3100	0.001	96.4	93.1	20	5000	−0.001	86.3	87.2
3	3100	−0.002	64.5	79.0	21	5000	−0.002	78.7	83.7
4	3100	−0.001	89.3	92.1	22	5000	−0.001	60.8	73.8
5	3100	0.000	82.6	83.9	23	5000	−0.002	93.9	89.2
6	3100	−0.002	56.6	74.4	24	5000	−0.003	90.1	88.6
7	3100	−0.002	29.2	58.6	25	5000	0.000	41.5	62.9
8	3100	0.000	90.8	89.3	26	5000	0.000	73.2	80.6
9	3100	−0.002	73.2	84.0	27	5000	0.001	92.9	88.0
10	4100	0.000	73.1	81.3	28	5600	0.001	59.0	71.2
11	4100	0.000	95.4	89.6	29	5600	0.002	65.2	74.7
12	4100	0.000	94.8	90.4	30	5600	0.002	93.0	87.0
13	4100	−0.002	91.2	87.2	31	5600	0.001	94.1	87.7
14	4100	0.000	87.0	88.0	32	5600	0.003	85.1	84.8
15	4100	0.000	80.1	84.9	33	5600	0.000	47.8	64.8
16	4100	0.000	58.6	73.2	34	5600	0.002	90.7	86.9
17	4100	−0.002	39.8	62.9	35	5600	0.002	79.1	81.9
18	4100	−0.001	66.1	77.5	36	5600	0.000	72.7	78.5

Source: TU Darmstadt.

For each colour temperature, there was a setting with the maximum possible colour-rendering index, which was used as a reference spectrum at each colour temperature, see Table 6.14. Three-dimensional real colour objects were used, which were arranged along the chromaticity circle (see Figure 6.40). Real objects increase the efficiency of the subject's cognitive processing of colour perception and ensure a better approximation to reality. The set of colour objects includes a red rose, an orange, a yellow lemon, a banana, green lettuce, a blue rose, and a purple onion as well as the respective test person's own hand (skin tone) (not shown in Figure 6.40).

A total of 34 young subjects with good colour vision were tested. The 36 spectra were presented randomly at each colour temperature and the order of the four colour temperatures was also randomised. For each spectrum, the spectrum under test was switched on for 15 seconds before the reference spectrum was also activated for 15 seconds. This process was repeated twice for each spectrum before the subject was asked to give a score between 0 and 100 with a cross on the scale in Figure 7.2 (right) based on the *perceived similarity of the colour appearance of the objects on the table* between the reference and test spectra. An adaptation time of two minutes was allowed for each change in colour temperature. The four reference spectra were always also compared with themselves. The latter comparisons achieved on average 'excellent' similarity on the scale of Figure 7.2 (right).

6.9.3 Results of the Experiment on the Semantic Interpretation of the Colour-Rendering Indices

Since the two colour-rendering indices R_a (CIE 1995) and R_f (CIE 2017) are numerically different, as shown above, their relationships to the subjectively determined semantic interpretation are analysed separately [60]. Figure 6.41 illustrates this relationship for the CIE CRI R_a colour-rendering index.

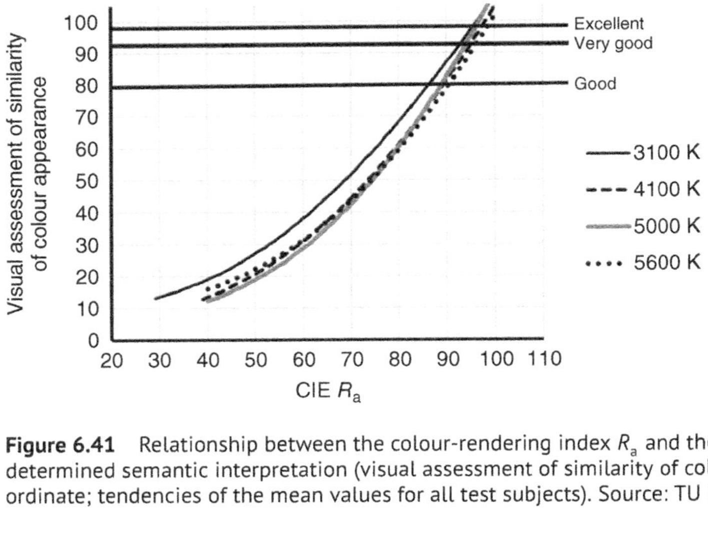

Figure 6.41 Relationship between the colour-rendering index R_a and the subjectively determined semantic interpretation (visual assessment of similarity of colour appearance, ordinate; tendencies of the mean values for all test subjects). Source: TU Darmstadt.

Table 6.15 CIE R_a values for specific values of the subjectively determined semantic interpretation scale (see Figure 6.41) corresponding to the categories 'excellent' (97.9), 'very good' (91.6), and 'good' (79.6).

T_{cp} (K)	'Excellent' (97.9)	'Very good' (91.6)	'Good' (79.6)
3100	95.4	92.3	86.2
4100	97.3	94.6	89.3
5000	96.1	93.7	88.8
5600	98.5	95.8	90.4

Source: TU Darmstadt.

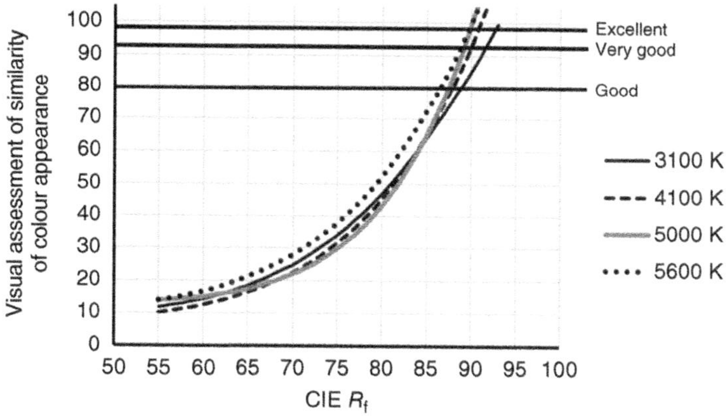

Figure 6.42 Relationship between the colour-rendering (colour fidelity) index R_f and the subjectively determined semantic interpretation (visual assessment of similarity of colour appearance, ordinate; tendencies of the mean values for all test subjects). Source: TU Darmstadt.

Figure 6.41 shows that the colour-rendering value $R_a = 80$, as recommended in DIN EN 12464 as the minimum value for most indoor lighting applications, corresponds to a semantic interpretation of 60–70. This is between 'moderate' and 'good' according to Figure 6.41. A semantic interpretation of just '*good*' requires a general colour-rendering index of at least 86.2 at CCT = 3100 K, see Table 6.15. The empirical visual experience of some experienced museum lighting designers that at least a general colour-rendering index of $R_a = 95$ is necessary has proven true according to Table 6.15.

In the same way, Figure 6.42 shows the relationship between the subjectively determined semantic interpretations and the colour-rendering (colour fidelity) index R_f. The CIE R_f *values* at the specific values of the subjectively determined semantic interpretation scale (similar to Table 6.15) can be seen in Table 6.16.

Table 6.16 CIE colour fidelity index R_f values for specific values of the subjectively determined semantic interpretation scale (see Figure 6.4) corresponding to the categories 'excellent' (97.9), 'very good' (91.6), and 'good' (79.6).

CCT (K)	'Excellent' (97.9)	'Very good' (91.6)	'Good' (79.6)
3100	92.7	91.4	88.8
4100	90.7	89.8	87.8
5000	89.8	89.1	87.4
5600	89.7	88.7	86.5

Source: TU Darmstadt.

6.10 Summary: Preliminary Consequences for Indoor Lighting

If we want to design for the best interior lighting conditions in office rooms, industrial plants, rooms in the education sector, or similar lighting situations during the day, the knowledge in this chapter provides the following key points:

- The best colour temperature range is between 4000 and 5000 K (see Figure 6.13).
- The chromaticity coordinate to be selected for the illumination spectra between 4000 and 5000 K lies below the Planckian locus with a chromaticity difference of approximately $\Delta u'v' = 0.000–0.003$ (see Figure 6.20).
- For the colour temperature range 4000–5000 K, illuminance should be selected higher than 1000 lx to ensure that the lighting installation obtains a 'good' subjective rating (see Figure 6.17).
- The colour-rendering index values should be dimensioned in the range of better than 86 (see Figure 6.41) in order to obtain a good *colour naturalness* of lighting and object representations.
- For rooms with creative and motivational activities (gallery, museum, sales room, and nursing home), an additional saturation increase in the range of $\Delta C^* = 2–3$ should be foreseen.

References

1 Khanh, T.Q. and Bodrogi, P. (2018). Nutzerpräferenzen in der heutigen Innenraumbeleuchtung (User preferences in contemporary indoor lighting). *Z. Licht* 8: 64–71.

2 Boyce, P.R. (1970). The influence of illuminance level on prologed work performance. *Light. Res. Technol.* 2 (2): 74–94.

3 Muck, E. and Bodmann, H.W. (1961). *Lichttechnik* 13: 512.

4 Söllner, G. Report No I. 14/22, Philips Lighting Engineering Laboratory.

5 Westhoff, J.M. and Horemann, H.W. (1963). Contribution to Bodmann, H.W. *Lichttechnik* 14: 24.

6 Bodmann, H.W., Söllner, G., and Voit, E. (1963). Evaluation of lighting level with various kinds of light. In: *Proc. CIE 15th session*, 502–509. Vienna: CIE.

7 Begemann, S.H.A., Aarts, M.P.J., and Tenner, A.D. (1994). *Daylight, Artificial Light and People*. Australia: IES.

8 Seidl, M. (1978). Ausreichendes Tageslicht und akzeptable Fensterabmessungen in Wohnräumen (Sufficient Daylight and Acceptable Window Dimensions in Living Rooms). Dissertation, TU Berlin.

9 Deutsche Lichttechnische Gesellschaft (LiTG) (2016). *LiTG-Publikation 33: Tageslicht kompakt – Tageslichttechnik und Tageslichtplanung in Gebäuden (Daylighting Compact – Daylighting Technology and Daylighting Planning in Buildings)*, 1e, 1–58.

10 Tops, M., Tenner, A.D., Van den Beld, G.J. and Begemann, S.H.A. (1998). *The Effect of the Length of Continous Presence on the Preferred Illuminance in Offices*, CIBSE National Lighting Conference.

11 Juslen, H. (2007). Lighting, productivity and preferred illuminances–field studies in the industrial environment. PhD thesis, Helsinki University of Technology, Espoo, Finnland.

12 Moosmann, C. (2015). Visueller Komfort und Tageslicht am Büroarbeitspalt – eine Felduntersuchung in neun Gebäuden (Visual comfort and daylight at the office workspace – a field study in nine buildings). Promotionsarbeit (PhD thesis), KIT Scientific Publishing. ISBN 978-3-7315-0312-52014.

13 Duff, J., Kelly, K., and Cuttle, C. (2017). Perceived adequacy of illumination, spatial brightness, horizontal illuminance and mean room surface existence in a small office. *Light. Res. Technol.* 49: 133–146.

14 Loe, D.L., Mansfield, K.P., and Rowlands, E. (1994). Appearance of lit environment and its relevance in lighting design: experimental study. *Light. Res. Technol.* 26 (3): 119–133.

15 Van Ooyen, M.H.F., van de Weijgert, J.C.A., and Begemann, S.H.A. (1987). Preferred luminances in offices. *J. Illum. Eng. Soc.* 16 (2): 152–156.

16 Tregenza, P.R. et al. (1974). Consistency and variation in preferences for office lighting. *Light. Res. Technol.* 6: 205–211.

17 De Boer, J.B. and Fischer, D. (1978). *Interior Lighting, Kap. 2*, 2e, 43–63. Philips Technical Review.

18 Balder, J.J. (1957). Erwünschte Leuchtdichten in Büroräumen (Desirable luminance levels in offices). *Lichttechnik* 9 (9): 455–461.

19 Guo, X., Bodrogi, P., and Khanh, T.Q. (2018). Nutzerpräferenz am Bildschirmarbeitsplatz – neue Experimente (User preference at VDU workstations – new experiments). *Z. Licht* 6: 78–82.

20 Littner, M.R. et al. (2005). Standards of practice committee of the American Academy of Sleep Medicine: practice parameters for clinical use of the multiple sleep latency test and the maintenance of wakefulness test. *Sleep* 28 (1): 113–121. PMID 15700727.

21 Veitch, J.A. and Newsham, G.R. (2000). Preferred luminous conditions in open plan offices: research and practice recommendations. *Light. Res. Technol.* 32 (4): 199–212.

22 Khanh, T.Q. and Bodrogi, P. (2016). Bevorzugte Farborte in der Innenraumbeleuchtung: Human-Centric-Lighting, visuelle Aspekte (Preferred chromaticities in indoor lighting: human-centric lighting, visual aspects). *Z. Licht* 3: 68–72.

23 Han, S. and Boyce, P.R. (2003). Illuminance, CCT, décor and the Kruithof curve: Proceedings of the 25th Session of the CIE. San Diego (25 June to 2 July 2003). Vienna: CIE, D3 282–285.

24 Khanh, T.Q. and Bodrogi, P. (2019). Beleuchtungsstärke und Farbtemperatur in der Innenraumbeleuchtung – Nutzerpräferenzbereiche (Illuminance and colour temperature in indoor lighting – user preference ranges). *Z. Licht* 1: 78–85.

25 Schweitzer, S., Schinagl, C., Djuras, G. et al. (2016). Investigation of gender- and age-related preferences of men and women regarding lighting conditions for activation and relaxation. SPIE Proceedings vol. 9954, Fifteenth International Conference on Solid State Lighting and LED-based Illumination Systems; 99540L.

26 Vos, J.J. (1978). Colorimetric and photometric properties of a 2° fundamental observer. *Color Res. Appl.* 3: 125–128.

27 Porkony, J. and Smith, V.C. (1994). *Scientific Basis for Visual Performance*, 80–88. Wien: Proceedings of the CIE symposium on Advances in Photometry.

28 Bodrogi, P., Vinh, Q.T., and Khanh, T.Q. (2019). Correlations among lighting quality metrics for interior lighting. *Light. Res. Technol.* 51 (8): 1192–1207.

29 Fotios, S.A. and Levermore, G.J. (1998). Chromatic effect on apparent brightness in interior spaces, II: SWS lumens model. *Light. Res. Technol.* 30 (3): 103–106.

30 Kruithof, A.A. (1941). Tubular luminescence lamps for general illumination. *Philips Tech. Rev.* 6 (3): 65–73.

31 Fotios, S. (2017). A revised Kruithof Graph based on empirical data. *Leukos* 13 (1): 3–17. (Taylor & Francis-Verlag).

32 Viénot, F., Durand, M.-L., and Mahler, E. (2009). Kruithof's rule revisited using LED illumination. *J. Mod. Opt.* 56 (13): 1433–1446.

33 CIE (2014). Guidance towards best practice in psychophysical procedures used when measuring relative spatial brightness. Vienna (Austria): CIE. CIE Publ. 212:2014.

34 Khanh, T.Q., Bodrogi, P., Guo, X., and Anh, P.Q. (2018). Towards a user preference model for interior lighting – Part 1: concept of the user preference model and experimental method. *Light. Res. Technol.* 51 (7): 1014–1029.

35 Khanh, T.Q., Bodrogi, P., Guo, X., and Anh, P.Q. (2018). Towards a user preference model for interior lighting – Part 2: experimental results and modelling. *Light. Res. Technol.* 51 (7): 1030–1043.

36 Bodrogi, P., Guo, X., Stojanovic, D. et al. (2018). Observer preference for perceived illumination chromaticity. *Color Res. Appl.* 43: 4. https://doi.org/10.1002/col.22210.

37 Rea, M.S. and Freyssinier, J.P. (2013). White lighting. *Color Res. Appl.* 38 (2): 82–92.

38 Liu, X.Y., Li, H., and Luo, M.R. (2015). Assessment of white perception in a real lit room. In: *Proceedings of 28th CIE Session*. Manchester, United Kingdom.

39 Perz, M., Baselmans, R., and Sekulovski, D. (2016). Perception of illumination whiteness. In: *Proceedings of the 4. CIE Expert Symposium on Colour and Visual Appearance*. Prague 2016. ISBN 978-3-902842-59-6.

40 Whitehead, L. (2013). Interpretation concerns regarding white light. *Color Res. Appl.* 38 (2): 93–95.

41 Ohno, Y. and Fein, M. (2014). Vision experiment on acceptable and preferred white light chromaticity for lighting. In: *Proceedings of CIE 2014 "Lighting Quality and Energy Efficiency."* Kuala Lumpur, Malaysia.

42 Bodrogi, P., Jungnitsch, K., and Khanh, T.Q. (2015). *Farbqualität: Definition und Anwendungen (Colour Quality: Definition and Applications)*. LiTG-Publikation 31 ISBN 978-3-927787-49-0.

43 International Commission on Illumination (CIE) (1987). International lighting vocabulary. CIE Publ. No. 17.4.

44 DIN 6169. (1968). Bl. 1, August 1962, Bl. 2. Farbwiedergabe. Allgemeine Begriffe (Colour rendering; general terms).

45 International Commission on Illumination (1995). CIE Publication No. 13.3, Method of measuring and specifying colour rendering properties of light sources.

46 David, A., Fini, P.T., Houser, K.W. et al. (2015). Development of the IES method for evaluating the color rendition of light sources. *Opt. Express* 23 (12): 15888–15906.

47 Illuminating Engineering Society of North America. (2015). *IES TM-30-15, IES Method for Evaluating Light Source Color Rendition*.

48 CIE (2017). CIE 2017 Color Fidelity Index for accurate scientific use. CIE Publication 224:2017 (CIE, 2017).

49 Fairchild, M.D. and Pirrotta, E. (1991). Predicting the lightness of chromatic object colors using CIELAB. *Color Res. Appl.* 16 (6): 385–393.

50 Davis, W. and Ohno, Y. (2010). Colour quality scale. *Opt. Eng.* 49: 033602.

51 Smet, K., Ryckaert, W.R., Pointer, M.R. et al. (2012). A memory colour quality metric for white light sources. *Energ. Build.* 49: 216–225.

52 Babilon, S. and Khanh, T.Q. (2018). Color appearance rating of familiar real objects under immersive viewing conditions. *Color Res. Appl.* 43: 551–568.

53 Khanh, T.Q., Bodrogi, P., Vinh, Q.T. et al. (2018). Colour preference, naturalness, vividness and colour quality metrics, Part 4: experiments with still life arrangements at different correlated colour temperatures. *Light. Res. Technol.* 50 (6): 862–879. https://doi.org/1477153517700705.

54 Khanh, T.Q., Bodrogi, P., Guo, X. et al. (2019). Colour preference, naturalness, vividness and colour quality metrics, Part 5: a colour preference experiment at 2000 lx in a real room. *Light. Res. Technol.* 51 (2): 262–279.

55 Krause, N., Bodrogi, P., and Khanh, T.Q. (2012). Bewertung der Farbwidergabe. Reflexionsspektren von Objekten unter verschiedenen weißen Lichtquellen (Evaluation of colour rendering. Reflectance spectra of objects under different white light sources). *Z. Licht* 5: 62–69.

56 Royer, M.P., Wilkerson, A., Wei, M. et al. (2017). Human perceptions of colour rendition vary with average fidelity, average gamut, and gamut shape. *Light. Res. Technol.* 49: 966–991.

57 Vinh, Q.T. and Khanh, T.Q. (2018). Beleuchtung von Hauttönen mit spektral veränderlichen LED-Lichtarten unterschiedlicher ähnlichster Farbtemperaturen (Illumination of skin tones with spectrally variable LED lights of different correlated colour temperatures). *Z. Licht* 3: 71–75. (Richard Pflaum Verlag GmbH & Co. KG.).

58 Quellman, E.M. and Boyce, P.R. (2002). The light source color preferences of people of different skin tones. *J. Illum. Eng. Soc.* 31 (1): 109–118.

59 Jost-Boissard, S., Avouac, P., and Fontonont, M. (2016). Preferred color rendition of skin under LED sources. *Leukos J. Illum. Eng. Soc.* 12: 79–93.

60 Khanh, T.Q. and Bodrogi, P. (2019). Über Farbwiedergabeindizes – Wesen und semantische Bedeutung (On colour rendering indices – essence and semantic meaning). *Z. Licht* 6: 73–79.

7

New Light-Quality Models from Laboratory Experiments and their Validation in Field Trials

7.1 Introduction

The history of lighting technology and lighting science from the beginning of the twentieth century until today has been accompanied by a series of efforts and attempts to determine, through experiments and modelling approaches, a reliable relationship between indoor lighting parameters and the ophthalmo–physiological, psychophysical as well as work-related responses of room users. Over time, the criteria for the above responses have been shifted and redefined in lighting science. The aim of these lighting science investigations was and is to define the respective threshold values for the lighting parameters in guidelines and in standards by establishing an acceptable or a preferred threshold value.

The first milestones of lighting research were marked by the pioneering work of M. Luckiesh and Moss in the period 1930–1940 in the USA [1], which was then continued by Blackwell in the four decades from about 1950 to the 1980s [2]. The criterion in Luckiesh's study was the boundary between 'seeing' and 'not seeing' (a threshold for visibility) in visual tasks, relative to a standard 8-point, '*Bondi Book*' font size task, read from a distance of 35.5 cm on a white paper page with a reflectance of 0.7 at an illuminance of 108 lx on the paper. The contrast studies by H.R. Blackwell [2] as well as numerous other studies on the visibility of reading objects as well as tools (such as the visibility of screws in industrial sorting) led to the minimum illuminance levels in today's standards (e.g. 500 lx in European EN-DIN standards, etc.).

In the 1970s, perceptual psychologists, lighting architects, and building designers began to look at user behaviour in interior lighting from the perspective of psychology and user satisfaction, rather than through visual performance alone. When viewing and perceiving an illuminated space in a particular context – with a variety of lighting options – there are a number of adjectives, attributes, and descriptors that describe different aspects of the viewer's emotions.

At the beginning of the 1970s, lighting research began to investigate interpretative dimensions of interior scenes and their subjective spatial impression. The first work in this field was carried out by Flynn et al. in 1973 [3] in a meeting room with 6 different lighting configurations, where 96 test persons were given a questionnaire with 34 different scales to assess these lighting configurations with the following

Human Centric Integrative Lighting: Technology, Perception, Non-Visual Effects, First Edition.
Tran Quoc Khanh, Peter Bodrogi, and Trinh Quang Vinh.

pairs of adjectives, such as 'clear–hazy', 'pleasant–unpleasant', 'bright–dim', or 'stimulating–subduing'. Flynn et al. [4] conducted a multidimensional analysis of the visual scales to establish the photometric properties for assessing similarity and differences and found the following three basic characteristics: (i) homogeneity (uniform/uneven); (ii) brightness (bright/dark); and (iii) '*light comes from above – light comes peripherally*'.

In 1998, Veitch and Newsham conducted subjective experiments to assess the appearance of a room with 292 test persons and 9 lighting systems using 27 semantic scales [5]. The most important characteristics for describing the impression of the room's appearance were – after a detailed data analysis – the so-called '*visual attraction and complexity*' and the '*brightness*'. Veitch et al. also analysed characteristics of good room lighting in further numerous studies and established lighting parameters such as luminance distribution in the room, colour rendering, light direction, and glare. Nevertheless, a modelling approach with a mathematical background was lacking in order to determine the limit values of the lighting parameters for a certain level of user preference.

With the dynamic development of LED technology, first with coloured LEDs at the beginning of the twenty-first century and the subsequent development of LED phosphors from around 2005 until today, a new era of colour research began with the aim of investigating colour quality in indoor lighting with the focus on such areas as retail lighting, museum and gallery lighting, and TV and film studio lighting. From about 1965 to 2017, scientists tended to study 'colour rendering'. In numerous studies on colour quality, the more complex attributes such as colour preference (CP), colour memory, colour gamut, colour harmony, naturalness, and colour attractiveness have been analysed by several research groups in China, in the USA, France, Belgium, and in Germany [6–11], the results of which led to two CIE workshops in South Korea in 2018 and in the USA in 2019 with the leading participation of the authors of the book. Nevertheless, there are still no unified opinions within the CIE on a possible unified colour quality model that can be used in practice. According to the studies from Germany, colour quality depends on lighting level, colour temperature, and colour saturation [12].

Thus, until about 2015–2017, we had a difficult situation: on the one hand, we knew very well the correlations between visual performance (reading speed, reaction time, and error rate) and lighting level. Nevertheless, visual performance goes into saturation from about 300 or 500 lx and thus loses its importance. The psychology of perception in the 1990s greatly helped scientists and lighting engineers to analyse the complexity of perception in indoor lighting. Nevertheless, a mathematical correlation between the perceptual characteristics and the lighting parameters was missing.

The research on colour quality in the last 15 years raised the research quality significantly, on which lighting science could build a practical lighting-quality model in the near future, which would be necessary in the age of the Internet of Things (IOTs), connectivity, and smart lighting – for the automatic control of lighting quality. If the technological prerequisites such as controllable LED luminaires, sensor technology, internet protocols, cloud, and artificial intelligence methods are already and soon available, one has to ask oneself according to which criteria one should control and regulate the building systems and LED luminaires and how one should implement these criteria in order to optimise user satisfaction in different weather conditions, geographical locations, or user groups by considering individuality according to age, country, cultural group, and gender.

Therefore, the motivation of this chapter is to model light quality in indoor lighting on the basis of experiments in real rooms and to establish the mathematical relationship between the important parameters of user satisfaction and the lighting parameters that have a direct effect on user perception in order to enable optimal control of LED luminaires in the future with the help of '*smart lighting computers*' [13, 14]. Accordingly, various model approaches for lighting quality in interior lighting [14] form the content of this chapter.

7.2 Input and Output Parameters of the Light-Quality Models

7.2.1 Input Parameters

A distinction shall be made between the different levels of origin of the lighting parameters:

(a) *Physical level*: these are spectral radiant flux values of the light source, spectral radiance factors of room objects (textiles, furniture, and food), or spectral irradiances at the working plane or directly in front of the eye of the room users.
(b) *Lighting parameters derived from the spectra*: these are luminance levels, illuminance levels on the working plane; colour temperature, white point of the walls or ceiling, colorimetric parameters such as colour rendering; and hue and saturation of the coloured objects in the room.
(c) *Other eye-physiological or psychophysical parameters* are e.g. colour shadings on objects, the ratio of direct to indirect light components, and the spatial luminance distribution in the room.

At the beginning of the 1970s, the level of illumination (*illuminance*) was the focus of investigations [15, 16] but in the last 15 years, this has tended to be *correlated colour temperature* (CCT) [17, 18] as well as the correlation between illuminance and CCT for an acceptable room impression [19]. In a study from 2018 [12], the *colour saturation* of coloured objects placed in the room was considered.

7.2.2 Output Parameters

The initial parameters of a model approach to lighting quality are the responses of room users to changes in the lighting situation. According to the research of Flynn et al. [3, 4] and Veitch and Newsham [5], there are the following two perceptual attributes: (i) brightness (light–dark), which is still more *physiologically* attributable; and (ii) '*visual attraction and complexity*', which emerges at a higher, *psychophysical* level, together with *room preference* (or *scene preference* [SP], see below) of furnished rooms in a certain weather situation and viewing context, combined with a certain expectation about the appearance of the room. Considering the history of lighting science and analysing existing literature, the authors of this book identified the following four output parameters as particularly relevant [13]:

(a) *Brightness*: the so-called *spatial* brightness of *extended* visual fields is decisive as a visual stimulus.
(b) Visual clarity (VC).
(c) *CP*: an important aspect of the general term '*colour quality*'.
(d) The term *room preference* or SP has been introduced recently as the observer's judgement of how a complex space is accepted or preferred. The strength of the three-dimensional perception of the room (table, floor, walls, furniture, ceiling, and objects in the room) and the three-dimensional recognition features (structure, texture, shadiness, and modelling) of the faces of the other persons or the objects in the whole room are taken into account.

7.3 Experimental Set-Ups for the Light-Quality Models

For the experimental investigation of the four output parameters (visual attributes), brightness, VC, CP, and room preference mentioned in the previous Section 7.2.2, a laboratory room – in the lighting context of a 'modern office' – was set up with white diffuse walls. The room was lightproof from the outside and the windows were covered with white diffuse curtains. Figure 7.1 (top left) shows a table with a white diffuse tablecloth to which the test subjects adapted and rated their brightness perception with a scale from 0 to 100 according to an empirical study [20] (see Figure 7.2, left). In the subsequent test set-up for VC (Figure 7.1, top right), objects with fine structures and patterns (coloured jumpers, hair structure of a doll, thin strokes of a reticle, and fine structure from an artificial flower) were observed.

In the test set-up for CP, distinctive colour objects such as coloured books and the two Macbeth® ColourCheckers with colours in different hues and colour saturations were made available for evaluation. For SP, spatially extended (i.e. not predominantly flat, so-called '3D') objects such as vases of flowers, plates of fruit, and a painting were used. The test subjects rated VC, CP, and SP with a different scale, see Figure 7.2 (right).

The four test set-ups were diffusely illuminated – one after the other – with two thermally stabilised large LED luminaires (so-called *SkyPanels*, Company ARRI/Munich) with RGB LEDs and warm white LEDs that provided stable illumination on a short-term and also on a long-term basis. The light was carefully mixed in advance to 36 different spectra, divided into 4 different illumination levels in the centre of the table (200, 500, 1000, and 1800 lx) as well as three different colour temperatures (3000, 4100, and 5600 K) and three different colour saturation levels (three ΔC^* levels, ΔC^* represents the mean value of the CIELAB chromaticity difference to the reference spectrum for the 15 – rather saturated – CQS test colours according to [10]). These 36 spectra were measured in the centre of the white

Figure 7.1 Four test set-ups [13] for brightness (top left), visual clarity (top right), colour preference (bottom left), and scene preference (bottom right). Source: Reproduced with permission from Khanh et al. [13], SAGE PUBLICATIONS, INC.

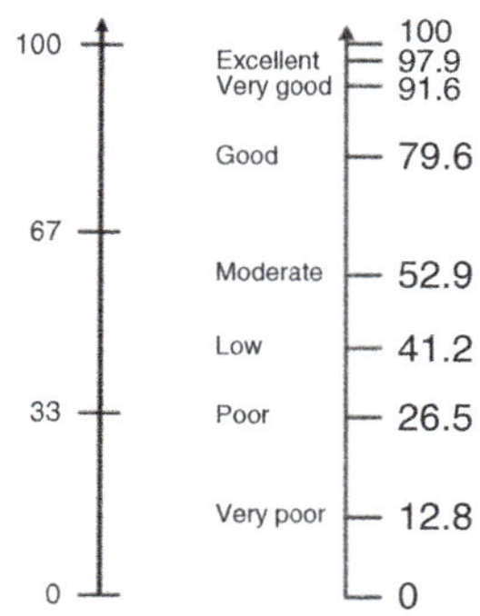

Figure 7.2 Rating scales (left: for brightness, right: for visual clarity, colour preference, and scene preference) with the anchor points shown (left: 0, 33, 67, and 100, right: 'excellent', 'very good', … to 'very poor') – after [13, 20]. Source: Reproduced from [13] with permission from Lighting Research and Technology.

tablecloth with a calibrated radiance-type spectroradiometer. The relative spectra calculated from this are shown in Figure 7.3. Table 7.1 contains the values of illuminance, colour temperature, colour saturation, and colour rendering for the 36 spectra.

A total of 30 subjects (14 men, 16 women) with normal colour vision (tested with 'The Standard Pseudoisochromatic Plates for Acquired Color Vision Defects' [21]) participated. The age of these subjects ranged from 19 to 32 years with an average age of 25.8 years (standard deviation: 3.3 years). Before the main experiment, training was conducted with different light settings and the same questionnaire. In the training phase, extreme light settings such as very bright, very dark and very high,

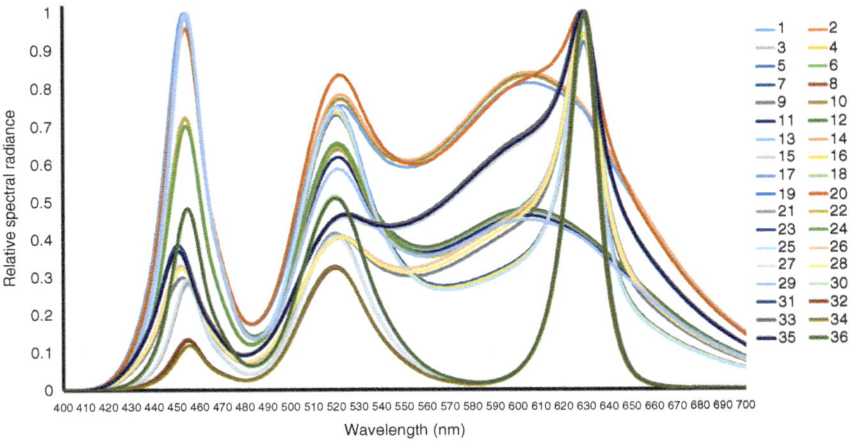

Figure 7.3 Spectra of the 36 light settings during the 4 visual experiments [13]. Source: Reproduced from [13] with permission from Lighting Research and Technology.

Table 7.1 Illuminance values (E_v in lx), correlated colour temperatures (CCT in K), colour saturations ΔC^* (ΔC^* represents the mean value of the CIELAB chroma difference to the reference spectrum for the 15 CQS colours according to [10]) and colour-rendering indices R_a of the $i = 1$–36 light settings in randomised order (see also Figure 7.3).

i	E_v	CCT	ΔC^*	R_a	i	E_v	CCT	ΔC^*	R_a	i	E_v	CCT	ΔC^*	R_a
33	197	2980	0.0	94	14	489	4077	−0.1	92	12	1009	5504	0.7	95
9	198	2992	1.9	91	22	491	4082	2.6	87	7	1000	5570	4.1	79
34	204	2995	10.7	21	4	502	4110	12.0	26	2	1005	5616	11.6	34
20	199	4082	0.5	96	11	495	5580	0.8	95	31	1791	2982	0.2	94
24	193	4068	2.4	88	16	496	5584	4.3	79	18	1796	2983	1.7	93
21	199	4101	11.8	27	30	499	5612	11.4	35	1	1818	2995	11.2	22
3	196	5545	0.7	95	35	997	2979	0.0	94	17	1793	4090	0.1	93
25	192	5602	4.4	78	26	999	2988	1.7	93	6	1791	4083	2.5	88
23	201	5587	11.5	35	8	1001	2991	11.2	22	27	1814	4107	12.2	25
15	491	2975	0.2	94	10	1000	4077	0.0	92	29	1789	5606	1.0	95
28	496	2988	1.9	92	19	990	4069	2.5	88	5	1801	5586	4.2	79
32	500	2996	11.3	21	13	1006	4110	11.8	27	36	1825	5617	11.6	34

Source: Reproduced from [13] with permission from Lighting Research and Technology.

and very low colour saturations were also set to create so-called anchor stimuli. The training results were not included in the final evaluation of the results. In the main experiment, the 36 spectra mentioned above were presented randomly.

For a better understanding of the four initial parameters of the light-quality models, it is beneficial to describe the instructions given by the experimenter to the subjects (cf. with Figure 7.1):

(a) *Brightness*: 'The term brightness is used in an everyday sense in that you evaluate the whole table with the tablecloth – compared to the reference situation – according to brightness. How bright does the table appear compared to the reference situation? The reference situation is shown several times in the training phase and corresponds to a brightness value of 100. Complete darkness corresponds to a value of 0. Please evaluate your impression of brightness (H) by looking at the table, i.e. the white tablecloth. In the training phase, you can memorise your brightness impression of the reference scene with $H = 100$, which is shown repeatedly. You can tick the scale – according to your brightness impression. After each adjustment of a new light source, please look at the table for 80 seconds first. After this adaptation, you can then evaluate the brightness. Please do not look at your own hand, nor at the faces of the other persons, only at the table, i.e. the white tablecloth. Please first decide in which third the current situation is (top, middle, or bottom) and then tick the scale – corresponding to your impression of brightness – within this third'.

(b) *VC*: 'VC is defined as the clear visibility of continuous colour transitions, of subtle shades of colour on the surfaces of objects, and the clear visibility of contrasts between the colours of different coloured objects. Thus, VC corresponds to the clear visibility of visual details, structure (e.g. the individual hairs of the doll) and texture (e.g. the coloured textile patterns of the jumper), as well as the clear visibility of colour gradients and coloured textures on the surfaces of the coloured objects (such as the coloured texture of the inside of the water lily), or the lines or letters on the black and white test patterns. Please assess the VC of the objects on the table. You are to look at all the objects in turn and then also at all the objects at the same time. Please form your judgement according to your general impression of the VC for all the objects in the scene. You can make your judgement by ticking the scale of VC. You should orientate yourself according to the categories of the scale (e.g. '*poor*', '*moderate*', '*good*', and '*very good*'). You can also mark the scale at any point between two categories. You should therefore always consider all the above aspects of VC and all objects. After each adjustment of a new light source, please first look at the table or walls for 80 seconds. After the adaptation, you can evaluate VC'.

(c) *CP*: 'In this experiment, you will evaluate the colour appearance of the coloured ColorChecker® boards and coloured books on the table. These two ColorChecker boards and the coloured books are illuminated with different light sources, creating different colour appearances (with 36 light sources in total). Consider the following point of view: Do you like the colour appearance of the coloured samples of the ColorChecker and the books? You should look at all the colours at the same time and evaluate your overall impression. The value of 100 corresponds to the maximum rating, i.e. when the ColorChecker and books appear as you would like to see their colour appearance from the point of view of your CP. You should tick the scale according to your impression. You should orientate yourself according to the categories of the scale, e.g. '*bad*', '*moderate*', '*good*', or '*very good*'. You can also mark the scale at any point between two categories. After each adjustment of a new light source, please look at the table for 80 seconds first. After adaptation, you can rate the colour appearance of the ColorChecker and the books according to your CP'.

(d) *Room preference (or SP)*: 'Room preference is defined as the strength of the three-dimensional perception of the space (table, floor, walls, furniture, ceiling, and objects in the room) and the three-dimensional recognition features (structure, texture, shading, and modelling) of the faces of the other persons or objects in the whole room and especially, at the table. Evaluate your overall impression of the room preference of the different scenes. We are curious to see how your impression of the room preference changes. You can give your assessment by ticking the scale. You are to orientate yourself according to the categories of the scale (e.g. '*bad*', '*moderate*', '*good*', and '*very good*'). You should express your impression honestly, if it is bad, then it is bad. We are really curious about your own impression. You can also tick the scale at any point between two categories. You should always consider all the above aspects of room preference. Please look not only at the table, but also at the two shutters, the walls, and the ceiling. After each adjustment of a new light source, please first look at the shutters for 80 seconds. After this adaptation, you can evaluate room preference. You can also '*wander*' around the room (as in a museum, for example) and look at all the objects on the table, but sometimes please also look at the shutters, the walls, and the ceiling'.

7.4 Equations of the Light-Quality Models

In this section, the four initial parameters brightness, VC, CP, and SP – as the mean responses of all test subjects on the scales of Figure 7.2 – are evaluated and modelled one after the other.

7.4.1 Brightness

The best model approach for the spatial brightness B (Brightness) was achieved with the following two equations [14]:

$$B = 27.058 \ln(E_{\mathrm{v,eq}}) - 105.25 \tag{7.1}$$

The symbol $E_{\mathrm{v,eq}}$ means the so-called equivalent illuminance, which is calculated according to Fotios and Levermore [22] in Eq. (7.2).

$$E_{\mathrm{v,eq}} = E_{\mathrm{v}}\,(\mathrm{S/V})^{0.24} \tag{7.2}$$

In Eq. (7.2), the symbol E_{v} means the illuminance measured on the table at a certain spectrum (i = 1–36, see Figure 7.3), S is the signal of the blue S-cones and V is the signal of the luminance channel V (calculated according to the well-known $V(\lambda)$-function). With a Pearson correlation coefficient of $r^2 = 0.97$, the correlation between the equivalent illuminance and the perceived brightness B according to Eq. (7.1) is very high, see Figure 7.4.

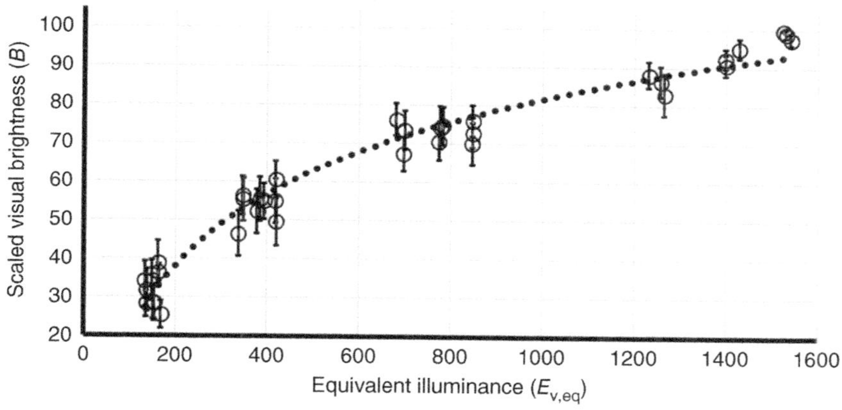

Figure 7.4 Mean visually scaled subjective brightness B as a function of equivalent illuminance $E_{v,eq}$; intervals represent 95% confidence intervals of mean subjective responses. Source: Reproduced from [14] with permission from Lighting Research and Technology.

This is proof that the brightness depends not only on the illuminance level E_v (lx) but also on the blue part of the spectrum of the light source. The blue component of the spectrum is modelled here by the signal of the (blue-sensitive) S-cones. Light sources with a high blue spectral component have a correspondingly high brightness (at the same illuminance level).

7.4.2 Visual Clarity (VC)

The best model approach for VC could be achieved by Eq. (7.3) [14] (see Figure 7.5):

$$VC = [19.804\ \ln(E_{v,eq}) - 60.496]\ [-0.0004\ \Delta C^{*2} - 0.011\ \Delta C^{*} + 1.0708] \tag{7.3}$$

In Eq. (7.3) there is on the one hand a dependence on the magnitude of $E_{v,eq}$ and an additional second term in the rectangular brackets, which represents a dependence on the magnitude of ΔC^*. (The quantity ΔC^* is, as mentioned earlier, an index for the colour saturation of the object colours under a certain spectrum.) The first term has the same form as for brightness (Eq. (7.1)) and the second term with ΔC^* means that excessive colour saturation of the object colours in the lit environment (caused by inappropriately chosen illumination spectra such as very saturating RGB LED spectra) reduces VC. This is an important indication – its significance was studied several times in the last 10 years (see [23]).

Applying the model equation of VC (Eq. (7.3)) to different illuminance levels and CCT values at a fixed oversaturation level of $\Delta C^* = 1.1$ (the level of maximum SP), we obtain the following minimum illuminance levels to ensure 'good' VC (VC = 79.6): between 1100 lx (with 6500 K) and 1300 lx (with 2800 K), see Figure 7.6. If in lighting practice the goal is set to achieve at least '*good*' VC, the CCT must first be selected depending on the engineer's intention and the application, and then the degree of chroma enhancement must be set to about $\Delta C^* = 1.1$. Finally, the required (horizontal) illuminance (for '*good*' VC) must be read from Figure 7.6.

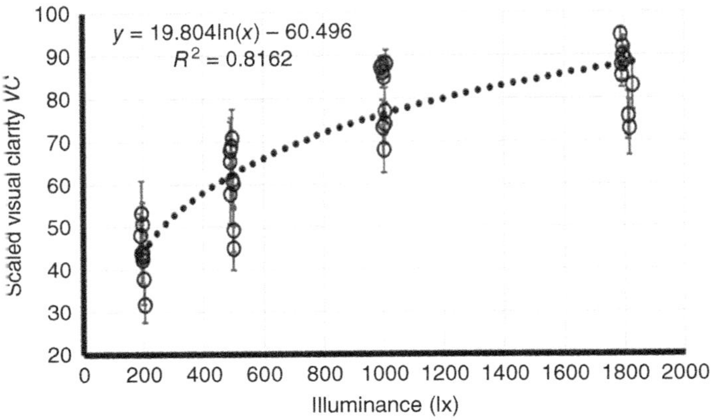

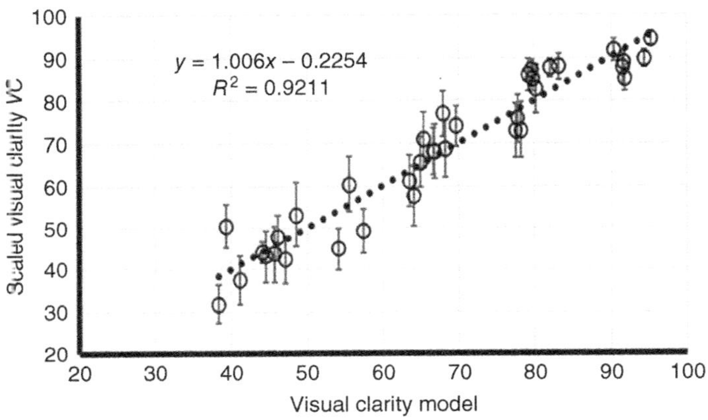

Figure 7.5 Visual clarity as a function of illuminance in lx (left) and the correlation between the modelled VC values of the visual clarity model in Eq. (7.3) and the visual VC values of the test subjects. Source: Reproduced from [14] with permission from Lighting Research and Technology.

7.4.3 Colour Preference (CP)

The subject's visually scaled mean values for CP were modelled using an equation with three terms (Eq. (7.4)) [14]:

(a) The first term is like brightness (Eq. (7.1)).
(b) The second term – with ΔC^* as a parameter – is like VC (Eq. (7.3)).
(c) The third term contains the quotient $(S/V)^{0.24}$ and takes light sources with different blue components in their spectrum into account.

$$\begin{aligned}\mathrm{CP} = {} & (14.089\ \ln(E_{v,eq}) - 25.397)\ [-0.003\ \Delta C^{*2} + 0.0252\ \Delta C^* + 1.0192] \\ & + [-518.554\ ((S/V)^{0.24})^2 + 864.872\ (S/V)^{0.24} - 356.578]\ (\text{with } r^2 = 0.84)\end{aligned} \tag{7.4}$$

The CP thus depends in turn on the brightness, on the colour saturation of the coloured objects in the room, and on the blue component of the spectrum. The dependence of the second term of Eq. (7.4), $[-0.003\ \Delta C^{*2} + 0.0252\ \Delta C^* + 1.0192]$,

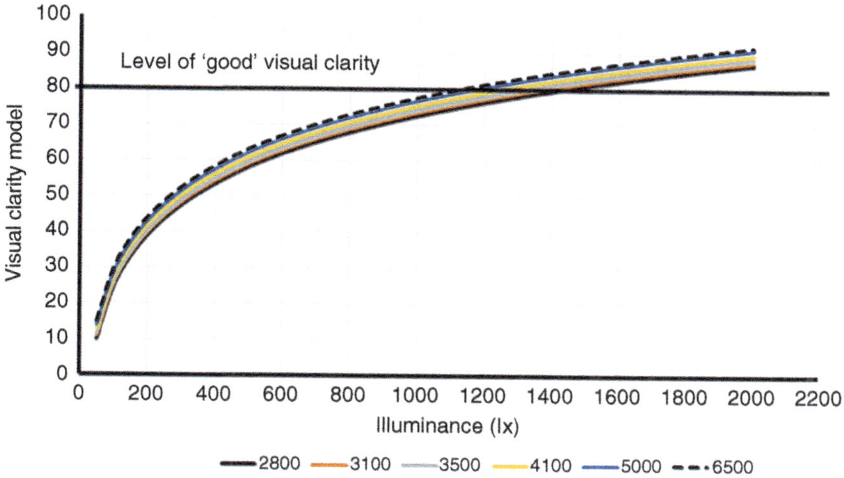

Figure 7.6 Metric of visual clarity (VC) according to Eq. (7.3) as a function of illuminance (lx) and CCT (K; legend) at a fixed oversaturation level of $\Delta C^* = 1.1$ (the level of maximum scene preference). The level for 'good' visual clarity (VC) is 79.6 (see Figure 7.2, right). Source: Reproduced from [14] with permission from Lighting Research and Technology.

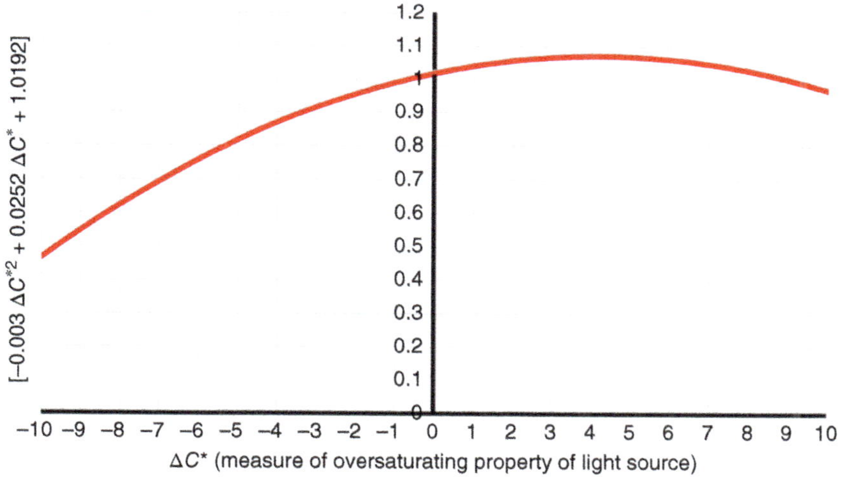

Figure 7.7 Dependence of the second term of Eq. (7.4), $[-0.003\ \Delta C^{*2} + 0.0252\ \Delta C^* + 1.0192]$, on the index of the colour-saturating property ($\Delta C^* > 0$: oversaturation; $\Delta C^* < 0$: desaturation) of the light source. Source: TU Darmstadt.

on the measure of the colour-saturating property (ΔC^*) of the light source is shown in Figure 7.7.

Figure 7.7 shows that the second term of Eq. (7.4) of CP, which depends on the colour-saturating property (ΔC^*) of the light source, has a maximum at approx. $\Delta C^* = 4.2$. This optimal value corresponds to a *moderate oversaturation* of the object colours compared to the reference light source (blackbody radiator or phase of daylight of the same colour temperature). Increasing the saturation of the object

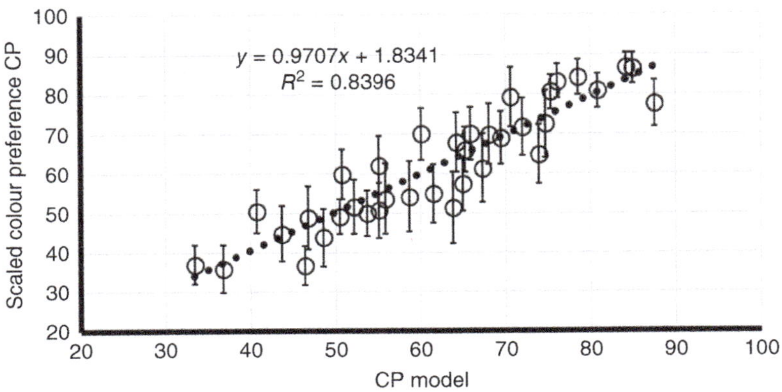

Figure 7.8 Correlation between the CP model according to Eq. (7.4) and the subjects' visually scaled mean colour preference; the intervals represent 95% confidence intervals of the subject's mean responses. Source: Reproduced from [14] with permission from Lighting Research and Technology.

colours (e.g. by increasing the RGB components of a multichannel LED luminaire) beyond approx. $\Delta C^* = 4.2$ leads to a decrease in subjective CP. If the object colours are desaturated by $\Delta C^* < 0.0$, then the CP curve (on the left side of Figure 7.7) goes downhill rapidly. Figure 7.8 shows the correlation between the CP model according to Eq. (7.4) and the subject's visually scaled mean value for CP.

If the model equation of CP (Eq. (7.4)) is applied to different illuminance levels and CCT values at a fixed oversaturation level of $\Delta C^* = 4.2$ (the value of maximum CP), we obtain the following minimum illuminance levels to ensure the '*good*' level (CP = 79.6): between 1100 lx (with 6500 K) and 1700 lx (with 3500 K), see Figure 7.9.

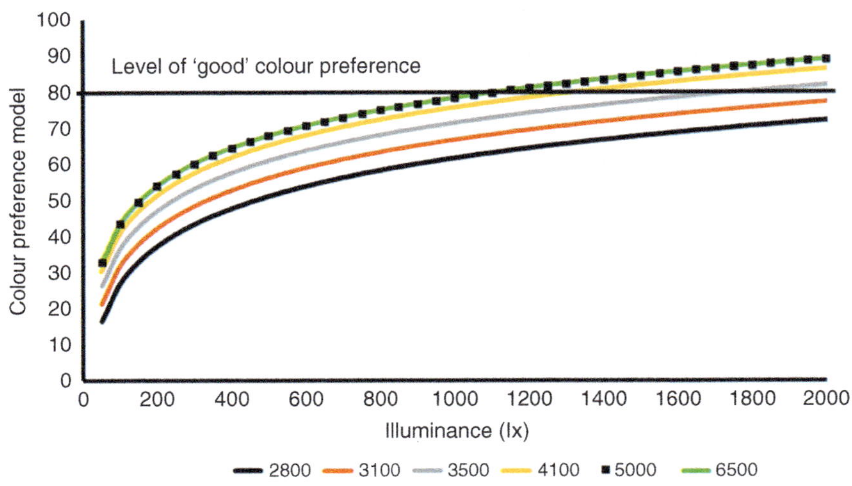

Figure 7.9 Colour preference (CP) metric of Eq. (7.4) as a function of illuminance (lx) and CCT (K; legend) at a fixed oversaturation level of $\Delta C^* = 4.2$ (the level of maximum colour preference). The value for a 'good' colour preference (CP) is 79.6 (see Figure 7.2, right). Source: Reproduced from [14] with permission from Lighting Research and Technology.

At lower CCTs, the 'good' CP level can only be achieved at E_v >2000 lx. This is only an extrapolation, as such high illuminances were not investigated in the experiments (Section 7.3). If in lighting practice the goal is to achieve at least a 'good' CP, the CCT must first be selected depending on the engineer's intention and the application, then the degree of oversaturation (i.e. increasing the chromaticity of coloured objects by changing the light source spectrum) must be set to about $\Delta C^* = 4.2$. Finally, the required (horizontal) illuminance can be read from Figure 7.9.

7.4.4 Scene Preference (SP)

The data evaluation for SP was like that for the CP mentioned above. The corresponding model approach in Eq. (7.5) [14] contains the similar three terms. Figure 7.10 shows the visually scaled SP as a function of illuminance and, in

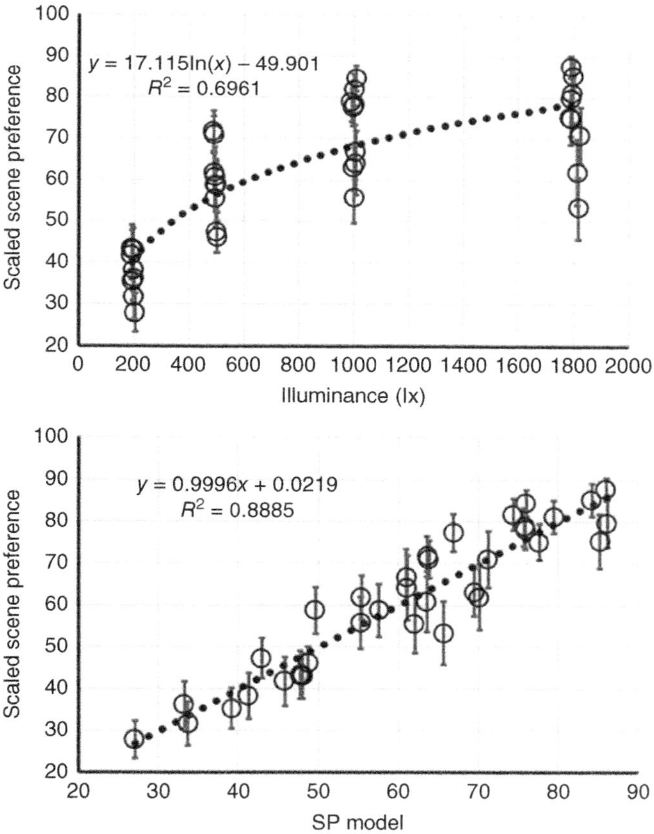

Figure 7.10 Mean values of visually scaled scene preference (or room preference) SP as a function of illuminance in lx (left) and correlation between the SP model according to Eq. (7.5) and the mean value of visually scaled scene preference (right); the intervals represent 95% confidence intervals of the mean responses of the test subjects. Source: Reproduced from [14] with permission from Lighting Research and Technology.

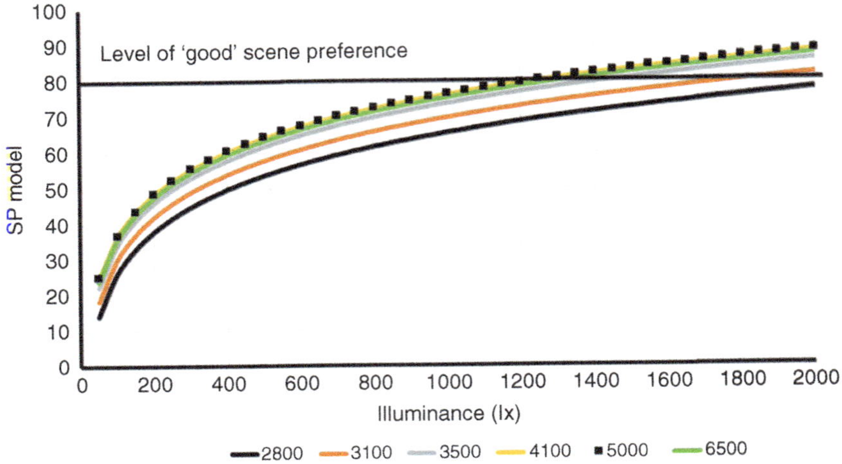

Figure 7.11 Scene preference values (SP) from the SP model according to Eq. (7.5) as a function of illuminance (in lx) at different correlated colour temperatures.
Source: Reproduced from [14] with permission from Lighting Research and Technology.

comparison, as a function of the SP model of Eq. (7.5).

$$\text{SP} = 17.127\ \ln(E_{\text{v,eq}}) - 41.844 - 0.1325\ \Delta C^{*2} + 0.2797\ \Delta C^{*} + [-622.378\ ((\text{S/V})^{0.24})^2 + 980.843\ (\text{S/V})^{0.24} - 382.535](r^2 = 0.89) \quad (7.5)$$

When evaluating the SP modelled according to Eq. (7.5) (with a constant ΔC^* value of 1.1; this value corresponds to the maximum SP) as a function of illuminance at different colour temperatures, a diagram (Figure 7.11) important for lighting design could be derived [14]. Figure 7.11 shows that a '*good*' rating for a scene in a room can only be ensured if a little more than 1200 lx is achieved at 4100 and 5000 K. At CCT = 3100 K, an illuminance of 1800 lx must be achieved for this.

7.5 Modelling with the Circadian Stimulus (CS)

In this section, the subject's mean responses on the scales of Figure 7.2 for brightness, VC, CP, and SP – as output parameters – are modelled using an *alternative* model. This alternative model [24] is based on a different set of three input parameters (as independent variables): (i) CCT; (ii) colour saturation level ΔC^* and (iii) the CS [25]. Another objective of this section is to numerically specify the input parameter ranges required to achieve the 'good' level of the output parameters, similar to Figures 7.6, 7.9, and 7.11.

The CS [25] was developed to predict the extent of nocturnal melatonin suppression induced by lighting. It was found that CS can also be used to predict other effects of light on humans. For example, receiving high levels of circadian light in the morning and throughout the day was associated with reduced latency to fall asleep, increased daytime alertness, reduced depression, and increased sleep quality [26, 27].

The ability of the CS parameter to predict the above effects can be attributed to the fact that the pathways responsible for the different effects of light on humans are intertwined in the brain [28]. The CS parameter combines and compresses the retinal photoreceptor signals, including the intrinsically photosensitive retinal ganglion cells (ipRGCs), the rods, and the three cone types (L, M, and S).

These advantageous properties of the CS parameter led to the idea of using it for the alternative model in this section (in combination with the above two other independent variables, CCT and ΔC^*) in order to predict the mean values of the four subjective variables (brightness *B*, VC, CP, and SP). It should be mentioned that the S-cone signal included in the CS parameter is also involved in the perception of brightness. In fact, CS correlates well ($r^2 = 0.97$ at 500 lx) with the magnitude of $(S/V)^{0.24}$ (S denotes S-cone signal and V denotes relative luminance) of the Fotios and Levermore brightness model [22].

The input parameter ΔC^* expresses the object saturation property of the light-source spectrum, see Section 7.3. In Section 7.4, the mean subjective ratings of SP and CP peaked at a low ($\Delta C^* = 1.1$) and moderate ($\Delta C^* = 4.2$) ΔC^* level, respectively. This result will also be expressed in the ΔC^* term of the alternative model in this section. The input parameter CCT significantly influences the output parameters CP and SP [29–31]. This is because the perception of the white tone (e.g. warm white, neutral white, and cool white) that illuminates the coloured objects and achromatic surfaces (e.g. white walls or grey furniture) in the scene interacts with the colour appearance of the coloured objects. The output parameters CP and SP reached their maximum in the CCT range between 4000 and 5600 K [29–31]. This effect is expressed using the CCT dependence of the alternative model in this section.

The model variables for B, VC, CP, and SP are scaled in this section in the same way as in Section 7.4 (see Figure 7.2): the interval scales for VC, CP, and SP have been labelled with the categories, 'excellent' (97.9); 'very good' (91.6); 'good' (79.6); 'moderate' (52.9); 'low' (41.2); 'poor' (26.5); and 'very poor' (12.8). These categories are important for interpreting the numerical scales of the model variables, especially the 'good' level (about 80) and the 'very good' level (about 90), which represent a *level of acceptance* and a *high-quality level*, respectively as optimisation targets of lighting design. The brightness scale (*B*) was not assigned such categories (see Figure 7.2, left). The *B* scale thus represents the magnitude of brightness perception of the illuminated scene without additional categorisation.

The input parameter ranges required to achieve the 'good' or 'very good' level of the output parameters (except B) were calculated. They will be presented in illustrative contour plots (ΔC^* vs. CS), and for CP and SP, also as a function of CCT. These contour diagrams (see below) are like the so-called *Kruithof diagram* [19], which illustrates the joint effect of the input parameters illuminance E_v (ordinate, in lx) versus CCT (abscissa, in K) on the output parameter 'pleasantness of the lighting situation'.

7.5.1 Calculation Method

The mean responses of all observers in the same experimental data set as in Section 7.3 were analysed in this section with respect to the combination of the input parameters CCT, ΔC^*, and CS as follows:

1) The input parameters (CCT, ΔC^*, and CS) were calculated for each of the 36 light-source spectra (see Figure 7.3). When calculating CS, the illuminances in Table 7.1 were used and *not* the illuminance at the cornea, the original input quantity of CS. The reason for this is that in lighting design, the level of horizontal illuminance measured on a *design plane* (where the coloured objects are placed, e.g. on a table) is *more often* used than the illuminance at the cornea, which depends on the seating position of the individual users, the room geometry, and the spatial luminous intensity distribution of the luminaires. It is important to use this horizontal illuminance when applying the model equations (see below).
2) The three input parameters (CCT, ΔC^*, and CS) were combined into four new model equations for B, VC, CP, and SP, see below.
3) The error defined by the square root of the sum of the squared differences between the model quantities and the mean subjective experimental data for each spectrum was minimised by varying the parameters of the model equations.

7.5.2 Brightness

Equation (7.6) contains the brightness model (*B*). The Pearson correlation coefficient between the model and the mean visual data was $r^2 = 0.92$. The dependence of the mean visually scaled brightness on CS is shown in Figure 7.12.

$$B = 202.89\,(\mathrm{CS})^2 - 11.835\,(\mathrm{CS}) + 26.624 \tag{7.6}$$

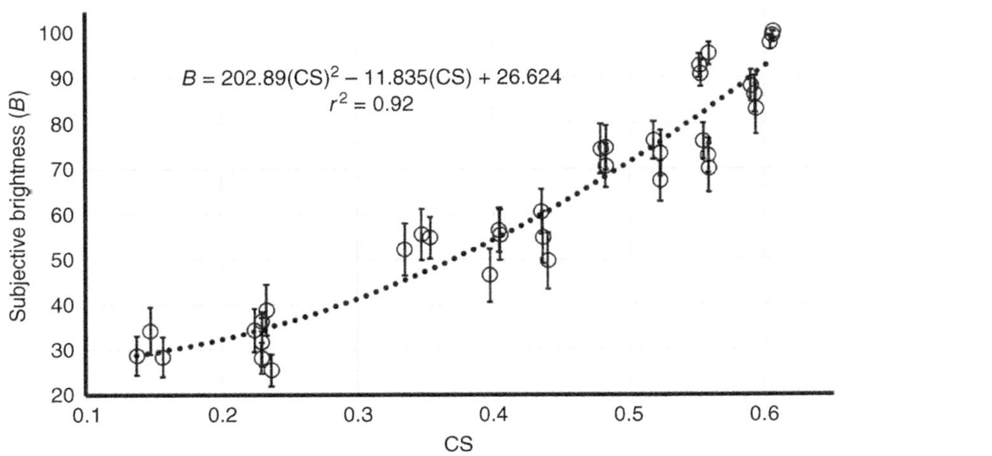

Figure 7.12 Dependence of the mean subjective brightness (*B*) on the circadian stimulus (CS) [25]. Dotted curve: The Brightness model curve of Eq. (7.6). Error bars are 95% -confidence intervals of the mean subjective B scale values. Source: Reproduced from [24] with permission from Lighting Research and Technology.

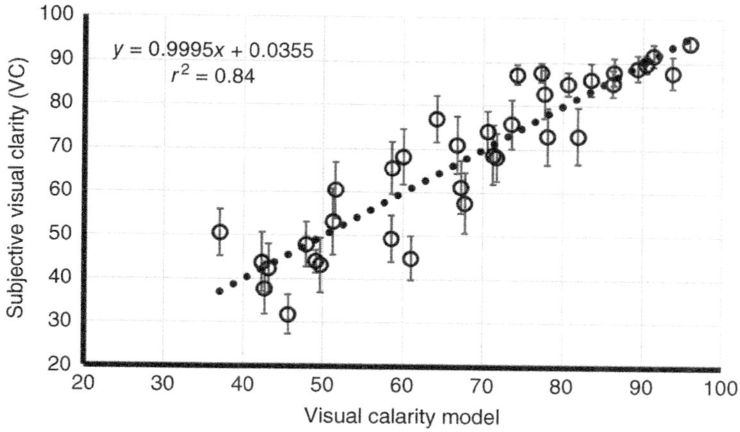

Figure 7.13 Dependence of mean subjective visual clarity (VC) on the model parameter of Eq. (7.7). Error bars are 95% confidence intervals of mean subjective VC scale scores. Source: Reproduced from [24] with permission from Lighting Research and Technology.

As can be seen from Figure 7.12, the mean subjective brightness ratings can be approximated by a quadratic function of CS. According to the definition of the signal compression function [25] built into CS, the absolute possible maximum of CS is the value of $CS_{max} = 0.7$, which, according to Eq. (7.6), leads to $B_{max} = 117.8$. The latter value (>100) allows for a further increase in perceived brightness at illuminances that were not investigated in the experiment (see Section 7.3) (i.e. $E_v > 1800$ lx).

7.5.3 Visual Clarity (VC)

The VC model with $r^2 = 0.84$ is shown in Eq. (7.7). As can be seen, VC was modelled by the summation of two quadratic terms of the input parameters ΔC^* and CS, respectively. The dependence of the mean visually scaled VC on the model of Eq. (7.7) is illustrated in Figure 7.13. The different levels of VC values are shown in a ΔC^* – CS contour plot according to Eq. (7.7) in Section 7.5.5.

$$VC = [0.00054037\,(\Delta C^*)^2 - 0.025396\,(\Delta C^*) + 1.00403]$$
$$[78.70534\,(CS)^2 + 56.74763\,(CS) + 32.15342] \quad (7.7)$$

7.5.4 Colour Preference (CP)

The CP data are modelled by Eq. (7.8) ($r^2 = 0.88$). As can be seen, the CP model is a combination of three quadratic terms: an additive CS term plus a ΔC^* term and a CCT term (CCT in K). As mentioned above, the CCT term expresses the subjective effect of the CCT of the light source on CP. The dependence of the mean visually scaled CP on the model parameter of Eq. (7.8) is shown in Figure 7.14. The different levels of CP values are shown in Section 7.5.5 in contour plots (ΔC^*– CS) according to Eq. (7.8) at different CCT values.

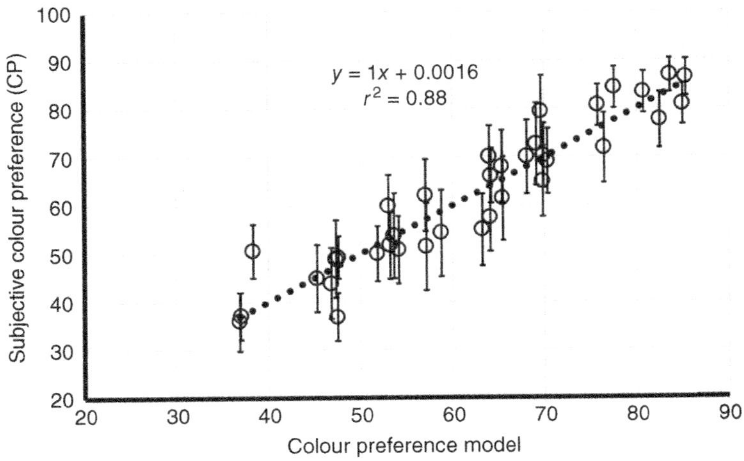

Figure 7.14 Dependence of the mean subjective colour preference (CP) on the model parameter of Eq. (7.8). Error bars are 95% confidence intervals of the mean subjective CP scale values. Source: Reproduced from [24] with permission from Lighting Research and Technology.

$$
\begin{aligned}
\mathrm{CP} = {} & [-6.5672441\,(\mathrm{CCT}/1000)^2 + 61.470144\,(\mathrm{CCT}/1000) - 110.68624] \\
& + \{[-0.00109479\,(\Delta C^*)^2 - 0.01126749\,(\Delta C^*) + 0.93721933] \\
& [-15.24209\,(\mathrm{CS})^2 + 110.9376\,(\mathrm{CS}) - 0.552997]\}
\end{aligned}
\tag{7.8}
$$

7.5.5 Scene Preference (SP)

The mean subjective responses of SP are modelled by Eq. (7.9) ($r^2 = 0.91$). As can be seen, the SP model is the sum of three additive quadratic terms: a CS term, a ΔC^* term, and a CCT term (CCT in K). Again, the CCT term expresses the subjective effect of the CCT of the light source on SP. The dependence of the mean visually scaled SP on the model of Eq. (7.9) is shown in Figure 7.15. The additive constants in the three quadratic terms are actually very similar (i.e. approximately equal to −39.39) for the mathematical reason of fitting to the visual SP scale. The different levels of SP values are visualised in this section as examples in selected ΔC^*-CS contour plots according to Eq. (7.9) corresponding to different CCT values.

$$
\begin{aligned}
\mathrm{SP} = {} & [-6.8586193\,(\mathrm{CCT}/1000)^2 + 61.480834\,(\mathrm{CCT}/1000) - 39.38999] \\
& + [-0.08361851\,(\Delta C^*)^2 - 0.66008326\,(\Delta C^*) - 39.389986] \\
& + [-71.276698\,(\mathrm{CS})^2 + 161.824364\,(\mathrm{CS}) - 39.389986]
\end{aligned}
\tag{7.9}
$$

7.5.6 Visualisation of the VC, CP, and SP Models in Contour Diagrams

Figure 7.16 shows the values of the VC model in a contour plot (ΔC^* versus CS) according to Eq. (7.7).

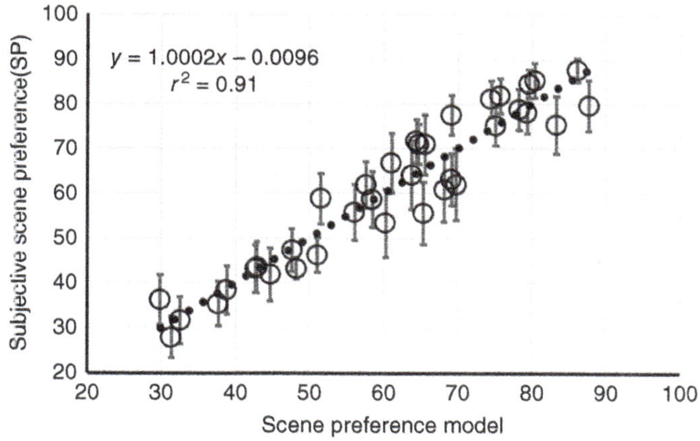

Figure 7.15 Dependence of mean subjective scene preference (SP) on the model parameter of equation (7.9). The error bars represent the 95% confidence intervals of the mean subjective SP scale scores. Source: Reproduced from [24] with permission from Lighting Research and Technology.

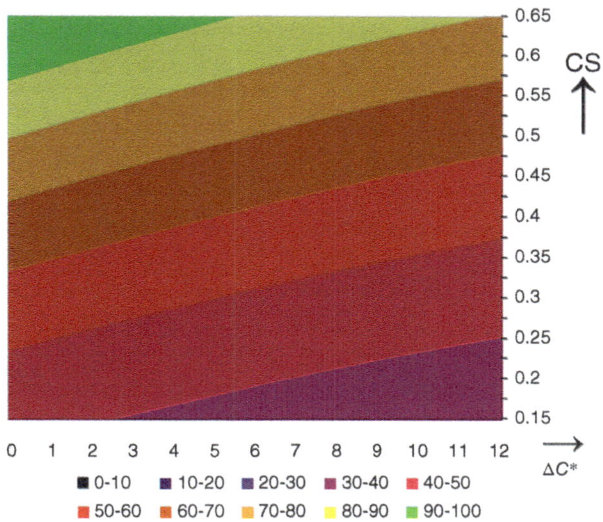

Figure 7.16 VC values of the visual clarity model in a ΔC^*-CS contour plot according to Eq. (7.7). Source: Reproduced from [24] with permission from Lighting Research and Technology.

In Figure 7.16, the two upper left areas (yellow and green) roughly correspond to the VC of the categories 'good' and 'very good' respectively. For an intuitive visualisation, the 'very good' value (91.6) was approximated by the value (90) and the 'good' value (79.6) by the value (80) in Figure 7.16. As can be seen from Figure 7.16, VC increases with increasing CS, while it decreases with increasing ΔC^*. The reason for the latter decrease is that the oversaturation of object colours by the spectrum of the

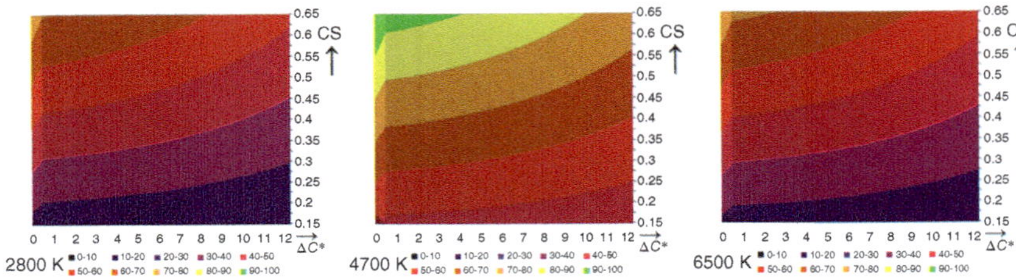

Figure 7.17 Different CP values in ΔC*-CS contour plots according to Eq. (7.8). Three examples of the following CCTs: 2800, 4700, and 6500 K. Source: Reproduced from [24] with permission from Lighting Research and Technology.

light source (with e.g. $\Delta C^* > 10$) destroys the reproduction of fine colour shades on object surfaces [23, 32] and reduces the ability to distinguish colours between adjacent coloured surfaces. Coloured surface structures are therefore less visible and VC is reduced [24]. Accordingly, with $\Delta C^* > 5.5$ it is not possible to achieve 'very good' VC, and with $\Delta C^* > 11.5$ it is not possible to achieve 'good' VC. With regard to CS, it is also evident from Figure 7.16 that it is not possible to ensure 'good' and 'very good' VC with CS <0.5 and CS <0.575, respectively.

Figure 7.17 shows the different levels of CP values according to Eq. (7.8) in three contour plots (three examples, similar to Figure 7.16) at three selected CCTs, 2800, 4700, and 6500 K.

As can be seen from Figure 7.17, we cannot achieve a 'good' CP at 2800 and 6500 K. The reason is the CCT dependency of CP already described in the introduction: for this type of lighting application, i.e. the context of a 'modern office', 2800 K are (subjectively) too warm and 6500 K too cold. In order to determine the optimal CCT for CP for this lighting application (modern office) using Eq. (7.8), we maximised the two upper left areas (green and yellow) in the ΔC*-CS diagram by changing the CCT value and achieved the optimal CCT of 4700 K, see the middle diagram in Figure 7.17 As can be seen from this middle diagram, it is not possible to achieve a 'very good' CP with $\Delta C^* > 4$ because the object colours become oversaturated when illuminated by such a light-source spectrum. To achieve a 'good' and a 'very good' CP at 4700 K, at least CS = 0.45 and CS = 0.55, respectively, are required.

Figure 7.18 visualises the different levels of SP values according to Eq. (7.9) in three contour plots (examples; similar to Figure 7.17) at three selected CCTs, 2800, 4500, and 6500 K.

As can be seen in Figure 7.18, we cannot achieve a 'good' SP at 2800 and 6500 K. The reason is – as in Figure 7.17 – the CCT dependence of the SP. To find the optimal CCT for SP for this type of lighting application (i.e. a 'modern office') using Eq. (7.9), the two upper left areas (green and yellow) in the ΔC*-CS *diagram* were maximised by changing the CCT value and the optimal CCT of 4500 K was achieved, see the middle diagram in Figure 7.18. As can be seen from the latter diagram, it is not possible to achieve 'very good' with $\Delta C^* > 4.5$ for the same reason as with CP (object colours become oversaturated). To achieve a 'good' and a 'very good' SP at 4500 K, at least CS = 0.475 and CS = 0.6, respectively, are needed.

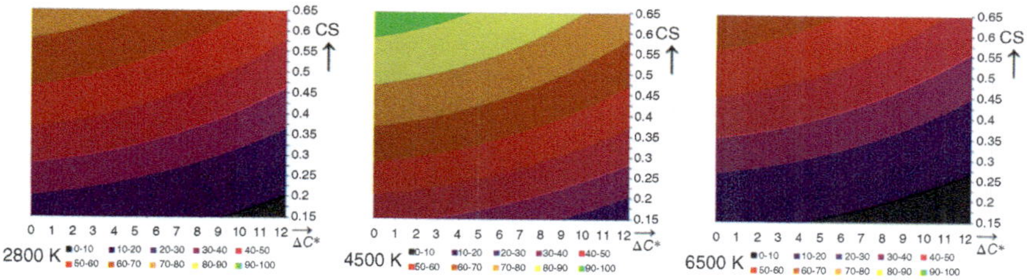

Figure 7.18 Different SP values in ΔC^*-CS contour plots according to Eq. (7.9). Three examples of the following CCTs: 2800, 4500, and 6500. Source: Reproduced from [24] with permission from Lighting Research and Technology.

7.6 Validation of the Light-Quality Models (in Section 7.4) in Three Museums in Japan

As a validation of the light-quality models (in Section 7.4), the results of an independent study [33] were used as follows. In these experiments, the effect of artificial lighting on observer's ratings in large museum spaces in three museums in Japan (National Museum of Western Art; The Aichi Prefectural Museum of Art; The Yamazaki Mazak Museum of Art) was investigated. The aim was to model the relationship between the observer's responses and the lighting parameters in these museums. In the study [33], 31 younger observers (aged 20–24) were invited to subjectively rate the lighting environment according to the four characteristics of 'comfort', 'clarity', 'preference', and 'warmth' on a scale between 1 and 6 (integers only; 1: very poor to 6: very good). The tests took place in the following three exhibition halls or with the following objects selected for focused viewing:

- $i = 1$: National Museum of Western Art: an exhibition hall for nineteenth and twentieth century paintings, selected painting: 'Morning on the Seine' (Claude Monet 1898).
- $i = 2$: The Aichi Prefectural Museum of Art: a permanent exhibition hall, selected painting: 'Colourful Flowers in Spring' (Rinen Hoshi 2016).
- $i = 3$: The Yamazaki Mazak Museum of Art: an exhibition hall for paintings, selected painting: 'Sailboat in the Morning' (Jean-Marc Nattier 1739).

The experimental process of subjectively rating 'comfort', 'clarity', 'preference', and 'warmth' had the following steps:

1. Introduction of the subjects.
2. Eye test, colour vision test.
3. Two minutes of adaptation in the respective exhibition hall.
4. Looking at the exhibition objects with 'moving, looking, and lingering' – focusing on the painting selected for the study (see above) at a viewing angle between 30° and 55°, for approx. five minutes.
5. Subjective ratings using the (ordinal) 6-point scales described above.

Changing the parameter 'museum or exhibition space' (1, 2, or 3, see above) had a significant influence (F-test, $p < 0.001$) on the mean values of all four subjectively

Table 7.2 Lighting parameters in the three test locations ($i = 1, 2,$ and 3) of the external study [33]. E_v: mean illuminance in the room (in lx). Co., Cl., Pr., Wa.: Mean ratings of all test subjects for 'comfort', 'clarity', 'preference', and 'warmth', respectively, and correlation coefficients with CP (Eq. (7.4)) and SP (Eq. (7.5)).

i	CCT (K)	x	y	R_a	R_9	ΔC^*	E_v	CP	SP	Co.	Cl.	Pr.	Wa.
1	2079	0.524	0.422	96	92	−0.08	84	7.8	9.0	3.77	3.39	3.74	5.03
2	2960	0.441	0.407	90	46	1.08	115	29.9	30.1	4.39	4.52	4.13	4.03
3	3568	0.413	0.420	93	88	0.41	538	61.2	64.0	4.87	5.00	4.74	3.52
Pearson correlation coefficient r with CP										0.99	0.95	1.00	−0.96
Pearson correlation coefficient r with SP										0.98	0.93	1.00	−0.95

Source: Darmstadt University of Technology.

scaled properties (comfort, clarity, preference, and warmth). Table 7.2 summarises the photometric parameters of the three test locations, the CP and SP values as well as the mean ratings of the test subjects [33].

Table 7.2 shows that the Pearson correlation coefficients between the model values CP (Eq. (7.4)) and SP (Eq. (7.5)) and the mean subjective ratings for 'comfort', 'clarity', and 'preference' are positive and high, respectively. A negative correlation was found with the 'warmth' property. There are also the following correlation coefficients [33] between the subjective attribute 'preference' and the other subjectively scaled attributes: preference-clarity ($r^2 = 0.94, p = 0.23$), preference-warmth ($r^2 = -0.95, p = 0.20$), and preference-comfort ($r^2 = 0.98, p = 0.13$). This means that 'warmer' scenes are not preferred, which is in line with the predictions of the model in this chapter.

7.7 Summary

In a historical view, the relationship between the illuminance level and contrast of the objects seen (at the input of the model) and visual performance (at the output of the model: reaction time, reading speed) could be modelled mathematically with closed formulae in the years around 1981–1991. The relationship between the perceptual psychological and physiological parameters such as 'visual attraction and complexity' or 'colour quality' (at the output) and the objective lighting parameters (at the input) used to be more qualitative and empirical.

In this chapter, the findings of the colour quality research of the last 15 years were combined with the possible model approaches from the literature of the last decades on brightness and VC. This has made it possible to set up a mathematically closed model approach for the four properties brightness B, VC, CP, and SP through extensive visual experiments in a real room.

An alternative modelling with the CS was presented in Section 7.5. In this alternative model, the output parameters B, VC, CP, and SP were modelled using three input parameters: CCT, a parameter of the object saturation property of the light-source spectrum (ΔC^*) and the CS. To foster lighting design, the horizontal illuminance in the working plane is to be used as input here – for the calculation of CS – and not the illuminance at the user's cornea.

Brightness was approximated by a quadratic function of CS in Section 7.5, with further experiments needed to cover the upper CS range (CS >0.6). VCwas modelled by the product of two quadratic terms of the input parameters ΔC^* and CS and showed an increasing trend with increasing CS and a decreasing trend with increasing ΔC^* (especially at high oversaturation levels). CP was modelled by an additive quadratic CS term plus the product of a quadratic ΔC^* term and a quadratic CCT term. The maximum CP occurred at 4700 K. Even at this CCT, it was not possible to achieve a 'very good' CP with $\Delta C^* > 4$ because then the object colours became oversaturated.

SP was modelled by the sum of three additive quadratic terms, a CS term, a ΔC^* term, and a CCT term. The maximum SP occurred at 4500 K. Even at this CCT, it was not possible – for the same reason – to achieve a 'very good' SP with $\Delta C^* > 4.5$. In the case of CP and SP, the CCT term expresses the subjective preference of the CCT of the white tone.

The ΔC^*-CS contour diagrams in Figs. 7.16–7.18, which result from the modelling Eqs. (7.7)–(7.9), are intended to support the lighting design as follows. For CP and SP, a CCT input value must first be specified in the interactive version of the diagram [34] (optimal: 4700 K for CP and 4500 K for SP). The ranges 'very good' or 'good' can be identified as green or yellow areas in the diagram. Subsequently, the required ranges of the input parameters ΔC^* and CS can be determined and the illumination spectrum (e.g. of a multichannel LED luminaire) and the horizontal illuminance in the working plane can be designed.

The VC Eq. (7.7) can be used for optimisation when the visibility of coloured surface structures is important, e.g. for highlighting structured coloured objects, e.g. in retail. The CP condition (designed for the highest colour quality), on the other hand, could play an important role, for example, in the illumination of colourful objects in an exhibition hall, while the SP criterion is to be applied in the most general case. After the lighting installation has been designed, the electrical power (P_{el}) becomes known. Thus, special key figures for a kind of 'specific electrical efficiency' can also be calculated for B, VC, CP, and SP by dividing the latter values by the value of P_{el}.

It should also be noted that the viewing conditions of the experiment (see Section 7.3) were chosen to represent the context of a formal or official situation or a working environment, e.g. a 'modern office', an exhibition hall, or a conference room. To design a more relaxed atmosphere, e.g. for creative mental activities, staying in the office with colleagues after working hours or for romantic evenings, dinners, and living rooms (in the evening), the models of this chapter are not valid and further experiments are needed. This is an important limitation for the applicability of all equations of this chapter as well as the contour plots in Figure 7.16–7.18.

Another limitation is that these equations ignore at least two other input parameters, which are known to influence the user preference of a lighting situation (in particular, the output parameter SP): (i) the ratio of direct and indirect illuminances

[35]; and (ii) a parameter of dynamic lighting (for example when the CCT or illuminance is deliberately changed during the working day). The effect of these parameters on SP should also be investigated in the future in interaction with the other input parameters.

An independent validation experiment in three museums in Japan [33] with three lighting types and contexts showed a high correlation between the visual perception and scaling of the preference property of (younger) East-Asian visitors of museums and the numerical values CP and SP predicted by the present model approach. (The authors of this book chapter and the authors of the study [33] did not know each other beforehand and did not communicate with each other.)

References

1 Luckiesh, M. and Moss, F.K. (1937). *The Science of Seeing*. New York: D Van Nostrand Co.
2 Blackwell, H.R. (1972). A human factor approach to lighting recommendations and standards. Technology for man 72. In: *Proceedings of the 16th Annual Meeting* (17–19 October 1972), 441–451. Los Angeles.
3 Flynn, J.E., Spencer, T.J., Martyniuk, O., and Hendrick, C. (1973). Interim study of procedures for investigating the effect of light on impression and behaviour. *J. Illum. Eng. Soc.* 3 (1): 87–94.
4 Flynn, J.E., AIA/FIES, Hendrick, C. et al. (1979). A guide to methodology procedures for measuring subjective impressions in lighting. *J. Illum. Eng. Soc.* 8 (2): 95–110.
5 Veitch, J.A. and Newsham, G.R. (1998). Lighting quality and energy-efficiency effects on task performance, mood, health, satisfaction and comfort. *J. Illum. Eng. Soc.* 27: 107–129.
6 Guo, X. and Houser, K.W. (2004). A review of colour rendering indices and their application to commercial light sources. *Light. Res. Technol.* 36: 183–199.
7 Khanh, T.Q., Vinh, Q.T., and Bodrogi, P. (2017). A numerical analysis of recent colour rendition metrics. *Light. Res. Technol.* 49: 561–573.
8 CIE (2017). *Color Fidelity Index for Accurate Scientific Use*, CIE Publication 224: 2017. Vienna: CIE.
9 David, A., Fini, P.T., Houser, K.W., and Whitehead, L. (2015). Development of the IES method for evaluating the colour rendition of light sources. *Opt. Express* 23: 15888–15906.
10 Davis, W. and Ohno, Y. (2010). Color quality scale. *Opt. Eng.* 49: 033602.
11 Houser, K.W., Wei, M., David, A. et al. (2013). Review of measures for light-source colour rendition and considerations for a two-measure system for characterizing colour rendition. *Opt. Express* 21: 10393.
12 Khanh, T.Q., Bodrogi, P., Vinh, Q.T. et al. (2018). Colour preference, naturalness, vividness and colour quality metrics, Part 4: experiments with still life arrangements at different correlated colour temperatures. *Light. Res. Technol.* 50: 862–879.

13 Khanh, T.Q., Bodrogi, P., Guo, X., and Anh, P.Q. (2018). Towards a user preference model for interior lighting - Part 1: concept of the user preference model and experimental method. *Light. Res. Technol.* 51 (7): 1014–1029.

14 Khanh, T.Q., Bodrogi, P., Guo, X., and Anh, P.Q. (2018). Towards a user preference model for interior lighting - Part 2: Experimental results and modelling. *Light. Res. Technol.* 51 (7): 1030–1043.

15 Bodmann, H.W., Söllner, G., and Voit, E. (1963). Assessment of illuminance levels for different types of illuminant. In: *Proceedings of the CIE*, vol. C, 502–509. Vienna.

16 Boyce, P.R. (1970). The influence of illuminance level on prolonged work performance. *Light. Res. Technol.* 2 (2): 74–94.

17 Fotios, S. (2017). A revised Kruithof graph based on empirical data. *Leukos* 13 (1): 3–17.

18 Viénot, F., Durand, M.-L., and Mahler, E. (2009). Kruithof's rule revisited using LED illumination. *J. Mod. Opt.* 56 (13): 1433–1446.

19 Kruithof, A.A. (1941). Tubular luminescence lamps for general illumination. *Philips Tech. Rev.* 6 (3): 65–73.

20 Bodrogi, P., Brückner, S., Krause, N., and Khanh, T.Q. (2014). Semantic interpretation of color differences and color-rendering indices. *Color Res. Appl.* 39: 252–262.

21 Ichikawa, H., Tanabe, S., and Hukami, K. (1983). *Standard Pseudoisochromatic Plates for Acquired Color Vision Defects, Part II*. Tokyo: Igaku-Shoin Medical Publishers.

22 Fotios, S.A. and Levermore, G.J. (1998). Chromatic effect on apparent brightness in interior spaces, II: SWS lumens model. *Light. Res. Technol.* 30: 103–106.

23 Royer, M.P., Houser, K.W., and Wilkerson, A.M. (2012). Color discrimination capability under highly structured spectra. *Color Res. Appl.* 37: 441–449.

24 Khanh, T.Q., Bodrogi, P., and Guo, X. (2020). Towards a user preference model for interior lighting - Part 3: an alternative model. *Light. Res. Technol.* 52 (2): 189–201.

25 Rea, M.S. and Figueiro, M.G. (2018). Light as a circadian stimulus for architectural lighting. *Light. Res. Technol.* 50: 497–510.

26 Figueiro, M.G., Steverson, B., Heerwagen, J. et al. (2017). The impact of daytime light exposures on sleep and mood in office workers. *Sleep Health* 3: 204–215.

27 Figueiro, M.G., Kalsher, M., Steverson, B.C. et al. (2018). Circadian-effective light and its impact on alertness in office workers. *Light. Res. Technol.* 51: 171–183.

28 Rautkylä, E., Puolakka, M., and Halonen, L. (2012). Alerting effects of daytime light exposure - a proposed link between light exposure and brain mechanisms. *Light. Res. Technol.* 44: 238–252.

29 Khanh, T.Q., Bodrogi, P., Vinh, Q.T., and Stojanovic, D. (2016). Colour preference, naturalness, vividness and colour quality metrics - Part 1: experiments in a real room. *Light. Res. Technol.* 49 (6): 697–713.

30 Bodrogi, P., Lin, Y., Xiao, X. et al. (2015). Intercultural observer preference for perceived illumination chromaticity for different coloured object scenes. *Light. Res. Technol.* 49 (3): 305–315.

31 Khanh, T.Q., Bodrogi, P., Guo, X. et al. (2017). Colour preference, naturalness, vividness and colour quality metrics, Part 5: a colour preference experiment at 2000 lx in a real room. *Light. Res. Technol.* 51 (2): 262–279. First Published 20 October 2017. https://doi.org/10.1177/1477153517737133.

32 Bodrogi, P., Guo, X. and Khanh, T.Q. (2016). Colour surface rendering: a new aspect of light source colour quality for museum lighting. In: *Proceedings of the 4th CIE Symposium on Colour and Visual Appearance* (5–7 September), 198–207. Prague.

33 Wang, Z., Nagai, Y., Liu, J. et al. (2020). Artificial lighting environment evaluation of the Japan Museum of Art based on the emotional response of observers. *App. Sci.* 10: 1121. https://doi.org/10.3390/app10031121.

34 Khanh, T.Q. and Bodrogi, P. (2019). Online spreadsheet with the contour map diagrams. TU Darmstadt. https://www.dropbox.com/s/wrlri0u5sc98nrf/deltaCs-vs-CS-diagram-for-download.xlsx?dl=0.

35 Houser, K.W. (2002). The subjective response to linear fluorescent, direct/indirect lighting systems. *Light. Res. Technol.* 34 (3): 243–264.

8

Correlation Analysis of HCL Parameters and Consequences for the Measurement Methods of Non-visual Effects

8.1 General Consideration of the Correlation of the Parameters for Visual Performance, Colour Quality, and Non-visual Effects

8.1.1 Introduction

Today's lighting technology for interiors such as offices, industrial halls, schools, hospitals, and salesrooms is in a dynamic transition from conventional luminaire systems via luminaires with white light-emitting diodes (LEDs) with a fixed luminous intensity distribution and a fixed colour temperature to the automated and intelligent LED luminaires with sensor-based and time-dependent lighting control and regulation. Lighting designers, lighting architects, and luminaire developers have talked about individual lighting quality features such as glare, luminance, colour rendering, and homogeneity in recent decades. Today, concepts for integrative lighting technology for user-dependent, individual, weather-dependent, and application-specific lighting applications are currently being created and practised on a trial basis. This complexity of tasks for *Human-Centric Lighting* (HCL) can no longer be described and fulfilled only by the application of lighting quality parameters. In principle, this complexity has been well understood within the German lighting industry [1], in application-oriented lighting science [2], and in modern concepts of lighting [3, 4]. The basic idea of placing humans with all their eye-physiological, psychological, and health-related characteristics into the centre of lighting technology comprises three main components, which are illustrated in Figure 8.1 (see [3]).

For the description of the photometric, colorimetric, and melanopic effects in Figure 8.1, scientists in lighting technology and medicine have defined a series of parameters of different significance for the different aspects. These parameters are described below. More comprehensive conceptual representations and interpretations were given in Chapter 2 of this book.

- In 1924, the $V(\lambda)$-function was defined for the spectral luminous sensitivity of the human visual system for daytime vision and a 2° visual field. This spectral function defines the photometric parameters such as illuminance, luminance, glare (uni-

Human Centric Integrative Lighting: Technology, Perception, Non-Visual Effects, First Edition.
Tran Quoc Khanh, Peter Bodrogi, and Trinh Quang Vinh.

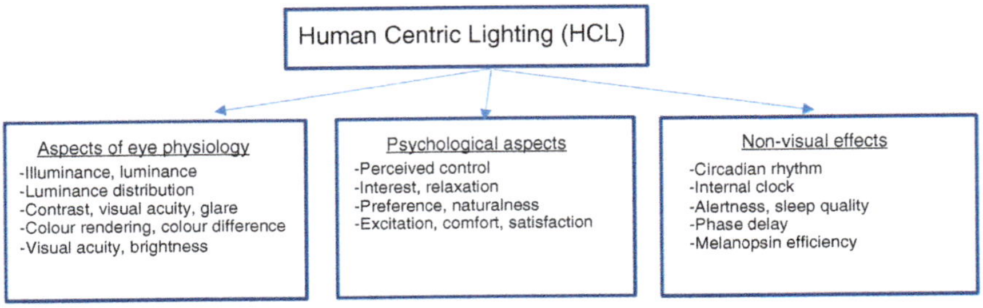

Figure 8.1 The three main components of human-centric lighting (HCL) technology. Source: TU Darmstadt.

fied glare rating (UGR) and TI formulae), visual acuity, contrast, and homogeneity, which form the foundation of today's photometric standards and regulations of international and national organisations (IEC, ISO, CIE, DIN, ANSI, etc.).

- Since this $V(\lambda)$-function only represents the achromatic part of the perceived scene, generations of scientists have tried to describe the *actual* perceived brightness with experimental and mathematical models. The most relevant results from today's point of view are presented in the publications [5–8]. The most recent investigations into the brightness and visual clarity aspects were carried out in [9] (see Chapter 7 of this book) and their results are modelled there. In principle, the processing chain of brightness perception in the visual brain can be illustrated as in Figure 8.2. According to this, when perceiving an illuminated room, the entire brightness signal consists of the luminance channel 'L + M' equivalent to the $V(\lambda)$-function channel, the two opposing channels 'L – M' and 'L + M – S', the channel of rod signals (R) as well as the signals generated in the intrinsically photosensitive melanopsin-containing retinal ganglion cells (ipRGC).

The following mathematical formulations have been proposed so far for the characterisation of perceived brightness (with R: rod signal; S: S-cone signal; Mel: ipRGC signal; A, A_S, A_{mel}, A_R, and γ: constants).

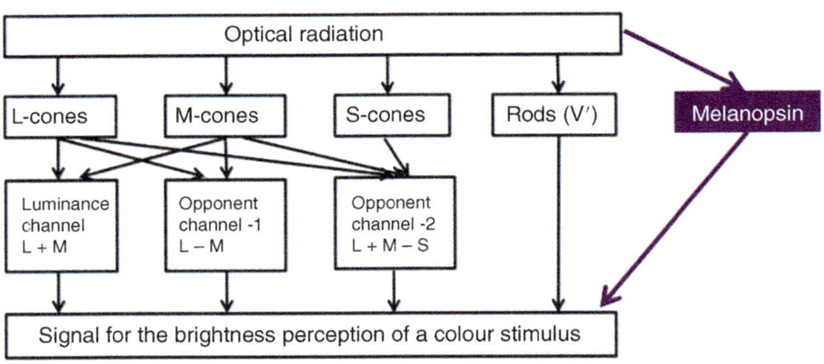

Figure 8.2 Processes of the formation of the signal responsible for brightness perception. Source: TU Darmstadt.

From Berman et al. [5]:

$$B = L_v\,(R/V)^{0.5} \tag{8.1}$$

By Fotios and Levermore [6]:

$$B = L_v\,(S/V)^{0.24} \tag{8.2}$$

From Bodrogi and Khanh [9]:

$$B = \log_{10}(L_v) + A\log_{10}\left[(A_S(S/V)^{\gamma} + A_{mel}(Mel/V)^{\gamma} + A_R(R/V)^{\gamma}\right] \tag{8.3}$$

Brightness and the retinal receptor's contributions to brightness are described in detail in Chapter 5 of this book. Analysing Eqs. (8.1)–(8.3) and Figure 8.2, the following mechanisms and signal contributions can be identified:

(a) The luminance channel is embodied here by (photopic) luminance L_v. This is only a part of the signal contributions to the perception of brightness (and visual clarity) and not the whole perceptual product;
(b) The signal of the (blue-sensitive) S-cones is weighted with the weighting factor A_S;
(c) The signal 'Mel' (i.e. 'melanopsin-active') of the light-sensitive ganglion cells is evaluated with the weighting factor A_{mel}
(d) The signal R ('rods') is evaluated with the weighting factor A_R. In the higher photopic range (daylight vision from approx. $L_v = 60\,cd\,m^{-2}$) these R signals are very weak or non-existent.
(e) The factor A weights the combined contribution of the S, Mel, and R signals compared to the luminance channel or the luminance contribution L_v.
(f) The exponent γ represents the *strength of signal compression* within the combination of the S, Mel, and R signals and equals 0.33 in the present Eq. (8.3).

The formulas of Berman et al. and Fotios and Levermore have in common that they start from the photopic luminance L_v and correct this signal by multiplying it by a further term, which in the case of Berman et al. consists of a power function of the ratio of the signals of the rods R to the photopic luminance signal V (see Eq. (8.1)) and in the case of Fotios and Levermore of a power function of the ratio of the signals of the blue-sensitive cones S to the photopic luminance signal (see Eq. (8.2)). Thus, Berman et al. or Fotios and Levermore assume that the opponent channels 'L − M' or 'L + M − S' or the photosensitive ganglion cells, which have become known since 2001, do not contribute to brightness perception. The formula of Bodrogi and Khanh [9] in 2017 tried to improve these two views formulated before 1998 by the signal of the photosensitive ganglion cells ('*Mel*').

In a PhD thesis [10] at the TU Darmstadt in 2017, W. Pepler was able to prove experimentally with white light spectra in a room that the formulas of Berman et al. and Fotios and Levermore are superior to all brightness models given by the CIE in [8] until 2011, as well as also superior to the $V(\lambda)$-weighted quantity (photopic) *luminance*.

The colour-rendering index R_a was defined for the first time in 1965 by the CIE and was improved several times until 1995 (see [11, 12]). Through a fundamentally new colour space (CIECAM02-UCS), the colour matching functions for a field of view of 10° and with the help of 99 representative colour samples, the new CIE colour-rendering index R_f was introduced for scientific purposes in 2017 (see [13]).

In 2015, DIN SPEC 5031-100:2015-08 [14] specified the spectral evaluation of optical radiation in the visible range for the assessment of melanopic light effects. These effects influence sleep quality, concentration, well-being, alertness, and *quality of life* of light users. In this work [14], terms and action spectra (e.g. $M_{mel}(\lambda)$) were introduced to characterise *melanopic light effects*. The factor a_{mel} was defined – in this context – according to Eq. (8.4):

$$a_{mel} = \frac{\int S(\lambda) \cdot M_{mel}(\lambda)}{\int S(\lambda) \cdot V(\lambda)} \tag{8.4}$$

The integrations in Eq. (8.4) range from 380 to 780 nm. In summary, from today's perspective, the parameters and metrics in Figure 8.3 are available for a comprehensive, HCL-oriented characterisation of light systems.

The following questions are important for the use of the parameters in Figure 8.3 for the assessment of future HCL-compatible intelligent luminaires and HCL-oriented lighting conditions by lighting designers, building managers, luminaire developers, and lighting companies:

(1) Which parameters should be used for the practical characterisation of colour rendering and how do the colour-rendering indices relate to the parameters for melanopic effects, brightness, and luminous efficacy?
(2) Which brightness parameters should be used and what is the relationship between these parameters and colour temperature as well as the melanopic effect? Does a lighting system that generates more melanopic effects provide more or less brightness?
(3) How does the melanopic effect relate to colour rendering? Is it possible to improve both the *melanopic effect* and the *colour rendering* within a single lighting system?

The authors of this book attempt to answer such existential questions for lighting technology and luminaire development. For this purpose, the authors were provided with a set of 387 *absolutely measured* spectra of lamps and luminaires from various European or national, leading state institutes, and light laboratories within the scope of a research project in 2017. The achievable measurement accuracy and reliability

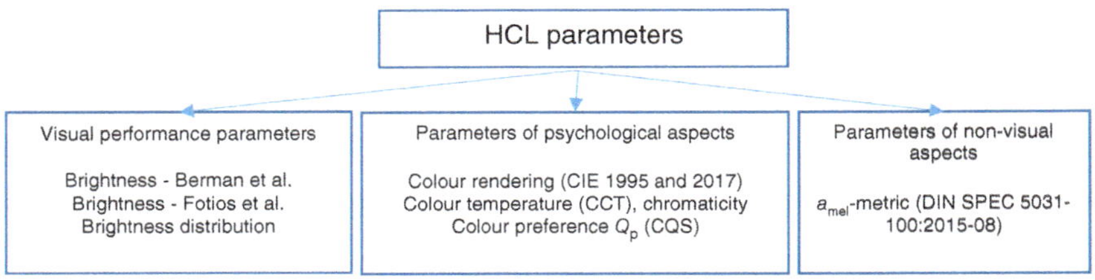

Figure 8.3 Parameters for a comprehensive HCL-oriented characterisation of lighting systems. Source: TU Darmstadt.

Table 8.1 Number of the different light source types in the database of 387 spectra in Figure 8.4.

Ord.	Lamp types	Count
1	Halogen–tungsten lamps	11
2	Compact fluorescent lamps	4
3	Linear fluorescent lamps	38
4	High-pressure discharge lamps	6
5	Phosphor-converted white LEDs	328

Source: TU Darmstadt.

of the light source data in these light laboratories are therefore very good. Table 8.1 shows the structure of these data.

Since the spectra of halogen incandescent lamps and fluorescent lamps were limited in their variety in the past and the transition from conventional light source technology to LED lighting technology has already been completed to a large extent, the phosphor-converted white LEDs formed the main part of the light sources examined. These relative spectra are illustrated in Figure 8.4. In this diagram, the variation of the peak wavelengths of the blue LEDs between 435 and 465 nm as well as the great variety of phosphor systems between 550 nm (yttrium aluminum garnet (YAG) phosphors) and 650 nm (red phosphors) can be seen when looking at the spectra of the white LEDs.

These 387 spectra and the corresponding electrical data were used in the time range of 2019 to calculate the relative spectra as well as the parameters of the melanopic effect, the brightness perception according to Berman et al. and Fotios and Levermore, the colour temperature and the colour-rendering indices. The correlation coefficients among these parameters are shown in Table 8.2.

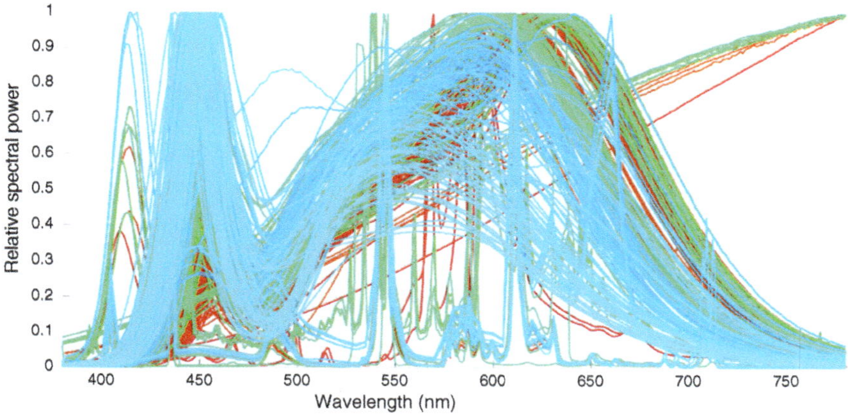

Figure 8.4 Relative spectra of the 387 light sources in the correlation analysis. Source: TU Darmstadt.

Table 8.2 Correlation coefficients between the parameters of the 387 light sources examined.

Parameter	R_a	R_f	R_g	a_{mel}	$L_{eq,Berman}$	$L_{eq,Fotios}$	CCT
R_a	1.000	0.880	0.083	0.099	0.075	0.137	0.178
R_f	0.880	1.000	0.200	0.083	0.075	0.129	0.250
R_g	0.083	0.200	1.000	0.007	0.007	0.074	0.021
a_{mel}	0.099	0.083	0.007	1.000	0.974	0.743	0.736
$L_{eq,Berman}$	0.075	0.075	0.007	0.974	1.000	0.772	0.782
$L_{eq,Fotios}$	0.137	0.129	0.074	0.743	0.772	1.000	0.736
CCT	0.178	0.250	0.021	0.736	0.782	0.736	1.000

Source: TU Darmstadt.

8.1.2 Evaluation of the Colour-Rendering Indices

Since 1995 until today, the general colour-rendering index R_a has been the dominant colour quality factor in all matters of lighting technology. After 22 years of scientific work, the International Commission on Illumination (CIE) introduced the colour-rendering index R_f in 2017. The mathematical relationship between these two *colour-rendering indices* is shown in Figure 8.5. With a correlation coefficient of $r^2 = 0.88$, they often yield similar results and can be assumed to be equivalent in case of those light sources that exhibit smooth spectra that do not contain sharp local maxima like in red green blue (RGB) LEDs.

Since the introduction of the term *colour gamut index* R_g by lighting scientists and engineers in TM 30-15 [15], colour quality has been substantially expanded to include the aspect of colour gamut. This aspect is a measure of whether a light source with its current spectral distribution can reproduce the various colour objects in the illuminated space more or less saturated (or with higher or lower colourfulness)

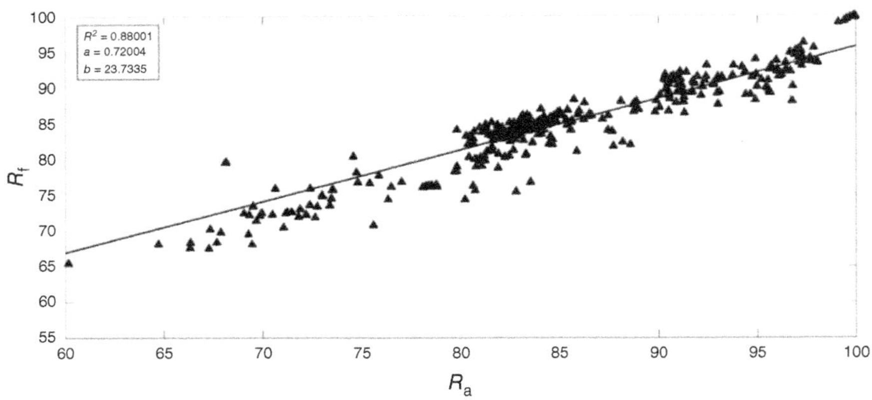

Figure 8.5 Correlation between the two colour-rendering indices CIE R_a (1995) and CIE R_f (2017). Source: TU Darmstadt.

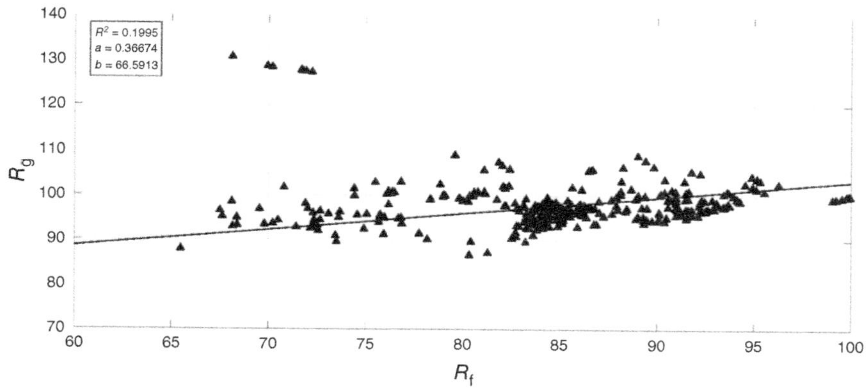

Figure 8.6 Correlation between the colour gamut index R_g and the colour-rendering index R_f (2017). Source: TU Darmstadt.

compared to the reference light source of the same colour temperature. This raises the question of whether a luminaire designer or an LED system designer can develop the colour rendering in favour of or against *colour gamut*. Based on the calculation of the 387 light sources, the correlation between the colour-rendering index R_f and the colour gamut index value R_g is shown in Figure 8.6. The correlation coefficient of only $r^2 = 0.2$ and the course of the 387 data points makes it clear that the two colour quality characteristics can be designed or optimised independently. The colour-rendering index can be varied over a wide range of values without substantially changing the colour gamut index.

8.1.3 Assessments of the Brightness Parameters

Figure 8.7 shows the correlation between the two *brightness metrics* according to Berman et al. (Eq. (8.1)) and Fotios and Levermore (Eq. (8.2)). The correlation coefficient $r^2 = 0.77$ shows the good linear relationship between the two metrics. Therefore, mathematically, these two indices are rather equivalent. Based on the study by Pepler [10], which proved a slightly better agreement between the formula of Fotios and Levermore with his own experimental data and, based on the fact that the formula of Fotios and Levermore (Eq. (8.2)) includes the signals of blue cones, it seems more sensible for the authors of this book to favour the formula of Fotios and Levermore when considering lighting technology in the photopic range indoors.

For many years, lighting designers, lighting users, and luminaire developers have reported their visual experience that light sources with a higher correlated colour temperature (CCT) appear brighter at the same luminance. This visual perception can be confirmed by the correlation between the brightness parameter according to Fotios and Levermore and CCT in Figure 8.8. The correlation coefficient $r^2 = 0.736$ clearly shows this tendency. A means of moderately increasing the brightness in an office room or a retirement home can thus be partially achieved by increasing CCT.

One of the biggest concerns of luminaire developers and serious luminaire manufacturer companies at the moment is how to increase both the melanopic effect

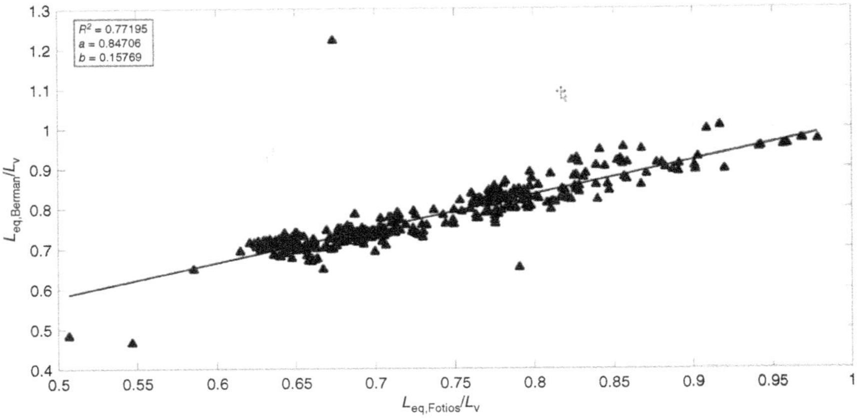

Figure 8.7 Correlation between brightness values according to the formulas of Fotios and Levermore (Eq. (8.2)) and Berman et al. (Eq. (8.1)). Source: TU Darmstadt.

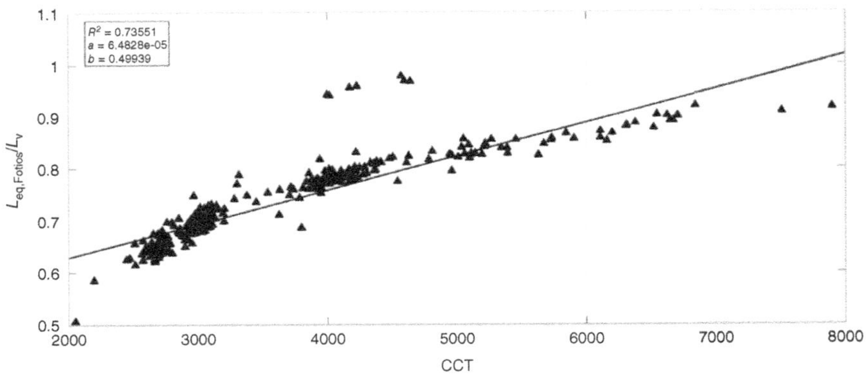

Figure 8.8 Correlation between correlated colour temperature (CCT in K) and the brightness metric according to Fotios and Levermore (Eq. (8.2)). Source: TU Darmstadt.

(a_{mel}-value) and brightness with a single luminaire system. In other way words, how to increase the melanopic effect and *visual performance* (brightness + visual clarity) at the same time? The concern is rooted in how to design an HCL luminaire with a high melanopic effect in the sense of a high a_{mel}-value (or also in the context of the new CIE metric MEDI, see Section 2.3.3 of this book) with a simultaneously high visual efficiency (brightness yield and luminous efficiency). It is also frequently asked how arguments could be made in the course of communication with potential building users and luminaire customers.

This is the key point of today's communication between the lighting industry and (professional) lighting users as well as the marketing strategy of modern HCL lighting technology. Figure 8.9 illustrates the correlation between the a_{mel}-value and *brightness* according to Fotios and Levermore for the 387 light sources. The high correlation coefficient r^2 =0.743 and the course of the 387 points show that

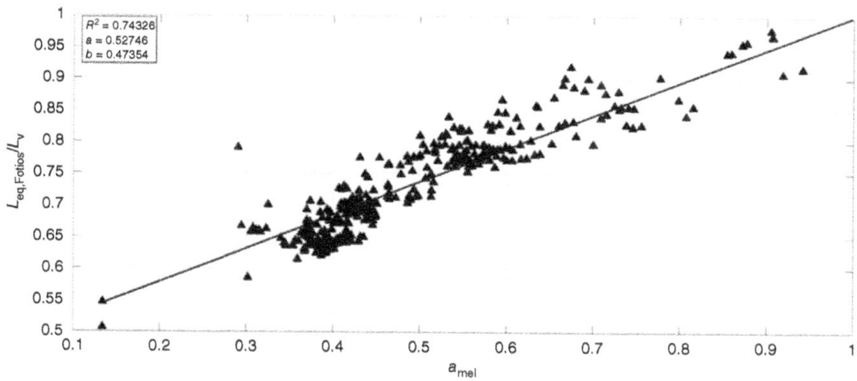

Figure 8.9 Correlation between the a_{mel}-value and brightness according to Fotios and Levermore for the 387 light sources in Figure 8.4. Source: TU Darmstadt.

the melanopic effect with the a_{mel}-value correlates well with the brightness in the sense of Fotios (Eq. (8.2)) and in the sense of Figure 8.2. If brightness is no longer evaluated *only* by the $V(\lambda)$-weighted photometric parameters such as luminance, illuminance, or luminous flux, but more according to the characteristics of the human visual system (Figure 8.2) then it will turn out that lighting systems with a high a_{mel}-value (and MEDI-value) also have a high *brightness yield* (i.e. *perceived brightness* divided by the electrical power of the luminaire).

8.1.4 Melanopic Effect and Colour Rendering

While the brightness according to Fotios and the a_{mel}-value show a good correlation, the a_{mel}-values of the melanopic effect and the colour-rendering index values R_a of the 387 light sources show no correlation ($r^2 = 0.099$), as can be seen in Figure 8.10. One can also formulate it differently, in that the variation of the a_{mel}-values (e.g. in

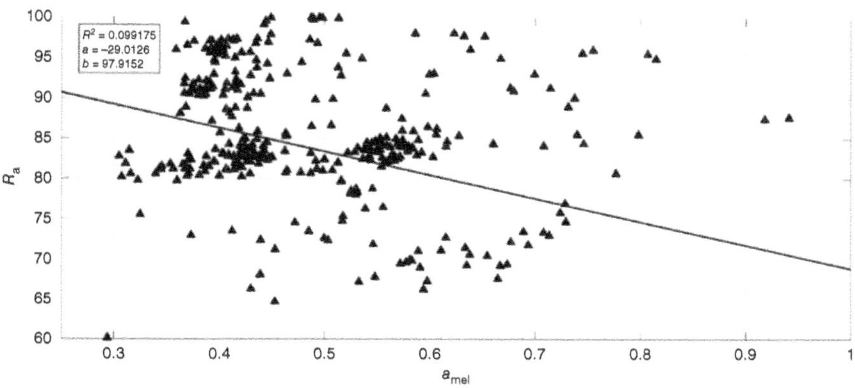

Figure 8.10 Correlation between the a_{mel}-value and the colour-rendering index R_a for the 387 light sources in Figure 8.4. Source: TU Darmstadt.

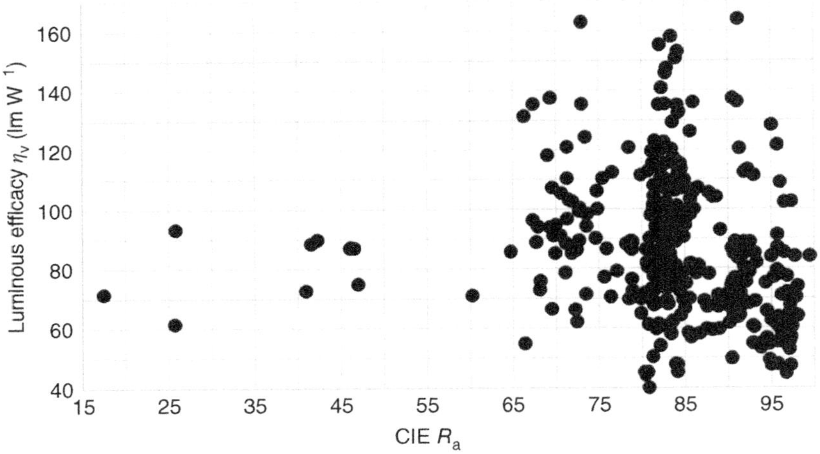

Figure 8.11 Luminous efficacy (in lm W^{-1}) and colour-rendering index CIE R_a of the 387 light sources in Figure 8.4. Source: TU Darmstadt.

the case of biodynamic light over the time of day or depending on the weather) has little influence on the colour-rendering property of the light source under consideration. One can also conclude that a light source with a high a_{mel}-value can have very good or low colour rendering. In a positive sense, one can also think that a light source with a high a_{mel}-value for daytime and a low a_{mel}-value for evening hours can be designed for high colour rendering in both cases.

The luminous efficacies (in lm W^{-1}) were determined from the spectra of the 387 light sources and their electrical data and compared with the colour-rendering indices of the same light sources in Figure 8.11.

Figure 8.11 shows the following:

- Luminous efficacy is between 40 and 160 lm W^{-1} – depending on the type of light source.
- The colour-rendering indices R_a are predominantly between 80 and 98 with a high concentration between 80 and 85.
- The most important finding is that there are light sources with a comparatively high luminous efficacy of around 120–160 lm W^{-1} and a high colour rendering of more than R_a = 90. *High luminous efficacy and simultaneously high colour rendering are technologically possible and do not contradict each other.* This is a very important statement for the lighting industry, facility managers, lighting designers, and politicians. The developers of light sources and luminaires must certainly optimise optics, electronics, *thermal management,* and the choice of light sources comprehensively, at the same time.

8.1.5 Correlation Between Further Parameters of Visual Performance, Colour Quality, and Non-visual Effects

In Section 8.1, the parameters brightness, colour rendering, colour gamut, and a_{mel}-factor (as they are generally anchored in the recommendations of CIE and

DIN-EN [12–14]) were considered in correlation to each other. In this Section, further parameters will be analysed. These parameters have not yet found their way into the standards and are taken from recent models of lighting literature.

The output parameters of the HCL system in Figure 8.3 can be classified into three groups according to the current understanding of HCL. The group of 'visual performance' includes *brightness perception* and *visual clarity* according to [16], as a function of the *signals* of the photoreceptor channels L, M, and S, and the *signal* of the intrinsically photosensitive retinal ganglion cells (G). The group of 'colour quality' (psychological aspects) includes the *colour-rendering* (R_a, R_f) metrics and the metrics of colour preference (colour quality scale (CQS) Q_p) according to [17] and colour gamut (R_g) according to Illuminating Engineering Society (IES) TM 30-15 [15]. The last two metrics represent an evaluation according to *colour preference* or according to the representation of saturated object colours mainly at the cognitive level of the human visual brain, where effects of personality and culture also come into play, but that cannot be described by these metrics. To calculate the values of these metrics, both the spectral distribution of the light source and the spectral reflectance of a number of selected object colours must be used.

The group of non-visual effects includes the Circadian Stimulus (CS, see Chapter 2, CS of version 2016 of Rea et al.) and the logarithmic parameter $\ln(a_{mel}\, E_v)$ defined by the authors of this book. By definition, the signals of the L-, M-, and S-cones and the intrinsically photosensitive retinal ganglion cells (ipRGCs) cells G are included in both parameters. This leads to the hypothesis that the metrics of visual performance and non-visual effects contain the same components (receptor signals according to Lucas [18] and according to the CIE [19]) and should therefore exhibit a good correlation. The factors CS and $\ln(a_{mel} \cdot 500\,\text{lx})$ are compared in Figure 8.12 at a vertical illuminance of 500 lx (in case of the light spectra in Figure 8.4). With a value of the correlation coefficient of $r^2 = 0.98$, one can speak of a high correlation or almost the same tendency of the two parameters of the non-visual effects. Instead of measuring the CS factors at a certain illuminance, the illuminance E_v and the a_{mel}-factor can therefore be measured and then the quantity $\ln(a_{mel} \cdot 500\,\text{lx})$ can be calculated at e.g. $E_v = 500$ lx.

Figure 8.13 compares the CS parameters and the brightness metric values according to Fotio's formula at the illuminance $E_v = 500$ lx. With a correlation coefficient of $r^2 = 0.97$, the conclusion is that brightness and CS-values show the same tendency. For lighting designers and lighting developers, this finding means that a scene that is brightly lit also produces a correspondingly high circadian effect. Combining the results in Figure 8.12 and 8.13, it can be stated that a scene with a high setting of the vertical illuminance E_v and at a high a_{mel}-value (i.e. with more accentuated short-wave components in the light spectrum around 480–490 nm while maintaining the white tone of the light) also appears *bright*.

If, on the other hand, one compares a colour quality parameter, here as an example the parameter CQS Q_p according to Davis and Ohno [17] with the parameters CS or the brightness value $(S/V)^{0.24}$, one obtains the results in Figure 8.14, which prove that the colour quality parameters do not correlate with the brightness and the CS-values

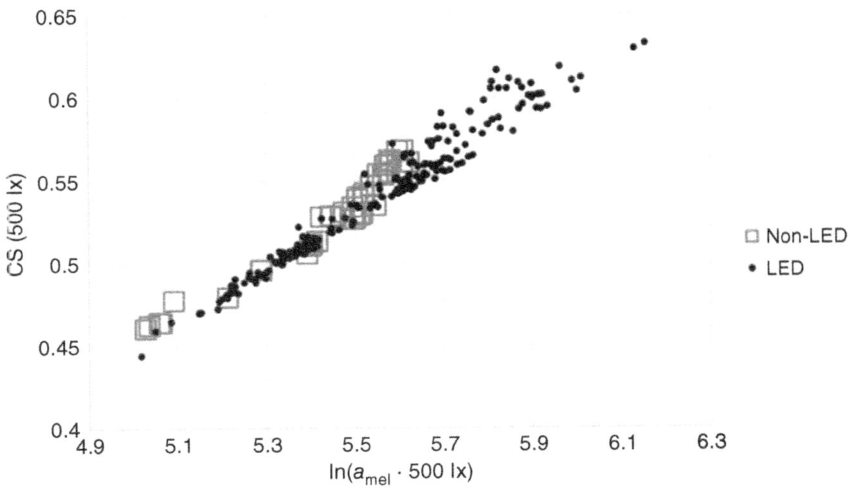

Figure 8.12 The parameters Circadian Stimulus (CS) [20] and $\ln(a_{mel} \cdot 500\ \text{lx})$ at a vertical illuminance of 500 lx for the 387 light spectra in Table 8.1. Source: Reproduced with permission from Lighting Research and Technology [21].

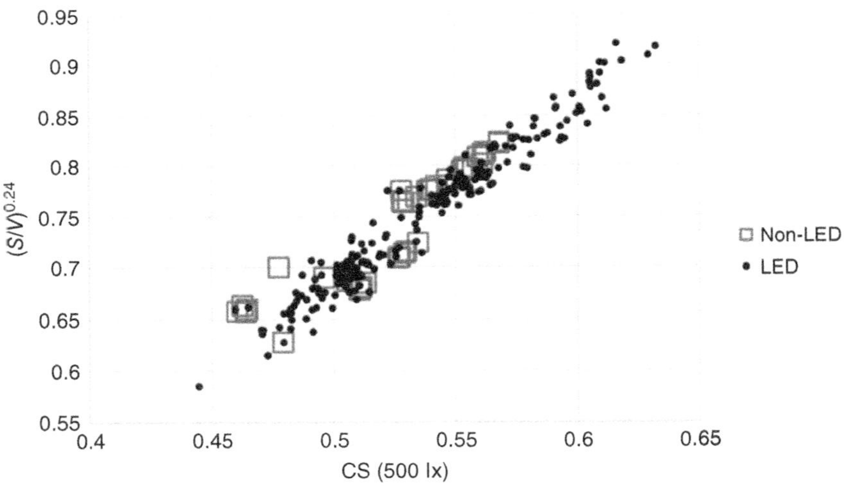

Figure 8.13 CS factor at an illuminance of $E_v = 500$ lx and the brightness multiplier values $(S/V)^{0.24}$ according to Fotios. Source: Reproduced with permission from Lighting Research and Technology [21].

(they are independent). A scene that is bright and provides a high circadian effect can have both high and low average colour quality.

Based on these computational findings [21], the numeric descriptors of the HCL system were divided into two independently acting groups of parameters. One group contains colour quality criteria. The other group contains the parameters for brightness and visual clarity (for visual performance) as well as for non-visual effects, see Figure 8.15.

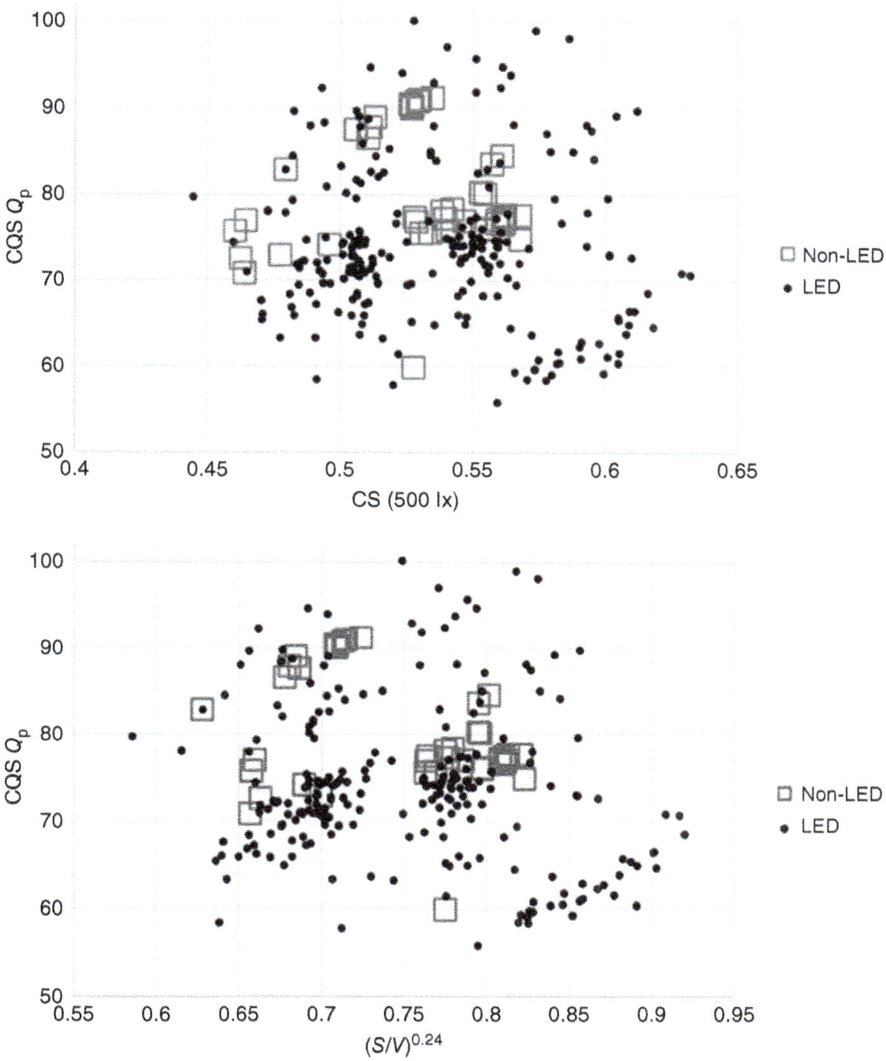

Figure 8.14 Comparison of CQS Q_p [17] with CS [20] (left) and the Fotios brightness metric (right). Source: Reproduced with permission from Lighting Research and Technology [21].

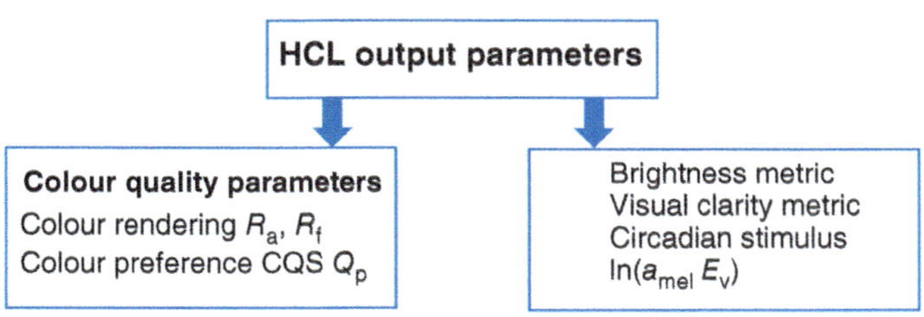

Figure 8.15 Structure of the output variables (numeric descriptors) of the HCL systems [21]. Source: TU Darmstadt.

8.2 Structure and Categories of the Input Parameters of the HCL System

The input parameters of the HCL system are important in lighting applications as they represent *control variables* for the intelligent lighting system. They can be divided into two groups. One group contains *physiologically based parameters* such as the receptor signals of the L-, M-, and S-cones and the ipRGC as well as the signals of the opponent channels (L – M) and (L + M) – S. These are fundamental parameters that contribute to the understanding of the relationships among higher-level parameters important for lighting design. They can be calculated by measuring the emission spectra of the light sources. However, they cannot be measured with conventional (i.e. non-spectral) measuring instruments generally used in today's practice (e.g. handheld colour meters, colour cameras, illuminance meters, and luminance meters).

In the second group are *photometrically* defined input parameters such as the illuminance E_v, the luminance L_v, the chromaticity coordinates x, y, and z, the a_{mel}-factor as well as the CCT. These parameters are mathematical constructs to describe a certain aspect of the visual system that is important for lighting design. Except for the a_{mel}-factor, these parameters can be measured with relatively inexpensive spectrally integrative measuring instruments such as handheld colorimeters, illuminance meters, or a luminance-type colour camera. Looking at Figure 8.16 in order to analyse the spectral sensitivity curves of the retinal photoreceptors and mechanisms (also compared with photometric functions and characteristic emission spectra), it becomes obvious that the *short* wavelength range (from 380 to about 520 nm) contains several physiological mechanisms:

- the S-cones, the rods R, and the ipRGC (ipRGC, G-signal);
- the spectral sensitivity to melatonin suppression according to Brainard et al. [22] and Thapan et al. [23], summarised by Gall [24];
- the colour-matching function $z(\lambda)$ for a field of view of 2° (CIE 1931);
- a mathematical function (defined here in this book) as a rectangular function from 380 to 520 nm, which is the so-called '*blue signal window*'.

On this basis, the following research questions are posed regarding the input parameters:

(a) Do these above-mentioned '*short-wave sensitive*' channels (Figure 8.16) show a high correlation?
(b) Are there correlations between the photoreceptor signals and the psychophysically derived colour quality parameters?
(c) Do the colour quality parameters depend on the photoreceptor signals of the L channel? This question is justified because – according to Figure 8.16 – the spectral sensitivity of the L channel dominates from 600 nm, which is almost the only channel still active from about 650 nm. From 600 nm, a large number of reddish surface colours (that are important in colour quality evaluations) have

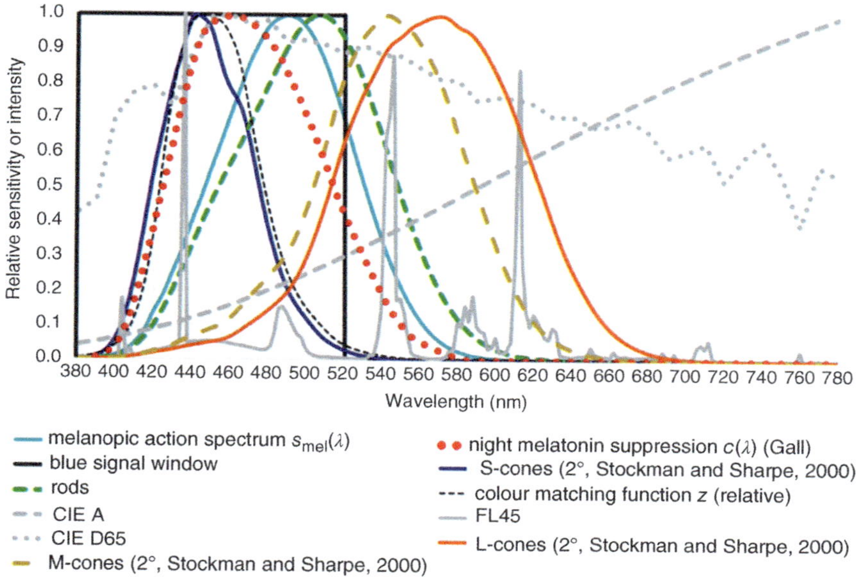

Figure 8.16 Spectral sensitivity of different photoreceptor types and retinal physiological mechanisms together with the relative emission spectra of the reference illuminants CIE A (CCT = 2856 K) and CIE D65 as well as a fluorescent lamp (FL45) at CCT = 4500 K. Source: TU Darmstadt.

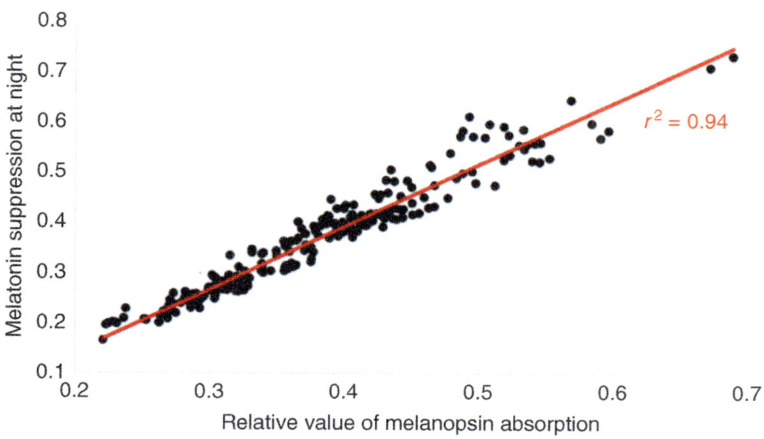

Figure 8.17 Comparison of the relative signal of melatonin suppression at night according to Gall [24] with the relative signal of melanopsin absorption [14]. Source: TU Darmstadt.

a high-spectral reflectance (e.g. fruits, flowers, skin tones, books, furniture, and clothing).

Figure 8.17 compares the relative signal of melatonin suppression at night according to Gall [24] with the relative signal of melanopsin absorption according to the standard DIN SPEC 5031-100:2015-08 [14] based on the light spectra in Table 8.1.

Figure 8.17 shows that these two types of signal correlate very well with each other ($r^2 = 0.94$). This is interesting from a lighting technological point of view because the result states that the signals generated by light absorption in the photosensitive retinal ganglion cells tend to also be used to describe melatonin suppression in the nocturnal hours.

Table 8.3 shows the correlation coefficients of all the above signal types active in the short-wave range. The correlation coefficients of all these signal types are very high. This means that the signals of the photoreceptors and the mechanisms of melatonin suppression as well as the photometric signal values exhibit a strong linear relationship (in this example for the representative light spectra in Table 8.1). Mathematically, they differ only by a linear scaling factor that can be easily determined by calibration in light measurement practice. They all thus belong to a single group of input parameters.

In the longer wavelength range up to 780 nm, the L-cones and the M-cones are active (in the photopic range) and contribute to visual performance, colour quality, and non-visual effects. The signals of the L-, M- and S-cones and the ipRGCs (G), the opponent channel (L − M) as well as the descriptor parameter of brightness, the CS-value, and the colour quality parameters Q_p, R_f, and R_g were calculated and analysed, in case of the light spectra in Table 8.1. The resulting correlation coefficients are shown in Table 8.4.

Table 8.4. shows that no receptor signals correlate significantly with the colour quality parameters Q_p, R_g, and R_f. This provides evidence that the physiologically identified receptor signals do not lead directly to a cognitive or psychologically obtained parameter such as the colour preference index CQS Q_p. Except for the L-signals (the signals of the L-cones, which are sensitive in the long-wave range), receptor signals correlate very well with the brightness values and the CS-values.

Table 8.3 Correlation coefficients of all signal types that are active in the short-wavelength range, see Figure 8.16.

	Melanopsin absorption [12]	Melatonin suppression at night [22]	Blue signal window	S-cones	Rods	*Z* (2° tristimulus value)
Melanopsin absorption	1.00	0.97	0.96	0.91	0.99	0.92
Melatonin suppression at night	0.97	1.00	0.98	0.98	0.98	0.99
Blue signal window	0.96	0.98	1.00	0.94	0.97	0.95
S-cones	0.91	0.98	0.94	1.00	0.93	1.00
Rods	0.99	0.98	0.97	0.93	1.00	0.94
Z (2° tristimulus value)	0.92	0.99	0.95	1.00	0.94	1.00

Source: TU Darmstadt.

Table 8.4 Correlation coefficients (r) of the colour quality metrics (CQS Q_p, IES R_f, and IES R_g), the brightness multiplier signal $(S/V)^{0.24}$ and the CS [13] at 500 lx with the relative signals of the different photoreceptor types.

	CS(500 lx)	CQS Q_p	IES R_f	IES R_g	$(S/V)^{0.24}$
L_{rel}	−0.896	0.308	0.582	0.243	−0.939
M_{rel}	0.946	−0.253	−0.511	−0.242	0.969
S_{rel}	0.959	−0.210	−0.482	−0.188	0.986
R_{rel}	0.978	−0.027	−0.199	−0.182	0.928
G_{rel}	0.984	−0.048	−0.229	−0.191	0.943
$\|L_{rel} - M_{rel}\|$	−0.930	0.273	0.537	0.243	−0.961

Source: Reproduced with permission from Lighting Research and Technology [21].

Table 8.5 Correlation coefficients between the chromaticity coordinates x, y, and z with the brightness value $(S/V)^{0.24}$, the CS-value and the colour quality parameters CIE R_a, CQS Q_p, and IES R_f, IES R_g.

	CS(500 lx)	R_a	Q_p	R_f	R_g	$(S/V)^{0.24}$
x	−0.960	0.424	0.233	0.504	0.224	−0.986
y	−0.945	0.269	0.153	0.383	0.188	−0.956
z	0.971	−0.375	−0.208	−0.469	−0.215	0.992

Source: TU Darmstadt.

Table 8.5 shows the correlation coefficients of the chromaticity coordinates x, y, and z with the brightness value, the CS-value, and the colour quality parameters. The results in Table 8.5 are important in that lighting designers or room users may come to the conclusion that these three chromaticity coordinates can be used as control variables for setting the brightness or the CS or the colour quality parameters of a specific lighting installation. The result from the analysis of Table 8.5 is that the x and y chromaticity coordinates do not correlate with any of the output parameters, whereas the z chromaticity coordinate correlates very well with the brightness signal $(S/V)^{0.24}$ and the CS-values.

In summary, the output parameters of the HCL system can be divided into the following two groups:

1. a group with the visual performance (brightness, visual clarity) and the non-visual effects (a_{mel}-factors, CS-values, and in the same context MEDI, see the Eqs. (2.12) and (2.13) in Chapter 2 of this book)
2. an independent group with the colour quality parameters (R_a, R_f, and Q_p).

The input parameters in the short-wavelength range correlate very well with each other (see Table 8.3), so that a parameter that can be measured *in practice* can be taken as a representative. This is the chromaticity coordinate z, which is suitable for controlling an intelligent luminaire. Chapter 10 describes the formulas and mathematical derivation of how the brightness and the CS values can be determined from measurable quantities such as vertical illuminance E_v and the chromaticity coordinate z (or from the CCT) with reasonable accuracy in the practical lighting design of a lighting system installation. The control of a lighting system with the CS-value, with MEDI-metric or with the brightness can then be carried out with the aid of the combination (E_v; CCT).

References

1 ZVEI (2016). Position paper: The use of Human Centric Lighting (HCL) enables the right light for every time of day, September 2016.
2 Boyce, P. (2016). Editorial: exploring human-centric lighting. *Light. Res. Technol.* 48: 101.
3 Khanh, T.Q., Vinh, Q.T., und Bodrogi, P. (2017). Visual performance, emotional and non-visual effects - The fundamental pillars of future lighting technology. In: Paper presented at the Biowi (LitG) conference (September 2017).
4 Bodrogi, P. und Khanh, T.Q. (2017). Human Centric Lighting (invited lecture). CIE 2017, W.S. 6: Temporal Light Artefacts in Automotive and General Lighting (October 2017). Jeju, Korea.
5 Berman, S.M., Jewett, D.L., Fein, G. et al. (1990). Photopic luminance does not always predict perceived room brightness. *Light. Res. Technol.* 22 (1): 37–41.
6 Fotios, S.A. and Levermore, G.J. (1998). Chromatic effect on apparent brightness in interior spaces, II: SWS lumens model. *Light. Res. Technol.* 30 (3): 103–106.
7 Sagawa, K. (2006). Toward a CIE supplementary system of photometry: brightness at any level including mesopic vision. *Ophthalmic Physiol. Opt.* 26: 240–245.
8 CIE Supplementary System of Photometry (2011). CIE Technical Report 200:2011. 978-3-902842-37-4.
9 Bodrogi, P. und Khanh, T.Q. (2017). Visual clarity and brightness in indoor and outdoor lighting: experiments and modelling. CIE 2017 (presented poster, October 2017). Jeju, Korea.
10 Pepler, W. (2017). Photopic brightness perception in indoor lighting with different light sources. Dissertation. TU Darmstadt, Darmstadt.
11 Bodrogi, P., Brückner, S., and Khanh, T.Q. (2013). Colour rendering for modern-light sources. Fachgebiet FG 05 der LiTG. ISBN 978-3-927787-30-8.
12 CIE (1995). *Method of Measuring and Specifying Color Rendering Properties of Light Sources*. CIE Publication 13.3-1995. Vienna: CIE.
13 CIE (2017). Colour fidelity index for accurate scientific use. *CIE* 224: 2017. ISBN 978-3-902842-61-9.
14 DIN SPEC 5031-100:2015-08: Optical radiation physics and illuminating engineering – Part 100: Melanopic effects of ocular light on human beings – Quantities, symbols and action spectra. Berlin: Beuth Verlag.
15 David, A., Fini, P.T., Houser, K.W., and Whitehead, L. (2015). Development of the IES method for evaluating the color rendition of light sources. *Opt. Express* 23: 15888–15906.

16 Bodrogi, P., Guo, X., and Khanh, T.Q. (2018). Brightness and visual clarity rather than luminance in the evaluation of indoor lighting. Light. Creative lighting solutions, 23. Europäischer Lichtkongress. Davos, Schweiz.

17 Davis, W. and Ohno, Y. (2010). Color quality scale. *Opt. Eng.* 49: 033602.

18 Lucas, R.J., Peirson, S.N., Berson, D.M. et al. (2014). Measuring and using light in the melanopsin age. *Trends Neurosci.* 37: 1–9.

19 International Commission on Illumination (2018). CIE S 026/E:2018 CIE System for Metrology of Optical Radiation for ipRGC-Influenced Responses to Light. International standard. ISBN 978-961-93733-4-7.

20 Rea, M. and Figuiero, M.G. (2016). Light as a Circadian stimulus for architectural lighting. *Light. Res. Technol.* 50 (4): 497–510.

21 Bodrogi, P., Vinh, Q.T., and Khanh, T.Q. (2019). Correlations among lighting quality metrics for interior lighting. *Light. Res. Technol.* 51 (8): 1192–1207. https://doi.org/10.1177/1477153518818856.

22 Brainard, G.C., Hanifin, J.P., Greeson, J.M. et al. (2001). Action spectrum for melatonin regulation in humans: evidence for a novel circadian photoreceptor. *J. Neurosci.* 21: 6405–6412.

23 Thapan, K., Arendt, J., and Skene, D.J. (2001). An action spectrum for melatonin suppression: evidence for a novel non-rod, non-cone photoreceptor system in humans. *J. Physiol.* 535: 261–267.

24 Gall, D. (2004). *The measurement of circadian radiation quantities.* Ilmenau: TU Ilmenau.

9

Psychophysical–Emotional Aspects – Visual Comfort and Non-visual Effects

In this chapter, the psychological–emotional aspects of lighting effects, visual comfort, and non-visual effects are analysed and summarised through an in-depth literature review. The discussion of the psychological–emotional aspects in Chapter 6 with preferred colours, preferred correlated colour temperatures (CCTs), *colour rendering,* and colour saturation led to a three-dimensional model of light quality in Chapter 7 with the model parameters '*illuminance-CCT-saturation*'. We accumulated knowledge about how the light emitted by the luminaires and reflected in the interior should be optimised to evoke a quality assessment of light in the observer's (room user) visual brain. Accordingly, the first part of the present chapter will deal with luminance distributions of different indirect and direct luminous flux components, whereby the light colours (e.g. warm white [ww] and cool white) in the direct and indirect components shall not necessarily be constant. These components will be adjusted depending on the weather conditions, time of day, or the context of the application (type of activity in the room). Lighting technology is thus moving in the direction of lighting design, interior design, and lighting architecture. Concerning visual comfort, new experimental results obtained by the authors of the present book regarding stroboscopic effects will also be described [1]. *Indoor glare* (characterised, e.g. according to the unified glare rating (UGR)-method [2]) has been researched in detail and included in international standards for lighting technology. It is therefore not the subject of this book. The various aspects of *non-visual effects* are described in the second half of this chapter on the basis of a comprehensive literature analysis.

9.1 Psychological–Emotional Aspects of the Effect of Light

9.1.1 Introduction

The Encyclopaedia of Psychology (https://www.psychomeda.de/lexikon/emotion.html) defines the term '*emotion*' as follows:

> Emotions (movements of mind) are understood to be feelings, moods, and effects. Emotions can be triggered by perceptions, thoughts and memories,

Human Centric Integrative Lighting: Technology, Perception, Non-Visual Effects, First Edition.
Tran Quoc Khanh, Peter Bodrogi, and Trinh Quang Vinh.

and express themselves in psychological and physiological changes of state. Emotions exert a great influence on thinking and acting and are closely linked to motivation and cognition (thoughts, beliefs, and expectations). In 1980, the American professor of medicine Robert Plutchik postulated a complex system of emotions with eight basic emotions. According to his idea, mixing, and combining these eight basic emotions results in more complex emotions, just like mixing colours. The basic emotions according to Robert Plutchik are:

- Joy–Sorrow
- Trust–Mistrust
- Fear–Anger
- Uncertainty–Certainty

Lighting research in recent decades has used the *PAD model* of Mehrabian et al. [3, 4] to characterise the emotions of a person and to analyse the effect of light on the emotional states of room users. Pleasure (***P***), arousal (***A***), and dominance (***D***) are the three factors that are considered. The PAD model is used in psychology to study non-verbal communication, such as body language [5] and follows the research thesis that physical environments (e.g. room architecture, light, and nature) influence people through their emotional impact. The combinations of the three components *PAD* lead, according to [6], to the emotional states shown in Figure 9.1.

Fleischer [6] applied the PAD model to investigate the psychological–emotional effects of light in field tests and in laboratory rooms. According to this model, the factors influencing people are divided into three groups: Outdoor space, indoor space, and activity (see Figure 9.2).

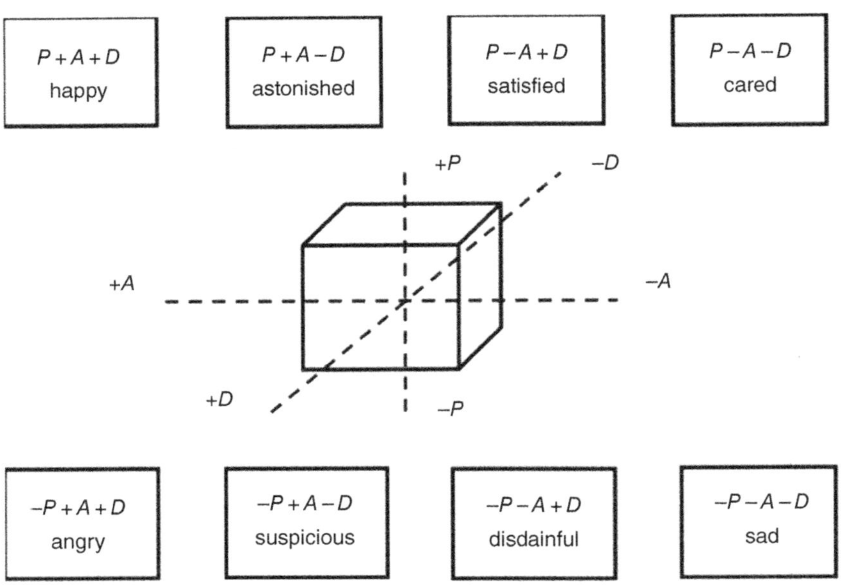

Figure 9.1 Characterisation of human emotions with the Pleasure–Arousal–Dominance (PAD) model [3, 6]. Source: TU Darmstadt.

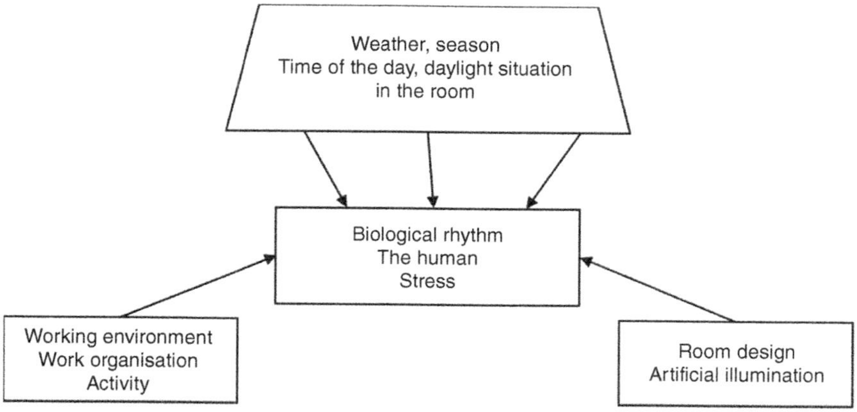

Figure 9.2 Emotional influencing factors from the perspective of Human-Centric Lighting [6]. Source: TU Darmstadt.

The main influencing factors in Figure 9.2 are the daylight situation, such as weather (rain, degree of cloud cover, and proportion of daylight), time of day, season (length of day, relative duration of sunshine, winter, and summer), type of activity (reading documents, concentrated work at a table, meetings, telephoning, and working at a computer screen), room design (windows, room furniture, and decoration), and artificial lighting (illuminance on the table, luminance distribution on the ceiling, on the walls, light colour, and indirect components), see Figure 9.3.

The totality of influencing factors with their respective physical properties (see Figure 9.3) evoke perceptions, evaluations, cognitions, and emotions in people's sensory organs and then in the brain. This information processing of environmental stimuli takes place on several levels. The level of perceptions and sensations is followed by the levels of affective evaluations and attributions [6], see Figure 9.4.

In [7], the term *attribution* is defined as follows: 'in social psychology, the term *attribution* refers to both the subjective and the social ascription of characteristics such as abilities or experiences, as well as to the assumed or presumed causes of one's own actions or those of other people, and to the causation of real external processes and situations (*causal attribution*). Consequences and effects for experience and behaviour resulting from these assumptions are also referred to by the term *attribution*, whereby different attributions of real situations can be regarded as constructs of reality and make different motivations for future behaviour'.

For several decades, a large part of lighting research and experimental psychology has been concerned with the effects of light in different weather (degree of cloudiness), room design, and different activities on the psychological–emotional states of light users. Among many well-founded literatures, the study by Fleischer [6] is one of the most comprehensive studies, the contents, and results of which are summarised in the following section.

Figure 9.3 Emotional influencing factors in an interior from the perspective of Human-Centric Lighting (example). Source: Technische Universität Darmstadt.

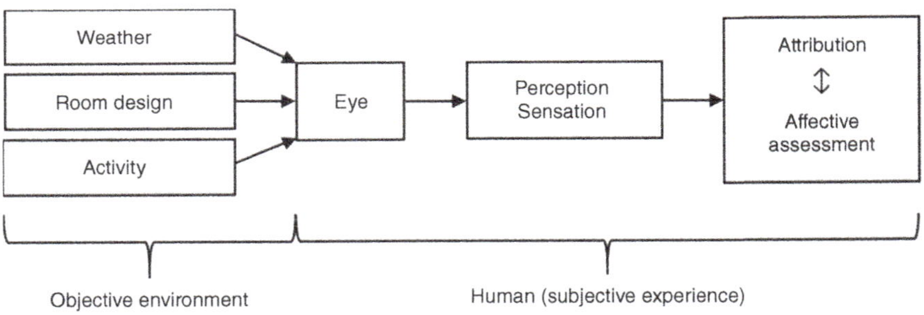

Figure 9.4 Effect of environmental stimuli [6]. Source: TU Darmstadt.

9.1.2 Psychological Effect of the Variable Lighting Situations, Spatial Effects

The study by Fleischer [6] pursued the aim of investigating the effects of different types of light distributions in the room by varying the indirect light component on the ceiling and on the upper half of the walls and, consequently, the effects of different luminance distributions on the emotional state in combination with either the illuminance level or the light colour (ww 3000 K, neutral white 4000 K, and daylight white [dw] 5600 K). The PAD model was used to describe the emotional state, whereby the following 18 pairs of adjectives were used to subjectively describe the lighting situation:

- satisfied–melancholic
- delighted–annoyed
- hopeful–desperate
- self-employed–dependent
- important–unimportant
- bored–relaxed
- dissatisfied–satisfied
- unhappy–happy
- slow running
- unexcited–excited
- lame–nervous
- calm–excited
- sleepy–wide awake
- stimulated–relaxed
- controlled–controlling
- influenced–influencing
- cared-for-self-determined
- inferior–superior

The study [6] sought to answer the following four research questions:

1. Whether the room users favour different lighting scenarios depending on the external measurable variables of time of day, season, and sky condition? The daylight situation was characterised by the length of the day (time between sunrise and sunset), the relative average sunshine duration and the shading (for the sky condition, degree of cloudiness).
2. Whether the motivation of light users is influenced by artificial lighting situations that are varied depending on the time of day or the state of the sky?
3. Whether the emotional state of humans (*PAD*) can be influenced by artificial light?
4. Whether – depending on the activity – different lighting situations that are not perceived as optimal can be derived?

The experiments were divided into field experiments and laboratory experiments. Their ideas, implementation, and results are presented in Sections 9.1.2.1 and 9.1.2.2.

9.1.2.1 Field Trial

In the field test, which was conducted over a year in two open-plan offices, the current daylight, and artificial light were combined to create four lighting scenarios. The artificial lighting systems consisted of indirect–direct pendant luminaires, with both luminous flux components equipped with *neutral white* fluorescent lamps with CCT = 4000 K and with *colour rendering* level 1B ($86 < R_a < 90$). The lighting scenes *A* and *B* were dependent on the time of day. The lighting scenes *C* and *D* were dependent on the state of the sky, see Table 9.1. The artificial lighting systems were dimmed depending on the *time of day* so that the illuminance on the work surface (500 lx) and the type of light distribution in the room were always maintained.

Table 9.1 Four light scenarios in the field trial by Fleischer [6].

Light scenarios	Description
A	Indirect light in the morning from 7 a.m. to 11 a.m., direct light in the afternoon from 2:30 p.m. to 6 p.m., 50 : 50 ratio between 11 a.m. and 2:30 p.m.
B	Direct light in the morning from 7 a.m. to 11 a.m., indirect light in the afternoon from 2:30 p.m. to 6 p.m., 50 : 50 ratio between 11 a.m. and 2:30 p.m.
C	Indirect light when the sky is overcast, direct light when the sky is clear
D	Direct light with overcast sky, indirect light with clear sky

Cf. with Figure 9.4.
Source: Adapted from Fleischer [6] by the TU Darmstadt.

Each scenario in Table 9.1 was carried out several times throughout the year, each time lasting three weeks, in order to be able to take the '*season*' factor into account in the evaluation. At the end of the three weeks, questionnaires were distributed on the impression of the room, the activity and the type and frequency with which the light users intervened in the settings of the lighting system. In addition, a mini-diary was kept in case of certain test persons. The diary documented how they felt in the morning and afternoon, their motivation and the effect of the lighting situation in the afternoon. The data were evaluated in connection with the length of the day and the degree of cloud cover. Table 9.2 contains those light scenarios of Table 9.1, which provided satisfactory results after the evaluation of the questionnaires and the mini-diaries.

Table 9.2 Light scenarios of Table 9.1, which provided satisfactory results.

Question	Lighting scenarios with satisfactory results
Questionnaires	
'How did you like the light in your office in the last three weeks' 'very bad … very good'	C and D
'The ceiling was too bright … just right … too bright'	C and D
'The light made the room seem predominant' (unpleasant … pleasant)	C and D
Mini-diary	
'How are you feeling this afternoon' (very bad … very good)	B and D
'What effect did the lighting situation at the workplace have on you? It makes me feel sleepy … awake'	D
'How satisfied were you with the lighting today?' (not at all … very much)	D

Evaluation of the questionnaires and the mini-diaries of the field trial.
Source: Adapted from Fleischer [6] by the TU Darmstadt.

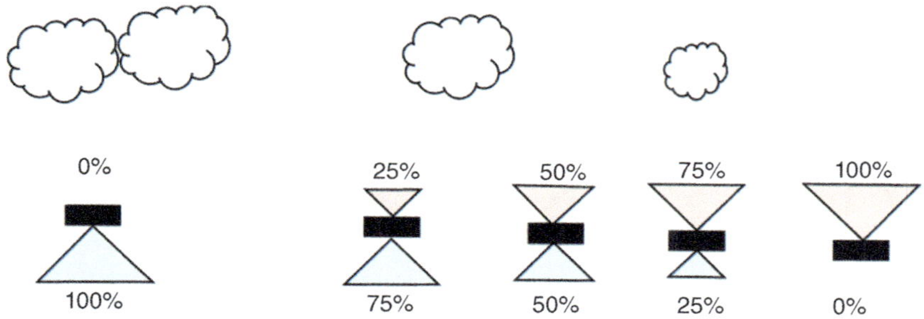

Figure 9.5 Functional principle of the sky-state-dependent lighting scenario D. Source: Adapted from Fleischer [6].

The result in Table 9.2 shows that the subject's state of health, activation, and satisfaction both in responding to the questionnaires after three weeks of testing and with the diary under the condition of constant illuminance of 500 lx on the work surface differ in different lighting scenarios and are mostly found in lighting scenario *D*. The results in Table 9.2 show that the subject's state of health, activation, and satisfaction differ from one lighting scenario to another. Figure 9.5 illustrates the functional principle of lighting scenario *D*, which depends on the state of the sky. It is apparently not desirable to vary the lighting scenarios with the '*indirect–direct*' lighting emphasis depending on the time of day.

9.1.2.2 Laboratory Experiment

The laboratory experiment [6] pursued the goal of realising different light distributions by varying the indirect component in combination with the illuminance level in test series 1 and with the light colour (3000 or 5600 K) in test series 2 and of determining the preferred lighting situations depending on the type of activity. The lighting installation consisted of three pendant luminaires with indirect and direct luminous flux components and was configured in such a way that the illuminance of 500 lx on the work surface could be achieved by both the direct and the indirect luminous flux components *alone*.

In test series 1, 10 lighting situations with 5 types of light distributions (i.e. five different *indirect-to-direct* ratios) were specified at two illuminance levels (300 and 500 lx) and at a constant colour temperature of CCT = 4000 K. In test series 2, 10 light colour – light distribution combinations and 2 additional light distributions were generated with 2 light colours (3000 and 5600 K) and 5 types of light distributions, in which both the indirect and the direct light components were configured with the same light colour (3000 or 5600 K). With these 12 lighting scenarios in total, the illuminance was always 500 lx on the working plane.

In another part of the experiment, lighting scenes for six predefined types of activity were set *by the test subjects*. Light distribution type and lighting level (max. 1050 lx) could be set. Here, these different experimental settings and their results will be described in detail in the context of optimising individuality-dependent lighting. For the values set by the test persons, Fleischer [6] formed confidence

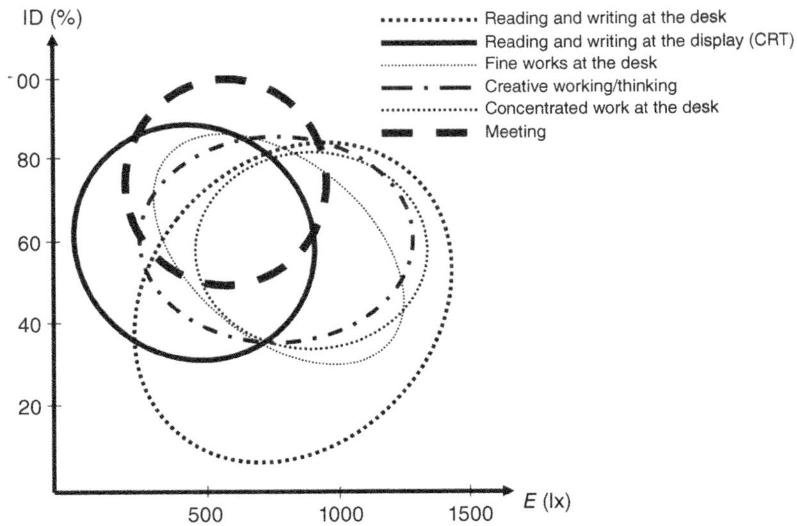

Figure 9.6 Confidence ranges at the 50% level about the setting of the lighting situation with the parameters illuminance level E_v (lx) and indirect component ID (in %) according to Fleischer [6]. Source: Reproduced from Fleischer [6] with permission of Susanne E. Fleischer.

ranges at the 50% level, which are shown in Figure 9.6. They differ in their size, the position of their centres and in the orientation of the main axes.

Figure 9.6 shows the following:

(a) The centres of the ellipses accumulate in the indirect component range ID = 50–70%.
(b) For the activity '*meeting*', a high indirect light component of over ID = 70% is desired. The so-called *modelling effect* (recognition of the fine visual details of faces and other vertically oriented objects by the lighting) is then higher – in favour of better visual communication.
(c) For reading and writing at a desk, some test persons require an illuminance of up to 1400 lx.
(d) For working with the screen, an indirect component of 60% is favoured at about 450–500 lx on the working plane.

In the context of psychological–emotional effects and non-visual lighting effects, it is also interesting to determine the lighting situations favoured by the test subjects depending on the time of day, whereby both the indirect light component and the light colours are varied in the direct and indirect luminous flux components. The lighting situation favoured for the morning between 8 a.m. and 12 p.m. is shown in Figure 9.7.

For the two light colours dw and ww in the direct and indirect light components, the ratings in Figure 9.7 are relatively similar. The indirect luminous flux component between about 50% and 75% provides the best rating in both cases.

For the daytime phase in the afternoon between 3 p.m. and 6 p.m. (Figure 9.8), the assessment of the lighting situation with ww light in the indirect part and with dw

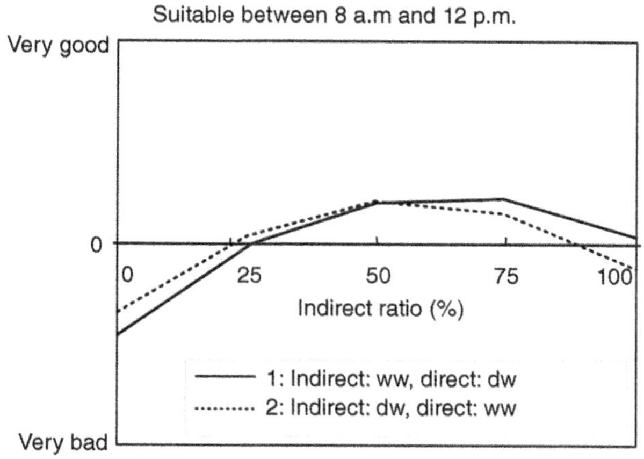

Variable	*p*-Value
ID	0.0001
FK	0.5747
FK*ID	0.8861
ID*ID	0.0001
FK*ID*ID	0.8576

Figure 9.7 Favourite lighting situation for the morning between 8 a.m. and 12 p.m. [6]. dw, daylight white; ww, warm white. Source: Reproduced from Fleisher [6] with permission of Susanne E. Fleischer.

light in the direct part is high, whereby the indirect luminous flux of 75% with the ww light of 3000 K (25% luminous flux with the direct light and with 5600 K) was most favoured.

The results in Figure 9.8 thus demonstrate that people in office contexts with screens and paper documents (offices, educational institutions) prefer ww light in the predominant indirect luminous flux component in the second-afternoon phase between about 3 and 6 p.m.

A *factor analysis* was carried out for the data of test series 1 and 2 with given lighting situations and with the 18 adjective pairs. The resulting three factors are consistent with the PAD model. In the factor *Pleasure*, adjective pairs such as

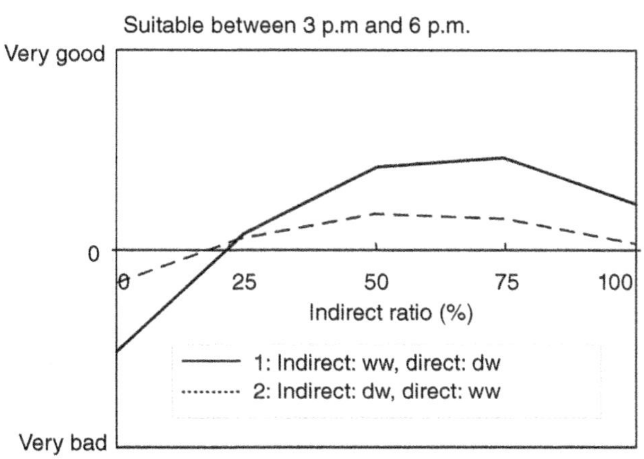

Variable	*p*-Value
ID	0.0001
FK	0.0310
FK*ID	0.0189
ID*ID	0.0001
FK*ID*ID	0.0895

Figure 9.8 Lighting situation favoured for the afternoon between 3 p.m. and 6 p.m. [6]. dw, daylight white; ww, warm white. Source: Reproduced from Fleischer [6] with permission of Susanne E. Fleischer.

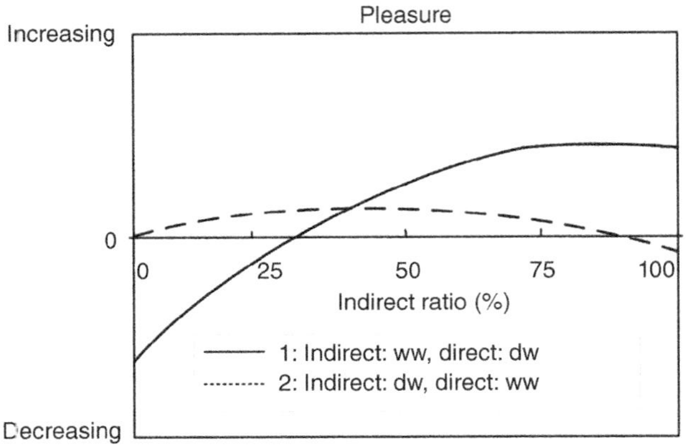

Variable	p-Value
ID	0.0031
FK	0.0059
FK*ID	0.0764
ID*ID	0.0298
FK*ID*ID	0.4557

Figure 9.9 Attribute 'Pleasure' for the two lighting situations ww/dw and dw/ww of the direct and indirect luminous flux components [6]. dw, daylight white; ww, warm white. Source: Reproduced from Fleischer [6] with permission of Susanne E. Fleischer.

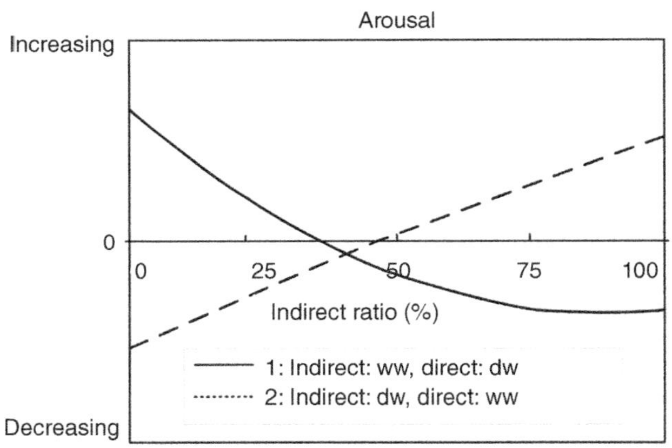

Variable	p-Value
ID	0.3100
FK	0.0001
FK*ID	0.0017
ID*ID	0.2639
FK*ID*ID	0.1673

Figure 9.10 Arousal factor (activation) for the two lighting situations ww/dw and dw/ww of the direct and indirect luminous flux components [6]. dw, daylight white; ww: warm white. Source: Reproduced from Fleischer [6] with permission of Susanne E. Fleischer.

content–melancholy, pleased–annoyed, hopeful–despairing, and happy–unhappy were collected. The results of test series 1 with the combination of the *type of light distribution* and *illuminance* show that the *pleasure* factor increases significantly with illumination level E_V (in lx). For test series 2 with the combination of the *type of light distribution* and *light colour*, the analysis of the results shows that the highest ratings are achieved with a high indirect proportion of luminous flux from about 65% to 100%, whereby the indirect proportion of light should contain *ww* and the direct proportion *dw*, see Figure 9.9.

Analysing the two curves in Figure 9.10 for the arousal factor, one comes to the conclusion that the greatest *arousal* can be achieved either with a high indirect component and *dw* or with a high direct luminous flux component and also with *dw*. In both cases, the *dw* light colour is the light source to increase arousal and activation.

The results of the evaluation of the psychological effects of variable lighting situations can be summarised as follows: With a constant illuminance of, e.g. 500 lx, the type of light distribution (indirect component), as well as the content of light colour in the indirect and direct luminous flux components, significantly influence the emotional states and the positive impressions of the room. The type of light distribution and the illuminance also have a lasting influence on the preferred lighting situations for different activities.

9.2 Space Impression, Space Brightness, and Visual Field Luminance

The study by Fleischer [6] also evaluated results on perceived *room brightness*. As already explained in Chapter 6, the assessed room brightness depends on the mean luminance of a 40° region (±20°) in the field of view (see Figure 9.11). According to the study by Fleischer [6], the room brightness is highest at a horizontal illuminance of 500 lx (in the luminance range investigated) when the luminance of this 40° region is around 120 cd m^{-2}, see Figure 9.12.

Under the condition of a constant horizontal illuminance of 500 lx on the working plane, the brightness of the wall perpendicular to the viewing direction

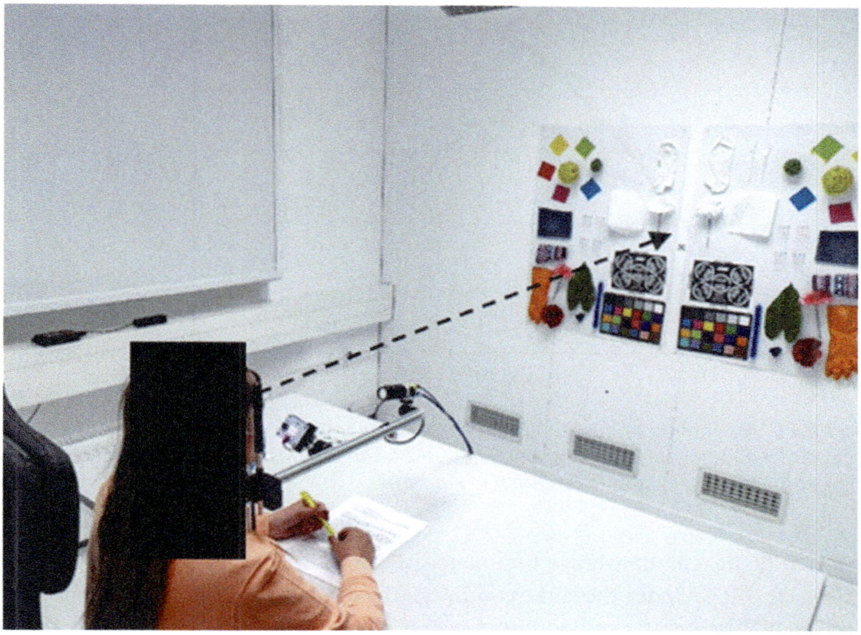

Figure 9.11 Field of vision of a seated person with the line of sight of the eye in the horizontal direction. Source: Technische Universität Darmstadt.

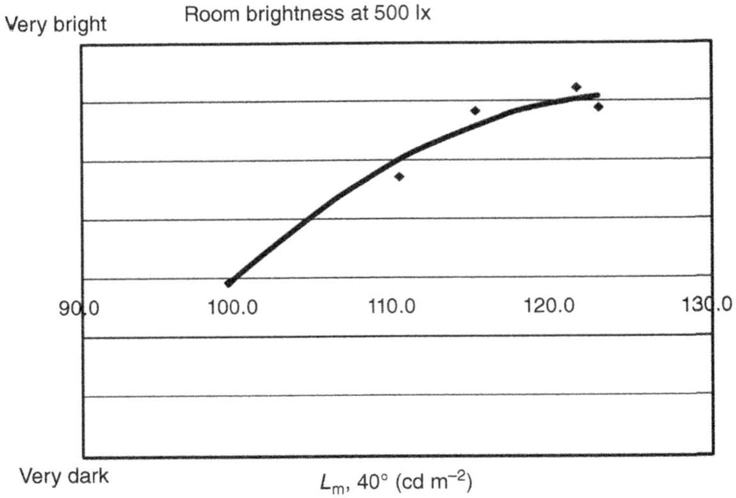

Figure 9.12 Perceived room brightness as a function of the mean luminance L_m of the 40° visual field region. Source: Reproduced from Fleischer [6] with permission of Susanne E. Fleischer.

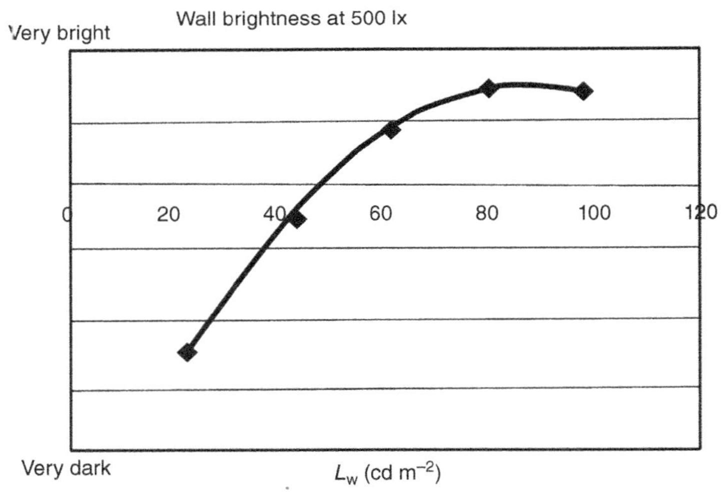

Figure 9.13 Brightness rating of the wall perpendicular to the viewing direction as a function of wall luminance L_W at a constant illuminance of 500 lx. Source: Reproduced from Fleischer [6] with permission of Susanne E. Fleischer.

was rated highest at an average luminance of the wall between 80 and 100 cd m^{-2} (see Figure 9.13). Under the same condition, the brightness of the ceiling was perceived to be highest at a ceiling luminance in the range of around 300–400 cd m^{-2} (Figure 9.14). Combining these two results, we arrive at a *ceiling-to-wall luminance ratio of* about 3 : 1 to 5 : 1.

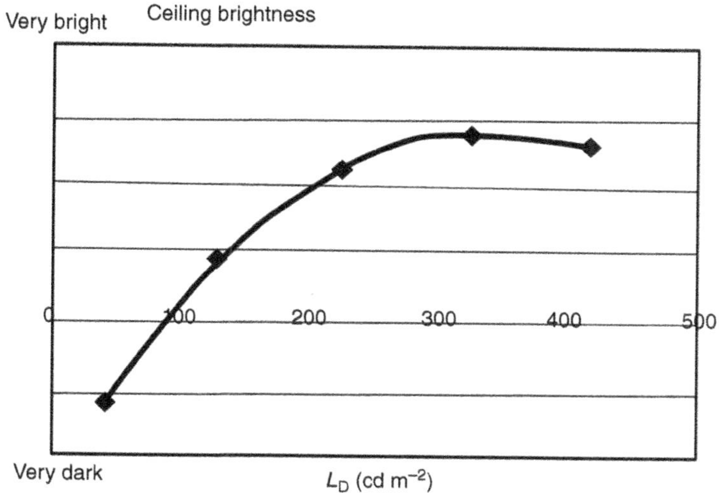

Figure 9.14 Brightness rating of the ceiling as a function of ceiling luminance L_D at a constant illuminance of 500 lx. Source: Reproduced from Fleischer [6] with permission of Susanne E. Fleischer.

9.3 Visual Comfort: Flicker and Stroboscopic Effects

With the development of light-emitting diode (LED) technology in recent years, the topic of '*visual comfort*' with the two aspects of '*flicker and stroboscopic effects*' and '*glare*' has become the subject of lighting research and vision science. In the context of this book, the author's own research results on the first topic [1] will be presented.

There are several reasons for dimming LED light sources including daylight-dependent and presence-dependent lighting control, dynamic lighting concepts that promote the well-being of users via non-visual mechanisms, and optimising LED operating parameters in terms of efficiency, colour rendering, spectrum, and white point. This is usually done via pulse width modulation (PWM). The LED is switched on and off in rapid succession. A research team from the Lighting Laboratory at the Technical University of Darmstadt conducted an experiment to investigate the influence of pulsed LED light on human visual perception.

9.3.1 Pulse Width Modulation and Constant Current Control

LEDs are a light source that can be dimmed reproducibly and without great technical effort. In practice, two methods have become established for this purpose [8].

9.3.1.1 Pulse Width Modulation (PWM)

The LED is switched on and off in rapid succession. Due to the very fast on-and-off times in the nanosecond to microsecond range, the emitted light follows the current pulses directly. At frequencies above the flicker fusion frequency (greater than 100 Hz), humans perceive a temporal average. The measured light emission (luminance, luminous flux) corresponds to the *duty cycle*, i.e. the ratio of the pulse duration to the period duration of the PWM signal. With the same switch-on and switch-off time (50% duty cycle), a measurable light intensity of 50% is produced.

9.3.1.2 Constant Current Regulation (CCR)

With constant current regulation (CCR), the variation of luminous flux is achieved by changing the *LED current levels* accordingly. As a rule, an electronic circuit for constant current control is more complex and consequently more expensive. In addition, the chromaticity shift with CCR *is* often greater than with PWM (see Figure 9.15), so PWM *is* preferred in many lighting applications.

9.3.2 Flicker and Stroboscopic Effects

The perception of flickering light can be perceived as distracting, disturbing, or irritating. Neurobiological studies show that flickering causes changes in the electrical activity of the brain [9]. Discomfort, headaches, eye sluggishness, reduction of cognitive performance or epileptic seizures in photosensitive people can be the result [10].

Due to the inertia of the light-sensitive receptors in the eye, humans are able to directly perceive flicker under certain lighting conditions up to about 100 Hz [11]. At frequencies above the flicker fusion frequency, the light appears as continuously luminous to the resting eye. However, irritating *stroboscopic* effects are still perceptible – especially in dynamic situations [12–14].

For PWM of high-power LEDs, currents in the range of 0.7 until 1 A must be switched. This generates electromagnetic fields that become stronger with

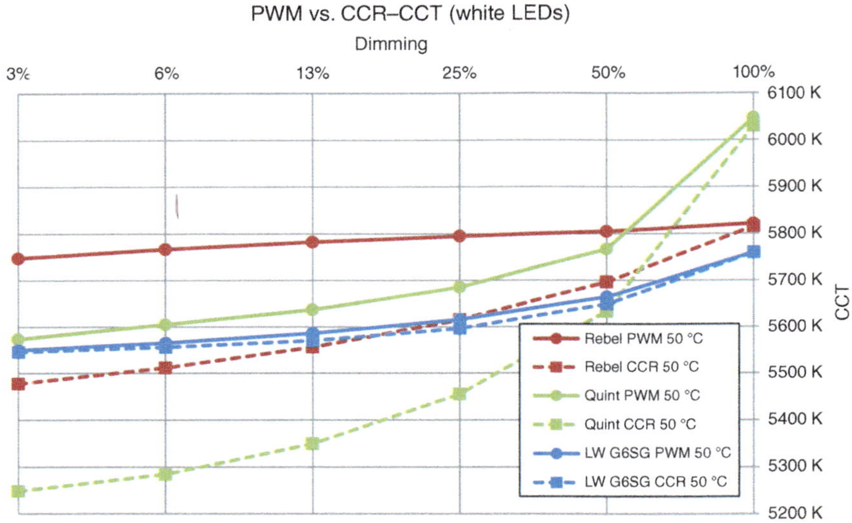

Figure 9.15 Change of the correlated colour temperature (CCT) of white LEDs with pulse width modulation (PWM) and constant current regulation (CCR) [8]. Source: TU Darmstadt.

increasing frequency. In order to keep the interference of these fields on surrounding electrical devices as low as possible and to meet the requirements for electromagnetic compatibility, PWM frequencies that are only slightly above the flicker fusion frequency are used. Typical frequencies are around 150 Hz.

In view of the rapid spread of LED technology, it is important to investigate how pulse-width-modulated LED lighting affects people's visual systems. In order to obtain objective data on this question under conditions that are as realistic as possible, a research team at the Technical University of Darmstadt conducted the experiment [1] described below.

9.3.3 State of Research

While there are numerous studies on the influence of light from fluorescent lamps on the visual system and on people's health, there are comparatively few studies on the indirectly perceptible flicker effect of LED lighting. When fluorescent lamps are operated directly, flicker occurs at twice the mains frequency (100 or 120 Hz). Although the flicker is not directly perceptible, it can lead to more frequent headaches [15]. In addition, such lighting impairs performance when reading text and increases the frequency and duration of rapid eye movements (REMs) [16]. Negative effects can be demonstrably reduced by means of an electronic ballast [15]. In this case, fluorescent lamps are operated with alternating current at a significantly higher frequency of 45 kHz. Studies [14] show that the probability of perceiving a stroboscopic effect increases with increasing modulation depth. This is defined as follows:

$$m = \frac{L_{\text{max}} - L_{\text{min}}}{L_{\text{max}} + L_{\text{min}}} \tag{9.1}$$

L_{max}: maximum luminance, L_{min}: minimum luminance.

In contrast to fluorescent lamps, the switching on and off processes of an LED are very fast. This means that the emitted light follows the PWM control signal almost without delay. The minimum luminance is therefore zero and the modulation depth is 100%, even at high frequencies.

From a visual point of view, the change between the light and dark phases in pulse-width modulated LED lighting is a scan to which the visual system reacts by interpolating the discrete temporal and spatial information in order to create a uniform impression of movement. Whether a continuous movement is perceived depends on the relative speed of the object. The higher it is, the higher the critical flicker frequency [11]. Figure 9.16 illustrates this dependency: the white circle symbolises the object whose temporal luminance curve describes a square-wave oscillation (PWM). The grey circles represent the after-images on the retina caused by the inertia of the receptors. While the afterimage of a slow movement is shown schematically on the left, the right image shows the effect of a faster object movement. The PWM signal is the same in both cases.

In order to avoid gaps between the after-images, the distance *s* travelled by the object between two light pulses must correspond at most to the object's extension *b*

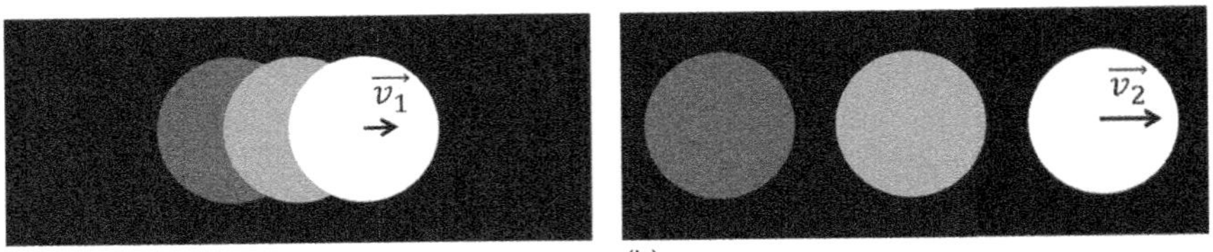

(a) (b)

Figure 9.16 Stroboscopic effect with slow (a) and fast (b) object movement. Source: TU Darmstadt.

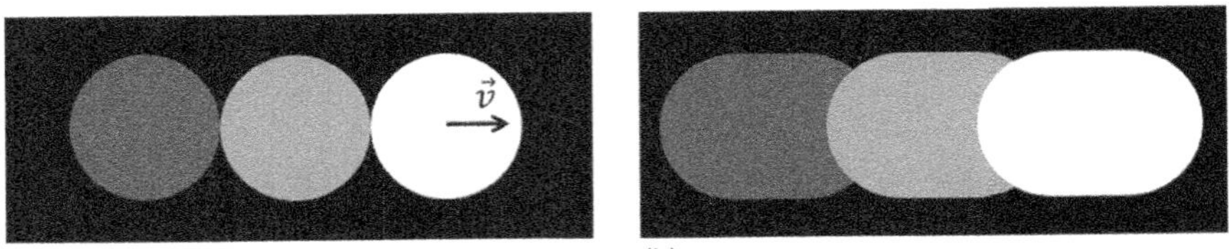

(a) (b)

Figure 9.17 Stroboscopic effect at critical flicker frequency and small duty cycle (a) or large duty cycle (b). Source: TU Darmstadt.

in the direction of movement, see Eq. (9.2).

$$b = s = vT = \frac{v}{f} \tag{9.2}$$

In Eq. (9.2), v is the object velocity, T is the time between two light pulses, and f is the frequency. If this condition is fulfilled, patterns like those in Figure 9.17 come into existence on the retina. From Eq. (9.2) we obtain the critical frequency at which the gap between after-images disappears according to Eq. (9.3).

$$f = \frac{v}{s} \tag{9.3}$$

Since the resolving power of the visual system is limited (for a person with normal vision it is about one angular minute), s can be correspondingly longer. At a distance from the observer d, Eq. (9.4) results.

$$s = b + \varphi \cdot d \tag{9.4}$$

In Eq. (9.4), the symbol ϕ means the resolving power of the observer in rad units. If one substitutes Eq. (9.4) into Eq. (9.3), the critical flicker frequency is given by Eq. (9.5),

$$f = \frac{v}{b + \varphi \cdot d} = \frac{v/d}{b/d + \varphi} = \frac{\omega}{\beta + \varphi} \tag{9.5}$$

In Eq. (9.5), the quantity ω (relative object velocity) is given in *rad/s* unit and the relative object expansion β is given in *rad* unit.

This stroboscopic effect is called *phantom array* [12] or *bead line* effect [13]. In contrast to another stroboscopic effect, the so-called *wagon wheel* effect, which simulates a slower movement or even standstill in the case of rapidly rotating wheels or working machines, the *bead string* effect occurs at significantly lower speeds and mostly in straight-line movements. The relative movement can be composed of the observer's eye movement, object movement, or a superposition of both.

Depending on the object geometry, this effect can also be perceived despite the overlapping of the after-images if a ripple occurs in the after-images at the edge. However, this ripple becomes smaller for large duty cycles (see Figure 9.17). Accordingly, it should be possible to reduce the stroboscopic effect both by increasing the PWM frequency and by increasing the duty cycle. The latter was checked in the investigation [1] described in Section 9.3.4 below.

9.3.4 Investigation

9.3.4.1 Settings

To ensure that the conditions and influencing parameters were as real as possible, the study was conducted in an office room. This had white walls and a light-proof blind at the window. An LED luminaire installed in the ceiling with a diffuse selective transmitting cover lens was used in the study. The LED *luminaire* consisted of eight high-power LEDs with a colour temperature of 4000 K. The luminaire was installed in the ceiling. Under this luminaire, there was an office table with some objects (see Figure 9.18).

Figure 9.18 Arrangement of the experiment [1] carried out in an office room with white walls and no daylight. The LED luminaire had eight high-power LEDs behind a diffuse cover. Source: Technische Universität Darmstadt.

The electronics used to control the luminaire allowed for step-less adjustment of the current. In addition, the power could be adjusted by means of PWM. The *PWM signal* was generated with a microcontroller. Both the duty cycle and the frequency could be varied. All LEDs were controlled with the same PWM signal.

9.3.4.2 Parameters Investigated

The experiment described below was conducted to investigate the conspicuousness of stroboscopic effects in a pulse-width modulated LED illumination depending on the frequency as well as the duty cycle. In addition, the test subjects were asked about the *disturbance factor* in relation to these effects. Three signal types with duty cycles of 20%, 50%, and 100% were used in the investigation. To ensure comparable conditions and to exclude the influence of the luminance level, the average luminance was kept constant for all settings by adjusting the maximum current accordingly. Figure 9.19 shows the three signal types.

Since the areas under the pulse signal curves in Figure 9.19 are the same, the time average results in the same value for the luminance. The values for the maximum current were determined using an illuminance meter and checked for all light settings. The PWM settings (frequency, duty cycle) were validated using a measurement with a photodiode on the oscilloscope.

With the present electronics, only *one* LED luminaire in the room could be controlled. The maximum achievable illuminance (at the lowest duty cycle of 20%) on the desk was 250 lx. Figure 9.20 shows the luminance distribution as seen by the test person during the test.

Figure 9.20 shows that the luminance was relatively homogeneously distributed both on the table (60 cd m^{-2}) and on the white wall in front of the table (25 cd m^{-2}). With a table luminance of 60 cd m^{-2}, the test subject's adaptation is in the photopic range. The luminance of the cable of the computer mouse was in the range of 5 cd m^{-2}. This results in the following contrast values for the mouse depending on the background: $K_{\text{desk}} = 70\%$, $K_{\text{wall}} = 67\%$.

Another test parameter was the frequency of the PWM. The duty cycles of 20% and 50% were tested at five different frequencies. In addition, there was a constant light setting-duty cycle of 100%. In total, there were 11 different settings, which are summarised in Table 9.3. Their order was randomised once. All tests were then carried out in this order.

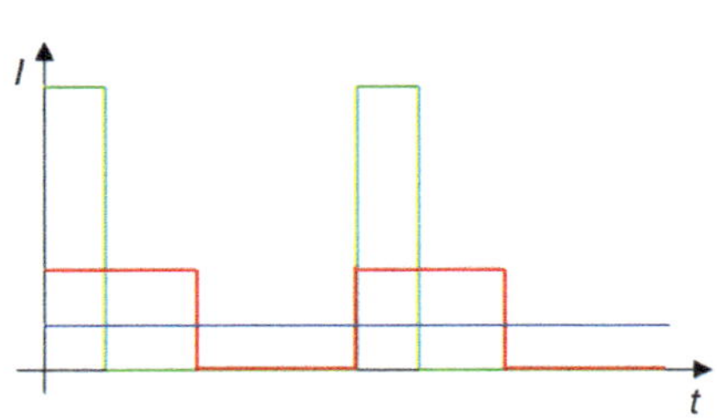

Figure 9.19 Three signal types used in the experiment [1]. Source: TU Darmstadt.

Figure 9.20 Luminance distribution as seen by the test person. Source: Polin and Khanh [1], Technische Universität Darmstadt.

Table 9.3 Settings of the tested PWM signals [1].

Duty cycle			**20%**					**50%**			**100%**
Frequency (Hz)	100	200	300	400	100	100	200	300	400	100	NA

Source: TU Darmstadt.

9.3.4.3 Experimental Procedure

The test procedure was recorded in a PC program developed for this purpose. It was operated by the test leader who was in the room during the entire test. In addition to controlling the lamp, the software was also used to record information about the test person (age, gender, and whether the person wears glasses or contact lenses). In addition, the ratings given by the test subjects were documented with the help of this programme. At the beginning of the examination, each test person was instructed about the procedure. In a '*training phase*', PWM frequencies of 50 and 80 Hz and a duty cycle of 20% were used to illustrate the stroboscopic effects to be evaluated. At the same duty cycle and 100 Hz, a test run was also carried out and the tasks were explained. Using fast hand movements, the difference to a setting with *constant light* was illustrated. The examination took place under conditions that were as realistic as possible. For this purpose, tasks were formulated for the test persons with partly fast but natural movements – at least with regard to the activities in an office. Each test person had to perform 5 tasks at each of the 11 light settings, see Figure 9.21. This took an average of 20 minutes per test person.

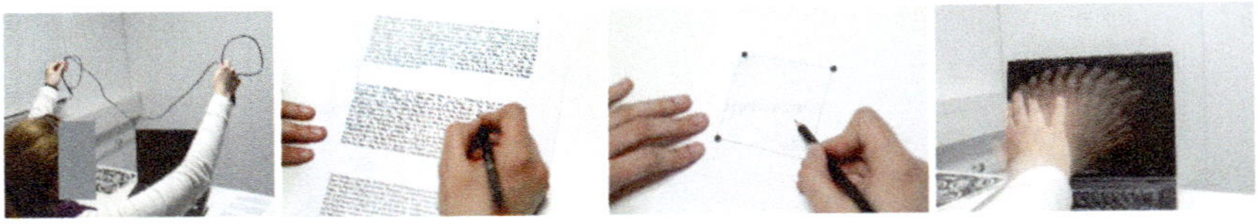

Figure 9.21 Subjects had to perform tasks with partly fast but natural movements typical for office activities. Source: Polin and Khanh [1], Technische Universität Darmstadt.

The subject had the following five tasks (cf. with Figure 9.21):

Task 1: Before each run, the subject stood about 1 m away from the desk. After the light was adjusted, the subject was asked to sit down at the desk. The aim of this task was to have the subject perform natural orientation movements.

Task 2: In this task, the movement of high-contrast objects were investigated. On the table, there was a computer mouse that was to be connected to the notebook so that the cable was behind the screen. The test person was asked to take the cable in their hand and swing it over the screen (Figure 9.21 left). The speed of the cable was estimated as 2–5 $m\,s^{-1}$.

Task 3: Next, the test subjects had to evaluate the lighting in a cognitive task. In a text printed on an A4 sheet, the subjects had to underline certain words. For each of the three paragraphs, the test leader said a number and the test person counted off the words and then underlined the corresponding word. A pen was used for support.

Task 4: The subject then turned the sheet over, connected pre-printed dots to form a square and hatched it, simulating fast typing.

Task 5: Finally, the subject had to move his hand back and forth as quickly as possible.

After each task, the subjects gave two ratings on a scale with five discrete levels from −2 to +2. First, the conspicuity of stroboscopic effects was rated. Here, the extreme values have the following meaning:

−2: The stroboscopic effect is not perceptible
+2: The stroboscopic effect is very noticeable.

Secondly, the disturbance of the perceived effect was assessed in relation to the task:

−2: The stroboscopic effect is not disturbing
+2: The stroboscopic effect is very disturbing.

9.3.5 Results

A total of 38 subjects aged 22–61 took part in the experiment [1]. Among them were 23 men (60%) and 15 women (40%). 13 people (34%) wore a visual aid: 10 people wore glasses, and 3 people wore contact lenses. The histogram of the age distribution is shown in Figure 9.22. The average age of the participants was 33 years. The oldest participant was 61 and the youngest was 22 years old.

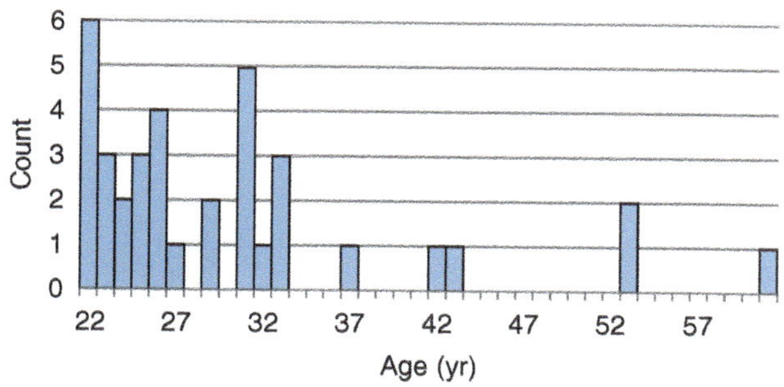

Figure 9.22 Histogram of the age distribution of the test subjects [1]. Source: TU Darmstadt.

9.3.5.1 Mean Subjective Values

Figure 9.23 (left) shows the mean subjective ratings of the conspicuity of the stroboscopic effect for the duty cycle of 20% as a function of frequency, sorted by task. It can be seen that conspicuity is higher for tasks with fast movements (hand movement, connecting the mouse, and hatching the square) than for tasks with slower movements (sitting down, counting words). It is also apparent in all tasks that the conspicuity decreases exponentially with increasing frequency.

Similar behaviour can also be observed with a duty cycle of 50%, see Figure 9.23 (right). As expected, the conspicuity here is lower than at the duty cycle of 20%. This also indicates that the conspicuity increases with decreasing duty cycle.

The same tendencies can also be found in the ratings of the *disturbance factor*, see Figure 9.24. There is, however, a clear difference in the level of the rating. Here, the mean values are generally lower. Furthermore, the slope of the curve is steeper in the range of 200–300 Hz.

Now the question arises, whether an optimal frequency for visual perception can be determined from these results. Of all the tasks tested in the study, task 5 (fast hand movement) was the most critically evaluated. If an optimal frequency is found for this task, it will also be optimal for tasks with slower movements. Based on these considerations, the results for task 5 at a duty cycle of 20% will be analysed in the

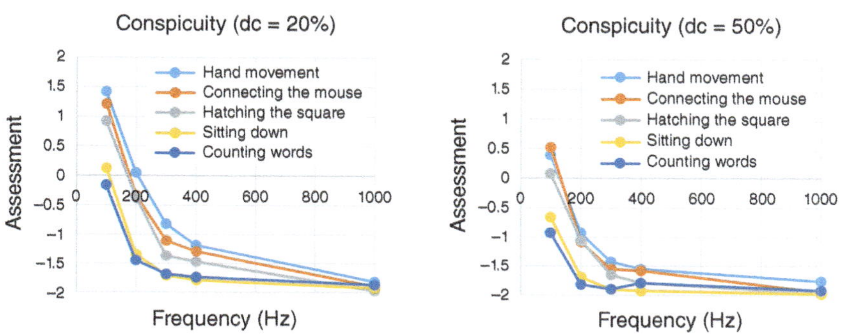

Figure 9.23 Mean subjective assessment of 'conspicuity of stroboscopic effect' by 'task' [1]. dc, duty cycle. Source: TU Darmstadt.

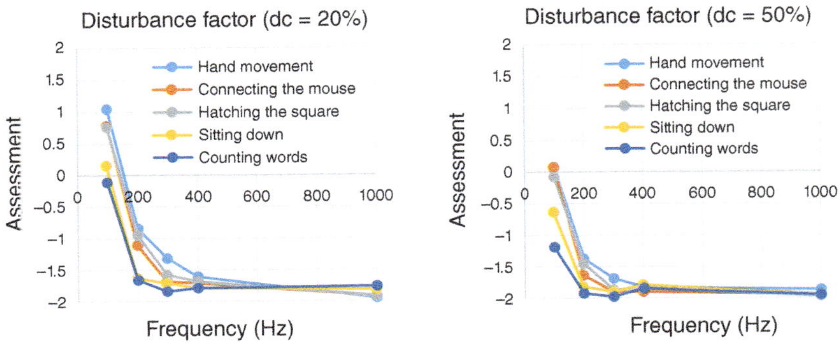

Figure 9.24 Mean subjective assessment of 'disturbance factor' by 'task' [1]. dc, duty cycle. Source: TU Darmstadt.

following. This duty cycle is of great importance for dimming in most lighting applications since a physical dimming of the maximum value of the luminance to about 18–20% according to the logarithmic brightness perception corresponds to a halving of the maximum value of the perceived brightness. As already mentioned, the test persons gave a discrete evaluation. The setting at which no stroboscopic effect was perceptible was to be rated −2.

Consequently, the value −1 can be considered as the threshold for the effect. From the results in Figure 9.23 (left), it can be seen that the mean rating for task 5 reaches the value −1 in the range between 300 and 400 Hz. In other words, on average, different people would not perceive a stroboscopic effect at 400 Hz and a duty cycle of 20%. But in any case, there are people who will certainly perceive it. Therefore, the ratings for task 5 will be looked at statistically in more detail. Figure 9.25 shows the mean value (blue line) as well as the curves for the lower quantile (grey dotted line) and for the upper quantile (grey dashed line). The error bars show the simple empirical standard deviation. It can be seen that the standard deviation is very large up to 400 Hz. The maximum value is 1.3 for 200 Hz. For 200 and 300 Hz, the quantile distance is also the largest. This indicates that different people evaluate the stroboscopic effect very differently.

One reason for the large scatter could be that the test subjects moved their hands at different speeds. However, this can be compensated for by averaging over all test

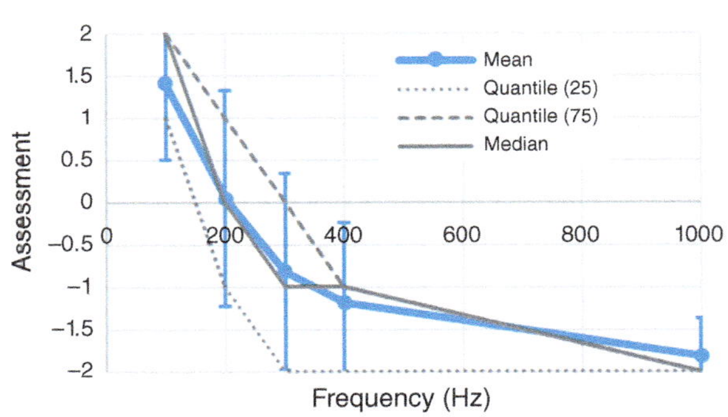

Figure 9.25 Conspicuity assessment (task 5), duty cycle 20% [1]. Source: TU Darmstadt.

subjects. On the basis of a measurement, speeds in the range of 2–5 m s^{-1} were determined for such movements. For the distance to the eye of about 0.4 m, this results in an angular velocity in the range of 250–700° s^{-1}.

Figure 9.26 shows the results of the *disturbance factor* assessment for task 5 at a duty cycle of 20%. The threshold of −1 is reached here on average at slightly more than 200 Hz. The differences in the conspicuity rating show the following: although the stroboscopic effect is perceptible, it tends *not* to be perceived as disturbing.

In the following, the conspicuity assessments of the stroboscopic effect at 200 and 400 Hz will be compared. The reason for this comparison is the widespread use of PWM frequencies in the 200 Hz range in lighting products. Figure 9.27 shows the histogram of the conspicuity ratings at these frequencies and a duty cycle of 20%.

The results in Figure 9.27 clearly show that the stroboscopic effect can be greatly reduced by doubling the frequency from 200 to 400 Hz. While at 200 Hz only 10% of the people did not perceive the effect, at 400 Hz it is about half of all participants. Clear differences can also be seen with ratings greater than −1: at 200 Hz it is 63%, and at 400 Hz it is 21%.

Furthermore, it was checked whether these conspicuity ratings differed by gender. Figure 9.28 shows the mean ratings of these two groups of test subjects for the task with the *rapid hand movement* at a duty cycle of 20%.

Figure 9.26 Assessment of the disturbance factor in task 5, duty cycle 20% [1]. Source: TU Darmstadt.

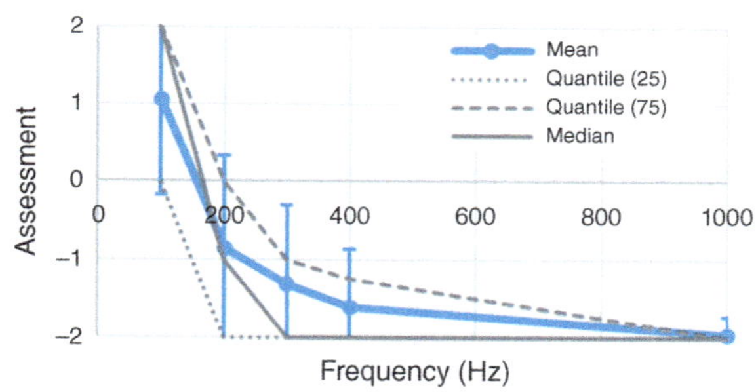

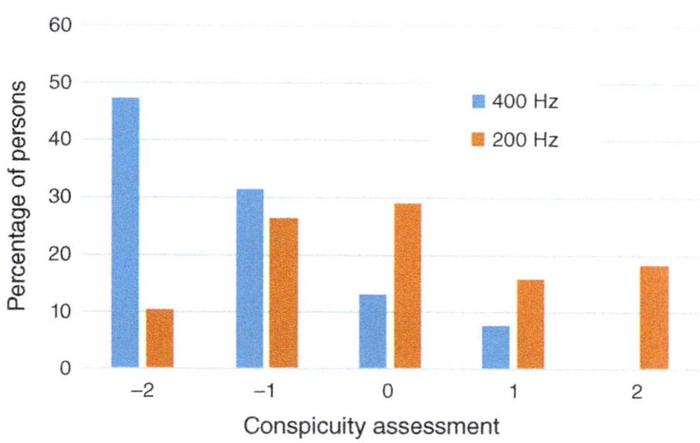

Figure 9.27 Histogram of the conspicuity assessments of the stroboscopic effect in task 5. Duty cycle: 20% [1]. Source: TU Darmstadt.

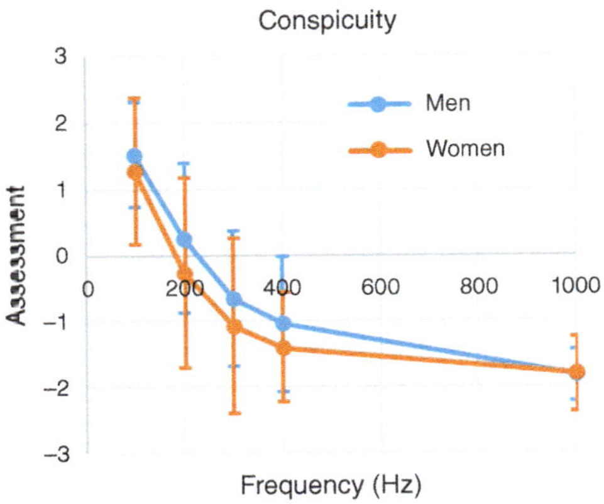

Figure 9.28 Mean conspicuity assessments grouped according to gender. Task 5, duty cycle: 20% [1]. Source: TU Darmstadt.

According to the results in Figure 9.28, women rated the conspicuity of the stroboscopic effect lower on average than men. In order to check whether the differences in the evaluation between these two groups are significant, a U-test was carried out with this data. It turned out that the differences seen in Figure 9.28 are not statistically significant across all frequencies.

9.3.6 Conclusion

The experiment [1] described here had the aim of investigating the conspicuousness of the stroboscopic effect with a pulse-width modulated LED illumination as a function of frequency and duty cycle. Another objective was to determine the degree of disturbance of the stroboscopic effect. The results of the investigation are based on the evaluations of 38 test persons.

The investigation shows that both the conspicuity and the disturbance factor of the stroboscopic effect decrease exponentially with increasing frequency. It is greater for small duty cycles. For movements with speeds around 500° s^{-1}, the PWM frequency should be in the range of around 1000 Hz or higher to avoid the stroboscopic effect. There are office activities where a similar object speed is achieved. For PWM settings with duty cycles higher than 20%, the frequency can be selected lower than 1000 Hz. *A pulse frequency higher than 400 Hz is therefore recommended and appropriate for office lighting.*

Since the examination was carried out with instructed subjects, the results show extreme values with regard to visual perception. In studies with naïve subjects, lower values for the conspicuity of the stroboscopic effect are to be expected.

9.4 Non-visual Light Effects During the Night Hours

In the current research, the non-visual effects of light are divided into two aspects: (i) aspects that come into play during night-time hours without daylight and (ii) aspects during the day with daylight. This makes sense because people's circadian rhythms are determined by the course of daylight and by the waking–sleeping rhythm. Sections 9.4–9.7 deal with the following topics:

(a) *Section* 9.4: Aspects of the effects of light during the night hours.
(b) *Section* 9.5: Psychological and non-visual aspects related to the interior use of daylight.
(c) *Section* 9.6: Influences of light intensity and time of light exposure on sleep behaviour through a literature review.
(d) *Section* 9.7: Effects of light on alertness from the analysis of various publications of the past years.

There is a wealth of information in the areas to be described. The authors concentrate on essential aspects in order to work out a lighting concept based on the current state of knowledge. For many other aspects of sleep research, neurobiology, and psychology, readers are referred to specific literature.

9.4.1 Introduction

Non-visual light effects on phase dynamics in the evening and night hours have been intensively researched over decades up to the present day and represent a potential for lighting technology applications on the one hand in the living area and on the other hand for shift work indoors from late evening from about 10 p.m. until about 6 a.m. the next day. This is a difficult task in today's global society with a 24/7 rhythm. On the one hand, lighting both in the home and in professional spaces should ensure that working people have the best possible alertness, concentration, mood, and well-being to carry out their tasks. On the other hand, after finishing their work, these people should get quality of sleep with sufficient sleep duration, short sleep latency, and less restlessness to generate vitality for the next day. This gives rise to many questions that should be asked of neurobiology, sleep research, occupational medicine and, ultimately, lighting designers as follows:

1. To what extent do the light effects in the evening and night hours depend on the wavelengths and on the light source spectra?
2. To what extent are the lighting effects related to lighting parameters such as illuminance levels at the eye, luminance distribution in the visual field, circadian-effective metrics, as well as the timing and duration of lighting, or more precisely, how should one design a time profile over the shift duration for lighting in the interiors of professional working areas?
3. What additional measures in terms of work technology and occupational hygiene must be applied to ensure a relatively good quality of sleep after the end of the shift?

Answering these and other questions has been the subject of intensive research in recent years. In the nocturnal hours, the light effects (output variables) in Figure 9.29 are generally relevant and useful for health and physiological assessment.

In the case of the photometric input parameters on the left-hand side of Figure 9.29, the time, exposure duration, and time sequence (continuous light,

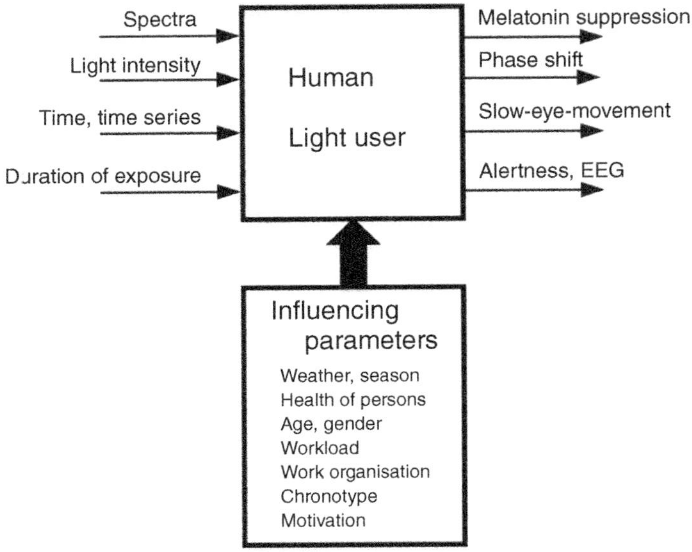

Figure 9.29 Lighting parameters, influencing parameters, and effects of light during the night-time hours. Source: TU Darmstadt.

intermittent light, and pulsed light sequences) are of great importance, in addition to the spectra and light intensity. In the following, the essential research results since the end of the twentieth Century until today are described and some lighting technology consequences for lighting technology will be derived.

9.4.2 Light Effects in Night Hours with Polychromatic White Light

In the year 2000, Zeitzer et al. [17] and Cajochen et al. [18] reported experimental results with a common set-up and with 23 young, healthy test subjects. Zeitzer's research group analysed the experimental data to investigate phase shift and melatonin suppression. Cajochen's research group examined alertness, slow eye movement due to sleepiness, and electroencephalography (EEG) activity. Their protocol and test procedure can be described as follows:

1. The test subjects were 22 men and 1 woman aged between 18 and 44 years (27.8 years ± 8.91). They were informed about the test content and signed a consent form. The test procedure and protocols were supervised by the clinic in charge.
2. Initially, two weeks were planned during which subjects stayed at home and maintained a regular wake–sleep rhythm by calling the clinic immediately before sleep time and immediately after waking to maintain an eight-hour sleep duration. One week directly before radiation, the subject's activity was recorded with an *actigraph* on their wrist.
3. This was followed by three days and nights in baseline mode, during which the test subjects habitually maintained their sleep rhythm in the laboratory, with an

illuminance of less than 150 lx set during the waking phase. In the 50 hours thereafter in a so-called first routine round, the circadian phase was determined by the core body temperature (CBT), whereby the illuminance at the eye was kept at less than 10 lx at the eye in the waking phase and less than 0.03 lx in the sleeping phase.

4. Exposure occurred 6.75 hours before the time of the individually determined minimum of CBT (at approximately 4 a.m. in the early morning) and lasted 6.5 hours. Each subject was exposed in a randomised manner to a single illuminance between 3 and 9100 lx. The light sources were neutral white fluorescent lamps with CCT = 4000 K, which illuminated the test room homogeneously. The light level 4.75 hours before and after the 6.5-hour exposure was approx. 3 lx. The test person maintained a sitting position during the exposure and could always switch once for six minutes between a fixation on a point on the wall and free gaze. Reading and eye closures were not allowed.
5. After every 30 minutes, alertness and sleepiness were determined using the questionnaire KSS (Karolinska Sleepiness Scale). Melatonin was recorded twice an hour with blood samples. CBT was measured continuously. The EEG measurements and the slow eye movements were carried out on 18 test subjects.
6. The exposure was followed by a 30-hour period (second routine round) in which the light effect on the phase of the circadian rhythm was examined. After a final eight-hour sleep period, the subject was discharged on day 9.

9.4.2.1 Results

The data evaluation and the mathematical relationships between the illuminance at the eye and the three light effects or output parameters provide the following statements:

- With higher illuminance levels, alertness increases at night, while at the same time, the number of slow eye movements and EEG activity (the power density of the EEG signals in the frequency range 5–9 Hz) decreases.
- The mathematical relationships between illuminance and the three above-mentioned output parameters follow a logistic function (Figure 9.30) with the four parameters a, b, c, and d according to Eq. (9.6) and Table 9.4. The correlation coefficients are relatively high, between 0.84 and 0.94 (see Table 9.4).
- While the maximum light effects with the three output parameters were reached at 9100 lx, the 50% values for alertness were at approx. 100 lx, for slow eye movements at approx. 180 lx and for EEG activity in the theta–alpha frequency band at approx. 90 lx. This means that the illuminance in the living area in the evening hours as well as in professional indoor rooms during the night shift is capable of noticeably triggering circadian effects. A detailed analysis will follow in the course of this chapter.

$$f(x) = \frac{a-d}{1+\left(\frac{x}{b}\right)^c} + d \quad (9.6)$$

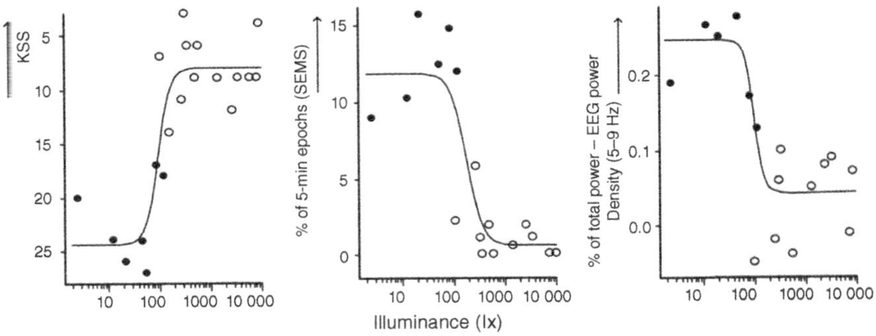

Figure 9.30 Data and mathematical relationships between the illuminance at the eye (lx) and the output parameters alertness, slow-eye movement, and EEG activity according to Cajochen et al. [18]. Source: Reproduced from Cajochen et al. [18] with permission from Behavioural Brain Research.

Table 9.4 Constants *a*, *b*, *c*, and *d* for the logistic functions describing the relationship between illuminance and the variables alertness, slow eye movement, and EEG activity.

Variable	*a*	*b* (lx)	*c*	*d*	r^2
KSS	-24.4 ± 2.1	94.8 ± 17.5	3.7 ± 2.5	-8.11 ± 1.2	0.94
SEMS	-11.9 ± 1.7	179.4 ± 62	2.8 ± 1.7	-0.52 ± 1.3	0.84
EEG	-0.25 ± 0.02	88.8 ± 1.1	98.2 ± 0.0	-0.04 ± 0.02	0.84

Source: Compiled by TU Darmstadt according to the data from Cajochen et al. [18]/with permission of Elsevier.

The phase shift was calculated from the shift of the phase in the first routine round to the phase in the second routine round. The melatonin suppression is calculated according to Eq. (9.7).

$$\text{suppression} = \frac{\text{AUC(baseline)} - \text{AUC(light)}}{\text{AUC(light)}} \tag{9.7}$$

In Eq. (9.7), area under the curve (AUC) (baseline) means the area under the melatonin profile for the time four hours before the maximum of melatonin content in the second melatonin cycle within the first routine round. AUC (light) is the area under the melatonin profile for the same four hours within the exposure phase when the subject was sitting and facing the wall. These results and the diagrams in Figure 9.31 provide the following knowledge:

- The phase shift has the 50%-value at about 120 lx and the 90%-value of saturation at 550 lx. The 50%-value corresponds to a phase shift of about 1.5–1.7 hours.
- The curve for melatonin suppression provides a 50% value in the range 50–130 lx and a 90% value of saturation at an illuminance around 200 lx.

This means that an illuminance with CCT = 4000 K at the eye of a light user of about 100–130 lx can cause a phase shift of about 1.7 hours and a melatonin suppression of 0.5 (or 50%) after an exposure time of 6.5 hours. The test conditions

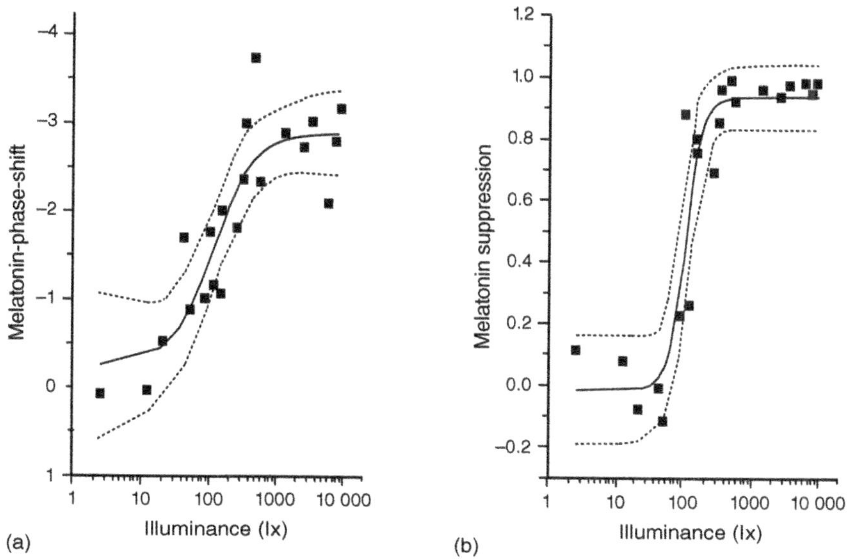

Figure 9.31 Data and relationships between the illuminance at the eye and the output parameters melatonin-phase-shift (in hours) (a) and melatonin suppression in % (b) according to Zeitzer et al. [17]. Source: Reproduced from Zeitzer et al. [17] with permission from the Journal of Physiology.

at the time of Zeitzer and Cajochen's work in 2000 were expressed in illuminance (in lx) and in colour temperature (in K). From Eq. (10.9) for the conversion of illuminance and colour temperature to a CS-value (Circadian Stimulus, see Chapters 2 and 10), we obtain the circadian-effective values in Table 9.5 for the various light effects in the night-time hours.

For a phenomenal description in the scientific and light planning sense, the high correlations between melatonin suppression on the one hand and alertness (KSS), slow eye movement, and *EEG theta–alpha activity* on the other (see Table 9.6) are

Table 9.5 Value of the Circadian Stimulus (CS, Version from 2018) for different categories of different light effects in the nocturnal hours for 1 hour exposure time.

Value category	E_v (lx)	CCT (K)	CS for 1 h exposure
50%-value of melatonin-suppression	120	4000	0.116
90%-value of melatonin-suppression	200	4000	0.2
90%-value of the phase shift	550	4000	0.369
50%-value of alertness	100	4000	0.085
50%-value of slow eye movement	180	4000	0.183
50%-value for EEG activity	90	4000	0.068

Source: Adapted from Cajochen et al. [18].

Table 9.6 Correlation coefficients and significance (p) between melatonin suppression and alertness (KSS), slow eye movement and EEG theta–alpha activity [18].

Variable	r^2	p
KSS	0.89	0.0001
Slow eye movement	0.90	0.0001
EEG	0.72	0.0001

Source: Adapted from Cajochen et al. [18].

revealing. Table 9.6 clearly states that high melatonin suppression also corresponds to high alertness and a higher reduction in slow eye movement and EEG activity. These physiological and psychological phenomena, which are triggered by different mechanisms and bodily functions, are mediated by the same cause, which is formed by light absorption in the photoreceptors on the retina, by signal transmission to suprachiasmatic nucleus (SCN) in the retinohypothalamic pathways, among others, and by signal projections from SCN to further brain areas.

With this work by Zeitzer and Cajochen, as well as through some later studies, the cause–effect relationships with *white polychromatic light* were experimentally determined. The scientific studies for the nocturnal hours were analysed in [19]. In addition to the knowledge gained through the research results of Cajochen and Zeitzer, there is also the knowledge that the concentration of melatonin increases again 15 minutes after the end of the exposure phase. A short exposure or a long exposure at night to suppress the melatonin concentration and to stabilise alertness may therefore be necessary.

9.4.3 Light Effects in Nocturnal Hours with Quasi-monochromatic Light

In addition to exposure parameters, such as timing, light intensity, and exposure duration, the question of which *wavelength* plays a role when considering the non-visual effects of light on people is of great interest, both for science and for practical lighting technology. In 2005, a research group (Cajochen et al. [20]) conducted a study with 10 young male subjects (21–29 years, mean 25.9 ± 3.8 years) who had no health problems and maintained a normal wake–sleep rhythm for one week before exposure. Daily activity was recorded with an *actigraph* on the wrist and with a diary. The test protocol was recorded as follows:

- The test procedure began at 6 p.m., first with a 1.5-hour period with an illumination level of 2 lx, followed by a two-hour phase with total darkness (no light). After this, the actual exposure took place with a duration of two hours either with a blue light (460 nm) with a green light (550 nm) or with no light. After this exposure period, there was a phase of 2 lx on the eye before the subject slept for 7.75 hours. The examination procedure was completed at approximately 10 a.m. the next day.

- The monochromatic radiation was generated with a xenon lamp and with interference filters each with a 10 nm-bandwidth and transmitted with optical fibres to two small integrating spheres. The two spheres were mounted in front of the eye and coated inside with a diffuse sphere cover colour for homogeneous eye illumination.
- The photon densities of the two radiations of 460 and 555 nm wavelengths were the same and equal 2.8×10^{13} photons $cm^{-1}\,s^{-2}$, corresponding to an irradiance of 12.1 $\mu W\,cm^{-2}$ for 460 nm and 10.05 $\mu W\,cm^{-2}$ for 550 nm.

The results of this investigation ([20, 21]) are shown in Figure 9.32.

In each diagram of Figure 9.32, the courses of the dependent variables are shown in the dark adaptation phase (0 lx), in the main phase of light exposure and in the two-hour phase after exposure to 2 lx at the eye. These results can be summarised as follows:

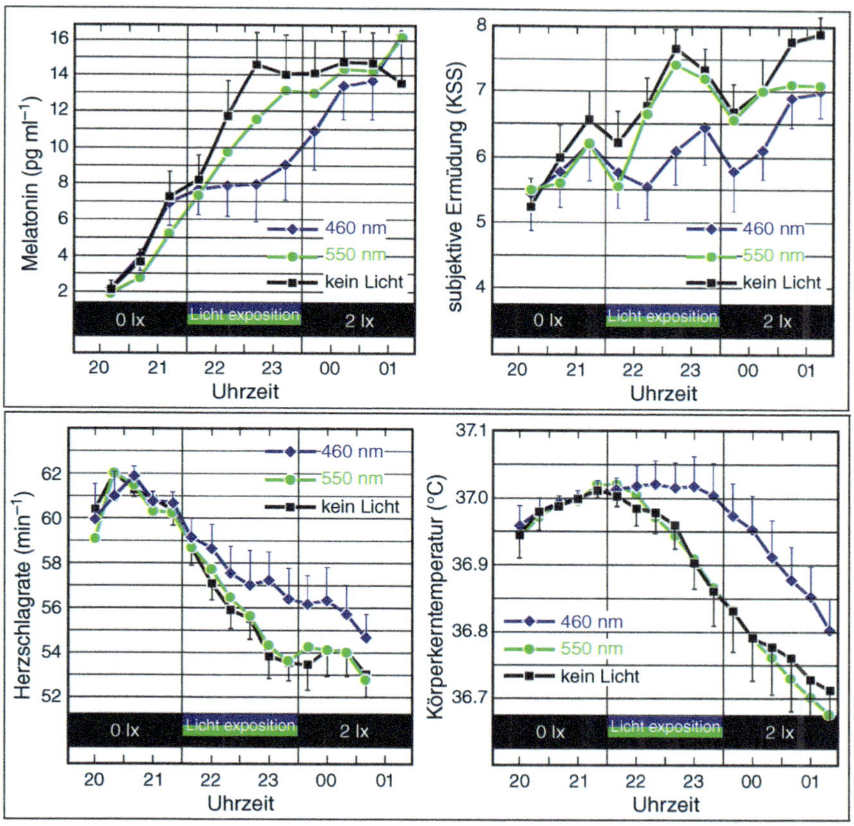

Figure 9.32 Time courses of the four variables 'melatonin suppression,' 'subjective fatigue' ('subjektive Ermüdung'), 'heart rate' ('Herzschlagrate'), and 'core body temperature' ('Körperkerntemperatur') over time ('Uhrzeit') in the dark phase ('kein Licht' = 'no light'), in the light exposure phase ('Lichtexposition') and two hours after exposure to 2 lx [20, 21]. Source: Reproduced from Schierz [21] with permission of the author.

(a) Melatonin suppression at 460 nm is comparatively significantly greater than in the case with 550 nm and with no light. This differentiation tendency started about 30 minutes after the onset of light exposure. Melatonin suppression at 550 nm is much smaller than at 460 nm, yet also significant over the time of exposure. About 1.5 hours after the end of the exposure, the melatonin concentrations in the three cases are again at the same level.
(b) Sleepiness (subjective fatigue as determined by the KSS questionnaire) from 460 nm exposure is much smaller than from 550 nm or no light. The difference between 550 nm and no light is not significant.
(c) The CBT at 460 nm is significantly higher than at 550 nm and with no light and remains higher about two hours after the end of light exposure. There is no significant difference between the results with 550 nm and with no light. The same results apply to the case with the heart rate. These test results lead to the conclusion that radiation of the wavelength 550 nm is not effective with regard to alertness (or subjective fatigue), heart rate and CBT under the given test conditions. The mechanisms that control these three functions are not sensitive or not sufficiently sensitive at 550 nm. For all four physiological and subjective variables described the effect of radiation at 460 nm is statistically significant. Since the 460 nm wavelength can induce sensitivity in both LMS cones, rods, and intrinsically photosensitive retinal ganglion cells (ipRGC) cells, the question of the contribution of the respective receptors to the formation of the behaviour of these four body functions is of substantial importance. For the case of melatonin suppression, some important research findings by Brainard et al. [22], Thapan et al. [23], Rea et al. [24, 25], and Nagare et al. [26] are known, which are described in the following section.

The two separately conducted experiments by Brainard et al. and Thapan et al. aimed to investigate and model the effectiveness of radiation of different wavelengths on melatonin suppression in the nocturnal hours (2:00 to 3:30 a.m. or 11:00 to 2.30 a.m., see Table 9.7) at an exposure duration of 0.5–1.5 hours. The wavelength selection ranged from 420 to 600 nm in both experiments. In Brainard's experiment, each subject was exposed to at least seven different irradiances at each wavelength, with a time interval of at least one week between each irradiation. The eye pupil of the test subject was artificially dilated during the irradiation. Melatonin concentrations were determined for blood samples before, during and after irradiation. Brainard had 627 observations and Thapan had 215 observations with monochromatic irradiations. These action spectra from the Brainard and Thapan experiments were presented in Chapter 2. The measuring points in Brainard's experiment have a maximum around 460–464 nm. Above 575 nm, the effectiveness of melatonin suppression in the data from Brainard et al. [22] is only very low.

The data from Brainard and Thapan were obtained in different laboratories and under different conditions. Nevertheless, they have some things in common. For each measurable irradiance (in $\mu W\,cm^{-2}$) at each wavelength (in nm), there was a measurable melatonin suppression (in %) after the respective irradiation duration. The respective irradiance at the respective wavelength of the two experiments can

Table 9.7 Test conditions for determining the action spectrum for melatonin suppression [22, 23].

First author	Year	Number of test subjects	Age (yr)	Irradiation ($\mu W\,cm^{-2}$)	Exposure	Time	Eyes	Wavelengths (nm)
Brainard	2001	72 (37 women, 35 men)	24.5 ± 0.3	0; 0.01; 1.5; 2.3; 3.1; 6.0; 12.1; 24.2; 42.2	1.5 h	2 a.m. to 3:30 a.m.	Pupil dilated	420; 440; 460; 480; 505; 530; 555; 575; 600
Thapan	2001	22 (4 women, 18 men)	27 ± 7	0.7–65.0	0.5 h	11 p.m. until 2:30 a.m.	Pupil dilated	424; 456; 472; 496; 520; 548; band-width: 13 nm

Source: TU Darmstadt, compilation based on data from Brainard et al. [23] and Thapan et al. [23].

be converted into an illuminance (in lx) using the $V(\lambda)$-function. This determination was carried out by Rea and Figueiro [25].

Figure 9.33 shows the individual data from Brainard and Thapan (melatonin suppression in % as a function of the illuminance at the eye of the test subjects) and a fitted curve. The data set shows a large scatter with $r^2 = 0.38$,which led to the conclusion that the illuminance on the x-axis of this diagram cannot be the correct independent variable. The $V(\lambda)$ function, on the basis of which the illuminance is formed, applies rather to a field of view of 2° and is composed by the cone signal combination L + M. Therefore, a different mathematical signal combination of the LMS cones, the rods, and the ipRGC cells had to be formed to minimise the data scatter in Figure 9.33. This work was carried out for an exposure time of one hour by Rea et al. [24, 25].

9.4.4 Formation of a Metric to Characterise Time-Dependent Melatonin Suppression

From the irradiances at the eyes of the test subjects at the selected wavelengths, Rea and Figueiro [25] calculated the so-called *circadian light* (CL_A) according to Eq. (10.3). The background to this is also given in Section 10.3.2 of the present book. The method according to Rea and Figueiro transforms the *Circadian Light CL_A* with the help of Eq. (10.4) into the so-called 'CS', which is proportional to the melatonin suppression in per cent. A value of CS = 0.4 thus corresponds to a nocturnal melatonin suppression of 40% compared to the state before irradiation. Eq. (10.4) applies in the publication by Rea and Figueiro [25] an exposure time of one hour. In Figure 9.34, the melatonin suppression data of Brainard and Thapan (in %) and the CS values are plotted as a function of the newly defined '*Circadian Light CL_A*'. It can be seen from this that the data scatter has become much smaller compared to

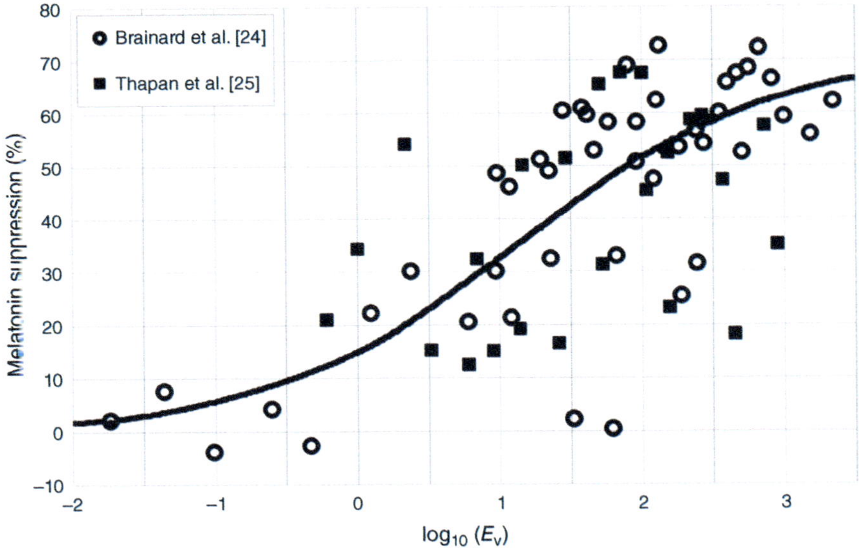

Figure 9.33 Melatonin suppression by Brainard and Thapan (in %) as a function of the illuminance (in log10-lx) at the eye of the test subjects. Source: TU Darmstadt based on data adapted from Rea and Figueiro [25].

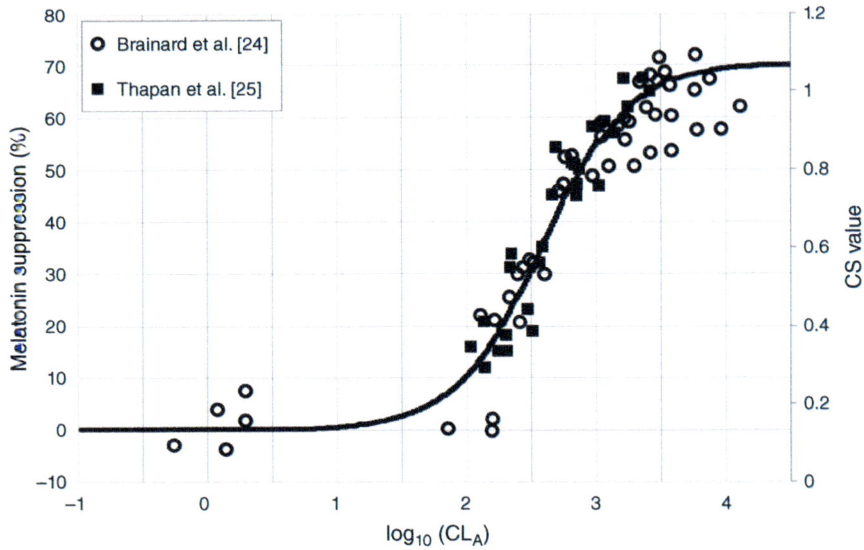

Figure 9.34 Melatonin suppression by Brainard and Thapan (left ordinate, in %) and the CS values (right ordinate) as a function of the newly defined quantity 'circadian light CL_A.' Source: TU Darmstadt based on the model adapted from Rea and Figueiro [25].

Figure 9.33. The mathematically adjusted S-curve in Figure 9.34 yields a correlation coefficient of $r^2 = 0.91$. In Figure 9.34 the model for calculating CS-values, which has been published in 2016 in [25], is used. Rea and Figueiro have improved and published new models in 2020 and 2021.

Since many people work or spend their leisure time in the evenings as well as in the nocturnal hours around or after midnight, an exposure duration of one hour is a special case. To address this, Nagare et al. [26] studied the course of melatonin suppression in the night as a function of light intensity, spectra, and exposure duration. Eighteen adolescents aged 13–18 years and 23 adult people aged 24–55 years were exposed to table lamps with two different spectra (CCT = 2700 and 6500 K), with different illuminance levels from 40 to 1000 lx at the eye and exposure duration between 0.5 and 3 hours. The circadian effective values CL_A, the CS values as well as the obtained results of melatonin suppressions for different exposure durations are listed in Table 9.8.

An analysis of the experimentally obtained dataset in Table 9.8 led to a calculation formula for the CS values by extending Eq. (10.4) by the new variable '*exposure time*' t (in hours).

One thus gets Eq. (9.8) with CL_A and t as new independent variables.

$$CS_t = 0.7 \cdot \left[1 - \frac{1}{1 + \left(\frac{CL_A}{355.7 \cdot t^{-1}} \right)^{1.1026}} \right] \tag{9.8}$$

For the exposure time $t = 1$ hour, Eqs. (9.8) and (10.4) are mathematically identical. For different CL_A values as well as for exposure times between 0.5 and 3 hours,

Table 9.8 Melatonin suppression (%) depending on CS- and CL_A-values measured with a measuring device at the eye in the experiment of Nagare et al. [26].

				Nocturnal melatonin decrease (%) after exposure duration in hours (h)					
Spectrum	**CS**	**CL_A**	**$\log_{10}$ (CL_A)**	**0.5 h**	**1.0 h**	**1.5 h**	**2.0 h**	**2.5 h**	**3.0 h**
2700 K	0.06	41.6	1.62	−1	2	2	2	3	6
	0.11	77.5	1.89	2	7	7	5	11	13
	0.27	233.2	2.37	12	17	21	22	25	25
	0.45	606.2	2.78	26	39	45	50	53	58
6500 K	0.06	41.6	1.62	2	5	0	−1	5	8
	0.11	77.5	1.89	3	7	11	12	17	19
	0.28	246.3	2.39	19	28	33	37	40	44
	0.46	641.7	2.81	26	43	51	57	61	65

Source: TU Darmstadt, compilation based on data adapted from Nagare et al. [26].

the new CS_t values are shown as a function of $\log_{10}(CL_A)$ in Figure 9.35. A threshold value of CS = 0.1, at which melatonin suppression of approx. 10% is reached, which can lead to different CL_A values for different exposure times or to different illuminance levels (in lx) at the eye for a constant light source spectrum.

A CCT = 2700 K in Table 9.9 in the living room area is the typical colour temperature of earlier incandescent lamps and today's LED retrofit lamps. The colour temperature of 6500 K is typical for reading information on PC monitors, laptops as well as television (compare with Figure 9.37).

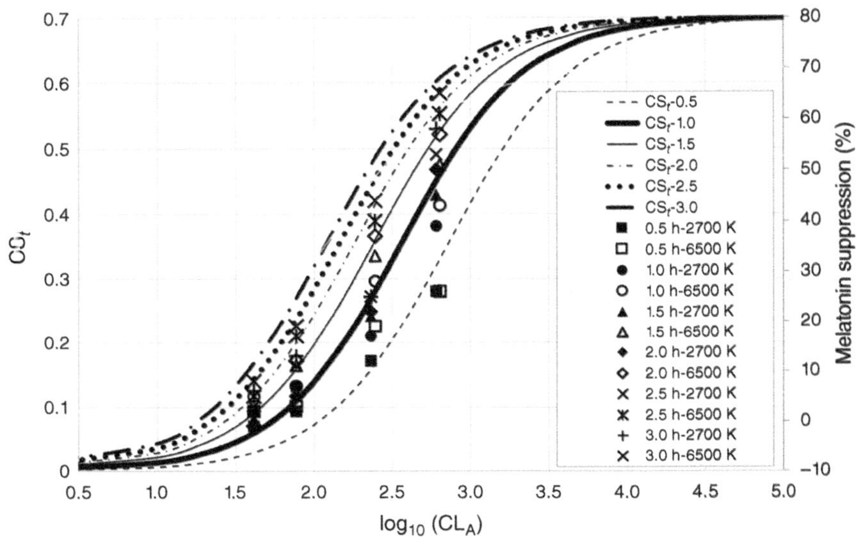

Figure 9.35 CS_t values (left ordinate) as a function of $\log_{10}(CL_A)$ compared with the percentage of melatonin suppression (%) [27] for different exposure times (hours) and CCT (2700 or 6500 K), compare with Table 9.8. Source: TU Darmstadt, based on the model Nagare et al. [27]/Springer Nature/CC BY 4.0.

Table 9.9 Illuminance levels at the eye (in lx) of adolescents and adults at two different correlated colour temperatures (CCT = 2700 and 6500 K) for the case of the threshold value CS = 0.1 or for a melatonin suppression of 10%.

Exposure time (h)	2700 K adolescents	2700 K adults	6500 K adolescents	6500 K adults
1	154	185	71	85
2	125	138	53	74
3	86	104	36	49

Source: TU Darmstadt, based on data adapted from Nagare et al. [26].

9.4.5 Determining the Potential Causes of Melatonin Suppression in Nocturnal Hours

In today's industrialised society, working indoors during night hours is necessary in many professions. In addition, there are other jobs such as home office or homework of pupils and students. At the same time, work with monitors such as PC monitors, laptops, and display devices (as in the surveillance rooms in power plants, ports, and airports) is widespread. Therefore, the question arises how high the exposure situation is when working with such displays. The authors of the book carried out measurements on PC screens at the operating modes 100, 75, and 50 for daytime work (*'normal mode'*, at 75 and 50, the *luminance* of the screen was dimmed from 100% to 75% and to 50% respectively). The operating modes for the night (*'night mode'*) with an increase in the yellow light component were also measured. Figure 9.36 shows the measurement set-up in which a computer monitor was set to different modes and spectrally measured by an absolute spectroradiometer at a measuring distance of 60 cm.

Figure 9.37 shows the measured illuminances and colour temperatures in the different operating modes of the computer display.

Figure 9.36 Measurement set-up for measuring spectral irradiance at a distance of 60 cm from a computer. Source: Technische Universität Darmstadt.

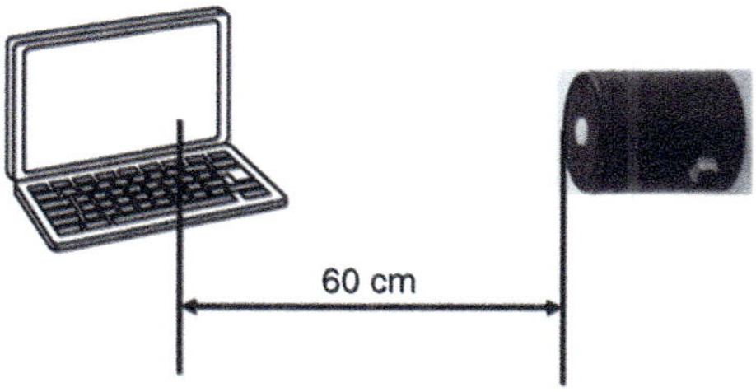

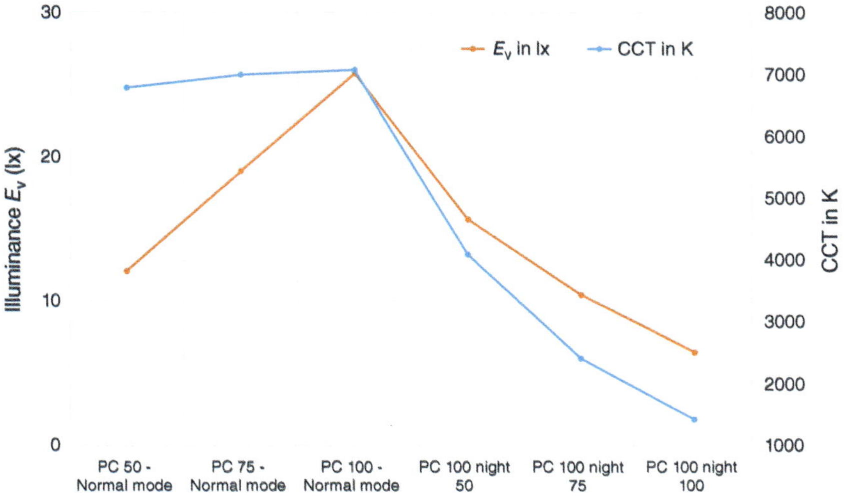

Figure 9.37 Illuminance and correlated colour temperature at a distance of 60 cm from a computer display in its 'normal' and 'night' operating modes (e.g. 'PC 75' means that the luminance of the display was dimmed to 75%, see text). Source: TU Darmstadt.

Table 9.10 Colorimetric, photometric, and CS-values of a measured PC monitor at a distance of 60 cm.

Name/parameter	PC 50-normal mode	PC 75-normal mode	PC 100-normal mode	PC 100 night 50	PC 100 night 75	PC 100 night 100
x	0.1360	0.1343	0.1349	0.1296	0.1280	0.1264
y	0.1472	0.1430	0.1415	0.2238	0.3743	0.5514
CCT (K)	6782	6993	7074	4094	2422	1440
E_v (lx)	12.08	19.02	25.81	15.70	10.48	6.53
a_{mel}	3.47	3.55	3.57	2.72	2.06	1.64
CS (exposure time 1 h, Eq. (10.4))	0.14	0.21	0.26	0.11	0.04	0.03
CS (exposure time 3 h, Eq. (9.8))	0.31	0.41	0.47	0.28	0.12	0.10

Source: TU Darmstadt.

The CCT of the daytime modes ('*normal mode*') was in the range between 6782 and 7074 K. The CS values determined from this for one hour exposure time (according to Eq. (10.4)) and for three hours (according to Eq. (9.8)) are given together with the colourimetric data in Table 9.10.

From the data in Table 9.10, it can be concluded that the CS values for a screen usage time of three hours definitely exceed the threshold value of CS = 0.1. This also applies to the PC monitor in the day modes. Only switching to the night modes PC 75 and PC 100 allows an exposure situation close to the threshold value. Shortening the working time with the monitor in the evening is a possible option, which is not always feasible. Similar measurement results have been reported by Gringras et al. [28] and Chang et al. [29].

9.4.6 Lighting Aspects for Shift Work

Many people work in normal office hours between about 7 a.m. and 6 p.m. and for this reason, their wake–sleep rhythm can be adapted to the light–dark rhythm of natural light in the normal case. Conditions for this are frequent stay outdoors during work breaks and good indoor lighting with artificial light and daylight (see Chapters 10 and 12). One result of this is a relatively high level of alertness during the day and a comparatively lower level of alertness between midnight and the morning hours (e.g. between 0 a.m. and 6 a.m.), which makes good sleep possible with a high probability.

In the German Working Hours Act § 2, the time from 11 p.m. to 6 a.m. is defined as night-time. An exception to this is bakeries and confectioneries with a period between 10 p.m. and 5 a.m. According to this law, night work exists if the work involves more than two hours of night-time. Working night shifts requires performance, vigilance, and concentration during a period when the body is not adapted to it. The workload for people on the early shift, which starts at about 6 a.m. and ends at 2 p.m., depends on the *chronotype*. '*Early birds*' can get up earlier and have fewer problems with the early shift. On the other hand, they have more difficulties with the night shifts. The *evening types* behave exactly the opposite way in this respect.

In Figure 9.38, the *circadian* (C) and *homeostatic* (S) components of people's alertness during the night shift are shown, as well as the sum of these, the so-called *total alertness*.

Initially, when the first night shift begins, according to Figure 9.38, the *circadian* component does not change, while the *homeostatic* component shifts with the night shift. The curves of the two alertness components and the alertness sum show the following:

(a) that the two alertness components and their sum decrease steadily during the night shift between 11 p.m. and about 6 a.m. The minimum is around 8 o'clock when there is no major light effect after the end of the shift.
(b) that during daytime sleep, around 8 a.m. to 4 p.m., alertness generally increases very strongly, which prevents falling asleep and shortens sleep length. In addition, there are other, external circumstances such as noises, light in the living area during the day, as well as children with school programmes.

In general, there are several *timer problems* for night shift workers. The timer's '*light–dark periodicity*' (light at night, darkness during daytime sleep) is strongly *out of phase* with the natural timer. The food intake timer (meals) is also shifted (meals during the night shift). The social timers (obligations, appointments) nevertheless remain in the usual course. Therefore, an adjustment to the new rhythm should be made by means of conceptualised lighting. The task of this adaptation is to ensure

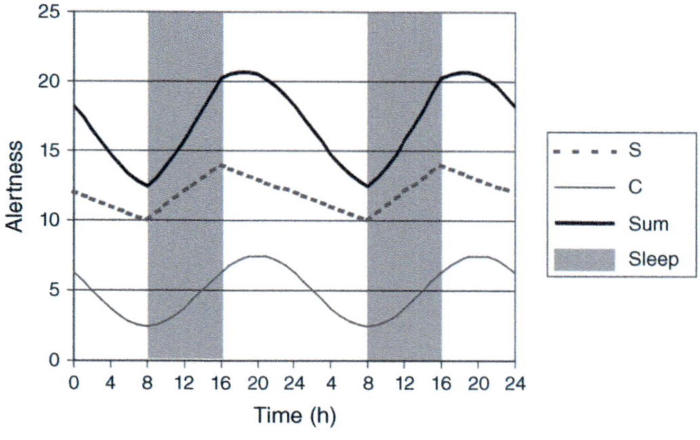

Figure 9.38 Homeostatic (S) and circadian (C) alertness components and the sum thereof ('total alertness') as a function of time of day in hours. Source: Reproduced from Vandahl et al. [30] with permission of the authors.

alertness, concentration, and vitality for people who work in *night shift mode* for longer periods of time and to enable good sleep after the end of the shift as much as possible.

This adaptation as a concept has several components. On the one hand, it is the time (timing) and duration of light exposure during the night shift, whereby these two parameters are closely related, as will be described later. On the other hand, there are the illuminance, the circadian-effective parameters, the spectrum or colour temperature, and the direction of light.

For the adjustment of the circadian rhythm or for a phase shift, it is first important to set the time. Setting light before the time of minimum body temperature, especially in the evening hours from *DLMO* (*'Dark Light Melatonin Onset'*, *i.e. start of melatonin secretion without light effect*), causes a phase extension, a delay of the rhythm, as the body is informed by the light that the person is still in the daytime [30]. The administration of light in the early morning hours shortens the rhythm. Figure 9.39 shows this in a diagram with the so-called PRC (*'Phase-Response Curve'*) [30, 31].

Figure 9.39 provides the following knowledge:

- Exposure to light between 12 noon and 8 p.m. does not cause a phase shift.
- Exposure to light between 8 p.m. and the minimum of the CBT (around 4:30 a.m.) prolongs the phase of the circadian rhythm.
- Setting the light after the minimum of the body's core temperature shortens this phase.
- Since the PRC curve is very steep shortly before and after the minimum of the body's core temperature and since each person has his or her own PRC curve, one should, on the one hand, precisely determine the time of the person-specific minimum and, on the other hand, not plan the time of light administration in this narrow time range, since a phase shift and a phase shortening cancel each other

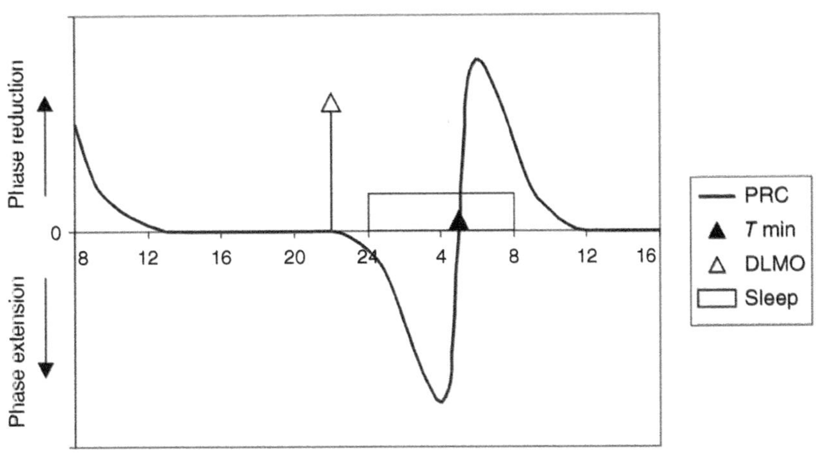

Figure 9.39 PRC (phase response curve) [30, 31]. Source: Reproduced from Vandahl et al. [30] with permission of the authors.

out in their effect. For these reasons, a light setting should take place approximately *before 3 a.m.*

- Since the phase position is individual, as shown above, the starting point of the light exposure should be determined individually in order to achieve the optimal phase shift. Because several people work in an open-plan office or in an industrial hall, for practical reasons the starting point should be fixed in the time range where the PRC curve is relatively flat. This is around 0:00 to 1:00 after midnight for the first night shifts. Due to the phase shift of the circadian rhythm, the starting point for the next night shifts is also shifted accordingly.

It is important to note when looking at Figure 9.39 that the end of the night shift at about 6 o'clock coincides with the maximum phase shortening. If night shift workers get a lot of light on the way home after the end of the shift (e.g. in summer, bright lights in public transport) or turn on bright lights in the living area before daytime sleep, adaptation is also hindered. The solutions are wearing dark sunglasses on the way home, having little light in the living area and having a short period of time between the end of the shift and the start of daytime sleep. The bedroom should be darkened during the day.

Table 9.5 has already shown the illuminance values at the eye of test subjects at the night for different categories of melatonin suppression and phase shift in case of white light with CCT = 4000 K. A 90% value of the maximum of melatonin suppression is possible at an illuminance of about 200 lx. The 90%-value of the maximum of the phase shift, which corresponds to about 2.0–2.5 hours, is reached at 550 lx. (A vertical illuminance at the eye of 550 lx corresponds to a horizontal illuminance of about 1100 lx for room lighting with a high indirect luminous flux component.)

The aim of *phase shifting* with the planned time and with the necessary illuminance and suitable colour temperature should ideally be place the previous time of minimum body core temperature at the beginning of daytime sleep and not at the time of returning home, because this can cause a possible *traffic accident* due to high fatigue or microsleep. The dimensioning of the amount of phase shift should therefore be done individually.

With a time of light setting around and shortly after midnight, which lasts about three hours and includes a vertical illuminance of around 300–550 lx at the eye of the workers with an appropriate spectral distribution of the lighting system, one is able to adapt the rhythm of the individual shift workers to the new sleep–wake rhythm. The results of this are that people are more awake, show higher performance, get more sleep, and have a better quality of daytime sleep [32–34].

With a permanent night shift system, one should strive for a complete adaptation to the new circadian rhythm. For this, as described above, the time of light setting, the duration, the illuminance, and the colour temperature of the light at the eye are important. Table 9.11 shows the results of the phase shift of night shift workers in experiments. Table 9.11 was constructed here on the basis of literature [30] by using only the data from the *field studies*.

After a detailed analysis of Table 9.11, it becomes clear that an exposure time of three hours and a high illuminance in the order of 2590–5000 lx were used there

Table 9.11 Data from three selected field studies on phase shift of night shift workers [30].

First author/year of publication	Subjects	Protocol	Lighting data	Result
Mitchell/1997	16 women 16 men Age: 24.7 yr ± 4.6	7 d baseline, 8 d night shift, Sleeping in dark rooms at home, Time changes with adjustment, and Duration from January to September	3 h of light with 5000 lx at eye level, less than 500 lx in the room	Suitable timing and light exposure of 3 h cause an adaptation to the night shift. Phase shift of 6 h is proven
Kripke/2007	50 young people (18–31 yr) 56 older persons (59–75 yr)	Duration of light exposure: 3 h on 3 d	3000 lx in horizontal viewing direction	Older people have a shorter phase shift, 1.8 h
James/2004	11 nurses (caregivers) as well as 6 men (40.7 ± 8.4 yr) and 5 women (36.9 ± 9.7 yr)	Full night shift, at least 8 d	Fluorescent lamps 4100 K illuminance in the field of vision, 2590 ± 1317 lx; indoors less than 100 lx	Phase shift: 11.04 ± 1.16 h

Source: TU Darmstadt, compiled on the basis of data adapted from Vandahl et al. [30].

to achieve a phase shift up to 11 hours, for example. If, however, another goal is pursued, that the minimum of the body's core temperature should be shifted from normally 4:30 a.m. to about 7 a.m., because night shift workers usually need an hour after the end of the shift at 6 a.m. to get home at about 7 a.m. and to start daytime sleep. According to Table 9.5, this phase shift of 2.5 hours is also possible with an illuminance at the eye of 550 lx. The spectra or colour temperatures can be dimensioned with LED luminaires with variable colour temperatures (see Chapters 11 and 12) between 4000 and 6000 K, depending on the age of the persons.

From the point of view of the circadian rhythm, it is generally complicated to optimise a suitable lighting technique for *rotating* shift workers. Shift workers who have completed their night shifts in a few nights already partially adapt their rhythm to work at night and to a daytime sleep from about 7 a.m. to 3 p.m. After a few days off, they must do their work in the morning shifts from 6 a.m. to 2 p.m., during which they have to concentrate and stay awake. The inner biological rhythm thus never adapts optimally to the outer rhythm and to the work routines.

The task of lighting technology is therefore help to adapt the circadian phase as well as possible to the new time condition. There are still few studies in the literature with long study times and with justifiable lighting profiles for such shift rotation.

Table 9.12 Summary of the study on rotating layers by Lowden and Åkerstedt [35].

Subjects	Protocol	Lighting data
7 subjects in the experimental group with the new dynamic light; mean age 50.3 yr (between 42 and 57 yr). 16 subjects in the control group; mean age 45.9 yr (between 30 and 62 yr)	*Subjective alertness* by questionnaire (Karolinska Sleepiness Scale [KSS], every 2 h) *Sleep* (objectively with actigraphy and subjectively with sleep diary) *Melatonin* (every 2 h during a night shift and the first 3 h during an early shift) to determine adaptation to shift work. Cross-over design between experimental group and control group with an interruption of 7 wk in between	*New dynamic lighting with three types of light:* white/blue, bright light with 745 lx, and 6000 K. moderate yellow light with 700 lx and 2700 K. dim yellow light with 650 lx and 4000 K. *Previous lighting:* CCT = 3000 K and E_v = approx. 200 lx (vertical at the eye)

Source: TU Darmstadt, Adapted from Lowden and Åkerstedt [35].

A notable study of rotating shift workers that proposed a lighting profile for phase stabilisation and examined its health effects on shift workers was conducted by Lowden and Åkerstedt in [35] in Sweden during the winter period to minimise the effect of daylight. Table 9.12 contains a summary of this study.

The time regime of layer rotation, the illumination profile, and the alternation of the three types of dynamic light are shown in Figure 9.40.

In the first three *night shifts* in Figure 9.40, which started at 0:00 on Saturday (N1–N2–N3), the *dynamic* light was applied. In the first night shift, N1, the bright blue–white light with 6000 K and 745 lx at the eye was used until 2:00 to accelerate the phase shift. In the night shift N2, this type of light was only used for one hour with a time delay starting at 1 o'clock and in the night N3, this type of light was no longer used to slow down the rhythm adaptation. In the remaining hours of these three night shifts, '*yellow light*' with 2700 K (light source Philips *TL5 827*) and with 650 lx was used. The illuminance was relatively high for visual performance, but the colour temperature was still in the '*ww–yellow*' range as with a previous incandescent lamp.

On the Monday after the night shift ended at 7 a.m. until 7 a.m. on Wednesday (i.e. in 48 hours), the subjects spent their free time. At 7 a.m. on Wednesday, they started the early shift M1 and the following day M2. On these two early shifts, the bright bluish-white light was used from 8 a.m. to 3 p.m. to maintain alertness. The phase delay could take place again (see also Figure 9.39 with the PRC curve). The late shift A1 on Friday started at 3 p.m. and the light was set according to the daylight setting in wintertime in Sweden. The bright blue–white light was maintained until 4 p.m. Then the moderate yellow light with 2700 K/700 lx at the eye was used between 17:00 (5 p.m.) and 19:00 (7 p.m.). From 20:00 to 23:00 (end of shift), the light with 2700 K and 650 lx was applied. At 24:00 on Friday, night shift N1 began

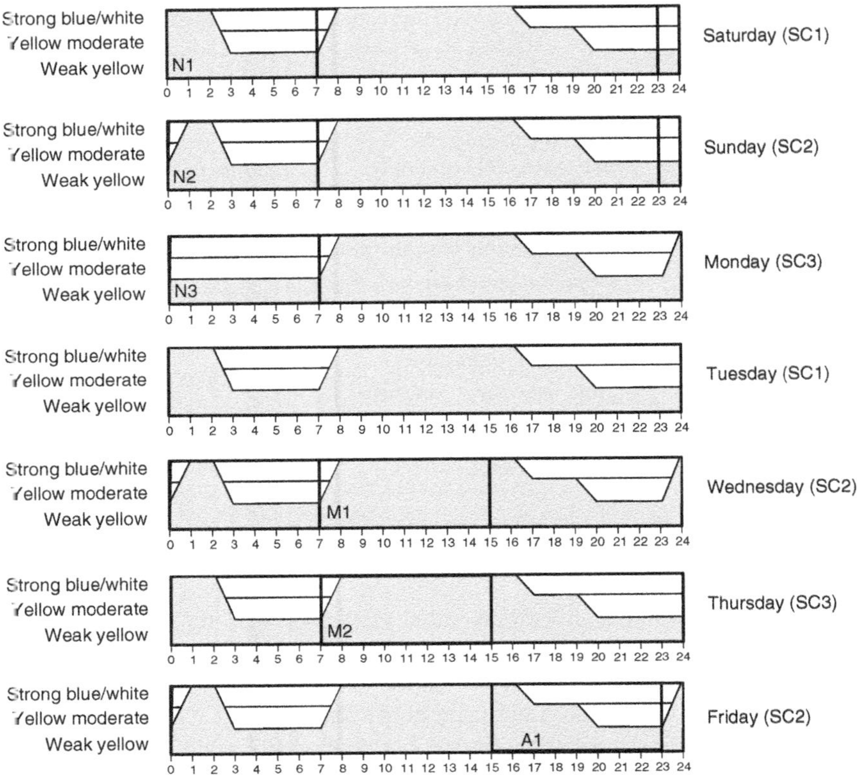

Figure 9.40 Time regime of layer rotation, illumination profile, and the alternation of the three types of dynamic light in the study by Lowden and Åkerstedt [35]. Source: Reproduced from Lowden and Åkerstedt [35] with permission of the journal Chronobiology International.

again. The normal light of the control group corresponded to 200 lx/3000 K and this light was kept statically constant over all shifts.

As for the results of the study [35], alertness was significantly increased during the second night shift. There was a significant interaction between the variables study group × light × time ($p < 0.01$). Wake-up time after the third night shift was significantly later due to dynamic lighting. As the sample (number of subjects) was small, the effect of dynamic lighting on melatonin levels was not clearly detectable. Nevertheless, the authors [35] considered their intervention as a possibility to favour the adaptation to a fast-rotating *shift system* (in winter).

In the dynamic lighting solution of Lowden and Åkerstedt [35] some noteworthy aspects can be seen. The light intervention in the first two night *shifts* N1 and N2 with bluish-white lights and with a relatively high illuminance before 2 and 3 o'clock is *effective for a phase shift*. Likewise high illuminance with the bluish-white light throughout the entire early shift is recommended for stabilising the wake–sleep rhythm during the day with alertness and concentration compared to current practice in many workplaces. Dynamic lighting with a reduction of colour temperature on the afternoon shift from 8 p.m. onwards is also a correct decision. The reader of this book will find more considerations on dynamic lighting in Chapter 12 (Section 12.5, Figures 12.10 and 12.11).

9.5 Psychological and Health Aspects of Daylight

The psychological and health aspects of daylight are shown in Figure 9.41.

Sections 9.5.1 and 9.5.2 are partly based on the studies by Galasiu and Veitch [36] and Aries et al. [37].

9.5.1 Psychological Aspects

Most research into the psychological aspects of daylight has been conducted to date in an era when the quality of light and colour was determined by fluorescent and incandescent lamps [36]. Cuttle [38] examined the relationship between people and windows at workplaces as early as 1983 by asking 471 office workers in England and in New Zealand about it with questionnaires,

(a) whether people considered the window to be part of a workplace and if so,
(b) how important is the window to the people seems and why this is so?

Ninety-nine per cent of respondents affirmed the use of windows and 86% considered daylight as the preferred source of light, as daylight at work causes less stress and more comfort than electric light.

Heerwagen and Heerwagen [39] conducted a survey in 1986 in an office building in winter and summer in the United States. More than half of the people surveyed felt that daylight was the better light source for psychological comfort, office appearance, health, and colour rendering of people and furniture. In general, people considered workplaces with daylight to be a comfortable working environments.

There have been a number of careful studies that have tried to determine the optimal window size using psychological methods. Generally, the openings and views to the outside provide the opportunity to get daylight with sunlight and skylight. In addition, visual access to the landscapes in front of the building and to weather

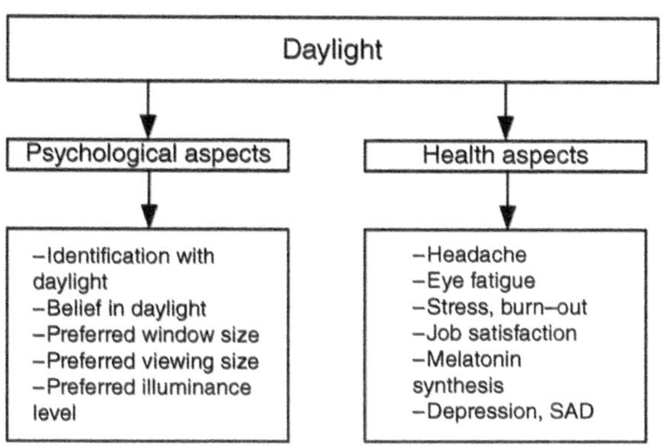

Figure 9.41 Systematics of the topics related to the use of daylight. Source: TU Darmstadt (partly based on the literature studies by Galasiu and Veitch [36] and Aries et al. [37]).

information is possible with the dynamics and the change of the landscape illuminated with daylight with the windows and skylights.

Ne'eman and Hopkinson [40] conducted research in 1970 on the critical minimum window size and on the provision of outward vision. The experiments were carried out in three office buildings in England with 318 test subjects between January and July under different sky conditions. The room geometry was 7.3 m length × 5.5 m width × 3 m height with the bottom edge of the windows 0.9 m above the floor. The result provided the knowledge that the minimum window size is determined more by the need for visual information than by daylight incidence. For two fixed window heights of 1.5 and 2.1 m, the minimum window widths were between 2.2 and 3.2 m. The ratio of window size to distance from the observer to the window was constant according to the test results. Table 9.13 lists the calculated percentages of the window area in relation to the façade area.

In [41, 42], Keighley carried out various investigations into window shapes, window areas, and arrangements, with the resultant finding that most subjects favoured a wide lateral view between the '*skyline*' and the horizon. Large horizontal window sizes of more than 25% of the wall size were favoured and window sizes smaller than 10% were considered insufficient.

There has also been a good amount of research using psychophysical methods to determine the best lighting condition (e.g. illuminance at the workplace) in daylight buildings. One notable study was the work of Laurentin et al. [43], which examined 30 test subjects in the two months of March and April to see how they adjusted to artificial light under different daylight conditions. The investigations took place at computer workstations at different distances from the window in two rooms facing east. The task of the test persons (each divided into groups of two) was to adjust the illuminance levels at different workstations. At the end of the adjustment, they were asked about their perception of the working environment by means of questionnaires. These results can be presented as follows:

1. At the workplaces close to the window, 57% of the subjects did not want any additional electric light, while the remaining 43% adjusted the electric light between 20 and 450 lx additionally. This range of settings is remarkably large. The lighting level with artificial light and daylight totalled approx. 1200 lx.
2. In the middle of the two rooms, 40% of the people did not set any additional light and the rest added an illuminance between 30 and 580 lx to the daylight amount.

Table 9.13 Percentage of window area in relation to façade area [40].

Facade size	Minimum window size	Percentage of window area with a minimum size
5.5 m width × 3 m height = 16.5 m^2	2.2 m × 1.5 m = 3.3 m^2	20%
	3.2 m × 2.1 m = 6.72 m^2	40%

Source: TU Darmstadt based on data adapted from Ne'eman and Hopkinson [40].

3. At the workplaces away from the window with a daylight contribution of about only 120 lx, 30% of the test subjects did not want any additional daylight. In contrast, the other test subjects wanted to add an additional 20–350 lx. This means that the workplaces furthest from the window were illuminated to the maximum with about 470 lx.
4. The workplaces at the window with about 1200 lx were preferred the most. The order of magnitude of 1200 lx for the preference at the window workplace is described in more detail in Chapter 12 (see Table 12.14).

An extensive field study was conducted by Moosmann [44] in nine office buildings in Germany. The results of the evaluation of almost 1000 questionnaires are summarised below:

- 67% of the respondents stated that they would not like any change in the lighting level at the mean value of the horizontal illuminance at the workplace of 966 lx.
- The lighting situations in which the lighting users did not express any wish to change the lighting level were subject to the influence of age. The horizontal illuminance at the workplace where the light users did not make any change requests was 1168 lx for people under 30 years of age, 1006 lx between 30 and 39 years of age, 883 lx between 40 and 49 years of age, and 834 lx for people over 50 years of age. These results were statistically significant.
- The desired horizontal illuminance was dependent on the season, higher in spring (1046 lx) than in summer (927 lx) autumn (875 lx), and winter (787 lx).

The results of Mossmann [44] and those of Laurentin et al. [43] are in the same order of magnitude between 966 lx (67% of the respondents in the case of Moosmann) and 1200 lx at the window (57% in the case of Laurentin et al.). Since natural light enters the room predominantly diffusely under varying weather conditions and sky cover, the vertical illumination component at the face of room users is large and can be estimated to be in the order of 480–600 lx at the eye.

9.5.2 Health Aspects of Daylight

For thousands of years, daylight (sunlight and skylight) has been linked to the physiological, psychological, and circadian-rhythmic health of the earth's inhabitants [37]. Daylight exhibits a high dynamic of spectral distributions, colour temperature, and illuminance in summer and winter (see Sections 10.4 and 10.5 for detailed information). This also changes the visual weather information (clouds, blue sky, cloud movement, sun and cloud, rain, and snowfall), the direction of light (sunlight irradiation over the day in summer, position of the sun, and shadow) as well as the scenic scene in front of the building mediated by the daylight, e.g. a park, meadow, or garden in winter or summer, in the morning or in the afternoon. Such phenomena also enter literature, philosophy, and people's consciousness, shaping their behaviour over centuries. All this forms the complexity of the term 'health', which is defined by the World Health Organisation (WHO) as follows:

> *Health is a state of complete physical, mental and social well-being and not merely the absence of disease or infirmity.* [45]

Thus, the WHO defines health as a broader concept that includes physical, mental, and social aspects of human existence and not the absence of disease alone. In a comprehensive study with a multilevel approach, Denissen et al. [46] analysed the effects of six weather parameters (temperature, wind, sunlight, precipitation, barometric pressure, and photoperiod) on daily mood with positive affect or negative mood and fatigue. The study results highlighted the importance of temperature, wind, and sunlight as mental mediators.

The effect of daylight on various components of people's health in an increasingly industrialised society must do first of all with the parameter '*daylight hours and daylight exposure*'. Bodis et al. [47] examined the relation between daylight hours and heart attacks and found a weak negative correlation, stating that an increasing number of hours spent in daylight is associated with fewer heart attacks. The number of available daylight hours depends on the time of year.

Alimoglu and Donmez [48] investigated the influencing parameters of the phenomenon of '*burn-out*', which is expressed in long-term exhaustion and absence of interest. The research methodology was based on the analysis of the daylight utilisation time of nurses at a university hospital. No direct correlation between daylight exposure and burnout was found, but an indirect correlation between job satisfaction and stress was found. More daylight led to less stress and more satisfaction.

Cowling et al. [49] investigated the lighting conditions in office buildings in Brisbane (Australia) and their consequences on health. Of the 310 questionnaires distributed in nine office buildings, 254 questionnaires were returned and statistically analysed. The results significantly showed the correlation that increased use of natural light is associated with less eye strain and eye fatigue.

Wilkins et al. [14] analysed the relationship between daylight exposure and headaches. Twenty test subjects worked in a building on different floors with an increase in daylight illuminance of 80 lx per floor upwards, measured on a sunny day. Statistical analysis of the results confirmed that headaches decreased with floor height ($z = 2.13$, $p < 0.02$). Robertson et al. established the relationship between light level and the frequency of headaches in a study in 1989 [50] in two buildings with a total of 106 people participating in the survey. In building with less daylight and generally with lower illuminance at the workplace, the frequency of headaches was statistically significantly higher ($p < 0.001$).

In general, the proper use of daylight not only has an effect on alertness, on reducing stress during the day or on reducing headaches and eye fatigue but also has positive effects on what happens at night. Good lighting at the workplace, combined with daylight, shapes the amplitude of the light–dark or wake–sleep rhythm. Cheung et al. [51] examined the relation of daylight exposure to physical sleep activity and quality of life. With the analysis of daylight exposure and with the questionnaire on sleep quality (Pittsburgh Sleep Quality Index), it was statistically proven that the 49 people with a window received significantly more daylight ($p < 0.005$) and on average 47 minutes more sleep duration (objectively recorded with *Actiwatch*, $p < 0.05$).

The topic of '*daylight*' is also directly related to the *season* because of the different day lengths, and times of sunrise and sunset. In summer, sunset in Central Europe takes place between 21:00 and about 22:00 and influences the increase of melatonin. Melatonin, in turn, is closely related to various diseases such as cancer, insomnia, depression, dementia, and diabetes. Breast cancer is a common disease. Women with malignant tumours have significantly lower concentrations of aMT6 (6-sulphatoxymelatonin) in urine over 24 hours than women with benign tumours. A study by Obayashi et al. [52] demonstrated that daylight at a minimum of 1000 lx for a period of 37–124 minutes (mean of 72 minutes) in an experiment with daylight during the day correlates positively with aMT6 excretion in elderly subjects.

The first scientific field of work that dealt intensively with the non-visual effect of light on people was light therapy. According to Wirz-Justice and Bromundt [53], observations made by neurologists in the nineteenth century on patients with affective disorders revealed that a main characteristic of depression is the change in biological rhythm (sleep disturbances, rapidly changing states between depression and mania, and seasonal affective disorder/SAD). This depression can be partially treated with light therapy using daylight and artificial light. Thus, light therapy specifically, and the effect of light in general, carries great significance for today's society. The Foundation for German Depression Aid issued the following balance sheet (https://www.deutsche-depressionshilfe.de/depression-infos-und-hilfe/was-ist-eine-depression/haeufigkeit):

> *Depression is one of the most common and most underestimated illnesses in terms of its severity. Currently, 11.3% of women and 5.1% of men in Germany suffer from depression. Women thus suffer from depression about twice as often as men. In total, 8.2% of the German population falls ill in the course of a year. This corresponds to 5.3 million.*

Table 9.14 shows various recommendations of illuminance levels or illuminations at the eye for light therapy, which will be discussed in more detail later. These recommended values should also be realised in the everyday life of many people, not only in therapeutic practice, but also in workplaces and in the private sphere, so that depression can be treated sustainably.

The importance of daylight on the psychological and health aspects is very high and more research is needed to document the daylight effects especially with more field studies and with more test subjects and to make them more popular. A lot of research on daylight has been done at a time when fluorescent lamps with significant radiation and colourimetric deficits were still widely used in buildings. With the advent of LED technology, the question arises whether daylight and the best LED spectra already have similar colourimetric, brightness perception, and non-visual effects. The authors of this book conducted research in this area. In Section 9.5.3, some results of the comparisons between daylight and artificial light with LEDs or discharge lamps are presented.

9.5.3 Quantitative Characteristics of Daylight and Electric Light – A Comparison

In the articles about daylight and the comparison of daylight to artificial light, quantitative and numerical statements are often missing with which light sources

Table 9.14 Recommendations of illuminance levels or illuminations at the eye for light therapy.

Disease pattern	Therapy recommendation	Literature
Seasonal affective disorder (SAD)	Light in the morning, individual timing according to chronotype (MEQ), 7000–10 000 lx for 30 min as initial dose, min. 2 wk. Depending on the patient's response to the light treatment, the dosage of light can be changed by gradually increasing the duration of light therapy by 15 min every 3–4 d	[53]
Alzheimer's dementia	Clear light–dark changes and sufficient light during the day (1000 lx), long-term treatment (months, years), if possible: start at an early stage	[53]
S-SAD (subsyndromal SAD)	Short-term irradiation with 1000 lx provides positive effects such as good moods	[54]
SAD	An illumination of 5000 lxh is necessary for light therapy	[55]
Dementia	In one study, longer sleep duration and less depression were achieved in a home for people with dementia with a light intervention with CS ≥ 0.6	[56]

Source: TU Darmstadt (compiled from the literature shown).

of different technologies and origins could be compared with each other. For example, a daylight spectrum and a so-called '*LED full spectrum*' that has often been described as '*similar to daylight*' *in* recent years can only be compared with each other if the photometric, colourimetric, and circadian data show similar value ranges at the same illuminance (such as E_v = 1000 lx) or at the same brightness. For the characterisation of daylight and artificial light, the following parameters are used here, see Tables 9.15–9.17.

Most readers are familiar with the colourimetric and brightness parameters in Tables 9.15 and 9.16. The *light quality* parameter SP (scene preference) in Table 9.17

Table 9.15 Colourimetric parameters for the comparison of daylight and artificial light.

Parameter	Designation	Literature
Colour rendering	*CIE* R_a, R_9, R_{1-14}	[57, 58]
Colour rendering	*IES TM-30-15* R_f	[59, 60]
Colour preference	*NIST CQS* Q_p	[61]
Colour gamut	*IES TM-30-15* R_g	[59, 60]
Correlated colour temperature (CCT)	CCT	[62]
Colour saturation	*NIST CQS* ΔC^*(CIELAB)	[61]
Chromaticity distance from reference curve	$\Delta u'v'$	[62]

Source: TU Darmstadt.

Table 9.16 Brightness parameters for the comparison of daylight and artificial light.

Parameter	Designation	Literature
Brightness (Fotios and Levermore)	$L_{eq,Fotios}$	[63]
Brightness according to Eq. (7.1)	B	[64–66]

Source: TU Darmstadt.

Table 9.17 Circadian parameters and light quality parameters for the comparison of daylight and artificial light.

Parameter	Designations	Literatures
Melanopic effect factor	a_{mel}	[67]
Circadian Stimulus	CS	[24, 25]
Light quality	SP	[64–66]

Source: TU Darmstadt.

according to [64] was obtained in visual experiments at the TU Darmstadt and experimentally proven in a series of experiments in three different Japanese museums by lighting scientists and found to be suitable [68]. This parameter includes saturation (emotional effect), colour temperature (light balance, partly circadian effect), and illuminance (visual performance). When considering circadian effects, there are currently open discussions among experts as to which parameters are useful for circadian lighting design. Therefore, both the melanopic effect factor a_{mel} and the 'CS' are used. In the opinion of the authors of this book, each parameter has its own disadvantages, so that lighting research should continue to work on a new relevant parameter.

For the daylight–artificial light comparisons, daylight and fluorescent lamps as well as thermal radiators (incandescent lamps, halogen lamps) are used here as artificial light sources in indoor spaces for the time period rather until about 2015 (Figure 9.42) and daylight and fluorescent-converted white LEDs as artificial light sources for present and future indoor lighting (Figure 9.43) are considered.

Figure 9.42 shows that the incandescent lamps and halogen tungsten lamps have a continuous spectrum with a high relative proportion in the wavelength range from approx. 600 nm and a comparatively low proportion in the blue-cyan range up to approx. 500 nm. The fluorescent lamps show some emission lines around 405, 435.8, 480–490, 546.1, and around 615–620 nm as well as some prominent spectral gaps. The measured daylight spectrum has a full continuous spectrum from about 410 nm (violet). The relative spectrum contains radiation values between about 440 and 600 nm, which are more than 80% of the radiation maximum.

In Figure 9.43, three standardised daylight spectra are shown (Commission Internationale de l'Éclairage (CIE) spectra with 5000, 5500, and 5800 K). The choice of white LEDs is varied and ranges from WW.pc LEDs (ww phosphor converted

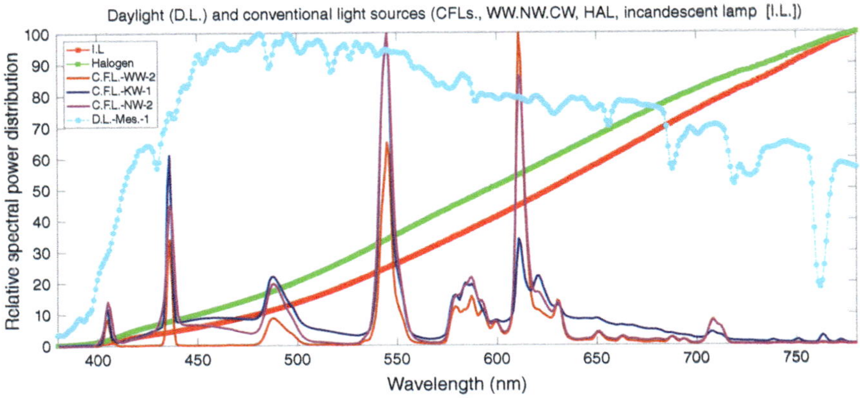

Figure 9.42 Daylight spectrum ('D.L.-Mes.-1,' measured in Darmstadt) and spectra of fluorescent lamps (warm white WW, neutral white NW, cool white KW), incandescent lamp, and tungsten halogen lamp. Source: TU Darmstadt.

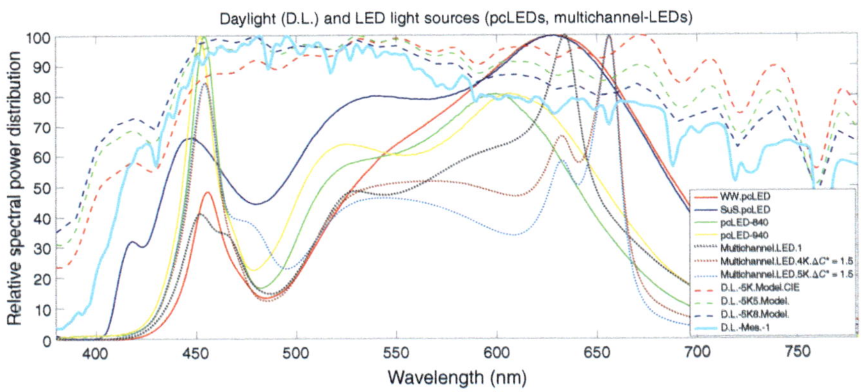

Figure 9.43 Daylight spectra (D.L.), spectra of phosphor converted and multichannel white LEDs (pcLEDs) with different phosphor components and colour properties. Source: TU Darmstadt.

LEDs), pc LEDs with 840 and 940 properties (R_a = 80 and 90 at CCT = 4000 K), multi-channel LEDs with 4000 and 5000 K and with increased saturation of ΔC^* = 1.5–1.8 (see Section 7.3). The red curve in Figure 9.44 shows a daylight-like LED full spectrum (designation of the authors: SuS.pcLED) with a considerable proportion of radiation from 420 nm, with a relatively remarkable proportion in the red wavelength range up to 750 nm and with a maximum around 625–630 nm. Figure 9.44 shows this daylight-like full LED spectrum together with a daylight spectrum (light blue curve; measured in Darmstadt on a summer day) as well as their ratio as a function of wavelength (green curve).

Compared to a measured daylight spectrum, the LED full spectrum in Figure 9.44 shows lower radiation components in the wavelength range from 390 to 580 nm (violet–blue–green areas) and from 670 nm (deep red areas). The loss of radiation

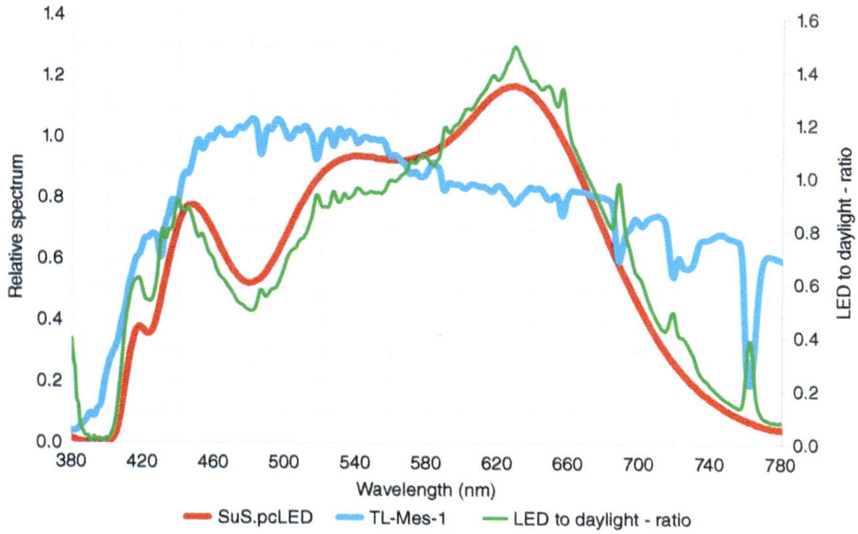

Figure 9.44 Red curve (SuS.pcLED): a full LED spectrum similar to daylight; blue curve (TL-Mes-1): daylight spectrum (measured in Darmstadt); green curve: the ratio of the two spectra. Source: TU Darmstadt.

components compared to daylight is pronounced in the wavelengths around 480 nm, where melanopic efficacy is the highest. Compared to the LED spectra, the daylight spectra in Figures 9.43 and 9.44 have a high proportion of radiation before 540 nm and after 660 nm. Table 9.18 contains the colour quality parameters of the light sources in Figures 9.42–9.44.

Table 9.18 reveals the following tendencies:

- The fluorescent lamps of type 850 and 840 considered here have a moderate colour rendering index around 80, with a desaturation of the light colour (negative ΔC^* values, see Section 7.3) and a colour preference index in the moderate range: National Institute of Standard and Technology Colour Quality Scale (NIST CQS) Q_p between 73 (for ww fluorescent lamps) and 85 (for cool white lamps).
- In contrast, certain fluorescent-converted, hybrid-multi-channel and so-called '*daylight-like*' white LEDs have a high colour rendering index (R_a or R_f) and a NIST CQS Q_p colour preference comparable to each other, which are very close to the corresponding characteristic values of daylight.
- Some fluorescent-converted and hybrid multi-channel white LEDs have a slightly larger colour volume (colour gamut) than daylight spectra. *Overall, LED technology has succeeded in developing white LEDs with different colour temperatures and with that colour quality that is similar to the colour quality characteristics of daylight.*

The brightness values of the light sources to be compared are determined with the help of the two brightness models by Fotios and Levermore [63] ($L_{eq,Fotios}$) and Khanh et al. [64–66] (B, see Eq. (7.1)) at the same illuminance of 1000 lx. The model

Table 9.18 Colour quality parameters of daylight and the artificial light sources in Figures 9.42–9.44.

Type	Non-LED		Fluorescent			pcLED				Multi-channel LED			Daylight			
Name	I.L.	Hal.	CFL-WW-2	CFL-KW-1	CFL-NW-2	WW-pcLED	SuS-pcLED	pcLED-840	pcLED-940	Multi-LED 1	4K-LED-DC1.5	5K-LED-DC1.5	D.L.-5K	D.L.-5K5	D.L.-5K8	TL-Mes-1
CCT	2439	2683	2656	5089	4081	2747	3965	3933	4018	3149	4000	5000	5002	5502	5800	5919
R_9	96	94	−20	56	7	57	98	12	58	79	91	49	100	100	100	95
R_a	98	98	81	83	79	89	97	84	94	95	92	93	100	100	100	98
R_{1-14}	98	97	69	76	68	84	96	78	90	92	88	89	100	100	100	98
Q_p	87	90	73	85	81	85	99	84	94	96	92	97	100	100	100	97
ΔC^*	−0.61	−0.60	−1.93	−0.97	−0.81	−0.19	0.72	−1.07	0.06	0.50	1.50	1.82	0.00	0.00	0.00	−0.50
R_f	98	98	76	86	79	88	96	84	91	91	87	90	100	100	100	97
R_g	99	99	100	98	98	100	103	96	101	102	104	104	100	100	100	98

Source: TU Darmstadt.

of Fotios and Levermore uses the signals of the L + M receptor channels and the S channels, whereas the model of Khanh et al. combines the L + M − S channels and the channel of the melanopsin-containing ganglion cells ipRGC. These brightness parameters are listed in Table 9.19.

Table 9.19 shows the following:

- With fluorescent lamps, white LEDs and daylight with a colour temperature CCT around approx. 5000 K, the brightness values of daylight and the white LEDs are the same and higher than those of the fluorescent lamp.
- The brightness values of fluorescent lamps, pcLEDs, and hybrid LEDs (multi-channel LEDs) are almost the same at the same colour temperature of approx. 4000 K.
- If we compare the brightness of white objects in the room (white ceilings, white walls, furniture) illuminated with fluorescent lamps or white LEDs at a typical colour temperature of 4000 or 5000 K with the brightness values of lighting with daylight coming in when the window in the room is open at 5500 or 5800 K, the brightness values of daylight are much higher in this case.

Table 9.20 shows the melanopic effect factor a_{mel}, the CS and the light quality parameter '*Scene Preference*' (SP, see Eq. (7.5)). At a CCT of approx. 5000 K, the corresponding values for daylight are highest, closely followed by white LEDs, with both types of light source showing significantly higher values than those of fluorescent lamps with 5000 K.

From the above considerations, it can be summarised that daylight and certain new high-quality white LED types already have colour quality parameters and melanopic factors of similar magnitude at the same illuminance level, such as 1000 lx. The advantage of daylight is clearly the much higher illuminance outdoors in general as well as indoors with well-equipped windows and skylights, the high luminance of the sky and clouds even on winter days, the dynamics of changes in daylight scenes in relation to the landscape in nature, and the integrative combination of daylight with other weather phenomena such as sun–clouds–wind–rain–heat, which have shaped human consciousness and emotion throughout evolution.

9.6 Influences of Light Intensity and Timing of Light Exposure on Sleep Behaviour

In Section 9.4, the knowledge about the effect of polychromatic white light and monochromatic light on the subjective and objective behaviour of humans in the evening and night hours were presented. Alertness, phase shift, melatonin suppression, EEG activity, CBT as well as heart rate are influenced by light exposures of different durations, at different times, with different amounts of light and different spectral distributions in the late hours. In parallel, there is also a large body of research aimed at investigating the light effects of different *amounts of light* (e.g. illuminance at the eye) and at different *times* distributed over the 24-hour day

Table 9.19 Brightness parameters of daylight and the artificial light sources in Figures 9.42–9.44. The parameter B is defined by Eq. (7.1).

Type	Non-LED		Fluorescent			pcLED				Multi-channel LED			Daylight			
Name	I.L.	Hal.	CFL-WW-2	CFL-KW-1	CFL-NW-2	WW-pcLED	SuS-pcLED	pcLED-840	pcLED-940	Multi-LED 1	4K-LED-DC1.5	5K-LED-DC1.5	D.L.-5K	D.L.-5K5	D.L.-5K8	TL-Mes-1
CCT	2439	2683	2656	5089	4081	2747	3965	3933	4018	3149	4000	5000	5002	5502	5800	5919
$L_{eq,Fotios}$ at 1000 lx	615	649	631	816	779	656	791	770	780	711	790	840	840	864	876	868
B at 1000 lx	69.74	73.86	63.95	73.93	60.75	71.49	69.19	64.75	68.35	78.88	69.27	82.08	81.07	84.87	86.67	86.18

Source: TU Darmstadt.

Table 9.20 Values of the melanopic effect factor a_{mel}, the Circadian Stimulus (CS), and the light quality parameter SP (scene preference) of daylight and the artificial light sources in Figures 9.42–9.44.

Type	Non-LED		Fluorescent			pcLED				Multi-channel LED			Daylight			
Name	I.L.	Hal.	CFL-WW-2	CFL-KW-1	CFL-NW-2	WW-pcLED	SuS-pcLED	pcLED-840	pcLED-940	Multi-LED 1	4K-LED-DC1.5	5K-LED-DC1.5	D.L.-5K	D.L.-5K5	D.L.-5K8	TL-Mes-1
CCT (K)	2439	2683	2656	5089	4081	2747	3965	3933	4018	3149	4000	5000	5002	5502	5800	5919
a_{mel}	0.35	0.40	0.288	0.651	0.496	0.367	0.613	0.552	0.605	0.468	0.572	0.743	0.77	0.82	0.85	0.86
CS at 1000 lx	0.49	0.51	0.46	0.51	0.44	0.50	0.49	0.46	0.48	0.54	0.49	0.55	0.55	0.57	0.57	0.57
SP at 1000 lx	53.6	62.0	56.96	81.90	76.42	62.20	79.27	77.86	79.68	73.46	78.89	83.96	85.0	81.2	77.1	75

Source: TU Darmstadt.

(early morning, mid-morning, afternoon, or late hours) on sleep behaviour [69]. In this context, the measured variables in the research on the investigations of sleep behaviour were divided into the following two blocks:

(1) *Subjectively assessed sleep behaviour*: Through sleep notebook, through questionnaires such as pittsburgh sleep quality index (PSQI), Leed's sleep evaluation, and self-reports;
(2) *Objectively assessed sleep behaviour*: By polysomnography of phases 1–4 or 2–4, *actigraph*.

The sleep data includes the following parameters: Time of sleep onset, sleep latency, sleep efficiency, total sleep length, sleep fragmentation (number of awakenings), REM, time, and frequency of awakenings in the individual sleep phases (first phase of sleep, third phase of sleep).

In 2019, Dautovich et al. published a literature review entitled: '*A systematic review of the amount and timing of light in association with objective and subjective sleep outcomes in community-dwelling adults*' [69]. This review found – out of 14 745 publications in scientific databases – 45 articles that met the search objectives and scientific criteria (e.g. the number of subjects, statement of research questions, hypotheses, measurement of light, and sleep parameters) and was published between 1986 and 2016. Of these, there were 32 publications that reported on experiments with the deliberate and systematic variation of light conditions (amounts of light over several light levels) and 13 publications that observed sleep behaviour and light change. Of the 32 publications mentioned above, 30 author collectives dealt with the changing amount of light at a specific time (morning, evening) and only 2 with the time of light exposure. The test subjects in the studies were healthy and had no sleep deficits. They all performed a socially relevant occupational task and were not dependent on medical assistance.

Regarding the lighting conditions, the illuminance levels were divided into three ranges [69]:

(a) $E_V < 100$ lx ('*dim light*'): the range frequently occurs in private living areas;
(b) $100\,\text{lx} < E_V < 1000$ lx ('*moderate light*'): the range occurs in professional interiors;
(c) $E_V > 1000$ lx ('*bright light*'): the range is often outdoors as well as in high-quality illuminated interiors.

For the objectively assessed sleep behaviour (e.g. via actigraph or polysomnography), there were two publications that compared the level of '*dim light*' with the level of '*moderate light*' in the morning and in the evening (e.g. 146 lx with 773 lx in the evening in [70]). There was no difference in sleep behaviour. In the opinion of the authors of the present book, these results can be explained by analysing Figure 9.31: an illuminance of 146 lx in the evening is already relatively high and, according to the curve in Figure 9.31, there is only a little difference.

In [71] Campbell and Dawson used objectively assessed parameters to investigate the sleep behaviour of older people by comparing light exposure between 300 lx (*moderate light*) and 3800–6000 lx (*bright light*) in the evening. These polysomnographic results demonstrated significantly longer sleep latency and later sleep onset at the higher light levels.

In 13 of 19 studies comparing the *dim light level* below 100 lx with the *bright light level* above 1000 lx, significant correlations were found between the light conditions and the objectively assessed polysomnographic sleep parameters. The tendency of the results depended on the time of light exposure as follows:

- The bright light (>1000 lx) *in the morning* caused a shorter sleep latency, less general waking frequency, decreased waking in the first third of sleep, increased waking in the third third of sleep, a shorter sleep duration, and an earlier sleep onset (*advanced sleep period*) compared to the *dim light*;
- Conversely, the bright light in the evening, compared to the '*dim light*', caused a delay in the sleep period (*delayed sleep period*), a greater sleep latency and a greater sleep delay in the first sleep phase.

The subjectively assessed sleep parameters in the self-reports of sleep quality also showed similar tendencies. Bright light in the evening caused longer sleep latency and overall self-reported poorer sleep quality. Bright light in the morning higher than 1000 lx enabled better sleep quality and sleep maintenance compared to '*moderate & dim light*'. In general, higher light levels throughout the day also produced positive sleep quality. Consequently, light levels such as illuminance higher than 1000 lx should be preferred in the morning and throughout the day and avoided in the evening.

This consequence for higher illuminance higher than 1000 lx, derived from research on non-visual light effects (sleep research), is consistent with research on the psychology of light, colour research (Chapters 6 and 7), and work performance/visual performance (Chapter 4), and leads to the principles of detailed lighting design conception presented in Chapter 12.

9.7 Light Effects on Alertness – Literature Analysis of Various Publications

The term '*alertness*' has already been explained in Chapter 3. Figueira et al. [72] formulated that 'Alertness is a construct associated with high levels of environmental awareness' and Posner [73] wrote that 'alertness is defined as achieving and maintaining a state of high sensitivity to incoming stimuli'. With these considerations, alertness is not exactly the opposite of sleepiness (latter includes a large component of the physiological state in which body functions enter sleep mode and fatigue). Even when *body functions* are in a good physiological and sensory state, there is still an additional level of *cognitive processing* required to reach a necessary level of alertness.

9.7.1 Alertness in the Evening and Night Hours

The effects of light on alertness and their consequences for the lighting design of shift work systems and activities in the evening were described in Section 9.4.

The mechanisms that cause melatonin suppression and the mechanisms that maintain alertness in the evening and night-time hours are, according to current knowledge, closely related, see Table 9.6. The 50% values of the maximum for alertness and for melatonin suppression are reached at the level around 100–120 lx (see Table 9.5). The research on nocturnal alertness has been continued with the studies by, among others, Cajochen et al. [18] and Zeitzer et al. [17].

9.7.2 Alertness in the Daytime

While the mechanisms of melatonin suppression and alertness at night are closely linked (high melatonin suppression means high alertness), the causes of people's alertness during the daytime are much more complex. Above all, the role of light exposure (light condition, time) in alertness analysis needs to be clarified in order to define one or more *optimal light conditions* during the day in the interest of maintaining alertness. Methodologically and bio-cybernetically, *alertness* represents a *system* with the input and output parameters shown in Figure 9.45.

The output variables of the alertness system can be divided into three groups according to Figure 9.45:

- *Subjective expression*: E.g. the KSS – questionnaire for sleepiness;
- *Physiological parameters*: These include the measured variables of the central nervous system such as EEG and the autonomic nervous system (skin temperature, CBT, heart rate, and the heart rate's variability);
- Performance indicators such as concentration tasks (performance tests such as psychromotor vigilance task (PVT), see Chapter 3) and executive control (N-back test, Go/NoGo test, see Chapter 3).

In the period 2017–2018, there were carefully executed literature studies to summarise the results obtained on the light effects on alertness, which are briefly presented as follows:

1. The literature review by Lok et al. [74] analysed 18 studies for daytime and pointed out that 14 out of 18 publications found a positive correlation between light measures and subjective characteristics such as alertness. Thus, intense

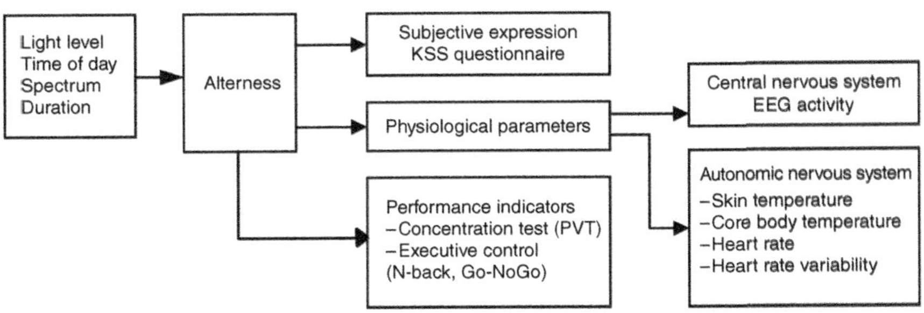

Figure 9.45 Input variables (Left: light level, time of day, spectrum, and duration) and output variables (see text) of the 'alertness' system in the bio-cybernetic sense. Source: TU Darmstadt.

light levels led to positive states of alertness. However, a positive correlation between the light measures and physiological and objectively measurable parameters such as EEG and skin temperature could only be found in a small number of publications. The reported 18 research experiments differed greatly in study design, light conditions, and time points (when and how alertness was assessed). The conclusion is that more intensive research is still needed with a uniform research paradigm as well as with considerably more measurement points (more lighting levels, more different spectra, and with more test subjects).

2. The literature review by Souman et al. [75] analysed the acute effect of lighting measures on alertness using 68 selected studies and found that a majority of the studies were able to establish a correlation between light intensity (lighting level) and alertness. The effect of colour temperature was not unambiguous.

From the point of view of the research result analysis of non-visual light effects, one can thus draw the following conclusion based on today's research:

- For the nocturnal and evening hours, the action spectrum of melatonin suppression according to *Brainard and Thapan* is known. The metrics 'CS' and MEDI for the calculation of melatonin – suppression as a function of the absolute spectra of the light sources and the exposure time, which was established and improved over several years by Rea et al. [27], and CIE [76, 77], could be proven and confirmed by many fundamental experiments of different international research groups at night and in the evening. For the night, there are several cause–effect functions (so-called '*dose–response relations*'), which have been demonstrated, among others, by the work of Zeitzer et al. [17] and Cajochen et al. [18]. The picture of the effects of light on people in the evening and at night is thus a relatively clear one (see Section 9.4).
- The effects of light on people's alertness during the daytime are to be further researched with a thoroughly considered plan, as outlined above. Relatively certainly, the connection between the light level and the subjective expression of alertness can be established by a majority of the literature studied. This subjective component was determined with the help of lighting research and sleep research as well as psychology with the aid of the KSS questionnaire.

This summary of the qualitative status of non-visual research formulated above leads to the idea of whether, in determining alertness with the KSS measurement variable (as the dependent variable) as a function of illuminance at the eye (as the independent variable), it is possible to combine all previous, relevant studies over the nocturnal hours to the daytime studies into a uniform '*dose–response curve*' by carefully analysing and compiling the underlying data. Hommes and Giménez pursued these goals in a publication in 2015 [78]. After a detailed data analysis of many publications, this author collective was able to compile a data set, which is reproduced in the present book as Table 9.21 [78].

The data in the eight selected research papers in Table 9.21 were obtained at different times, with different exposure durations (at least two hours of light exposure), different light histories before exposure, as well as different spectra

Table 9.21 Data compiled by Hommes and Giménez [78] for the reconstruction of a 'dose–response curve' for the nocturnal hours and for the daytime.

Literature source	Lamp type	Colour of light	Timing and duration of light exposure	Ref. condition	p-Illuminance (p-lx)	m-illuminance (m-lx)	ΔKSS	Moment of KSS reading
Cajochen et al. [18] (n = 23)	TL 840	White, 4000 K	6.5-h light pulse, centred 3.5 h, before minimum CBT	Dim light (about 3 lx)	20	11.53	0	N.A.: based on the fit
					100	57.66	0.9	
Phipps-Nelson et al. (2003) (n = 16)	TL 840	White, 4000 K	5-h light pulse, starting at noon	Dim light (55 lx)	1000	576.61	1.8	
					1000	576.61	2	2:00 p.m.
Phipps-Nelson et al. (2009) (n = 9)	LED	Blue, 460 nm, 25 nm FWHM	5-h light pulse, starting at 11:30 p.m.	(<1 lx)	1.14	13.38	0	N.A. overall does not signify different from control
Cajochen et al. [20] (n = 10)	Xe-Arc with narrow-band filter	Blue – 460 nm, 20 nm FWHM; Green – 550 nm, 20 nm FWHM	2-h light pulse, starting 3.5 h before habitual bedtime	Darkness	5	76.48	1.56	10:30 p.m.
Rüger et al. (2005) (n = 12)	TL 950	White, 5000 K	4-h light pulse starting at noon	Dim light (<10 lx)	68	23.39	0.27	2:00 p.m.
					5000	4309.09	1.6	

Viola et al. (2008) (n = 94)	TL 840 vs. TL Activiva	White, 4000 K (250 lx)	8.25-h light pulse starting at 8:30 a.m.	Blue enriched white, 17 000 K (200 lx)	200	287.84	1.8	N.A.: mean values reported
Santhi et al. (2011) (n = 22)	TL 827	White, 2700 K Blue depleted white Blue enriched white, 17 000 K Blue enriched and white, 17 000 K	4-h light pulse starting at 4.42 h before sleep onset	Dim light (<1 lx)	250	144.15	1.4	Last data point during light exposure 40 min before bedtime
					135	51.62	1.01	
Smolders et al. (2012) (n = 32)	TL 16	White, 4000 K	2-h light pulse, starting at 9 a.m. and at 1 p.m.	TL 840, 4000 K (1000 lx)	225	32.7	0.74	N.A.: mean values reported
	TL Activiva				135	194.29	1.29	
	TL Activiva				750	1079.41	1.19	
	TL 840				200	115.32	1.4	
					1000	576.61	1.8	

Compare with Figures 9.46, 9.47.
Source: Adapted from Hommes and Giménez [78].

and illuminances. The lighting conditions are represented in Table 9.21 by 16 measuring points of the illuminance weighted with the photopic $V(\lambda)$-function (so-called p-Illuminance, p-lux) as well as the melanopically effective illuminance (m-Illuminance, m-lux, see Chapter 2). The quantity ΔKSS is the difference of the sleepiness ratings at a certain light level compared to the reference condition. This resulted in Figures 9.46 and 9.47, for the *photopic* and *melanopically-effective* illuminance levels, respectively.

The measuring points in Figures 9.46 and 9.47 were taken from Table 9.21. The model functions (black, continuous curves) were calculated by Hommes and Giménez [78] with high correlation ($r^2 = 0.87$ for p-illuminance and $r^2 = 0.98$ for m-illuminance). The model function for photopic illuminance (Figure 9.46) reveals the following tendencies:

- An improvement in alertness goes into saturation as soon as the photopic vertical illuminance reaches a value in the range of about 500–1000 lx;
- The 50% value of the maximum increase in alertness (ΔKSS) is reached at an illuminance of around 150 lx at the eye.

In the context of interpreting these results, a maximum increase in alertness is interesting and useful *depending on the situation*. Therefore, an illuminance at

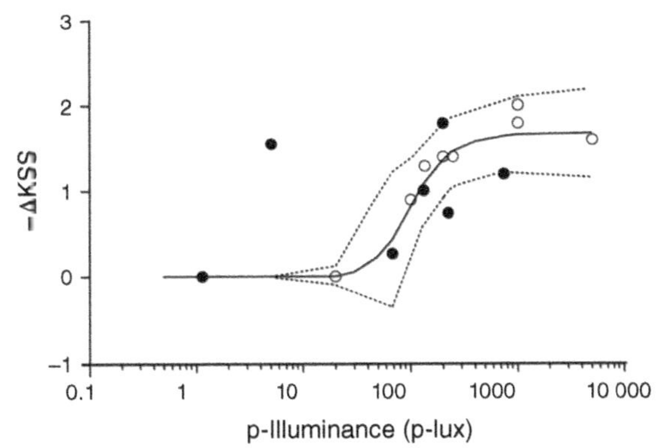

Figure 9.46 ΔKSS as a function of photopic illuminance (p-illuminance) according to [78]. Source: Reproduced from Hommes and Giménez [78] with permission from Chronobiology International.

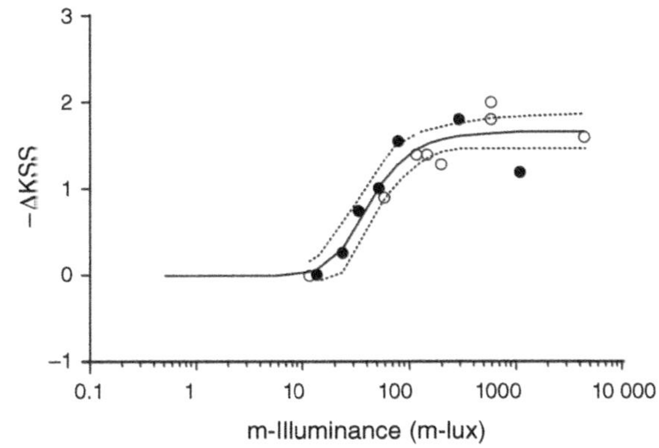

Figure 9.47 ΔKSS as a function of the melanopically-effective illuminance (m-illuminance) according to [78]. Source: Reproduced from Hommes and Giménez [78] with permission from Chronobiology International.

the eye of 500–1000 lx is desirable, corresponding to a horizontal illuminance of 1000–2000 lx in an indoor space with a high indirect luminous flux component.

The model functions in Figures 9.46 and 9.47 with the high correlation coefficients are heterogeneously composed of the data from the eight research studies with a relatively long exposure time of at least two hours (see Table 9.21). There was a research that attempted to determine the '*dose–response curve*' in the daytime on 60 subjects with several illumination levels between 20 and 2000 lx and with an exposure duration of one hour after a baseline exposure of 30 minutes at 100 lx in front of the eye (see [79]). These researchers were unable to obtain a '*dose–response curve*' of sufficient fitting quality with these data. There are several reasons for this, such as *too short* exposure time or *too high* baseline illuminance (100 lx). One crucial reason is that the time during the day includes a period when many people are already biologically *awake*, and if many of them had a good quality of sleep the night before the test day, then the alertness and concentration of these people will not be affected by the lighting condition. This statement has already been supported by the research results of Lok et al. [80] as well as in the research in Section 9.8.

In 2022, in a principal article, a group of scientists in neurophysiology, neuroscience, and sleep research recommended the following oriented values, expressed in MEDI metric, for lighting in indoor spaces [81]:

(a) '*Daytime light recommendations for indoor environments*: Throughout the daytime, the recommended minimum melanopic EDI is 250 lx at the eye measured in the vertical plane at approximately 1.2 m height (i.e. vertical illuminance at eye level when seated)'.

(b) '*Evening light recommendations for residential and other indoor environments*: During the evening, starting at least three hours before bedtime, the recommended maximum melanopic EDI is 10 lx measured at the eye in the vertical plane approximately 1.2 m height. To help achieve this, where possible, the white light should have a spectrum depleted in short wavelengths close to the peak of the melanopic action spectrum.'

(c) '*Night-time light recommendations for the sleep environment*: The sleep environment should be as dark as possible. The recommended maximum ambient melanopic EDI is 1 lx measured at the eye'.

9.8 Results of the Effect of Light on Alertness and Sleepiness During the Early Shift in an Industrial Company

Lok et al. [74] summarised and analysed a number of different studies on daytime sleepiness. Most of these studies were conducted under laboratory conditions. The question arises whether the non-visual effect would also be seen in a long-term field study in everyday conditions. This question was investigated by Truong and Khanh [82] in a typical early shift production as a *field study* on the effect of day-dependent and weather-dependent lighting conditions on *alertness* and *sleepiness*. The content of Section 9.8 is based on a publication by these authors [82].

The procedure of this field study [82] corresponds to a *repeated measure design*. While the lighting conditions are controllable and reproducible under laboratory conditions, they are dependent on the day of the field studies. The locations, the presence of the test subjects, the exposure duration, the eye movements, the physiological state of the test subjects (*light intake* before the start of the shift, intake of food) as well as the spectral composition are also highly variable compared to the laboratory conditions. In addition, in a field study under *real industrial* conditions, there are also possible organisational operational changes. Therefore, many factors had to be taken into account for the evaluation of the data. Due to the expected lack of data due to holidays, illness, etc., the so-called *multi-level analysis* according to [83–85] was selected for the statistical evaluation.

The study [82] investigated the influence of seasonal and weather-dependent lighting situations on the sleepiness of employees in a German industrial hall. Luminaires were manufactured in this industrial hall. The production of luminaires consisted of many different activities such as pre-assembly of components, wiring, and assembly of individual parts, which were carried out at different workplaces. The study took place in the period from 26 March 2019 to 28 June 2019 and was reviewed and approved by the ethics committee of the Technical University of Darmstadt (Germany). The participants of the study were recruited from the current manufacturing workforce.

The production management and the workers of the industrial hall were informed in advance about the study procedure and written consent was obtained from each study participant. Nine participants were given an identification number (ID) and were assigned to the age groups (in years) '*40–50*' (2×), '*50–60*' (6×), and '*60–70*' (1×). These participants thus belonged to the middle or older age group. In addition, the MCTQ and PSQI questionnaires were distributed monthly from August 2018 to June 2019 in order to determine the chronotypes MSF_{sc} and the PSQI values.

The built-in lighting consisted of static ceiling luminaires and workplace luminaires that were switched on manually at the start of work (05:30 a.m.) and switched off at the end of work (14:00, Fridays: 11:00). On the ceiling, the lighting was equipped with LED light sources of light colour 840 ($R_a > 80$, CCT = 4000 K). In contrast, *fluorescent lamps* with light colour 840 were used for the workplace luminaires. To describe the influence of static artificial lighting, *spectral measurements at the nights* were carried out at the workplaces. From these measurements, the vertical illuminance E_v, the melanopic daylight-equivalent illuminance $E_{v,mel,D65}$, and the CS (version 2016 of Rea and Figueiro [25]) could be determined from the measured spectra at a height of 1.5 m (mean value ± standard deviation: $E_v = 232 \pm 53$ lx, $E_{v,mel,D65} = 134 \pm 30$ lx, CS = 0.16 ± 0.03).

During the early shift, the lighting conditions were determined not only by artificial lighting but also by the changing variable daylight in the industrial hall. For the spectral measurements of the changing daylight, a measuring station with a spectroradiometer was set up. With the use of special input optics, the visual spectral range was measured. Every 15 minutes, a vertical spectral measurement was carried out at a height of 1.5 m with the following preset parameters: *integration time* (two seconds) and *measurement mean* (from 25 single measurements).

The sleepiness of the early shift workers was recorded with the aid of the subjective questionnaire KSS. Participants were instructed to complete the KSS questionnaire daily before their first break at work during the study period. The time of the KSS questionnaire varied slightly in the period 08:30 to 09:45 in the morning due to different break times. A mailbox was available at the workplace for the study to hand in the completed questionnaires. The collected documents were regularly collected by a study supervisor and digitised for evaluation.

9.8.1 Results of the Data Evaluation

In the following – for the respective chronotype groups '*late*' and '*early/normal*' – two separate linear models were built and examined. The *t*-test results for the model coefficients are shown in Table 9.22.

Table 9.22 shows that the coefficient of illuminance E_v is statistically significantly different from zero ($t(66) = -2.43$; $p < 5\%$). This means that the illuminance – in case of the 'late' chronotypes – has an influence on subjective sleepiness. The result from the '*late*' model can be set up as a linear model according to Eq. (9.9). It follows that an increase in illuminance by 200 lx results in a reduction of the KSS value by 1 unit (the person is therefore more awake by *one* KSS unit).

$$KSS_{\text{"late"}} = -0.005E_v + 5.650 \tag{9.9}$$

It is important to note that the linear model set up according to Eq. (9.9) is an estimate. The slope and the intercept are subject to certain uncertainties, which can be limited by the following 95% confidence intervals.

Furthermore, if the individual sleep types ('*bad*' or '*good*' sleep types) are considered separately from each other, then, according to the model analysis in Table 9.23, *illuminance* has a significant influence in case of the '*bad sleep*'-types ($t(107) = -2.92$; $p < 0.005$). However, no such statement is possible in case of the model of the '*good sleep*'-types, as the test strength, indicates a large error of the second kind ($t(249) = -0.58$; $p > 0.005$; $1 - \beta < 0.80$).

Table 9.22 *t*-Test results for the model coefficients for the chronotype groups 'late' and 'early/normal' [82].

Fixed effect	Value	*df*	*t*	*p*	$1-\beta$
		'Late'–model			
(Intercept)	5.650	66	6.63	0.00	1.00
E_v	−0.005	66	−2.43	0.02	0.68
		'Early/normal' model			
(Intercept)	3.152	245	14.58	0.00	1.00
E_v	0.000	245	−0.03	0.98	0.05

Source: TU Darmstadt.

Table 9.23 *t*-Test results for the model coefficients for the individual sleep types ('bad sleep' or 'good sleep').

Fixed effect	Value	*df*	*t*	*p*	$1-\beta$
		'Bad sleep' model			
(Intercept)	4.739	107	14.25	0.00	1.00
E_V	−0.003	107	−2.92	0.00	0.83
		'Good sleep' model			
(Intercept)	3.268	249	22.29	0.00	1.00
E_V	0.000	249	−0.58	0.56	0.09

Slope: −0.005 (−0.009; −0.001); intercept: 5.650 (3.973; 7.326).
Source: TU Darmstadt.

9.8.2 Summary and Discussion

In this field study [82], the influence of illuminance on the sleepiness of early shift workers in a German industrial hall was investigated. Two separate multilevel analyses showed that for the early/normal chronotypes and '*good sleep*' types, there was no effect of light intensity on subjective sleepiness. For the late chronotypes and '*bad sleep*' types, an increase in light intensity reduced subjective sleepiness. This could be an indication that higher light intensities only develop an effect when people are generally *sleepy*. This is because late chronotypes show a higher sleepiness in the morning hours (8:30 to 9:45 a.m.) than other chronotypes and a person's '*bad sleep*' type can also lead to increased sleepiness in the morning.

In other words, increased light intensity only has a reducing effect on subjective sleepiness when the exposed person is generally *tired*. The study was conducted without artificial light intervention (static lighting installation) so that the findings obtained here can be attributed to the seasonal and weather-dependent lighting situations. However, the study presented here had the limitation that the number of test subjects was relatively small (eight to nine persons) due to the company's organisation problem. Further studies are therefore necessary.

References

1 Polin, D. and Khanh, T.Q. (2014). Flimmern und Stroboskopeffekte Visuelle Aspekte der pulsweitenmodulierten weißen LEDs (Flicker and stroboscopic effects visual aspects of pulse-width modulated white LEDs). *Licht* 11–12: 80–86. ISSN: 0024-2861.

2 Geerdinck, L.M., Funke, C., Vissenberg, M.C.J.M., and Schierz, C. (2016). Discomfort glare of non-uniform luminaires – a literature review. *4th CIE Expert Symposium on Colour and Visual Appearance* (pp. 6–7), Prague,Czech Republic (September 2016). https://cie.co.at/publications/proceedings-4th-cie-expert-symposium-colour-and-visual-appearance-6-7-september-2016.

3 Mehrabian, A. and Russell, J.A. (1974). The basic emotional impact of environments. *Percept. Mot. Skills* 38: 283–301.

4 Mehrabian, A. (1995). Framework for a comprehensive description and measurement of emotional states. *Genet. Soc. Gen. Psychol. Monogr.* 121 (3): 339–361.

5 Mehrabian, A. (2007). *Nonverbal Communication*, 1e, 10–15. Routledge. ISBN 978-0-202-30966-8.

6 Fleischer, S.E. (2001). Die psychologische Wirkung veränderlicher Kunstlicht – Situationen auf den Menschen (The psychological effect of changing artificial light situations on humans). PhD thesis. ETH Zurich, No. 14033.

7 Stangl, W. (2021). Stichwort: 'Attribution'. Online Lexikon für Psychologie und Pädagogik (Keyword: 'Attribution'. Online encyclopaedia of psychology and education). https://lexikon.stangl.eu/2020/attribution (accessed 15 January 2021).

8 Brückner, S. and Khanh, T.Q. (2012). Dimmung von Hochleistung – LEDs (Dimming of high-power LEDs). *Licht* 4: 20–24.

9 Boyce, P.R. (2003). *Human Factors in Lighting*. CRC Press.

10 Hershberger, W. (1987). Saccadic eye movements and the perception of visual direction. *Percept. Psychophys.* 41 (1): 35–44.

11 Kelly, D.H. (1961). Visual response to time-dependent stimuli: I. Amplitude sensitivity measurements. *J. Opt. Soc. Am.* 51 (4): 422–429.

12 Bullough, J., Sweater Hickcox, K., Klein, T., and Narendran, N. (2011). Effects of flicker characteristics from solid-state lighting on detection, acceptability, and comfort. *Light. Res. Technol.* 43: 337–348.

13 Brückner, S. and Khanh, T.Q. (2007). *A Field Experiment on the Perception of Automotive Rear Lights Using Pulsed LEDs with Different Frequencies*. Darmstadt: ISAL.

14 Wilkins, A., Nimmo-Smith, I., Slater, A., and Bedocs, L. (1989). Fluorescent lighting, headaches and eyestrain. *Light. Res. Technol.* 21 (1): 11–18.

15 Kennedy, A., Brysbaert, M., and Murray, W.S. (1998). The effects of intermittent illumination on a visual inspection task. *Q. J. Exp. Psychol.* 51 (1): 135–151.

16 Watson, A.B., Ahumada, A.J., and Farrell, J.E. (1986). Window of visibility: a psychophysical theory of fidelity in time-sampled visual motion display. *J. Opt. Soc. Am. A* 3 (3): 300–307.

17 Zeitzer, J.M., Dijk, D.J., Kronauer, R.E. et al. (2000). Sensitivity of the human circadian pacemaker to nocturnal light: melatonin phase resetting and suppression. *J. Physiol.* 526 (3): 695702.

18 Cajochen, C., Zeitzer, J.M., Czeisler, C.A., and Dijk, D.J. (2000). Dose-response relationship for light intensity and ocular and electroencephalographic correlates of human alertness. *Behav. Brain Res.* 115: 75–83.

19 Tähkämö, L., Partonen, T., and Pesonen, A.-K. (2019). Systematic review of light exposure impact on human circadian rhythm. *Chronobiol. Int.* 36 (2): 151–170. https://doi.org/10.1080/07420528.2018.1527773.

20 Cajochen, C., Münch, M., Kobialka, S. et al. (2005). High sensitivity of human melatonin, alertness, thermoregulation, and heart rate to short wavelength light. *J. Clin. Endocrinol. Metab.* 90 (3): 1311–1316.

21 Schierz, C. (2006). Der Mensch im farbigen Licht (Humans in colourful light), Conference Licht 17: Lighting technology joint conference of the Swiss Lighting Society and the lighting technology societies of Germany, Austria and the Netherlands. Bern. 10 to 13 Sept. Conference CD.

22 Brainard, G.C., Hanifin, J.P., Greeson, J.M. et al. (2001). Action spectrum for melatonin regulation in humans: evidence for a novel circadian photoreceptor. *J. Neurosci.* 21: 6405–6412.

23 Thapan, K., Arendt, J., and Skene, D.J. (2001). An action spectrum for melatonin suppression: evidence for a novel non-rod, non-cone photoreceptor system in humans. *J. Physiol.* 535: 261–267.

24 Rea, M.S., Figueiro, M.G., Bierman, A., and Hamner, R. (2012). Modelling the spectral sensitivity of the human circadian system. *Light. Res. Technol.* 44: 386–396.

25 Rea, M.S. and Figueiro, M.G. (2016). Light as a circadian stimulus for architectural lighting. *Light. Res. Technol.* 50: 1–14.

26 Nagare, R., Rea, M.S., Plitnick, B., and Figueiro, M.G. (2019). Nocturnal melatonin suppression by adolescents and adults for different levels, spectra, and durations of light exposure. *J. Biol. Rhythms* 34: 178–194. https://doi.org/10.1177/0748730419828056.

27 Rea, M.S., Nagare, R., and Figueiro, M.G. (2020). Predictions of melatonin suppression during the early biological night and their implications for residential light exposures prior to sleeping. *Sci. Rep.* 10: 14114. https://doi.org/10.1038/s41598-020-70619-5.

28 Gringras, P., Middleton, B., Skene, D.J., and Revell, V.L. (2015). Bigger, brighter, bluer-better? Current light-emitting devices: adverse sleep properties and preventative strategies. *Front. Public Health* 3: 233. https://doi.org/10.3389/fpubh.2015.00233.

29 Chang, A.-M., Aeschbach, D., Dufy, J.F., and Czeisler, C.A. (2015). Evening use of light-emitting eReaders negatively affects sleep, circadian timing, and next-morning alertness. *Proc. Natl. Acad. Sci. U.S.A.* 112: 1232–1237. https://doi.org/10.1073/pnas.1418490112.

30 Vandahl, C., Bieske, K., Neuhäuser, S., and Schierz, C. (2009). Optimale Beleuchtung bei Schichtarbeit (Optimal lighting during shift work). Literature study, TU Ilmenau.

31 Burgess, H.J., Sharkey, K.M., and Eastman, C.I. (2002). Bright light, dark and melatonin can promote circadian adaptation in night shift workers. *Sleep Med. Rev.* 6 (5): 407–420.

32 Crowley, S.J., Lee, C., Tseng, C.Y. et al. (2004). Complete or partial circadian re-entrainment improves performance, alertness, and mood during night-shift work. *Sleep* 27 (6): 1077–1087.

33 Smith, K.A., Schoen, M.W., and Czeisler, C.A. (2004). Adaptation of human pineal melatonin suppression by recent photic history. *J. Clin. Endocrinol. Metab.* 89 (7): 3610–3614.

34 Czeisler, C.A., Johnson, M.P., Duffy, J.F. et al. (1990). Exposure to bright light and darkness to treat physiology maladaptation to night work. *N. Engl. J. Med.* 322 (18): 1253–1259.

35 Lowden, A. and Åkerstedt, T. (2012). Assessment of a new dynamic light regimen in a nuclear power control room without windows on rapidly rotating shiftworkers-effects on health, alertness, and circadian alignment: a pilot study. *Chronobiol. Int.* 29: 641–649.

36 Galasiu, A.D. and Veitch, J.A. (2006). Occupant preferences and satisfaction with the luminous environment and control systems in daylit offices: a literature review. *Energy Build.* 38 (7): 728–742. https://doi.org/10.1016/j.enbuild.2006.03.001.

37 Aries, M.B.C., Aarts, M.P.J., and van Hoof, J. (2015). Daylight and health: a review of the evidence and consequences for the built environment. *Light. Res. Technol.* 47: 6–27.

38 Cuttle, C. (1983). People and windows in workplaces. In: *Proceedings of the People and Physical Environment Research Conference*, 203–212. Wellington, New Zealand.

39 Heerwagen, J.H. and Heerwagen, D.R. (1986). Lighting and psychological comfort. *Light. Des. Appl.* 16 (4): 47–51.

40 Ne'eman, E. and Hopkinson, R.G. (1970). Critical minimum acceptable window size: a study of window design and provision of view. *Light. Res. Technol.* 2 (1): 17–27.

41 Keighley, E.C. (1973). Visual requirements, and reduced fenestration in offices: a study of window shape. *Build. Sci.* 8: 311–320.

42 Keighley, E.C. (1973). Visual requirements, and reduced fenestration in offices: a study of multiple apertures and window area. *Build. Sci.* 8: 321–331.

43 Laurentin, C., Berrutto, V., Fontoynont, M., and Girault, P. (1998). Manual control of artificial lighting in a daylit space. In: *3rd International Conference on Indoor Air Quality, Ventilation and Energy Conservation in Buildings*, 175–180. Lyon, France.

44 Moosmann, C. (2015). Visueller Komfort und Tageslicht am Büroarbeitspalt-eine Felduntersuchung in neun Gebäuden (Visual comfort and daylight at the office workspace – a field study in nine buildings). PhD thesis. KIT Scientific Publishing. ISBN: 978-3-7315-0312-52014.

45 WHO (1946). Definition of health. *Preamble to the Constitution of the World Health Organization as adopted by the International Health Conference*, New York (19–22 June 1946); signed on 22 July 1946 by the representatives of 61 States (Official Records of the World Health Organization, no. 2, p. 100) and entered into force on 7 April 1948.

46 Denissen, J.J., Butalid, L., Penke, L., and van Aken, M.A. (2008). The effects of weather on daily mood: a multilevel approach. *Emotion* 8: 662–667.

47 Bodis, J., Koppan, M., Boncz, I., and Kriszbacher, I. (2009). Time of sunrise and hours with daylight may have an effect on the seasonality and diurnal variation of heart attack. *Chin. Med. J. (English)* 122: 2107–2110.

48 Alimoglu, M.K. and Donmez, L. (2005). Daylight exposure and the other predictors of burnout among nurses in a university hospital. *Int. J. Nurs. Stud.* 42: 549–555.

49 Cowling, I., Coyne, S., and Bradley, G. (1990). *Light in Brisbane Office Buildings: A Survey*. Brisbane: Department of Physics, Queensland University of Technology, Centre for Medical and Health Physics.

50 Robertson, A.S., McInnes, M.F., Glass, D.F. et al. (1989). Building sickness: are symptoms related to the office lighting? *Ann. Occup. Hyg.* 33: 47–59.

51 Cheung, I.N., Reid, K.J., Wang, C. et al. (2013). Impact of workplace daylight exposure on sleep physical activity and quality of life. *J. Sleep Disord. Res.* 36: A422.

52 Obayashi, K., Saeki, K., Iwamoto, J. et al. (2012). Positive effect of daylight exposure on nocturnal urinary melatonin excretion in the elderly: a cross-sectional analysis of the HEIJO-KYO study. *J. Clin. Endocrinol. Metab.* 97: 4166–4173.

53 Wirz-Justice, A. and Bromundt, V. (2013). Light therapy. *Sleep* 2 (1): 20–29.

54 Aan het Rot, M., Moskowitz, D.S., and Young, S.N. (2008). Exposure to bright light is associated with positive social interaction and good mood over short time periods: a naturalistic study in mildly seasonal people. *J. Psychiatr. Res.* 42: 311–319.

55 Lam, R.W. and Levitt, A.J. (1999). *Canadian Consensus Guidelines for the Treatment of Seasonal Affective Disorders*. Vancouver BC: Clinical & Academic Publishing.

56 Figueiro, M.G., Hunter, C.M., Higgins, P.A. et al. (2015). Tailored lighting intervention for persons with dementia and caregivers living at home. *Sleep Health* 1: 322–330.

57 CIE (1974). *Method of Measuring and Specifying Colour Rendering Properties of Light Sources*. Paris: CIE Central-Bureau, CIE Publ. 13.2 (2nd edition, corrected reprint).

58 CIE (1995). *Method of Measuring and Specifying Colour Rendering Properties of Light Sources*. Vienna: CIE Central-Bureau, CIE Publ. 13.2.

59 David, A., Fini, P.T., Houser, K.W. et al. (2015). Development of the IES method for evaluating the colour rendition of light sources. *Opt. Express* 23 (12): 15888–15906.

60 CIE (2017). *Colour Fidelity Index for Accurate Scientific Use*, vol. 224. CIE. ISBN: 978-3-902842-61-9.

61 Davis, W. and Ohno, Y. (2010). Color quality scale. *Opt. Eng.* 49 (3): 1–16.

62 CIE (2006). CIE colorimetry 1931–2006. In: *Colorimetry – Understanding the CIE System* (ed. J. Schanda), 3–43. Wiley, Preprint Edition.

63 Fotios, S.A. and Levermore, G.J. (1998). Chromatic effect on apparent brightness in interior spaces, II: SWS lumens model. *Light. Res. Technol.* 30 (3): 103–106.

64 Khanh, T.Q., Bodrogi, P., and Guo, X. (2019). Towards a user preference model for interior lighting, Part 3: an alternative model. *Light. Res. Technol.* 52: 1–13. https://doi.org/10.1177/1477153519856056.

65 Khanh, T.Q., Bodrogi, P., Guo, X., and Anh, P.Q. (2019). Towards a user preference model for interior lighting. Part 1: Concept of the user preference model and experimental method. *Light. Res. Technol.* 51: 1014–1029.

66 Khanh, T.Q., Bodrogi, P., Guo, X., and Anh, P.Q. (2019). Towards a user preference model for interior lighting. Part 2: Experimental results and modeling. *Light. Res. Technol.* 51: 1030–1043.

67 DIN SPEC 5031-100:2015-08 (2015). *Strahlungsphysik im optischen Bereich und Lichttechnik – Teil 100: Über das Auge vermittelte, melanopische Wirkung des Lichts auf den Menschen – Größen, Formelzeichen und Wirkungsspektren (Radiation Physics in the Optical Field and Lighting Technology – Part 100: Melanopic Effect of Light on Humans Mediated via the Eye – Quantities, Formula Symbols and Effect Spectra).* Berlin: Beuth Verlag.

68 Wang, Z., Nagai, Y., Liu, J. et al. (2020). Artificial lighting environment evaluation of the Japan Museum of Art based on the emotional response of observers. *Appl. Sci.* 10: 1121. https://doi.org/10.3390/app10031121.

69 Dautovich, N.D., Schreiber, D.R., Imel, J.L. et al. (2019). A systematic review of the amount and timing of light in association with objective and subjective sleep outcomes in community-dwelling adults. *Sleep Health* 5 (1): 31–48.

70 Santhi, N., Thorne, H.C., van der Veen, D.R. et al. (2012). The spectral composition of evening light and individual differences in the suppression of melatonin and delay of sleep-in humans: artificial evening light suppresses melatonin and delays sleep. *J. Pineal Res.* 53 (1): 47–59. https://doi.org/10.1111/j.1600-079X.2011.00970.

71 Campbell, S.S. and Dawson, D. (1992). Aging young sleep: a test of the phase advance hypothesis of sleep disturbance in the elderly. *J. Sleep Res.* 1 (3): 205–210.

72 Figueiro, M.G., Bierman, A., Plitnick, B.A., and Rea, M.S. (2009). Preliminary evidence that both blue and red light can induce alertness at night. *BMC Neurosci.* 10 (1): 1–11.

73 Posner, M.I. (2008). Measuring alertness. *Ann. N.Y. Acad. Sci.* 1129: 193–199.

74 Lok, R., Smolders, K.C.H.J., Beersma, D.G.M., and de Kort, Y.A.W. (2018). Light, alertness, and alerting effects of white light: a literature overview. *J. Biol. Rhythms* 33 (6): 589–601.

75 Souman, J.L., Tinga, A.M., Te Pas, S.F. et al. (2017). Acute alerting effects of light: a systematic literature review. *Behav. Brain Res.* https://doi.org/10.1016/j.bbr.2017.09.016.

76 CIE S 026/E:2018 (2018). *CIE System for Metrology of Optical Radiation for ipRGC-Influenced Responses to Light.* Vienna, Austria: CIE Central Bureau.

77 CIE (2020). CIE S 026 alpha-opic Toolbox User Guide. https://doi.org//10.25039/S026.2018.UG; CIE S 026 alpha-opic Toolbox. https://doi.org//10.25039/S026.2018.TB.

78 Hommes, V. and Giménez, M.C. (2015). A revision of existing Karolinska Sleepiness Scale responses to light: a melanopic perspective. *Chronobiol. Int.* 32 (6): 750–756.

79 Smolders, K.C.H.J., Peeters, S.T., Vogels, I.M.L.C., and de Kort, Y.A.W. (2018). Investigation of dose-response relationships for effects of white light exposure on correlates of alertness and executive control during regular daytime working hours. *J. Biol. Rhythms* 33 (6): 649–661. https://doi.org/10.1177/0748730418796438.

80 Lok, R., Woelders, T., Gordijn, M.C.M. et al. (2018). White light during daytime does not improve alertness in well-rested individuals. *J. Biol. Rhythms* 33 (6): 637–648.

81 Brown, T.M., Brainard, G.C., Cajochen, C. et al. (2022). Recommendations for daytime, evening, and night-time indoor light exposure to best support physiology, sleep, and wakefulness in healthy adults. *PLoS Biol.* 20 (3): e3001571.

82 Truong, W. and Khanh, T.Q. (2021). Einfluss der Lichtintensität auf die Schläfrigkeit von älteren Arbeitern in der Frühschicht (Influence of light intensity on sleepiness of older workers on the early shift). *Z. Licht* 1: Pflaum-Verlag. https://lichtnet.de/artikel/beeinflusst-die-lichtintensitaet-die-schlaefrigkeit/.

83 Raudenbush, S.W. and Bryk, A.S. (2002). *Hierarchical Linear Models: Applications and Data Analysis Methods*, 2e. Thousand Oaks, CA, London: Sage Publications.

84 Field, A., Miles, J., and Field, Z. (2012). *Discovering Statistics Using R*. Los Angeles, London, New Delhi, Singapore, Washington, DC: Sage.

85 Snijders, T.A.B. and Bosker, R.J. (2012). *Multilevel Analysis: An Introduction to Basic and Advanced Multilevel Modeling*, 2e. Los Angeles, London: Sage.

10

Practical HCL Light Measurement Technology Indoors and Outdoors

10.1 Introduction

Alongside light source technology and lighting technology, light metrology is an important component of lighting research and lighting practice [1]. In general, light measurement technology, like all other measurement technologies such as electrical or thermal measurement technology, is used to record process and condition parameters to describe, control, or regulate a physical, biological, or chemical state or the relationship between input variables and output parameters of a system.

Until today, light measurement technology has been based on the spectral luminous efficiency function $V(\lambda)$ for daytime vision with a 2° field of view, which was defined in 1924 [2]. The photometric parameters derived from this function, such as luminance, luminous flux, or illuminance, are used to characterise the photometric properties of light sources, lamps, and luminaires and to describe a lighting situation such as the luminance distribution in a room or on a desk surface, the contrast perception when reading a document or to record reaction time or visual acuity. These photometric parameters are anchored in current international and national standards [3].

In 1931, the colour-matching functions $x(\lambda)$, $y(\lambda)$, and $z(\lambda)$ were defined by the CIE. These functions are used to initially calculate the chromaticity coordinates x, y, and z and later other colour perception attributes such as lightness L^*, hue h, or chroma C^* [4] from which colour difference metrics, the colour-rendering index (CRI) [5] or numeric descriptors of subjective colour preference [6] can be calculated.

This long historical development of light measurement technology and colorimetry produced many foundations, standardisations and, above all things, a range of light and colour measurement *devices* from 1924 until today. Generations of students, lighting designers, and lighting scientists worked with such parameters and measuring instruments to describe their teaching materials and research results. Despite major deficits, which were constantly improved by new corrections in the photometric and colorimetric models, these parameters and measuring instruments found their permanent place in research departments and in lighting practice.

Lighting technology has become much more complex and dynamic since the turn of the century. The reasons for this are the rapid and determined development

Human Centric Integrative Lighting: Technology, Perception, Non-Visual Effects, First Edition.
Tran Quoc Khanh, Peter Bodrogi, and Trinh Quang Vinh.

of LED technology with many technological and control advantages (including dimming, variety of spectra, speed, and flexible design) as well as the discovery of the so-called intrinsically photosensitive retinal ganglion cell (ipRGC) receptors (i.e. the intrinsically light-sensitive retinal ganglion cells), whereby daylight is used as a switching mechanism and a *clock* to control and stabilise the *circadian rhythm* with a great effect on wakefulness, sleep quality, or concentration [7, 8].

With the introduction of these above-mentioned '*non-visual*' effects, lighting technology, and its components such as lighting practice, luminaire technology, and light measurement technology became more complex. General light measurement technology therefore no longer consists solely of the previous photometric recording of luminous flux or luminance or the determination of colorimetric parameters such as the chromaticity coordinates or the correlated colour temperature (CCT). Light technology, sleep research, and neuroscience must also deal with the recording of biological output signals such as EEG, ECG, and heart rate as well as with the *melanopic-effective* input-side parameters, which this section will discuss in detail.

Considering the structure of the parameters at the input and output of the human light processing apparatus in the context of the analysis of the visual and non-visual effects of light on humans, a summary of system-related parameters is depicted in Figure 10.1.

Concerning Figure 10.1, the following questions should be asked:

- Which *output metrics* represent the visual performances, the emotional, and colorimetric effects as well as the non-visual effects? How can these output parameters be formed, modelled, and measured? Under which conditions and in which applications are they relevant? [9, 10]
- Which *input variables* of optical radiation are relevant for the formation of the signals of the visual system as the '*causes of the effect*'? How do they combine with

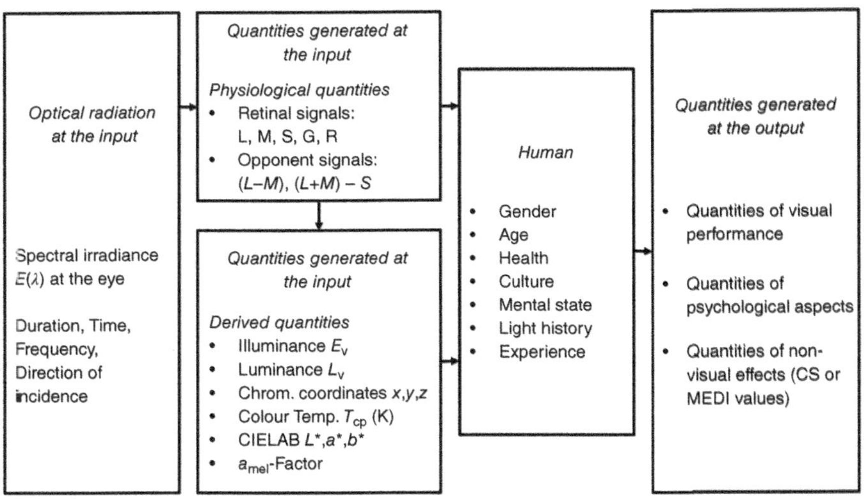

Figure 10.1 Structure of the parameters at the input and output of the human light processing apparatus. Source: TU Darmstadt.

each other to form a general *cause signal?* What is their effective or harmful duration, at different time points of the day? [10, 11]

- Do we know the relationships between causes and effects from previous research? Are we using the right descriptions for the causes and the effects? Which magnitude of the cause signal represents a threshold (50% probability of occurrence) or a certain effect (e.g. 90% or 99% probability of occurrence)? Are the effects detectable in the long term or are they more acute effects? [12]

Lighting technology in the twenty-first century thus forms a bridge between research in the fields of neuroscience, sleep research, medicine, and psychology on the one hand and lighting architecture, technological luminaire development, building technology, and lighting practice (lighting design, installation, acceptance, and logging) on the other. In the chain of problems explained above, the focus is always on the questions of the right parameters for describing visual phenomena and the right measurement technology for non-visual effects to support light and sleep research and practical lighting technology. In this context, the aim of this chapter is to propose a practice-relevant measurement technique for the non-visual effects and to document it with some field measurements. First, the authors go into the definition of the term before describing this new, practice-oriented measurement technique for recording non-visual effects.

10.2 Hypotheses and Questions for HCL Light Measurement Technology

Research in neurobiology, psychology, biomedicine, lighting technology as well as *sleep research* has developed greatly in the last two decades, both in depth and in the breadth of research topics in the laboratory field [13]. To a large extent, now it is possible to describe the effects of light in terms of its effective intensity, timing, and duration of intervention as well as its spectra. It has also been possible to explain the interaction of the various photoreceptors of the retina with the ipRGCs and the biological signal processing of various brain areas in response to a light impulse [14–16].

In this way, relationships between the light as a cause and its effect on sleep quality, alertness, mood, concentration, and well-being can be analysed qualitatively. To implement these initial approaches and results sustainably, comprehensibly, and quantitatively in the various applications of indoor lighting (hospitals, nursing homes, office workplaces, and industrial halls), the previous working methods, concepts, and their practical implementation must be recapitulated and improved. In the context of daylight research as well as general lighting research, Münch et al. [17] formulated three groups of deficiencies in our current knowledge, see Figure 10.2.

According to the authors of this book, groups 1 and 2 of deficiencies of our knowledge in Figure 10.2 are closely related to each other. In general, from the point of view of cybernetics, a purposeful cause–effect relationship should be found for every technical, biological, and physical system. For an optimal effect (e.g. good sleep quality,

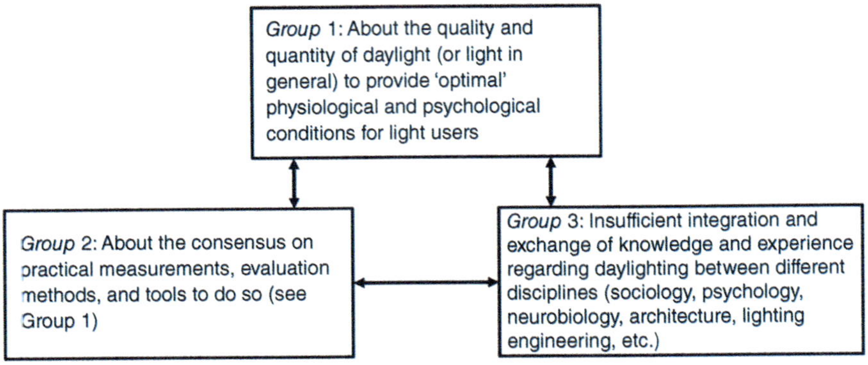

Figure 10.2 Münch et al. [17] formulated these three groups of deficiencies in our knowledge in daylighting research. Source: Münch et al. [17]/MDPI/CC BY 4.0.

good perception of contrast, and optimal pupil diameter), one or more interrelated causes (light intensity, timing, and spectra) must be found quantitatively, which so far has only been partially successful in case of the light effect '*melatonin suppression in the nocturnal hours*' [18].

For the effects during the day, this cause–effect relationship is more complicated. At present, light research and biomedicine are not yet fully capable of characterising the effects and the cause with suitable radiometric, Spectroradiometric, psychophysical, and biomedical parameters. There are currently some attempts and approaches to this. These are to be regarded conceptually and not as final steps, so that the resulting findings on cause–effect relationships as well as the measurement technique of integral radiation quantities for calculating the causes should rather be regarded as potential interim solutions. As effect variables for the non-visual effects of the '*human biological-physiological system*', many leading national and international research institutions are currently using [14–16] *biosignals* such as EEG, ECG, cortisol, melatonin as well as pupil diameter. Psychophysical variables such as sleepiness or alertness are also frequently tested (see Chapter 3). For psychophysical aspects, perceptual attributes such as colour quality, brightness and visual clarity are the current output variables (see Chapters 5–7 of this book).

The question of Group 2 in Figure 10.2, '*Gap in consensus on practical measurements and evaluation methods as well as tools for this*' is the subject of this chapter. The following research potentials and partly also problems have emerged in research practice so far in this respect:

(a) In current lighting measurement methods in indoor spaces, illuminance levels, colour temperature, uniformity, and colour rendering are specified and measured on a defined *horizontal* work surface (e.g. on the worktable), as visual tasks at this level (reading documents and books, operating workpieces, and tools on the table) have been assumed in recent decades. For the non-visual effects of light, illuminances, spectral irradiances on a *vertical* plane directly in front of the eye and the locally distributed spectral radiance or luminance in the viewing field are currently regarded as decisive parameters.

(b) The light emanating from the viewing field of the observer is imaged and distributed onto the retina by means of the *eye lens* at a certain pupil size, where it is absorbed by the various photoreceptors (rods, LMS-cones, and ipRGC-cells) and converted into nerve signals. Since the distribution of these receptors on the retina is not homogeneous, not all segments of the viewing field are equally evaluated, which leads to different contributions of the total light effect. In addition, the entire light stimulus of the room is averaged over a cosine dependency using integral measurement techniques. Neither the imaging properties of the eye nor the influence of different light sources in the room is considered. Therefore, knowledge of the spatial light effect is important and must be transferred to the analysis as well as to the measurement technique. The first approaches of spatial light characterisation were described by Knoop et al. [19].

(c) In lighting technology to date, illuminance, luminance, and colour temperature are the most important photometric and colorimetric parameters from a lighting design perspective. Circadian parameters, however, have not yet been finally defined due to the lack of definitive research results and international agreements. Nevertheless, in order to drive research forward, provisional, rather nationally agreed measures [20] or parameters proposed by certain research groups [12, 21] should be used to analyse measurement results quantitatively and comprehensibly.

(d) Due to the high acquisition costs of spectroradiometers and luminance cameras, these are often only used at large research institutes. In contrast, integral measuring devices, such as illuminance meters or handheld colour meters, have been available in the market for several decades and are widely used. These are classified in terms of their accuracy according to DIN standards such as DIN 5032-7 [22] and other international standards or technical papers of CIE. Therefore, the question arises whether and how accurately the circadian-effective quantities can be measured with these devices and whether measurement results with different types of measuring devices can be converted to the same relevant circadian-effective quantities.

(e) Up to now, there is a lack of metrological knowledge about the spectral, spatial, and temporal distributions of circadian-effective light outdoors and indoors as a function of weather, geographical location, time of day, and season as well as of *light-use profiles* with the *presence and movement* sequences of light users. The exact knowledge of the circadian-effective amount of light in connection with a registered course of sleep quality, work performance, or well-being provides information about which light quality and quantity are necessary for a certain quality of life and work. In this context, previous laboratory values are rather the first clues. Field measurements and *personal circadian-dosimetric* measurements over longer periods of time on the *vertical* working plane (e.g. at the eye) are therefore necessary.

(f) In addition to the viewing field, the measurement of lighting parameters should also consider the viewing behaviour and the head tilt of the observer. The line of vision of most office workers with paper documents and laptops tends to point downwards, whereas the light usually comes from the ceiling, the walls, or the sky.

Light measurement technology plays a major role in the realisation of the aforementioned research and offers the potential of spatially resolved measurement to record and describe light scenarios more precisely through cost-effective approaches. Based on a series of spectral and integral photometric measurements at real office workplaces, outdoors, and indoors (department stores, residential areas), this chapter attempts to convert the measurement results with different types of measuring instruments to the currently defined circadian-effective radiation quantities (CS 2016, MEDI) and to compare them. In addition, different viewing field sizes, head inclinations, and the contributions of the monitor at the work desk to circadian effectiveness will be analysed.

10.3 Light Measurement Aspects

10.3.1 Size of the Viewing Field

For further consideration of the light measurement techniques, it is important to first know the size of the human *viewing field* [13, 23, 24]. The visual range under the aspect of '*visibility*' for monocular vision, i.e. seeing with only one eye, is approx. 144–147° horizontally, provided the eye is immobilised and the head is not moved. This range is mainly limited to the nose. Binocularly, a viewing field of approx. 180° can be covered, which means that there is an overlapping area of approx. 120° horizontally. Depth perception is possible in this area. Vertically, it is approx. 120° binocularly. A distinction is made here as to the environment in which a person is located. Due to high solar radiation, the upper visual range, which normally lies at approx. 50°, can be reduced by the eyelid to as little as 20°. These results in the viewing field sizes are shown in Table 10.1 depending on ambient lighting [25, 26].

From the circadian efficacy perspective, studies of melatonin suppression in humans show that light sources located in the upper viewing field (mapped to the lower area of the retina) have a greater impact than sources located in the lower viewing field [27]. This is due to the density of ipRGCs, which, as shown in Figure 10.3, are mainly located in the posterior inferior region of the eye [28]. Furthermore, Rüger et al. showed [29] that the nasal areas of the retina are more sensitive than the lateral ones.

For the above-mentioned reason, the most recent literature classifies the viewing field according to the scheme described by Broszio et al. [24]. Accordingly, the upper

Table 10.1 Typical viewing field sizes corresponding to ambient lighting according to CIE S 026/E:2018 [25].

Ambient lighting	Vertical extension	Horizontal expansion
Interior	50° upwards, 70° downwards	180° (binocular)
Outdoor space	20° upwards, 70° downwards	180° (binocular)

Source: Adapted from CIE [25].

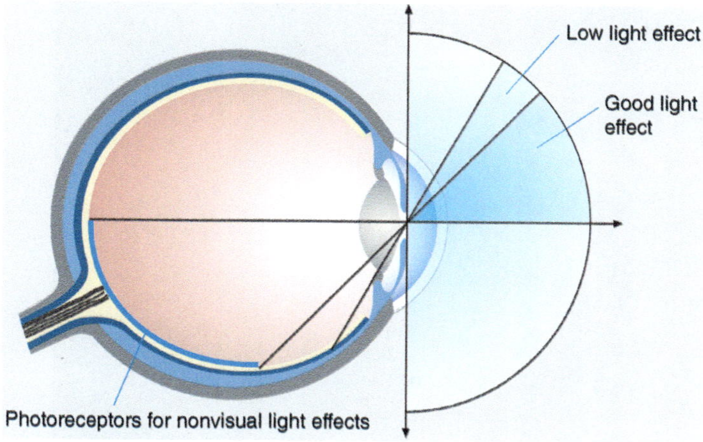

Figure 10.3 Schematic distribution of ipRGCs on the retina and the resulting areas of light effect. Source: Edited reproduction from [28] with permission from Licht.de.

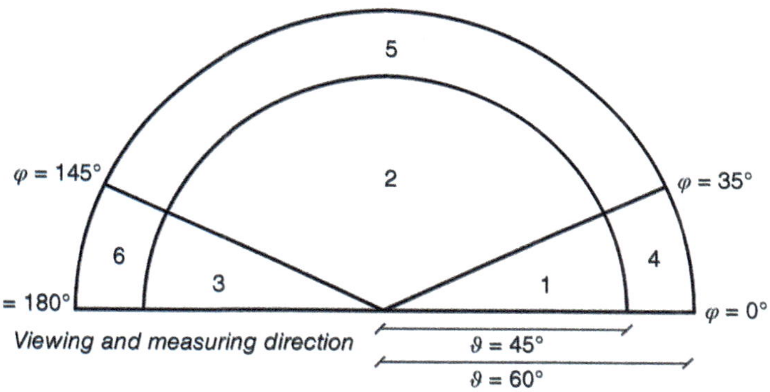

Figure 10.4 Schematic representation of the viewing field regions according to Broszio et al. [24], which serve as masking of the measurement data. The lower line represents the horizontal bisector of the image. Source: Edited reproduction [24] with permission from Light and Engineering.

half-space under consideration, to which a higher biological light effect is attributed, is divided into six different areas as shown in Figure 10.4. The regions are selected in such a way that the incident light can be divided into nasal, temporal, and central areas, depending on the angles φ and ϑ, the angles of incidence of the effective light [28] on the retina (see also Figure 10.3). An application example is shown in Figure 10.5 for the measurement data of a luminance camera at a measured office workplace.

10.3.2 Current Definitions of Circadian-Effective Irradiance

Since 2012, there have been three notable approaches worldwide on how to define circadian-effective irradiation parameters. On the one hand, scientists in neuroscience and sleep research start from the five receptor types on the retina,

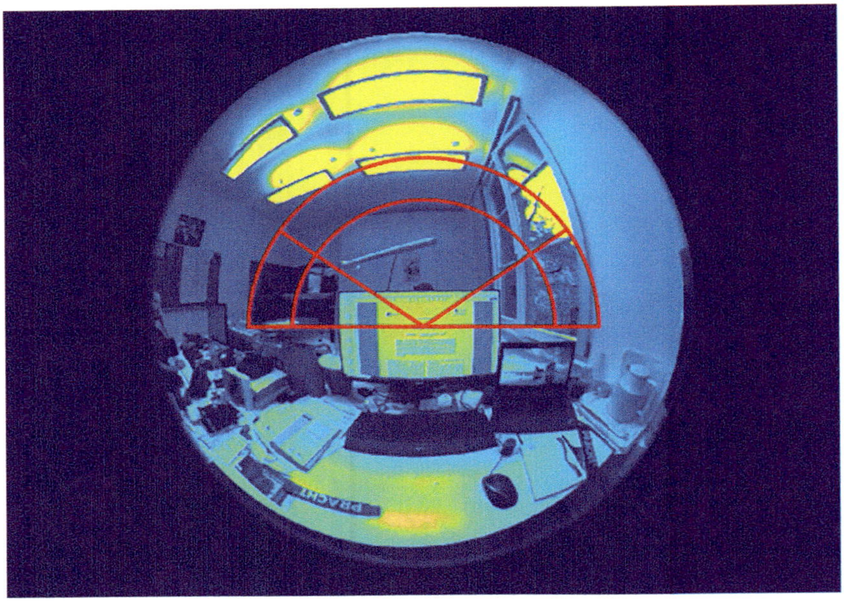

Figure 10.5 Sample application of viewing field masking according to Broszio et al. [24] from the measurement data of a luminance camera in case of a workplace. Source: Reproduced with permission from Broszio et al. [24], THE LIMITED LIABILITY COMPANY EDITORIAL BOARD LIGHTING.

each with an action spectrum (or normalised spectral sensitivity). These are the three types of cones (LMS-cones for colour vision and vision in the mesopic and photopic range), the rod type (for scotopic and mesopic vision), and the ipRGCs, whose signals are brought together in the brain apparatus and combined to the extent necessary to generate an ipRGC-influenced response to light. These ideas are mainly initiated by internationally known neuroscientists and sleep researchers and are illustrated in Figure 10.6 interpreted by the authors of this book. The fundamental remarks of Lucas et al. [30] in 2014 have been refined in the new CIE publication [25] of 2018, in which the five signals of the five receptor channels and their action spectra are defined. Each action spectrum of the five channels is a so-called '*α-opic spectral weighting function*', i.e. a so-called $S_\alpha(\lambda)$-function for a 10° field of view. These action spectra are shown in Figure 10.7.

10.3.2.1 DIN Evaluation Procedure

In the German DIN standardisation context, the standard DIN SPEC 5031-100:2015-08 was published in 2015 [20]. This standard describes the melanopically evaluated radiation quantities and the melanopic light effect. In Eq. (10.1), a general, melanopically evaluated radiation quantity is defined.

$$X_{\text{mel}} = \int_{380\text{ nm}}^{780\text{ nm}} X_e(\lambda) \cdot S_{\text{mel}}(\lambda) \cdot \mathrm{d}\lambda \tag{10.1}$$

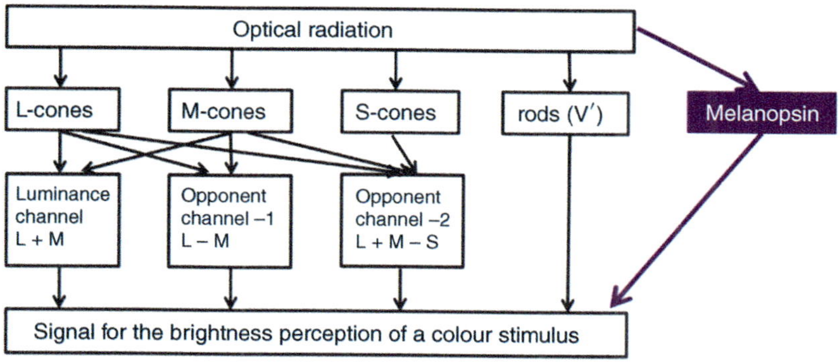

Figure 10.6 Components for the formation of a numerical parameter for brightness perception, for visual clarity perception and for the ipRGC-influenced system response. Source: TU Darmstadt.

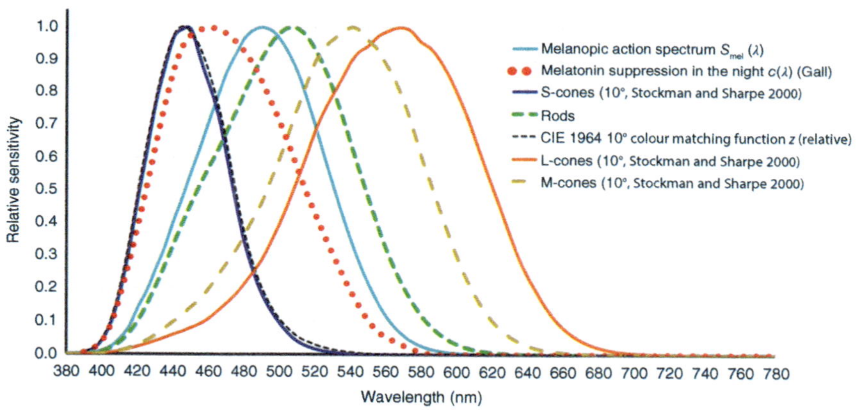

Figure 10.7 Action spectra of the five receptor channels, the colour matching function $z(\lambda)$, and the function according to Gall [31] for melatonin suppression. Source: TU Darmstadt.

The symbols in Eq. (10.1) have the following meaning:

- X_{mel}: the melanopically evaluated radiation quantity.
- $X_e(\lambda)$: the spectral radiometric quantity of the radiation source.
- $S_{\text{mel}}(\lambda)$: the action spectrum for the melanopic effect.

In the DIN standard, the melanopic effect factor $a_{\text{mel,v}}$ was defined, which is often used today. The factor $a_{\text{mel,v}}$ is the input variable of the melanopsin-effective light processing system and is defined – in this context – according to Eq. (10.2).

$$a_{\text{mel,v}} = \frac{\int_{\lambda_u}^{\lambda_o} X_\lambda(\lambda) \cdot S_{\text{mel}}(\lambda) \cdot \mathrm{d}\lambda}{\int_{\lambda_u}^{\lambda_o} X_\lambda(\lambda) \cdot V(\lambda) \cdot \mathrm{d}\lambda} \tag{10.2}$$

The integrations to be realised in Eq. (10.2) range from 380 to 780 nm. The factor $a_{\text{mel,v}}$ represents a ratio of the melanopsin-effective radiant quantity to the photometrically effective radiant quantity evaluated with the spectral luminous efficiency

Table 10.2 Melanopic effect factors of different light sources according to [20].

Light source	$a_{mel,v}$
LED, white, CCT = 3075 K	0.387
LED, white, CCT = 4250 K	0.669
LED, white, CCT = 6535 K	0.725
Standard Illuminant D65	0.906
Standard Illuminant D50	0.768
Light source type F12 according to CIE: Fluorescent lamp, CCT = 3000 K	0.366

Source: Adapted from DIN SPEC [20].

for photopic vision (2° field of view) and tells how high the melanopsin-effectiveness of the value of a photometric parameter (such as luminous flux or luminance) is. Different light sources with different spectral distributions cause different melanopsic effects on the same white wall or ceiling at the same luminance (L_V in cd m^{-2}), see Table 10.2.

From the present author's point of view, the above-mentioned publications have the following disadvantages in practice:

- From the theoretical, neuroscientific side, it is correct to calculate the five receptor signals (CIE publication [25]) or to determine the melanopic effect solely with the action spectrum of Lucas et al. (DIN SPEC 5031-100). However, the total effect of all five channels together for a non-visual assessment of lighting for a specific lighting situation cannot be calculated with this method. Even if all individual channel signals are calculated numerically correctly, it is not (yet) possible to quantify the entire effect.
- The action spectra in the CIE publication [25] are intended for a field of view of 10°. All commercially available integral measuring instruments such as illuminance meters or luminance meters as well as handheld colour meters have so far been manufactured for spectral sensitivity functions for 2° angles. This problem must be solved.

10.3.2.2 Procedure according to M. Rea and Figuiero

In parallel with the publications of Lucas et al. [30], CIE [25], and DIN SPEC 5031-100 [20] there is also modelling by M. Rea and Figuiero [12] in 2016 which has been improved in several steps in 2020 and 2021 [32, 33]. The improved models were built based on the experimental data, but the principle stays nearly unchanged since 2016. Therefore, this book describes the model and the measurement of circadian-effective light in the next parts of this chapter in view of the model from 2016. Based on the spectral distribution of the light source and with the help of the spectral sensitivity of the melanopic effect M_{mel} (λ), the S-cones, the $V(\lambda)$-function for photopic vision, and the $V'(\lambda)$- function for scotopic vision, the circadian-effective light CL_A, normalised to the standard illuminant A

(CCT = K2856), is first calculated. The final product, the '*Circadian Stimulus* CS', is calculated from CL_A and scaled so *that the CS-value is directly proportional to the melatonin suppression (in %)*. The CS-value is thus a *weighted* combination of the signals from the ipRGC, rod, L- and M-, and S-cone channels and is of attention for further consideration of the non-visual effects.

In doing so, M. Rea et al. had used the data in determining the spectral sensitivity of melatonin suppression by Brainard et al. [7] and Thapan et al. [8] and modelled the circadian-effective light, the so-called *circadian light* CL_A. Equation (10.3) shows the calculation of the CL_A parameter.

$$CL_A = \begin{cases} 1548 \cdot \left[\int M_{c\lambda}E_\lambda \mathrm{d}\lambda + \left(a_{b-y} \cdot \left(\int \frac{S_\lambda}{mp_\lambda}E_\lambda \mathrm{d}\lambda - k\int \frac{V_\lambda}{mp_\lambda}E_\lambda \mathrm{d}\lambda\right) - a_{\mathrm{rod}} \cdot \left(1 - e^{\frac{-\int V'_\lambda E_\lambda \mathrm{d}\lambda}{\mathrm{RodSat}}}\right)\right)\right] \\ \quad \text{if} \left(\int \frac{S_\lambda}{mp_\lambda}E_\lambda \mathrm{d}\lambda - k\int \frac{V_\lambda}{mp_\lambda}E_\lambda \mathrm{d}\lambda\right) > 0 \\ 1548 \cdot \int M_{c\lambda}E_\lambda \mathrm{d}\lambda \\ \quad \text{if} \left(\int \frac{S_\lambda}{mp_\lambda}E_\lambda \mathrm{d}\lambda - k\int \frac{V_\lambda}{mp_\lambda}E_\lambda \mathrm{d}\lambda\right) < 0 \end{cases} \tag{10.3}$$

According to Eq. (10.3), the circadian-effective light consists of the signal of the melanopsin-containing ganglion cells (ipRGC) alone when the signal of the opponent channel $B - (\mathrm{L} + \mathrm{M})$ is negative. This is approximately the case when the light source has a colour temperature of less than about 3500 K (warm white light). If the signal of the opponent channel $B - (L + M)$ becomes positive (CCT > 3500 K), the signals of this opponent channel are added to the signals of the ipRGCs. Since most lighting systems in the professional sector (schools, hospitals, offices, and industrial halls) have a colour temperature of CCT > 3500 K, the circadian-effective light in practical applications is composed of both above channels. This means that both the ipRGC components and the signals of the three cone types L, M, and S play an important role in modelling the circadian effect.

The method according to M. Rea et al. transforms the circadian-effective light CL_A with the help of Eq. (10.4) into a so-called 'circadian stimulus CS', which is proportional to *melatonin suppression* in percentage. A value of CS = 0.4 thus corresponds to a nocturnal melatonin suppression of 40% compared to the state before irradiation, see Figure 10.8.

$$\mathrm{CS} = 0.7 - \frac{0.7}{1 + \left(\frac{\mathrm{CL_A}}{355.7}\right)^{1.1026}} \tag{10.4}$$

The authors of this book are of the opinion that the circadian light effect should be a combined effect of the signals from all five channels in the eye and the visual brain processing system, which has been confirmed in [34]. Therefore, for further field measurements and evaluations in this chapter, the present authors use the parameter *CS 2016* for the calculation of the circadian effectiveness of light. Since the calculation of CS *is* determined by the Eqs. (10.3) and (10.4), this requires spectral radiation data from light sources and therefore a relatively cost-intensive spectroradiometer. Therefore, the present authors determined an empirical formula

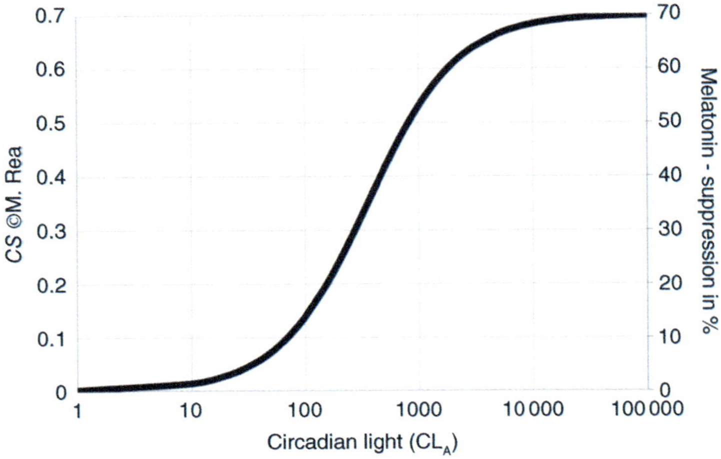

Figure 10.8 The path from Circadian Light CL_A to the CS (circadian stimulus, version 2016–2018) value (or melatonin suppression in percentage). Source: TU Darmstadt.

for the calculation of *CS* (version 2016) from the vertical illuminance E_V and the chromaticity coordinate *z* or the CCT by analysing hundreds of light source spectra. These approaches form the content of Sections 10.3.3 and 10.3.4.

10.3.2.3 Use of the Definitions for the Metrics MDER and MEDI according to CIE, Which Have Been Described in Chapter 2 (Section 2.2.3) of this Book

In Sections 10.3.3 and 10.3.4, the mathematical transformation from the chromaticity *z* or the CCT and the vertical illuminance *E* to the CS-metric (Version 2016–2018 of Rea at al.) will be formulated in detail [25]. In Sections 10.4–10.6, the practical field measurements of illuminance, luminance, chromaticity coordinates, CS, and spectra of daylight and light conditions in the indoor spaces (sales department, offices) will be described before the authors will go in the Section 10.7 to the mathematical conversion from the chromaticity coordinate *z* and the vertical illuminance E_V to MDER and MEDI metrics. In Figure 10.38 at the end of this chapter, the possibilities and measuring systems for the determination of visual and non-visual parameters are illustrated, for laboratory and practical field measurements and for measurements for feedback controlling the luminaries in the context of smart lighting for different weather conditions and applications.

10.3.3 Calculation of the Circadian Stimulus CS from Vertical Illuminance and Chromaticity Coordinate *z*

The authors of this book have been trying to answer the following three questions [1]:

(1) Is it possible to link the five receptor signals in the ocular apparatus mathematically according to some correlations among them in case of a set of lamp spectra and to describe the product of this linkage for the melanopic and visual effects during the day? This question is answered in good parts in [9, 10, 35].

(2) Does the product of this linkage correlate with the CS-value for a high number of known lamp spectra?
(3) Is it possible to replace the result of this linkage or the CS-value with a mathematical combination of photometric (such as illuminance, and luminance) and colorimetric parameters (such as colour temperature or chromaticity coordinates) so that the melanopic effect can be measured and documented with good accuracy using light measuring instruments that are readily available and affordable today?

To answer questions (1) and (2), the authors of this book computed the CS-values for 292 lamp spectra (white LED light sources, fluorescent lamps, thermal emitters) according to Eq. (10.4) and also the values of the quantity $\log_{10}(a_{\text{mel,v}} \cdot E_\text{v})$, see Eq. (10.2). The correlation (see Figure 10.9) was high, $r^2 = 0.96$. This means that the CS-value can be determined with a good approximation by determining the $a_{\text{mel,v}}$ value and the vertical illuminance E_v at the eye. However, if one analyses the curves and the data point collections in Figure 10.9, it can be seen that the difference between the data points and the mean red curve is considerable in some areas. The explanation may lie in the fact that the CS-value contains a combination of the signals of the ipRGC-cells, the rods, and the LMS-cones, and in the term, $\log_{10}(a_{\text{mel,v}} \cdot E_\text{v})$, the melanopsin-containing channel of the ipRGC-cells and the (L + M)-component (for the illuminance E) are taken into account.

Even in the ideal case that both the correlation between the CS-values and the $\log_{10}(a_{\text{mel,v}} \cdot E_\text{v})$ values are high and the difference between the mean red curve in Figure 10.9 and the individual data points is small, the result found is practically not very useful, because $a_{\text{mel,v}}$-values cannot be measured with today's integral (thus '*affordable*') light meters, for 2°-fields of view. A common integral colorimeter usually displays the illuminance E_v, the chromaticity coordinates x, y (and z), and the CCT. The practical implementation of the idea with the $a_{\text{mel,v}}$-value in the sense of DIN SPEC 5031-100:2015-08 is therefore difficult.

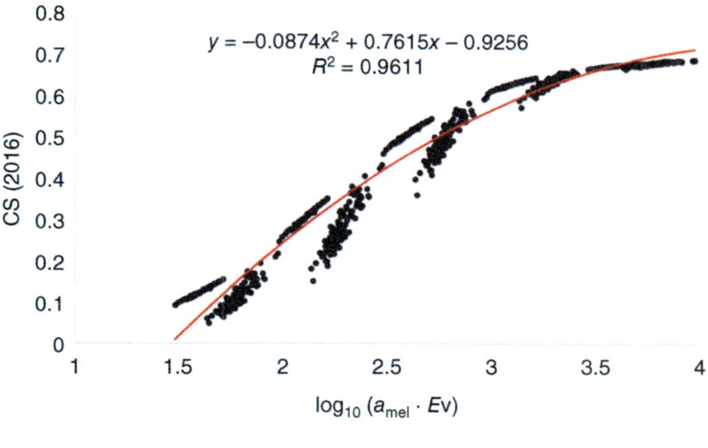

Figure 10.9 Relationship between CS and the quantity $\log_{10}(a_{\text{mel,v}} \cdot E_\text{v})$. Source: TU Darmstadt.

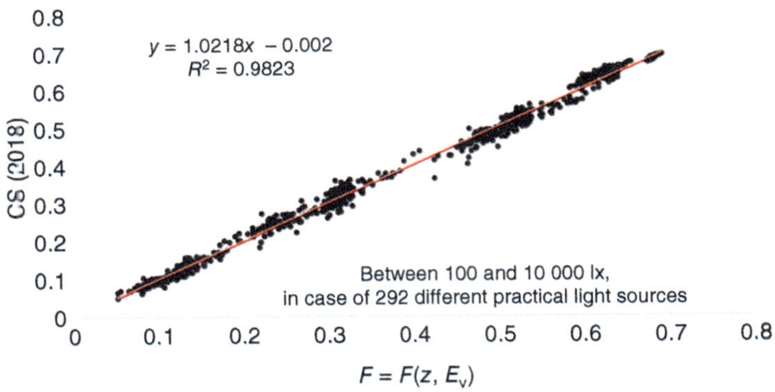

Figure 10.10 Relationship between the 'Circadian Stimulus CS' and the construct $F = F(z, E_v)$. Source: TU Darmstadt.

Table 10.3 Mean, maximum, and standard deviation values when determining the correlation between the CS-values and $F = F(z, E_v)$ in Figure 10.10.

Mean deviation	0.009
Max. Deviation	0.060
Standard deviation	0.009

Source: TU Darmstadt.

For this reason, using the 292 light spectra mentioned above, the authors found a mathematical relationship between the 'Circadian Stimulus CS' and a mathematical construct between the chromaticity coordinate z (for a 2° viewing field) and the vertical illuminance E_v, which is shown in Figure 10.10. There, it is clear that the CS-values sought to have a linear relationship (with good approximation) to the construct $F = F(z, E_v)$ and the differences between the mean linear red straight line and the 292 data points are small, see Table 10.3.

The actual mathematical relationship $F = F(z, E_v)$ is defined as follows.

- *Step 1*: the following quantity (D) *is* first to be calculated as a 'decision criterion' from the measured chromaticity coordinate z (2°):

$$D = 3.5585 \cdot z^2 + 0.9482 \cdot z - 0.3334 \tag{10.5}$$

- *Step 2*:

$$\text{If } D > 0, F(z, E_v) = a + b \cdot \log_{10}(z \cdot E_v) \tag{10.6a}$$

$$\text{otherwise } F(z, E_v) = c + d \cdot \log_{10}(z \cdot E_v) \tag{10.6b}$$

The parameters a, b, c, and d are four different poly-6 functions of the variable $\log_{10}(E_v)$ with the poly-6 coefficients in Table 10.4.

The coefficients in Table 10.4 make it possible to compute the circadian effect (CS-value) by measuring the vertical illuminance E_v and the chromaticity coordinate z for a 2° viewing field.

Table 10.4 Coefficients (a_0–a_6) of the four sixth-degree polynomial functions for the parameters a, b, c, and d, which define the dependence of these parameters on the variable $\log_{10}(E_v)$ in Eqs. (10.6a) and (10.6b).

	a_0	a_1	a_2	a_3	a_4	a_5	a_6
a	27.501	−84.373	101.064	−60.154	18.667	−2.882	0.175
b	−13.034	41.095	−50.827	31.475	−10.208	1.654	−0.106
c	4.484	−14.214	17.759	−11.133	3.688	−0.612	0.040
d	−2.493	7.996	−10.081	6.376	−2.115	0.351	−0.023

Source: TU Darmstadt.

The idea of calculating the parameter CS with a mathematical function of the vertical illuminance E_v and the chromaticity coordinate z was also realised in an equivalent way in a publication by Truong et al. [36]. For this (alternative) calculation according to Eq. (10.7), the z-coordinate is required in addition to the vertical illuminance E_v, which can be determined via a colorimetric measurement of an integral handheld colorimeter.

$$\mathrm{CS}(z, E_v) = \begin{cases} 0.7 - \dfrac{0.7}{1 + 0.016781\left(z \cdot E_v^{0.509265}\right)^{2.268904}}, & z > 0.195 \\ 0.7 - \dfrac{0.7}{1 + 0.011376(z \cdot E_v)^{1.109998}}, & z \leq 0.195 \end{cases} \tag{10.7}$$

When using Eq. (10.7), please note that it is only useful for polychromatic light sources in the range of 10–10 000 lx [36] in which most applications in indoor lighting are involved.

10.3.4 Computation of the Circadian Stimulus CS from Vertical Illuminance and Correlated Colour Temperature CCT

Since the CCT is a very familiar term for lighting designers and architects in the practice of human-centric lighting (HCL) computations, Truong et al. [37] used a data set of 7641 synthetic LED spectra at 19 illuminance levels to model a mathematical relationship between CS and the vertical illuminance E_v and the CCT and validated it with 838 spectra typical of the market (white phosphor-converted LEDs, discharge lamps, fluorescent lamps, incandescent, and halogen lamps). The model is described in Eq. (10.8).

$$\mathrm{CS}(E_v, \mathrm{CCT}) = \begin{cases} 0.7 - \dfrac{0.7}{1 + \dfrac{a}{10^6} \cdot \left(\dfrac{(\mathrm{CCT}/[K])^c}{1000} \cdot \dfrac{E_v}{[\mathrm{lx}]}\right)^b} & \text{for CCT} > 3710\,\mathrm{K} \\ 0.7 - \dfrac{0.7}{1 + \dfrac{a}{1000} \cdot \left(\dfrac{(\mathrm{CCT}/[K])^c}{1000} \cdot \dfrac{E_v}{[\mathrm{lx}]}\right)^b} & \text{for CCT} \leq 3710\,\mathrm{K} \end{cases} \tag{10.8}$$

The fitting parameters a, b, and c in Eq. (10.8) are shown in Table 10.5.

Table 10.5 Fitting parameters *a*, *b*, and *c* in Eq. (10.8) according to Truong et al. [37].

Parameter	CCT > 3710 K	CCT ≤ 3710 K
a	3.636 ± 0.0562	3.389 ± 0.0995
b	1.15000 ± 0.000238	1.10122 ± 0.000202
c	1.404 ± 0.0015	0.7831 ± 0.00329

Source: Reproduced with permission from Build. Environ. [37].

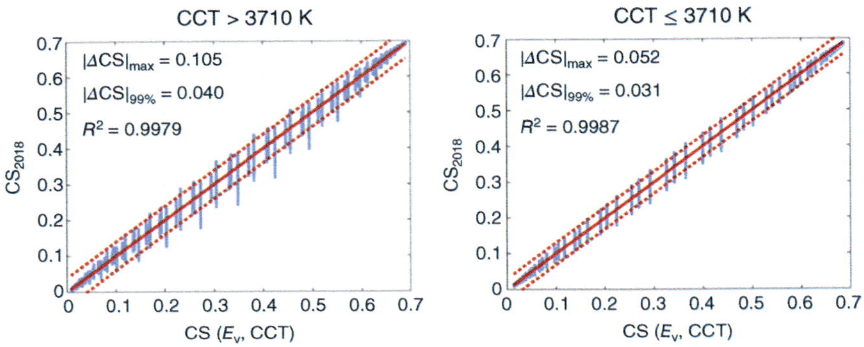

Figure 10.11 Correlation between the CS-values calculated from the original spectra (CS_{2018} on the ordinate) and the CS-values calculated from E_v and CCT, according to Eq. (10.8). Source: Reproduced with permission from Build. Environ. [37].

Figure 10.11 shows the relationship between the CS-values calculated from the original spectra (CS_{2018} on the ordinate) and the CS-values calculated from E_v and CCT according to Eq. (10.8). The maximum difference $|\Delta CS|_{max}$ equals 0.105 at CCT > 3710 K and 0.052 at CCT < 3710. The difference with a 99% confidence interval of the prediction ($|\Delta CS|_{99\%}$) equals 0.04 and 0.031, respectively.

For light sources with a general colour-rendering index (CIE CRI R_a) greater than 80, Eq. (10.8) can be transformed into Eq. (10.9).

$$\mathrm{CS}(E_v, \mathrm{CCT}) = \begin{cases} 0.7 - \dfrac{0.7}{1 + \dfrac{3.65}{10^6} \cdot \left(\dfrac{(\mathrm{CCT}/[K])^{1.404}}{1000} \cdot \dfrac{E_v}{[\mathrm{lx}]} \right)^{1.1500}} & \text{for (CCT > 3710 K) and } (R_a > 80) \\ 0.7 - \dfrac{0.7}{1 + \dfrac{3.40}{1000} \cdot \left(\dfrac{(\mathrm{CCT}/[K])^{0.783}}{1000} \cdot \dfrac{E_v}{[\mathrm{lx}]} \right)^{1.1012}} & \text{for (CCT} \leq \text{3710 K) and } (R_a > 80) \end{cases} \tag{10.9}$$

Figure 10.12 shows a three-dimensional representation of the relationship $CS = f(E_v, CCT)$ according to Eq. (10.8). As can be seen, a higher colour temperature CCT and a higher illuminance E_v result in a higher CS-value.

With the above findings on how to calculate the CS from vertical illuminance level, chromaticity coordinates *z* or CCT or from the spectra measured in the lighting installations, Sections 10.3.3 and 10.3.4 go into the practice of measurements

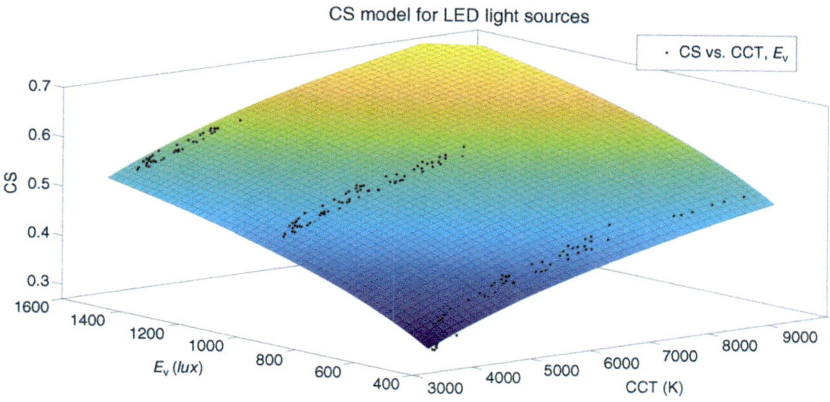

Figure 10.12 Three-dimensional representation of the relationship CS = $f(E_v$, CCT) according to Eq. (10.8). Source: TU Darmstadt.

in the field throughout the day at different places where the light users are, also in the evening in the living area as well as at professional workplaces in offices. Such measurements are necessary to determine the lighting conditions for the light users and to derive a relation from the measurement results to the non-visual lighting effects such as alertness, sleep quality, or mood. In a similar manner, MEDI and MDER quantities can also be calculated from the colour coordinate z and the vertical illuminance E_v (in lx) and this method will be described in Section 10.7 of this chapter.

10.4 Circadian-Effective Irradiation Outdoors and Indoors by Integral Field Measurements

In order to obtain circadian-effective irradiance data (CS) *in the version* 2016–2018 as well as colorimetric data (chromaticity coordinates, colour temperatures), or photometric data such as illuminances, long-term mobile field measurements were carried out at the TU Darmstadt (Germany) at different times of the year, times of the day, locations (in the building, in the park, on the streets in the flow of traffic, in the lecture hall, in the office, and in flats, until midnight) [38].

The test subjects wore a modified bicycle helmet with an integral colour-measuring device and a laptop computer in a backpack on the back of the person carrying out the measurement. The measurements were triggered, controlled, and monitored by a computer programme. The colour-measuring head with the *XYZ* colour channels with the full filtering method was mounted vertically on the helmet parallel to the face, see Figure 10.13, so that the measurement of the actual vertical circadian-effective irradiances became possible. The sky conditions as well as the localities were recorded with a digital camera and documented along the timeline in seconds of precision. The transmitted colorimetric data such as CCT and illuminance E_v were later checked mathematically for consistency and then converted into CS-values using Eq. (10.8).

Figure 10.13 Mobile field measurement of circadian-effective irradiation with a bicycle helmet and with a colour measurement detector, the face of the person is not shown. Source: Technische Universität Darmstadt.

In Sections 10.4.1 and 10.4.2, the measurement data on some days in summer and winter are presented. From a large amount of data, some data sets and their interpretations were selected. Since the data are dependent on location, weather, and people, they are not systematised here, but only presented as examples to illustrate the trends.

10.4.1 Field Measurements in Winter

One data set was recorded on a cloudy winter day (6 January 2018) at midday in Frankfurt am Main (Germany). The locations are partly outside on the way from the flat of the person carrying out the measurement to a large department store and inside the department store at various windowless sales departments (Figure 10.14).

From Figure 10.15, we learn that the vertical illuminance in the *outdoor* area (brown curves) was around 2000 lx and the colour temperature was around 6500 K. In the department store, in the windowless departments, there were 222 lx and 3050 K on average. An employee of the department store who is supposed to be there for eight hours thus only gets an exposure of 8 h × 222 lx = 1776 lxh. These

Figure 10.14 Mobile field measurement of circadian-effective irradiation: places to stay outdoors and in a department store with windowless sales departments. Source: Technische Universität Darmstadt.

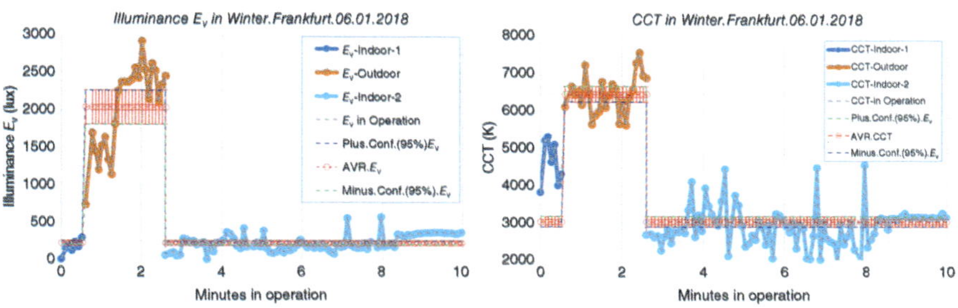

Figure 10.15 Vertical illuminance and colour temperature at the eye (ordinate) along the timeline (abscissa: minutes of observation with the detector, see Figure 10.13). Source: TU Darmstadt.

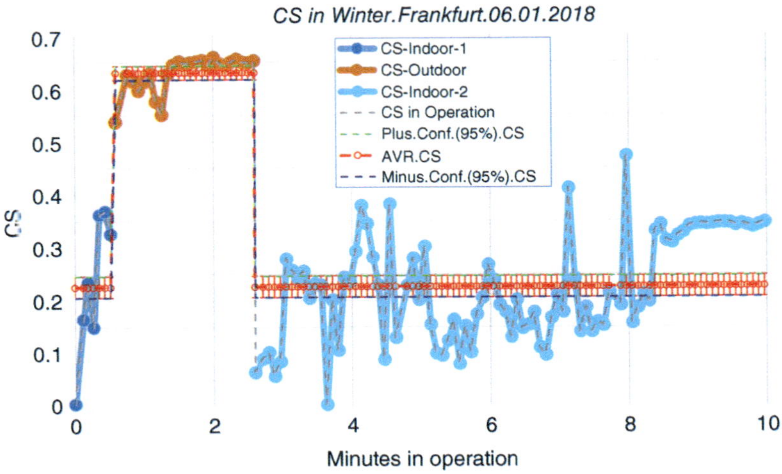

Figure 10.16 CS-values in the outdoor area (brown curve) and in the department store (blue curve). Source: TU Darmstadt.

sales rooms mainly use warm white light sources. The CS-values were converted from the illuminance and colour temperature into CS-values and are shown in Figure 10.16.

According to the above, the CS-values outdoors in winter on a cloudy day at midday are around 0.66. Thus, they exceed the recommended values for light therapy for people suffering from dementia. Therefore, it is recommended to spend more time outdoors (also) in winter. In contrast, the CS mean values in the department store (blue line) are only around 0.22. According to Figuiero et al. [39], the CS-value should be greater than 0.3 to obtain a *low to medium* circadian effect. At CS = 0.22, it is highly probable that a positive light effect will *not* occur. Department store employees who work there eight or more hours a day should therefore be outdoors more often during breaks and before or after work, and for longer periods, to provide more advantageous lighting. The vertical illuminance value of 240 lx at CCT = 6500 K recommended in DIN SPEC 5031-100 [20] as a planning value, provides a CS-value of 0.35.

10.4.2 Field Measurements on a Summer Day

On 3 July 2017, measurements were taken from 8:54 to 23:05 at various locations in Darmstadt (Germany). From Figure 10.17, it can be seen, that the weather was (slightly) cloudy at first, and then sunny from 12:00 to 18:00. The scenes at the visited locations are shown in Figure 10.18. In the photos, blue skies can be seen on more occasions. The outdoor locations are both shady due to trees and buildings and sunny in the open places. The indoor locations are the typical indoor areas with areas by the window as well as areas in the evening with little light.

Figure 10.19 shows mobile vertical illuminance recordings on 3 July 2017. In Figure 10.19, it is possible to locate the temporal areas where the person taking the

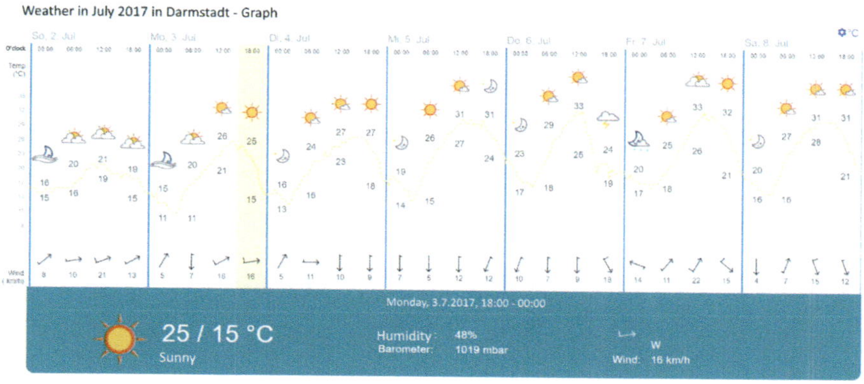

Figure 10.17 Weather patterns on the measurement day 3 July 2017. Source: www.timeanddate.de/wetter/deutschland/darmstadt/rueckblick?month=7&year=2017. Reproduced with permission of the editors of timeanddate.de.

Figure 10.18 Photos on 3 July 2017. Source: Technische Universität Darmstadt.

measurements was indoors (before 10 a.m., after 12:30 p.m. until 2:20 p.m., 3 p.m. until 4:30 p.m., before 6 p.m. until about 7 p.m., from 8 p.m. until 11:05 p.m.) and the illuminance levels were reduced. In the outdoor area, vertical illuminance peaked around 10 000 lx, about five times that of a cloudy winter day (see Figure 10.15). The colour temperature extended to about 11 500 K in the outdoor area when the head of the observer was oriented towards the blue sky, see Figure 10.20.

It can be seen from Figure 10.21 that the CS-values in the outdoor area reach the maximum value of 0.7 in summer. The CS-values in the residential area before 6 p.m.

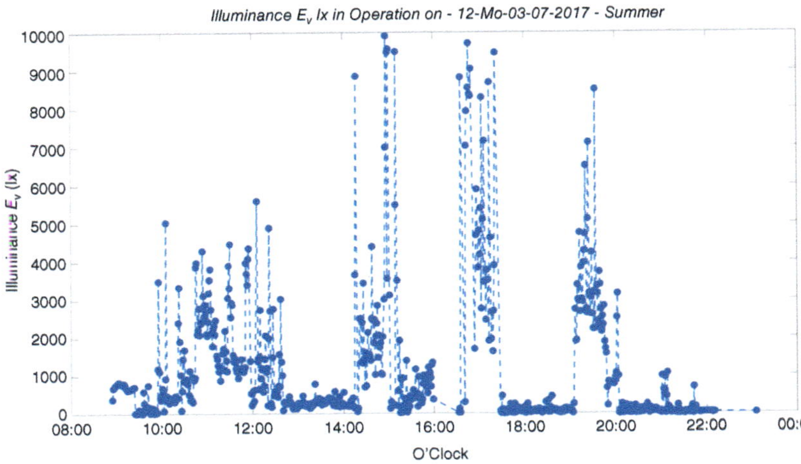

Figure 10.19 Recording of vertical illuminance from 8:54 to 23:05 hours on 3 July 2017. Source: TU Darmstadt.

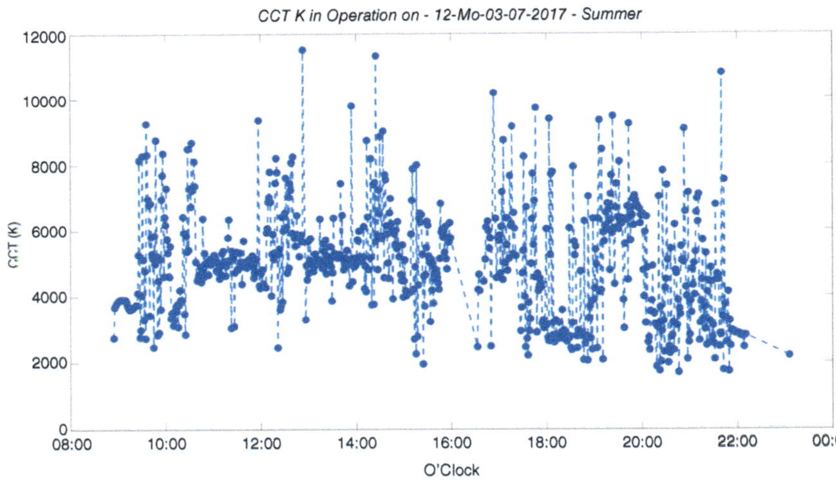

Figure 10.20 Recording of the colour temperature from 8:54 a.m. to 11:05 p.m. on 3 July 2017. Source: TU Darmstadt.

until about 7 p.m. are partly below the value of 0.1. In the early afternoon between 12:30 and 2:20 p.m., the CS-values are around 0.3.

10.4.3 Field Measurements on the Evening of an Autumn Day

It is of great interest to measure not only the circadian-effective irradiation during the day in the professional area such as at the workplace, in a department store or, in an outdoor area (on the street, in the park), but also in a *residential* area in the *evening* hours, as this irradiation also significantly determines the sleeping

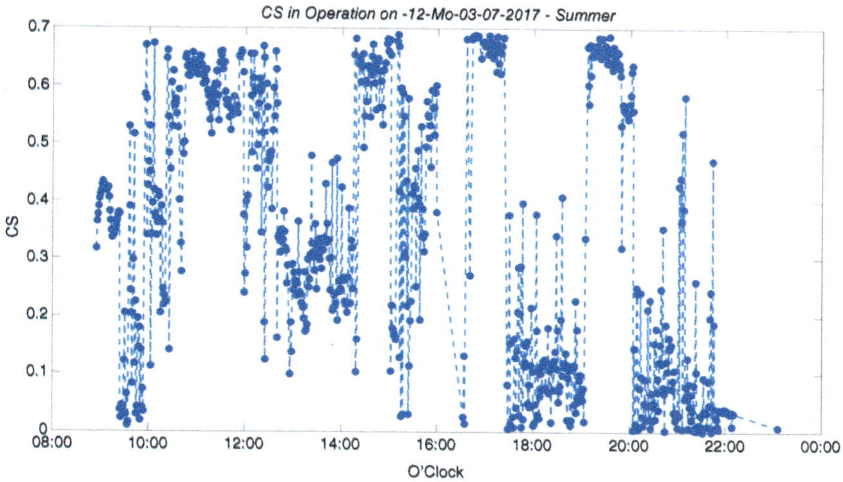

Figure 10.21 Circadian Stimulus CS-value over the day from 8:54 a.m. to 11:05 p.m. on 3 July 2017. Source: TU Darmstadt.

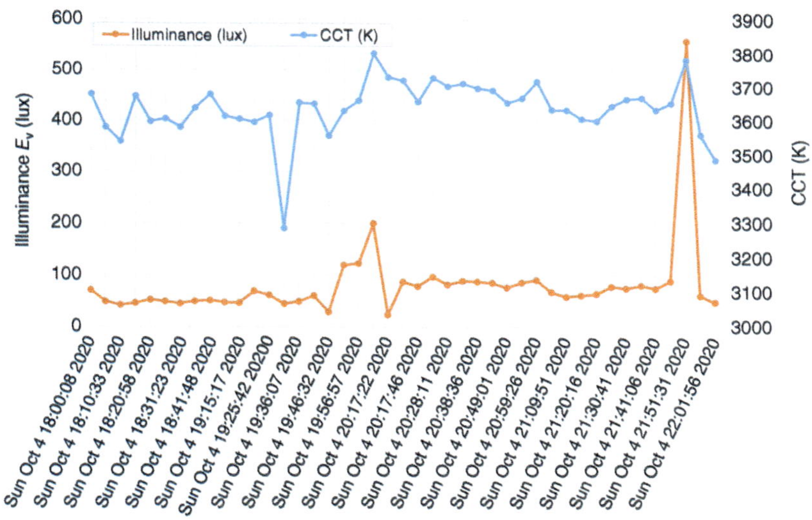

Figure 10.22 Time course of vertical illuminance and correlated colour temperature (CCT) from 18:00 hours in a student flat on 4 October 2020 in Darmstadt (Germany). Source: TU Darmstadt.

behaviour in the following night. Therefore, field measurements with a bicycle helmet and a colorimeter were also carried out shortly before and after sunset until the onset of sleep. An example of this is shown here for the autumn day 4 October 2020 in Darmstadt, carried out by a student, in the student's flat. Sunset occurred at 18:57.

Figure 10.22 shows the temporal curves of vertical illuminance and CCT from 18:00. With a few exceptions, the illuminance levels were in the range between 50

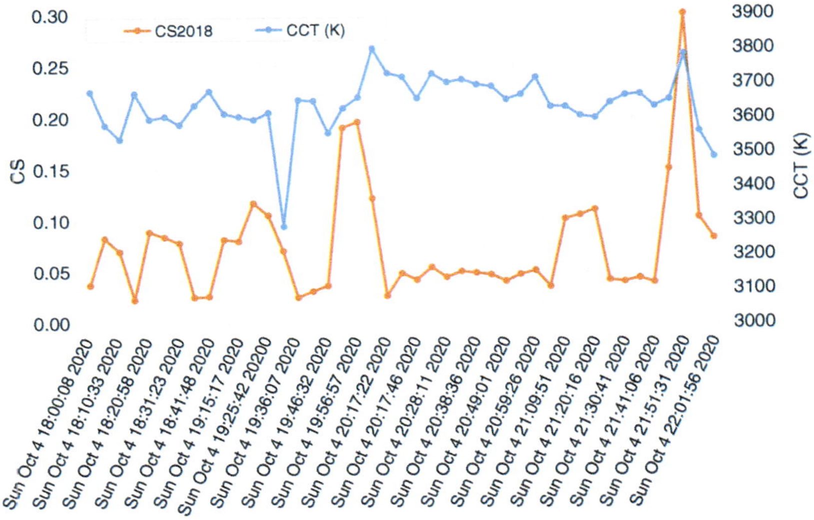

Figure 10.23 Time course of vertical illuminance and CS-values from 18:00 hours, in a student flat on 4 October 2020 in Darmstadt (Germany). Source: TU Darmstadt.

and 100 lx and the colour temperatures in the range around 3600 K. In late autumn, due to the weather, the influence of daylight shortly before and after sunset is no longer dominant. The lighting event is rather determined by the artificial light sources. Figure 10.23 shows the temporal course of the vertical illuminance and the CS-values. With two exceptions of short duration, the CS-values remained below 0.1 and, according to the study by Figuiero et al. [39], this low value did not cause noticeable circadian effects in the evening.

10.5 Daylight Measurement–Spectral Measurement and Practical Approaches

10.5.1 Spectral Measurement of Daylight Spectra

'*Spectral daylight measurements*' are usually carried out in the following way:

1. Measurement of horizontal spectral *irradiance*.
2. Measurement of vertical spectral *irradiances* in four cardinal directions at a shadow-free, relevant location over a long period of time (several years, months, or weeks, as appropriate).
3. Sky spectral *radiance* at many measurement points [40].
4. Mobile measurement of spectral *irradiances* on the horizontal and/or vertical plane (e.g. at the eye of a person) as a function of individual locations over several days in different weather conditions for certain selected persons and defined occupational groups (e.g. construction workers, nurses, and students).

From the spectral data, integral radiation quantities such as illuminance levels, chromaticity coordinates, colour temperatures, and Circadian Stimuli CS [12] in the version 2016 of Rea et al., melanopic effect factors $a_{mel,v}$ [20], MEDI and MDER can be calculated. *Spectroradiometers* are usually cost-intensive and maintenance-intensive, depending on the desired measurement accuracy. They are rather not for mobile use – therefore usually reserved for research laboratories.

At the TU Darmstadt (Germany), *spectral daylight measurements* on a horizontal plane with cosine evaluation have been carried out for a few years regularly, each measurement over a period of a few days. Daylight radiation penetrates into indoor spaces via skylights, and windows, and contributes to the illumination of living spaces and workplaces. From the spectral distributions, photometric (illuminance, luminance), colorimetric (colour rendering, chromaticity coordinates, colour temperatures) and melanopic parameters ($a_{mel,v}$ [20], CS [12], MEDI, MDER) were calculated. For analytical purposes, the measured data from a summer day (19 August 2020, in Darmstadt, see Figure 10.24) will be presented here. These measurements were carried out at a large recreational park without buildings.

Figure 10.24 shows the temporal course of illuminance (filled rectangles) and CCT (filled circles) 10 minutes after sunrise (at 6:22 a.m.) until the end of the measurement at 8:45 p.m. The colour temperature reached values up to over 15 000 K in the early morning and persisted at the level of just below 6000 K through a long sunny phase with clear skies. Illuminance increased steadily until midday, reaching maximum values of just over 90 000 lx shortly after 12 noon. In some short cloudy periods between 12:14 and 14:30 hours, illuminance dropped to about 30 500 lx, whereas colour temperatures rose to about 7000 K.

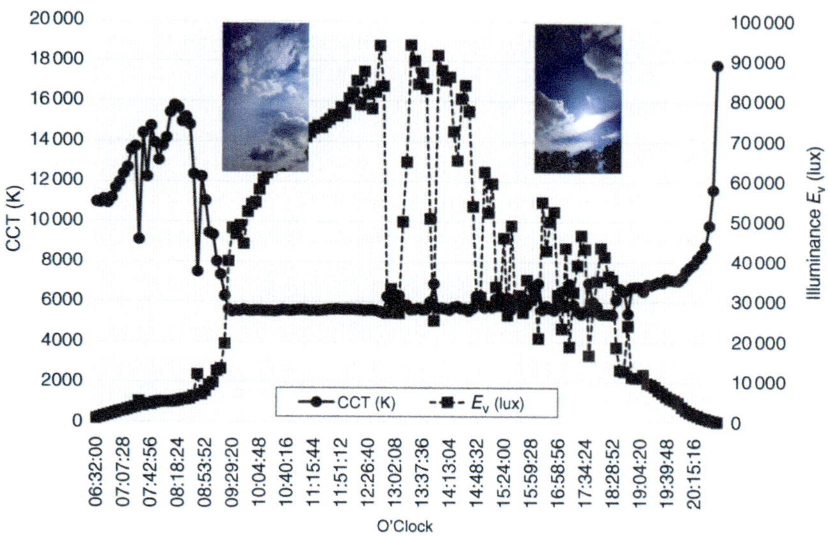

Figure 10.24 Horizontal illuminance (filled rectangles) and correlated colour temperature (filled circles) from sunrise to 20:45 on a summer day in a park in Darmstadt. Source: Technische Universität Darmstadt.

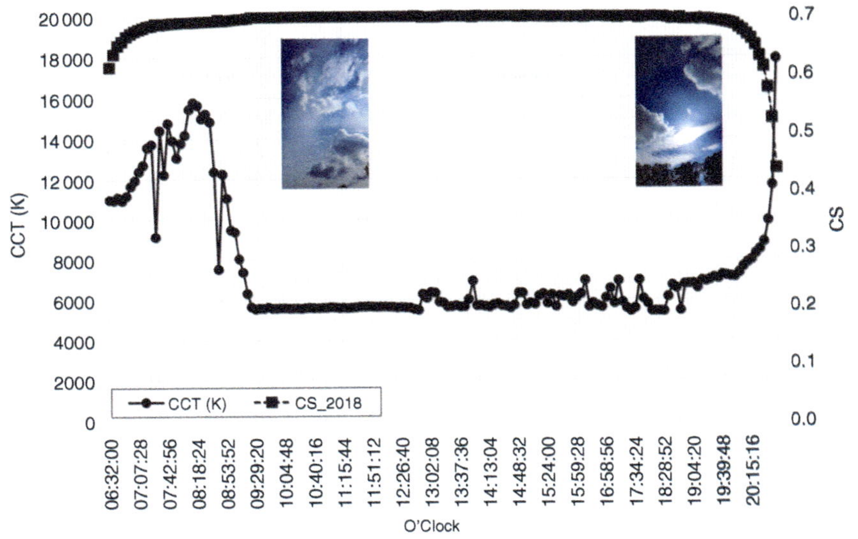

Figure 10.25 Course of the CS-value (filled rectangles) and correlated colour temperature (filled circles). Source: Technische Universität Darmstadt.

Figure 10.25 shows the temporal course of the CS-values (CS, filled rectangles) and CCT (filled circles). Shortly after sunrise, CS reached a high value of about 0.61 and from about 9 a.m., it went into saturation with CS = 0.7 for the outdoor lighting condition on a horizontal plane.

From the analysis of these daily patterns (Figures 10.24 and 10.25), we can conclude that there are *no* typical daylight patterns, in accordance with what we can often read in the literature on HCL lighting technology. Daylight is ***very dynamic*** depending on the time of day, the weather, and the geographical location. Therefore, programming dynamic light for LED luminaires of the future according to a '***general daylight pattern***' is *not* possible. However, it is possible to ***dynamically readjust*** and adapt the light spectrum of LED luminaires according to a specific concept using sensors and weather monitoring equipment.

Figure 10.26 shows daylight irradiance spectra (expressed in $\mathrm{W\,nm^{-1}\ m^{-2}}$ units) from 6:32 a.m. to about 20:45 o'clock at certain intervals of approx. 30 and 60 minutes. As can be seen, these spectra differ during the day, as expected. They exhibit typical absorption gaps between 380 and 430 nm and after 700 nm.

From the spectral daylight irradiances (see Figure 10.26), one can calculate the chromaticity coordinates x, y, z, u', and v' as well as the CCT according to the CIE publication [4], for the case of this particular location (Darmstadt, Germany), for the selected measurement period with different weather conditions at different times.

If these calculated CCTs are used as input variables in the CIE model for the spectral reconstruction of daylight spectra [4], then the daylight spectra modelled in this way deviate from the actual measured spectra. The CIE model for the spectral reconstruction of daylight spectra [4] is described by Eqs. (10.10)–(10.15).

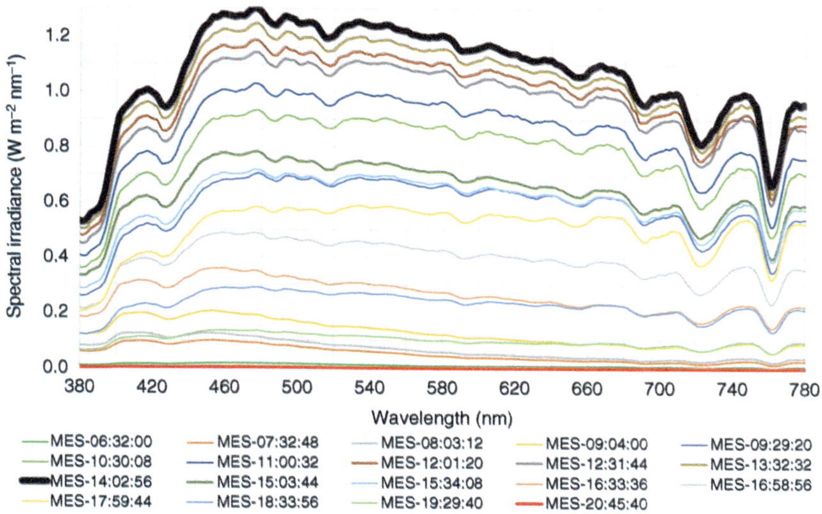

Figure 10.26 Daylight irradiance spectra from 6:32 to about 20:45 o'clock in approx. 30 and 60-minute intervals. Source: TU Darmstadt.

For the case of 5000 K ≤ CCT ≤ 7000 K, the chromaticity coordinates x_D of daylight are calculated according to Eq. (10.10).

$$x_D = \frac{-4.607 \cdot 10^9}{CCT^3} + \frac{2.9678 \cdot 10^6}{CCT^2} + \frac{0.09911 \cdot 10^3}{CCT} + 0.244063 \tag{10.10}$$

In the case of CCT > 7000 K, the chromaticity coordinates x_D of the daylight are calculated according to Eq. (10.11).

$$x_D = \frac{-2.0064 \cdot 10^9}{CCT^3} + \frac{1.9018 \cdot 10^6}{CCT^2} + \frac{0.24748 \cdot 10^3}{CCT} + 0.23704 \tag{10.11}$$

Then, the calculation steps according to Eqs. (10.12)–(10.15) should be carried out for all CCTs.

$$y_D = -3 \cdot x_D^2 + 2.87 \cdot x_D - 0.275 \tag{10.12}$$

$$M_1 = \frac{-1.3515 - 1.7703 \cdot x_D + 5.9114 \cdot y_D}{0.0241 + 0.2562 \cdot x_D - 0.7341 \cdot y_D}; \tag{10.13}$$

$$M_2 = \frac{0.03 - 31.4424 \cdot x_D + 30.0717 \cdot y_D}{0.0241 + 0.2562 \cdot x_D - 0.7341 \cdot y_D} \tag{10.14}$$

$$S_{CIE,model}(\lambda, CCT) = S_0(\lambda) + S_1(\lambda) \cdot M_1 + S_2(\lambda) \cdot M_2 \tag{10.15}$$

Figure 10.27 shows the ratios of real measured spectra (see Figure 10.26) to the corresponding calculated spectra, according to the CIE daylight model [4] (Eqs. (10.10)–(10.15)). The two data sets differ strongly in terms of radiation physics in the wavelength range before 430 nm and from 620 to 780 nm, and the differences also indicate a strong difference in weather conditions and geographical conditions between Darmstadt and the locations whose measurement data, the CIE had used

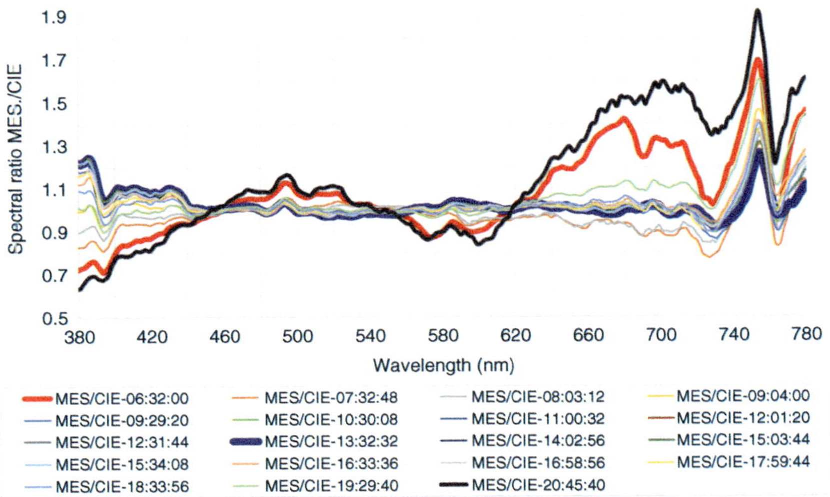

Figure 10.27 Ratios of the real measured spectra ('MES', see Figure 10.26) to the calculated spectra according to the CIE model ('CIE', Eqs. (10.10)–(10.15)). Source: TU Darmstadt.

for its daylight modelling (99 data sets measured at Budde in Ottawa, Canada; 249 data sets at Condit in Rochester, USA; and 274 data sets at Henderson and Hodgkiss in Enfield, UK) [41].

Table 10.6 shows the melanopic and colorimetric data of the measured spectra (the upper three rows) as well as the corresponding modelled spectra according to the CIE model (the next three rows) with the same colour temperatures.

As can be seen from Table 10.6, despite the differences in the shape of the spectra, the *melanopic and colorimetric descriptors* of the real measured spectra and the spectra modelled according to the CIE specifications do not differ greatly from each other. Except for the sunrise phase before 8:02 a.m., the colour differences $\Delta u'v'$ of the measured and modelled spectra at the same colour temperature are in the order of 0.003, i.e. about 3 MacAdam ellipses. Note that Table 10.6 contains $(1000{\cdot}\Delta u'v')$ data.

The melanopic factors $a_{\mathrm{mel,v}}$ of the measured and modelled daylight spectra also differ very little from each other. Hence, it can be stated that the CIE model can be used for the spectral reconstruction of outdoor daylight spectra for lighting and building applications from a practical point of view, although the relative courses of the modelled and measured spectra are different. If the illuminance and the CCT of the entire radiation are known or measurable by a handheld colorimeter or by an RGB colour sensor (see Chapter 11 of this book), then the spectrum can be reconstructed to the extent that the spectral, colorimetric, and circadian-effective radiation quantities can be calculated on the one hand and the spectrum of the LED luminaires can be adapted to it on the other hand, in order to obtain a desired spectrum for indoor applications.

Table 10.6 Melanopic and colorimetric data of the measured spectra (upper three rows) and data of the corresponding modelled spectra according to the CIE model (Eqs. (10.10)–(10.15), next three rows). Chromaticity difference $\Delta u'v'$ between measured and modelled spectra (last row).

Name	**MES-06:32:00**	**MES-07:32:48**	**MES-08:33:36**	**MES-09:29:20**	**MES-10:30:08**	**MES-11:30:56**	**MES-12:31:44**	**MES-13:32:32**	**MES-14:33:20**	**MES-15:34:08**	**MES-16:33:36**	**MES-17:34:24**	**MES-18:33:56**	**MES-19:29:40**	**MES-20:45:40**
$R_{1.14}$	94	96	97	99	99	98	99	99	99	99	98	99	98	97	92
a_{mel}	1.17	1.22	1.22	0.83	0.83	0.83	0.83	0.84	0.84	0.83	0.94	0.83	0.88	0.95	1.32
Name	CIE-06:32:00	CIE-07:32:48	CIE-08:33:36	CIE-09:29:20	CIE-10:30:08	CIE-11:30:56	CIE-12:31:44	CIE-13:32:32	CIE-14:33:20	CIE-15:34:08	CIE-16:33:36	CIE-17:34:24	CIE-18:33:56	CIE-19:29:40	CIE-20:45:40
$R_{1.14}$	100	100	100	100	100	100	100	100	100	100	100	100	100	100	100
a_{mel}	1.13	1.21	1.22	0.82	0.83	0.83	0.83	0.83	0.83	0.83	0.94	0.82	0.87	0.94	1.26
1000 $\Delta u'v'$	4.26	4.75	2.87	1.88	2.66	3.18	2.72	2.89	2.74	2.46	2.25	1.74	1.31	0.449	5.68

Source: TU Darmstadt.

10.6 HCL – Light Measurements at Office Workplaces

In this section, the measuring instruments of HCL measurements at office workplaces, the handling of these measuring instruments and the measurement results are described [13, 23]. The measured parameter is CS (Version 2016–2018) but the measuring method can also be applied for the metrics MDER and MEDI in a similar manner.

10.6.1 Measured Variables and Measurement Technology

In this series of measurements, a spatially resolved luminance measurement camera, and a spectral radiance measurement device were used to measure individual spectral radiances at specific points. The spatially resolved luminance images were recorded with a luminance measuring camera, a '*LMK 5 Color*' from the company *Technoteam*, see Figure 10.28. This is a digital camera with a CCD sensor whose beam path is equipped with a filter wheel.

The filter wheel allows the spectra to be weighted with different sensitivities, e.g. with the luminous efficiency function $V(\lambda)$. From six different glass filters, the filters of the colour channels $V(\lambda)$, X_1, X_2, and z were used in the present measurements. Using the company's own software, these measurement data were transformed into commonly used colour spaces such as *XYZ*. Calibration was built in the image evaluation software [42] thus the measured values were equivalent to real luminance and colorimetric descriptors of the scene with high accuracy.

Another advantage of the LMK-5 camera, besides spatially resolved *XYZ* values, is the possibility to use different lenses. For the measurements, a half-space lens was used to measure the workplace. According to the manufacturer, this half-space lens (so-called *fisheye*) makes it possible to record a field of view of up to ±92°, which means that this attachment lens (FCE08-027) can be used to image large scenes.

The point-like spectral radiance measurements were carried out with the spectroradiometer CS-2000 from Konica Minolta. With this measuring device, it was possible to manage a measuring angle between 0.1° and 1.0°. In this way, the light spectra of the lighting system at the workplace as well as the spectral reflectance on the table surface could be determined. Especially for the evaluation of the biological effect of light, the spectral distribution of the *vertical* irradiance at the eye plays a *major* role. It can be used to directly determine the CS-value, MEDI, or the

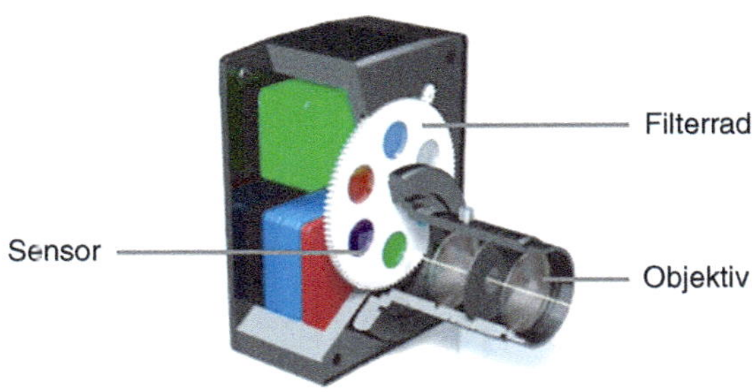

Figure 10.28 Construction scheme of the luminance measuring camera 'LMK 5 Color'. Filterrad: filter wheel; Objektiv: front entrance lens. Source: Reproduced with permission of the Technoteam Image Processing GmbH.

Figure 10.29 Measurement set-up consisting of the CSS-45 spectroradiometer (mounted on the tripod, to be seen on the left-hand side of the figure) and the LMK-5 colour luminance measurement camera (right-hand side) with a wide-angle lens. Source: Technische Universität Darmstadt.

melanopic factor $a_{mel,v}$ from the spectrum and thus provide information on *biologically effective* light.

The *spectrally* resolved *irradiance* was determined using a CSS-45 spectroradiometer from the company Gigahertz-Optik, see Figure 10.29, left. The latter measuring instrument measures the optical radiation via a diffusing disk with a diameter of 10 mm, which enables a cosine-adjusted field of view ($f_2 \leq 1.5\%$, according to DIN 5032- Part 7). This allows the entire half-space in front of the observer (sensor) to be measured. In addition to the spectrum, the calculated illuminance and the CCT can be read out via the software. The dynamic range of the sensor enables the detection of illuminance levels from 1 to 350 000 lx in the wavelength range from 360 to 830 nm [43].

10.6.2 Measurement Set-Up

The measurements were subject to the idea, that the observer, sitting at the workplace, directs their *line of sight* – at a certain sitting height – straight ahead, and parallel to the floor surface, without changing their gaze behaviour. When arranging and aligning the measuring devices, care was therefore taken to ensure that the arrangement of the devices corresponded to the range of vision of the eyes of a person sitting at the workplace.

The luminance measurement camera and the spectroradiometer, as shown in Figure 10.29, were mounted on a tripod centrally to the monitor, at a vertical height of 120 cm, which is, according to DIN EN 12464 [3], the average viewing height of a seated person. The distance to the computer screen was chosen so that the lens and the sensor were at the same height as the edge of the table (approx. 18 cm between the longitudinal axis of the tripod and the edge of the table). With some exceptions, the measuring devices were directed towards the half-space in front of the observer without any inclination. Figure 10.30 shows this arrangement with an inclination of −10° to the monitor and a measuring height of $h_{meas} = 120$ cm. Figure 10.31 illustrates the spectral radiance measurement of a light source (here a monitor) with the spectroradiometer CS-2000.

10.6.3 The Rooms in which the Measurement Took Place

The nine office workplaces provided for the measurements were located in medium-sized offices of about 4.5 m × 4.5 m, each with two large windows. The

Figure 10.30 Use of the measuring devices shown in Figure 10.29 at an office workplace at the angle of −10°. Source: Technische Universität Darmstadt.

Figure 10.31 Spectral radiance measurement of a light source (here monitor) using the spectroradiometer CS-2000. Source: Technische Universität Darmstadt.

workplaces were equipped with computer monitors, laptops, and white desks, see Figures 10.30 and 10.31. The electrical lighting was provided by pendant office luminaires mounted above the desks with fluorescent lamps of a CCT around 6000 K and a high CRI around CIE CRI $R_a = 95$. The fluorescent lamps were not dimmable and were installed in luminaire housings with prismatic output optics so that the light distribution provided glare-free and rather homogeneous room lighting.

Figure 10.32 shows the spectral radiances obtained at one of these office workplaces from the monitor screen with the typical fluorescent spectra, from the cloudy sky (for comparison with Figure 10.26), from the ceiling lighting as well as the

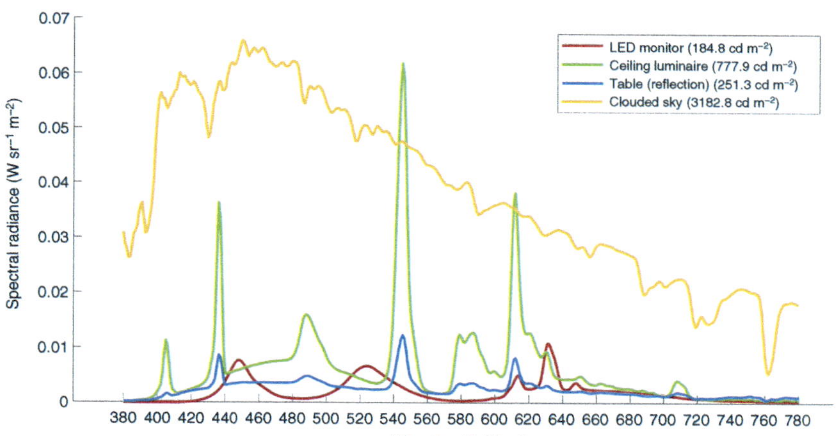

Figure 10.32 Spectral radiance of different objects and light sources at an office workplace, together with associated luminance values. Source: TU Darmstadt.

reflection from the table surface exhibiting typical spectra of white fluorescent lamps with secondary maxima at 404, 435.8, 490, 546.1, 577–579, and around 615 nm.

10.6.4 Measurement Results at Different Office Workplaces

In the following, the above office workplaces will be evaluated, regarding their biological lighting effect, see Table 10.7. The vertical illuminance levels equalled 383.7 lx on average. The 500 lx level, required by DIN EN 12464 as *horizontal* illuminance for an office workplace, was thus covered because 1 lx *vertical* illuminance caused about 1.8–3 lx *horizontal* illuminance at workplaces illuminated from above [44].

The average CS-value (*in the version of Rea and Figuiero in* [12]) of all workplaces equalled 0.351, which provided a melatonin suppression of approximately 35% when this lighting was used in the evening hours. The lowest CS-value was found to be 0.247 at workplace No. 9. The highest CS-value was provided by workplace No. 2 with 0.379. Apart from workplace No. 9, the other office workplaces differed only by a maximum of 3% melatonin suppression and could therefore be regarded as almost identical. They were above CS = 0.3 thus exhibiting circadian-effective light. The reason for this is that these workplaces were located close to the windows, and suitable luminaires were selected for office lighting during the day (CCT = 6000 K, CIE CRI $R_a = 95$).

With the help of some further LMK-5 recordings (Figure 10.28), additional observations were made regarding the light effect caused by the monitor alone. If the

Table 10.7 Illuminance and CS-values of the nine office workplaces calculated from the measurement with the spectroradiometer CSS-45.

Workplace (meas. height, cm)	Illuminance, lx (by CSS-45)	CS (by CSS-45)
#1 (120)	372.4	0.359
#2 (120)	414.3	0.379
#3(120)	372.7	0.342
#3 (98)	397.3	0.351
#4 (120)	370.0	0.343
#4 (107)	389.8	0.357
#5 (120)	387.0	0.360
#6 (120)	399.4	0.370
#6 (125)	399.9	0.370
#7 (120)	421.8	0.372
#7 (105)	416.5	0.364
#8 (120)	389.7	0.350
#9 (120)	257.3	0.247
Average	383.7	0.351
SD	39.9	0.032

Source: TU Darmstadt.

monitors under consideration do not have adaptive brightness adjustment, the biological light effect at the darkened workplace or in late evening hours without other light sources can be examined. For this purpose, the area of the monitor was cut out of the image, as shown in Figure 10.33.

Table 10.8 shows that the average illuminance at the eye caused by the monitor equals 60 lx and the corresponding average CS-value equals 0.1, which corresponds

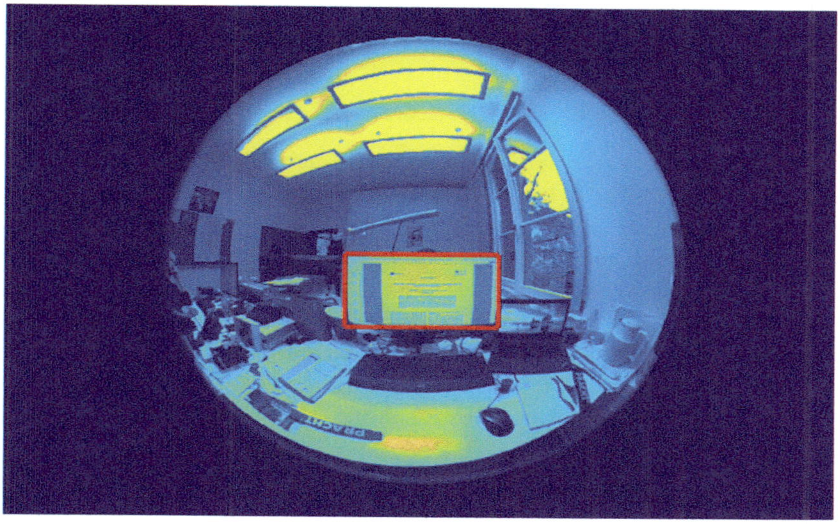

Figure 10.33 Cutting the monitor area out of a LMK-5 recording. Only the inner area of the red masking rectangle was used for evaluation. Source: Reproduced with permission from [24], THE LIMITED LIABILITY COMPANY EDITORIAL BOARD LIGHTING.

Table 10.8 Evaluation of the monitor area extracted from the LMK-5 image (see Figure 10.33). Illuminance and CS-values caused by the monitor at the eye are shown.

Workplace (meas. height, cm)	Illuminance (lx)	CS
#1 (120)	62.7	0.117
#2 (120)	86.9	0.140
#3 (120)	56.4	0.090
#4 (120)	67.1	0.119
#5 (120)	49.6	0.092
#6 (120)	62.8	0.115
#7 (120)	73.5	0.131
#8 (120)	35.8	0.073
#9 (120)	43.4	0.022
Average	59.8	0.100
SD	14.7	0.034

Source: TU Darmstadt.

to a melatonin suppression of about 10% in the evening hours. This value represents exactly the limit (CS = 0.1) above which a biological light effect can be assumed [39]. Such measurement results depend on the brightness setting of the monitor and the colours currently displayed on the monitor as well as the distance of the monitor to the measuring device or the viewer.

10.7 Calculation of the Metrics MDER and MEDI from Vertical Illuminance and Chromaticity Coordinate *z*

10.7.1 Definition of MDER and MEDI According to CIE-Publication

In Chapter 2 of this book, the authors used the CIE-publication [25] to define the non-visual metric MEDI [25]. In this section some basic formulas for MEDI and MDER will be formulated and repeated.

The aim of this mentioned CIE publication is to define spectral response curves, quantities, and metrics to describe optical radiation for each of the five photoreceptors in the human eye that may contribute to non-visual processes in humans. The spectral sensitivity functions of the five receptor types are illustrated as follows (see Figure 10.34):

1. S-cone-opic, $s_{10}(\lambda)$
2. M-cone-opic, $m_{10}(\lambda)$
3. L-cone-opic, $l_{10}(\lambda)$
4. Rhodopic, $V'(\lambda)$, rods
5. Melanopic, mel(λ), ipRGCs

The CIE defined the α-opic quantities in Table 10.9 by means of the spectral sensitivities in Figure 10.34, the spectral power $\Phi_{e,\lambda}(\lambda)$, the spectral radiance $L_{e,\lambda}(\lambda)$, and the spectral irradiance $E_{e,\lambda}(\lambda)$ (α represents one of the five photoreceptor types).

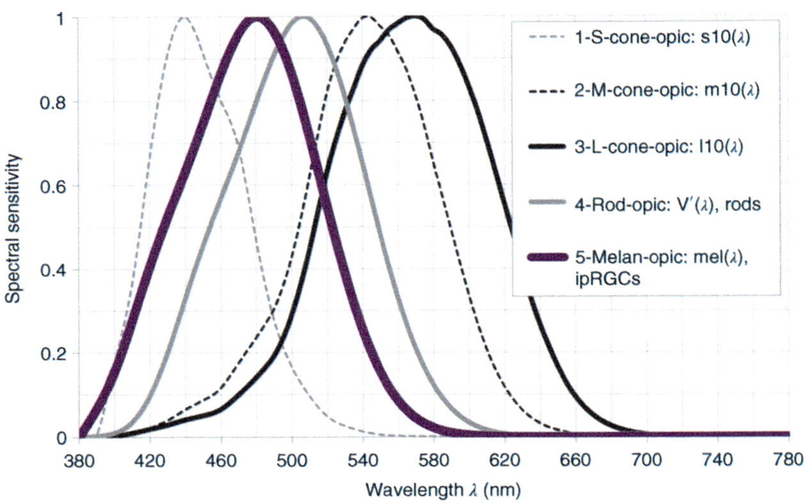

Figure 10.34 System of relative photoreceptor sensitivities in CIE S 026/E:2018.

Table 10.9 α-opic quantities in CIE S 026/E:2018.

α- opic quantities		
Parameter	**Equation (Equation No.)**	
α-opic power	$\Phi_{e,\alpha}$ or $\Phi_{\alpha} = \int_{\lambda_u}^{\lambda_o} \Phi_{e,\lambda}(\lambda) \cdot s_{\alpha}(\lambda) \cdot d\lambda$ in W	(10.16)
α-opic radiance	$L_{e,\alpha}$ or $L_{\alpha} = \int_{\lambda_u}^{\lambda_o} L_{e,\lambda}(\lambda) \cdot s_{\alpha}(\lambda) \cdot d\lambda$ in $\frac{W}{sr \cdot m^2}$	(10.17)
α-opic irradiance	$E_{e,\alpha}$ or $E_{\alpha} = \int_{\lambda_u}^{\lambda_o} E_{e,\lambda}(\lambda) \cdot s_{\alpha}(\lambda) \cdot d\lambda$ in $\frac{W}{m^2}$	(10.18)

Based on Table 10.9, further parameters characterizing the non-visual light effect can be derived, where $s_{\alpha}(\lambda)$ is the spectral sensitivity function of one of the five receptor types, K_m is the photometric radiation equivalent (=683 lm W^{-1}) and $\Phi_{D65,\lambda}(\lambda)$ is the spectrum of standard daylight type D65, see Eqs. (10.19) and (10.20). The quantity $K\alpha$, v is the α-opic efficacy of luminous radiation.

$$K_{\alpha,v} = \frac{\int_{\lambda_u}^{\lambda_o} \Phi_{e,\lambda}(\lambda) \cdot S_{\alpha}(\lambda) \cdot d\lambda}{K_m \cdot \int_{\lambda_u}^{\lambda_o} \Phi_{e,\lambda}(\lambda) \cdot V(\lambda) \cdot d\lambda} \tag{10.19}$$

$$K_{\alpha,V}^{D65} = \frac{\int_{\lambda_u}^{\lambda_o} \Phi_{D65,\lambda}(\lambda) \cdot S_{\alpha}(\lambda) \cdot d\lambda}{K_m \cdot \int_{\lambda_u}^{\lambda_o} \Phi_{D65,\lambda}(\lambda) \cdot V(\lambda) \cdot d\lambda} \tag{10.20}$$

For melanopic efficacy ($\alpha = mel$) the parameter values ${K_{mel,V}}^{D65}$ equals 1.3262 mW lm^{-1}.

The MEDI and MDER metrics can be formulated according to Table 10.10. For interpretation of the meaning of MEDI or MDER for lighting design and engineering, two main aspects should be pointed out:

a. MEDI is the illuminance of standard daylight illuminant D65 that has as much melanopic efficacy as the test light source at a given illuminance E_v (lx) caused by the test light source, see Eq. (10.21) in Table 10.10.

Table 10.10 Melanopic equivalent D65 quantities of CIE S 026/E:2018.

Melanopic equivalent D65 quantities		
Parameter	**Equation (Equation No.)**	
Melanopic Equivalent Daylight Illuminance (MEDI) in lx	$\text{MEDI} = \frac{\int_{380}^{780} S_{mel.}(\lambda) \cdot E_{e,\lambda}(\lambda) \cdot d\lambda}{K_{mel,V}^{D65}}$	(10.21)
Melanopic daylight efficacy ratio (MDER)	$\text{MDER} = \frac{\frac{1}{K_{mel,v}^{D65}} \cdot \int_{380}^{780} S_{mel.}(\lambda) \cdot E_{e,\lambda}(\lambda) \cdot d\lambda}{E_v}$	(10.22)

b. MDER is the ratio of the illuminance of standard illuminant D65 (MEDI) to the illuminance of the test light source E_v (in lx) when the absolute melanopic efficacy of both light sources is set equal, see Eq. (10.22) in Table 10.10.

The melanopically effective metrics *MDER* and MEDI can be calculated from the equations mentioned above, if the spectral irradiance $E_{e,\lambda}(\lambda)$ on the face of the light users is known from a spectroradiometric measurement. The key question is whether these two parameters can also be determined with sufficient accuracy in practical lighting technology using chromaticity coordinates (*xyz*) and illuminance E_v (vertical). If this is the case, practical field measurements in offices, industrial halls, in schools, or in outdoor areas can be done in the same manner like the measurements of CS-metric in Sections 10.4–10.6 of this chapter.

10.7.2 Mathematical Transformation for Calculation of MEDI and MDER

According to Eqs. (10.21) and (10.22), to obtain the melanopic equivalent daylight illuminance MEDI, one needs to know the MDER value and the illuminance E_v (in lx). Therefore, the first task is to calculate the dimensionless input quantity MDER (see Eq. (10.22)) from the chromaticity coordinates x, y, and z (z turned out to be the most predictive, hence this will be used).

For this mathematical conversion from z to MDER, a large number of measured and simulated light source spectra were analysed (see Table 10.11). They are real measured light sources from incandescent lamps (thermal radiators, 28 light sources), compact and linear fluorescent lamps (252 light sources), different LED types (419 light sources), and 185 measured daylight spectra. These 884 light source spectra in total are shown in Figure 10.35.

For the simulation of the combination of daylight-to-white LEDs or daylight-to-white fluorescent lamps, which are real lighting conditions of indoor spaces with white LEDs or with white fluorescent lamps with windows over a longer period of use, the authors of this book took 185 measured daylight spectra and mixed them with a market-typical white LED spectrum or to a typical fluorescent lamp spectrum in nine different mixing ratios from 10% to 90%. From the 185 daylight spectra

Table 10.11 Measured and simulated spectra (4214 altogether).

No.	Type of light sources and their parameter ranges CCT = 2201–17815 K; $-1.467 \times 10^{-2} < \text{Duv} < 1.529 \times 10^{-2}$; $0.0797 < z < 0.4792$; $80 < \text{CIE } R_a < 100$; $0.2784 < \text{MDER} < 1.46$
1.	Conventional incandescent lamps + filtered incandescent lamps, 28 spectra
2.	Fluorescent tubes + Compact fluorescent lamps, 252 spectra
3.	LED lamps + LED luminaires, 419 spectra
4.	Daylight (CIE-model + measurements), 185 spectra
5.	Mixtures (Daylight + LED) – [mixture ratio = 10–90%], 1665 spectra
6.	Mixtures (Daylight + FL) – [mixture ratio = 10–90%], 1665 spectra

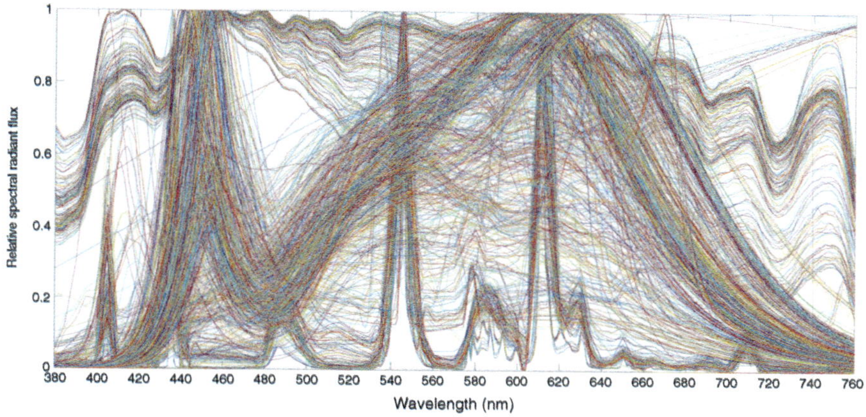

Figure 10.35 Spectra of 884 real measured light source spectra.

with nine mixing ratios each, 1665 simulated spectra result, each for LED and for fluorescent lamps ($185 \times 9 = 1665$).

With the 4214 light source spectra summarised in Table 10.11, 4214 MDER values and 4214 z values were calculated, from which their relationship was determined. The formula for MDER is shown in Eq. (10.23).

$$\mathrm{MDER} = a \cdot e^{b \cdot z} + c \cdot e^{d \cdot z} \tag{10.23}$$

The parameters a, b, c, and d for the fitting function and the correlation coefficients R^2 are shown in Table 10.12. The correlation between MDER (ordinate) and the chromaticity coordinate z (abscissa) is shown in Figure 10.36, where the fitting function follows the green course of the curve.

For estimation of the predicting quality of Eq. (10.23) and the transformation from z to MDER, the 4214 spectra in Table 10.11 were first taken to calculate the 4214 MDER values from the spectra themselves and then also another 4214 MDER values indirectly via Eq. (10.23) with z as the independent variable, see Figure 10.37 with 97%-confidence range. The goodness-of-fit measure (R^2) equalled 0.978. The difference ΔMDER between the fitted function and 97%-limits is about 0.0678.

Finally, this chapter delivers a formula for a mathematical conversion from the chromaticity z to the metric MDER, with which one can calculate the metric MEDI by knowing the vertical illuminance E. The illuminance E_v and the chromaticity z can be measured with handheld colorimeters, which are available on the market at moderate prices or with RGB sensors. The methods of practical measurements

Table 10.12 Parameter a, b, c, and d for the fit function in Eq. (10.23) and the correlation coefficient R^2.

Parameter	a	b	c	d	R^2
Value	0.53806	2.3212	−0.37585	−0.85808	0.968

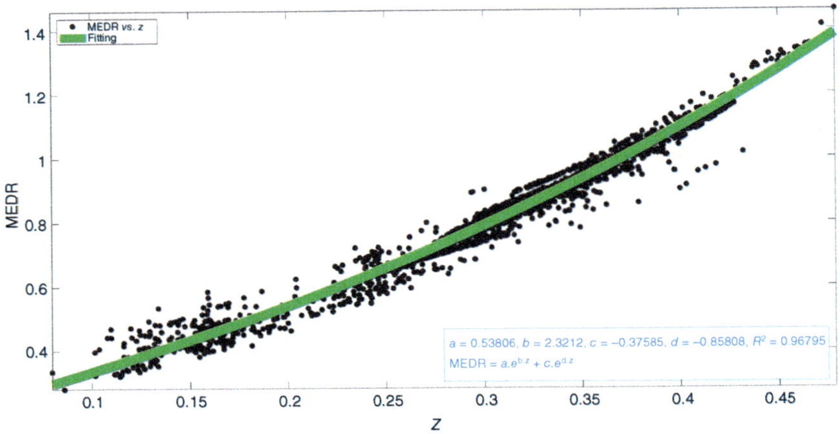

Figure 10.36 Correlation between MDER (ordinate) and the chromaticity coordinate z (abscissa).

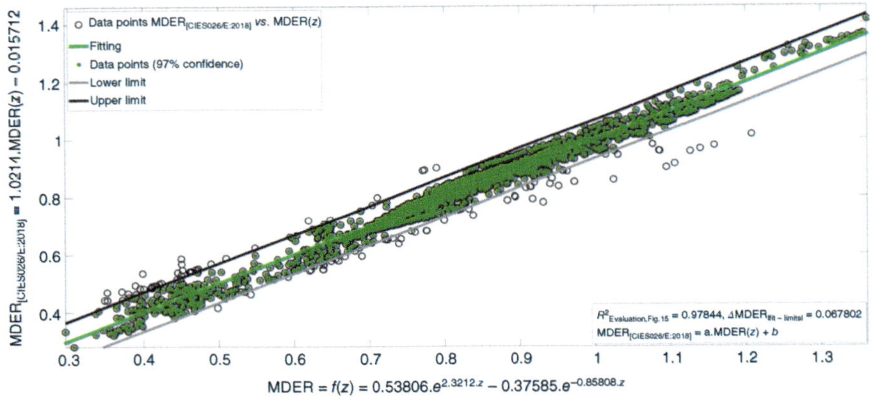

Figure 10.37 Comparison of the 4214 MDER values on the ordinate, calculated from the spectra themselves with the 4214 MDER values calculated by the use of Eq. (10.23).

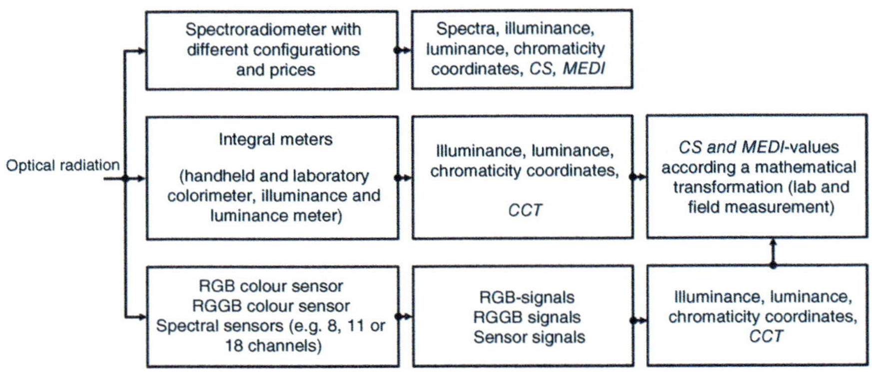

Figure 10.38 Spectroradiometer, Integral meter, and RGB colour sensors.

of visual and non-visual metrics for lighting engineers and lighting designers are illustrated in Figure 10.38 below.

References

1 Khanh, T.Q. and Bodrogi, P. (2019). Eine Brücke in die Praxisanwendungsorientierte Messtechnik zur Erfassung nichtvisueller Effekte in der Beleuchtungstechnik (A bridge to practical application-oriented metrology for the detection of non-visual effects in lighting technology). *Z. Licht* 5-2019: 65–71. ISSN: 0024-2861.

2 CIE (1983). *The Basis of Physical Photometry*, 2e. CIE Publ. 018.2-1983.

3 DIN Deutsches Institut für Normung (2019). *DIN EN 12464-1 Light and lighting - Lighting of work places - Part 1: Indoor work places*. https://www.din.de/en/getting-involved/standards-committees/fnl/drafts/wdc-beuth:din21:302583817?destinationLanguage=&sourceLanguage= (accessed 21 April 2023).

4 CIE (2018). *Colorimetry*, 4e. CIE Publ. 015:2018.

5 CIE (1995). *Method of Measuring and Specifying Color Rendering Properties of Light Sources*. CIE Publication 13.3-1995. Vienna: CIE.

6 Khanh, T., Bodrogi, P., Guo, X. et al. (2019). Colour preference, naturalness, vividness and colour quality metrics, Part 5: a colour preference experiment at 2000 lx in a real room. *Light. Res. Technol.* 51 (2): 262–279.

7 Brainard, G.C., Hanifin, J.P., Greeson, J.M. et al. (2001). Action spectrum for melatonin regulation in humans: evidence for a novel circadian photoreceptor. *J. Neurosci.* 21: 6405–6412.

8 Thapan, K., Arendt, J., and Skene, D.J. (2001). An action spectrum for melatonin suppression: evidence for a novel non-rod, non-cone photoreceptor system in humans. *J. Physiol.* 535: 261–267.

9 Vinh, Q.T. and Khanh, T.Q. (2018). Lichtsystem-basierte Lichtqualitätsparameter – Farbwiedergabe, Helligkeit, melanopische Wirkung, Lichtausbeute (Lighting system-based light quality parameters – colour rendering, brightness, melanopic effect, luminous efficacy). *Z. Licht* 2-2018: 118–123. ISSN: 0024-2861.

10 Bodrogi, P., Vinh, Q.T., and Khanh, T.Q. (2019). Correlations among lighting quality metrics for interior lighting. *Light. Res. Technol.* 51 (8): 1192–1207.

11 Vinh, Q.T., Bodrogi, P., and Khanh, T.Q. (2018). Preliminary measure for the characterization of the usefulness of light sources. *Optics Express* 26: 14538–14551.

12 Rea, M. and Figuiero, M.G. (2016). Light as a circadian stimulus for architectural lighting. *Light. Res. Technol.* 50 (4): 497–510.

13 Kunkel, J., Beck, S., Babilon, S., and Khanh, T.Q. (2020). Ideales Licht in Innenraum – Messtechnische Untersuchung Zirkadian Wirksamer Beleuchtungsgrößen in Büro. (Ideal light indoors – Metrological investigation of circadian effective lighting quantities in offices). *Light* 8: 70–76.

14 Lok, R., Smolders, K.C., Beersma, D.G., and de Kort, Y.A. (2018). Light, alertness, and alerting effects of white light: a literature overview. *J. Biol. Rhythms* 33 (6): 589–601.

15 Soumana, J.L., Tinga, A.M., Te Pas, S.F. et al. (2017). Acute alerting effects of light: a systematic literature review. *Behav. Brain Res.* http://dx.doi.org/10.1016/j.bbr.2017.09.016.

16 Wirz-Justice, A. and Bromundt, V. (2013). Lichttherapie (Light therapy). *Sleep* 1 (2013): 1–10.

17 Münch, M., Wirz-Justice, A., Brown, S.A. et al. (2020). The role of daylight for humans: gap in current knowledge. *Clock Sleep* 2: 61–85.

18 Cajochen, C., Zeitzer, J.M., Czeisler, C.A., and Dijk, D.-J. (2000). Dose-response relationship for light intensity and ocular and electroencephalographic correlates of human alertness. *Behav. Brain Res.* 115 (1): 75–83.

19 Knoop, M., Broszio, K., Diakite, A. et al. (2019). Methods to describe and measure lighting conditions in experiments on non-image-forming aspects. *Leukos* 15 (2, 3): 163–179. https://doi.org/10.1080/15502724.2018.1518716.

20 DIN SPEC 5031-100:2015-08 (2015). *Strahlungsphysik im optischen Bereich und Lichttechnik – Teil 100: Über das Auge vermittelte, melanopische Wirkung des Lichts auf den Menschen – Größen. Formelzeichen und Wirkungsspektren (Optical Physics and Photometrics – Part 100: Eye-Mediated Melanopic Effects of Light on Humans – Quantities. Formulae and Effect Spectra).* Berlin: Beuth Verlag.

21 Rea, M.S., Figueiro, M.G., Bierman, A., and Hamner, R. (2012). Modelling the spectral sensitivity of the human circadian system. *Light. Res. Technol.* 44: 386–396.

22 DIN 5032-7:2017-02 (2017). *Lichtmessung – Teil 7: Klasseneinteilung von Beleuchtungsstärke- und Leuchtdichtemessgeräten (Light Measurement – Part 7: Classification of Illuminance and Luminance Meters).* Berlin: Beuth-Verlag.

23 Kunkel, J. (2019). Untersuchung des alltäglichen Lichteinflusses durch Evaluation der Blaulichtschädigung sowie der biologischen Lichtwirkung unter Berücksichtigung von Blickdaten (Investigation of everyday light exposure by evaluating blue light damage and biological light effects considering gaze data). Master thesis, TU Darmstadt.

24 Broszio, K., Knoop, M., Niedling, M., and Völker, S. (2018). Effective radiant flux for nonimage forming effects – is the illuminance and the melanopic irradiance at the eye really the right measure? *Light. Eng.* 26 (2): 68–74. ISSN 0236-2945.

25 CIE (2018). Ed. CIE S 026:2018: CIE System for Metrology of Optical Radiation for ipRGC-Influenced Responses to Light. https://doi.org/10.25039/S026.2018.

26 Lauber, L.H. (2013). Das Gesichtsfeld. Untersuchungsgrundlagen. In: Physiologie und Pathologie, Bd. 3. Augenheilkunde der Gegenwart). The field of vision. Basics of investigations. In: *Physiology and Pathology*, Contemporary Ophthalmology, vol. 3. Berlin and Heidelberg: Springer. ISBN: 9783662019405 https://10.1007/978-3-662-02235-1.

27 Lasko, T.A., Kripke, D.F., and Elliot, J.A. (1999). Melatonin suppression by illumination of upper and lower visual fields. *J. Biol. Rhythms* 14 (2): 122–125. ISSN: 0748-7304. https://doi.org/10.1177/074873099129000506.

28 Fördergemeinschaft Gutes Licht (2014). Wirkung des Lichts auf den Menschen (The effect of light on humans) Vol. 19. Licht.wissen. Frankfurt am Main: Licht.de Fördergemeinschaft Gutes Licht. ISBN: 9783926193988.

29 Rüger, M., Gordijn, M.C.M., Beersma, D.G.M. et al. (2005). Nasal versus temporal illumination of the human retina: effects on core body temperature, melatonin, and circadian phase. *J. Biol. Rhythms* 20 (1): 60–70. ISSN: 0748-7304. https://doi.org/10.1177/0748730404270539.

30 Lucas, R.J., Peirson, S.N., Berson, D.M. et al. (2014). Measuring and using light in the melanopsin age. *Trends Neurosci.* 37: 1–9.

31 Gall, D. (2004). *Die Messung Circadianer Strahlungsgrößen (The Measurement of Circadian Radiation Quantities)*. Ilmenau: TU Ilmenau.

32 Rea, M.S., Nagare, R., and Figueiro, M.G. (2020). Predictions of melatonin suppression during the early biological night and their implications for residential light exposures prior to sleeping. *Sci. Rep.* 10 (1): 1–12.

33 Rea, M.S., Nagare, R., and Figueiro, M.G. (2021). Modeling circadian phototransduction: quantitative predictions of psychophysical data. *Front. Neurosci.* 15: 615322.

34 Rautkylä, E., Puolakka, M., and Halonen, L. (2012). Alerting effects of daytime light exposure – a proposed link between light exposure and brain mechanisms. *Light. Res. Technol.* 44 (2): 238–252.

35 Bodrogi, P., Guo, X., and Khanh, T.Q. (2018). Helligkeit und Sehklarheit statt Leuchtdichte in der Bewertung der Beleuchtung von Innenräumen (Brightness and visual clarity rather than luminance in indoor lighting assessment). Light 2018, Creative Lighting Solutions, 23rd European Lighting Congress. Davos, Switzerland.

36 Truong, W., Trinh, V., and Khanh, T.Q. (2020). Circadian stimulus – a computation model with photometric and colorimetric quantities. *Light. Res. Technol.* 52 (6): 751–762. https://doi.org/10.1177/1477153519887423.

37 Truong, W., Zandi, B., Trinh, V.Q., and Khanh, T.Q. (2020). Circadian metric – computation of circadian stimulus using illuminance, correlated colour temperature and colour rendering index. *Build. Environ.* 184: 107146.

38 Khanh, T.Q. and Vinh, Q.T. (2020). Zirkadian Wirksame Bestrahlung im Innen- und Außenbereich – Empfehlung für Lichtplanung und Leuchtenentwicklung auf Basis des aktuellen Wissensstands (Circadian effective irradiation indoors and outdoors – recommendation for lighting design and luminaire development based on current knowledge). *Light* 1: 70–76.

39 Figueiro, M.G., Hunter, C.M., Higgins, P.A. et al. (2015). Tailored lighting intervention for persons with dementia and caregivers living at home. *Sleep Health* 1: 322–330.

40 Knoop, M., Abdelmageed, A.A., Luo, T. et al. (2017). Spatially resolved spectral sky daylight measurement data: Methods of collection, evaluation, and representation. In: *Proceedings of the CIE 2017* Midterm Meeting, 1036–1046. Jeju, South Korea.

41 Diakite, A.K., Weber, N., and Knoop, M. (2018). Optimierung des Verfahrens zur Rekonstruktion von Tageslichtspektren aus Farbörter auf dem Daylight Locus

(Optimisation of the method for reconstructing daylight spectra from chromaticities on the Daylight Locus). In: *Proceedings of LICHT*. Davos, Switzerland.

42 Porsch, T., Krüger, U., and Schmidt, F. (2012). Einsatz von bildgebenden Leuchtdichtemesskameras bei der Bewertung von Blendungssituationen (Use of imaging luminance measurement cameras in the assessment of glare situations). *Licht 2012 - Berlin* 196–203. ISBN 978-3-927787-37-7.

43 Gigahertz Optics (2022). CSS-45. https://www.gigahertz-optik.com/de-de/produkt/css-45/ (accessed 10 January 2022).

44 Schierz, C. (2018). Blaulichtschädigung der Augen-Netzhaut – Stand der wissenschaftlichen Erkenntnisse. (Blue light damage to the ocular retina – state of scientific knowledge). www.tu-ilmenau.de/fileadmin/public/lichttechnik/Publikationen/2018/LICHT2018_Schierz_Christoph_Langfassung.pdf.

11

Technological Aspects of Human Centric Lighting in Buildings

11.1 Introduction to the Topic '*Smart Lighting*'

Today's world is currently experiencing dynamic social and technological development [1]. On the one hand, the digitalisation and automation of processes in production (Industry 4.0), in communication and in the home (Smart Home) are taking place, which means that decision-making and the content of processes are becoming increasingly complex but also shorter in time. On the other hand, new social challenges are emerging, which can be formulated as follows:

- *Formation of urban living spaces and large cities*: These cities are the new living centres of young and older generations of people with temporally and locally condensed supply chains, traffic situations, and medical provisions. The structure of '*Smart City*' includes the interconnections of digital infrastructures of buildings, transport systems, road infrastructures, logistics centres, as well as units of the education and health care systems of a city.
- *New division of working time and leisure time*: In a previous, industrially oriented society, there is a clear division between working hours and rest periods in the companies, at the workplaces, in the offices, or on the factory floors. In the information technology-oriented society, people work for the value creation of the labour-providing institutions at various locations networked by the Internet, including from home. The consequence of this is that living spaces also become working spaces (home offices). Therefore, '*very good and human-oriented*' lighting on one's own four walls will also become more important than previously.
- *An increasing demographic development*: the proportion of older people and people in need of care is growing rapidly. For this reason, the number of care facilities is increasing strongly, e.g. in Germany, see Table 11.1. This social development requires great responsibility, foresight, and care, but also innovation from human medicine, pharmacy, and product development to improve the life and vitality of the elderly. Besides medical aspects, light and lighting play a major role in this process. Since 1995 until today, thanks to intensive sleep research, human medicine, and lighting technology, we have concluded, that light – in the right measure – positively influences well-being, concentration, vitality, and the

Human Centric Integrative Lighting: Technology, Perception, Non-Visual Effects, First Edition.
Tran Quoc Khanh, Peter Bodrogi, and Trinh Quang Vinh.

Table 11.1 Current development of care facilities (homes and outpatient care).

Feature | Year →	2009	2011	2013	2015	2017	2019
Number of nursing homes	11 634	12 354	13 030	13 596	14 480	15 380
Of which with full inpatient permanent care	10 384	10 706	10 949	11 164	11 241	11 317
Outpatient care services	12 026	12 349	12 745	13 323	14 050	14 688

Source: Adapted from Statistisches Bundesamt [2].

healing process. Therefore, new, innovative lighting products are important for living areas, nursing homes, and hospitals, as well as for other professional areas such as offices and industrial halls.

During these processes, lighting industry and lighting design are undergoing a rethinking process. The previous argument for the use of LED products, that they have a long service life and high energy efficiency, will no longer be decisive in the foreseeable future with the current luminous efficacy of the best-LED luminaires around 160–180 lm W^{-1} and with increasingly reliable LED technology. Possible future progress must therefore be seen in connection with the development of intelligent and '*human-centred*' lighting solutions with lighting design that is close to the user, but this is currently partly the subject of discussion in industry and lighting research. The motivation of the authors of this book is therefore to describe technological and computing aspects, as well as lighting engineering-related aspects of the subject of '*Smart Lighting*'. The following questions are in the foreground:

(a) How shall we structure the terms '*Smart Lighting*' *and* '*Connected Lighting (Connectivity)*'?
(b) What advantages and challenges can be drawn from the *Internet of Things* (IoTs) platform for the realisation of *Human Centric Lighting* (HCL) solutions?
(c) Which basic technical components of networking and data processing can be worked out, and how can *sensor data* (especially data from colour sensors) be used for lighting control?

These are complex issues that cannot be dealt with in the final depth and breadth in just one chapter. Therefore, the readers of this book may wish to consider the content of Sections 11.2–11.5 as a first possible systematic attempt or contribution to the discussion. The focus of the following considerations will be on interior lighting.

Until about 2013–2014, indoor luminaires for professional applications were mostly produced with gas discharge lamps (fluorescent lamps, high-intensity discharge lamps) with a maximum luminaire luminous efficacy of about 80–90 lm W^{-1}, which could be switched on and off with wall switches, analogue signals, or digital formats, or dimmed to a defined value at a given time. LED technology with phosphor-converted high- and mid-power LEDs with a correlated colour temperature (CCT) and a general colour rendering index around R_a = 80–86 enabled the luminaires to be connected to transmission formats such as KNX, DALI, and LON and dimming between about 1–2% and 100%, with the first standard sensors such as presence sensors, motion sensors, or daylight sensors being applied. These were developments up until about 2017.

In 2017–2018, the first LED luminaires with variable colour temperatures between 2700 and around 6500 K were used by mixing the colours of the warm white and cool white LED light sources. Control and regulation were carried out with special apps and daylight sensors or presence sensors. Each luminaire system had its own transmission formats, such as KNX or DALI, to interconnect the luminaires in a building or on a floor. Figure 11.1 summarises the evolutionary stages of luminaire development up to Smart Lighting.

Today's lighting technology and lighting research are in the process of developing the upper level, i.e. '*Smart Lighting*'. In recent years, terms such as '*Smart Lighting*', '*Connected Lighting*', '*Connectivity*', and 'HCL' have often been the subject of discussion in lighting industry and lighting research institutions. From the present authors' point of view, these terms and the platforms behind them, can be characterised as shown in Figure 11.2.

According to Figure 11.2, '*Smart Lighting*' is a generic term that includes the technological platform '*Connectivity*' and HCL lighting technology. '*Connected Lighting*' or '*Connectivity*' refers to the systems or platforms that *network* LED

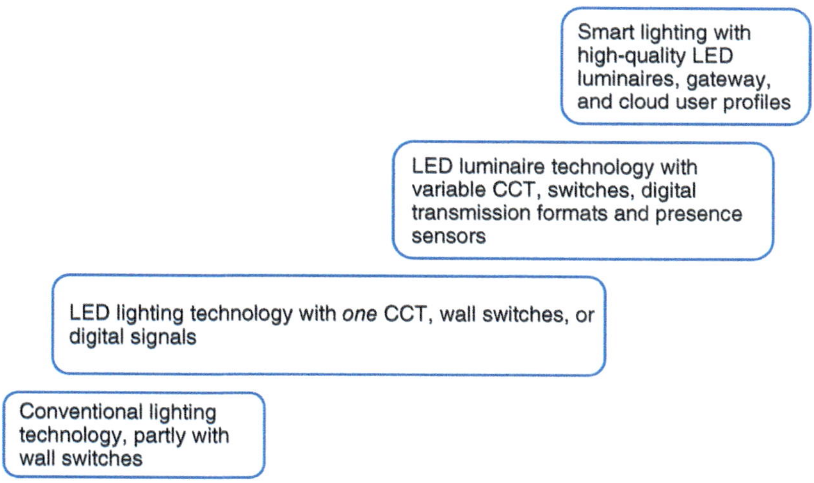

Figure 11.1 Evolutionary stages of luminaire development up to Smart Lighting. Source: TU Darmstadt.

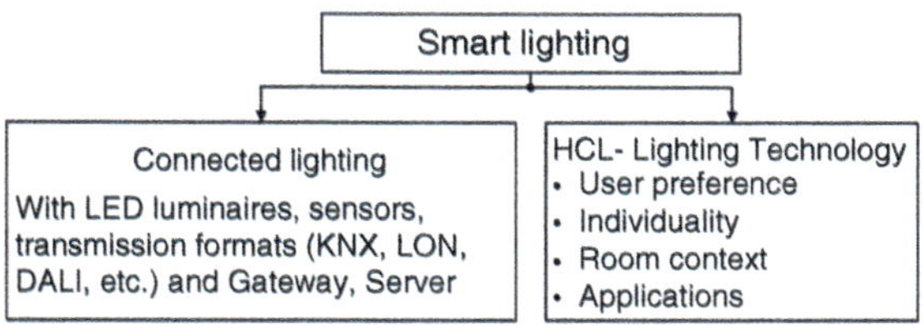

Figure 11.2 Smart Lighting and HCL Lighting. Source: TU Darmstadt.

luminaires of various degrees of intelligence (dimmable, non-dimmable, or one or more colour temperatures can be set) with various real and virtual sensors using various transmission formats such as DALI, LON, KNX, and Zigbee. In the context of the 'IoT', these LED luminaires at different locations in a building are connected via gateways to a specific Internet form, either to a *cloud server* or to several cloud servers of different providers. There, the physically generated luminaire data (electrical data, temperatures of the LEDs, temperatures of the electronics, humidity, time, season, and location of the luminaire), building data (so-called '*BIM data*'), data of other building technologies such as heating, air conditioning, hot water, and sun protection technology (curtains, blinds), as well as weather data or GPS data are collected, managed, and processed (Figure 11.3).

Depending on the level of difficulty, the following three levels of data processing exist for '*Connected Lighting*', see also Figure 11.4.

1. *Recording of the energy balance*: over a certain period of time, usually over several weeks within a season, the electrical energy consumption and the energy structure for the lighting, for the building services (lifts, automatic doors, sanitary facilities), and for the air-conditioning technology are recorded and brought into relation with user occupancy, user density, time of day and year, and weather data. This level is currently of interest for building management and the lighting

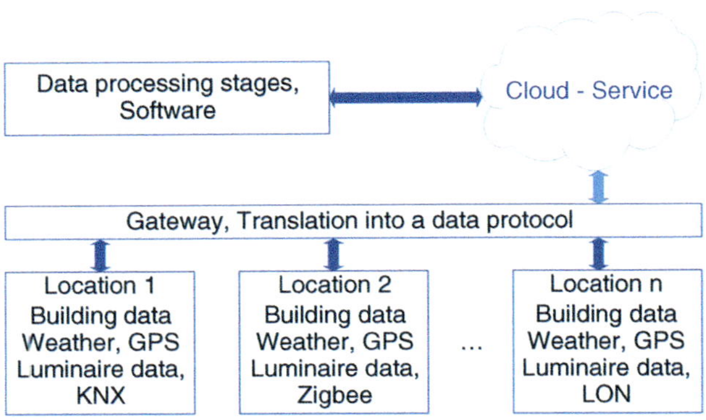

Figure 11.3 Structure of 'Connected Lighting'. Source: TU Darmstadt.

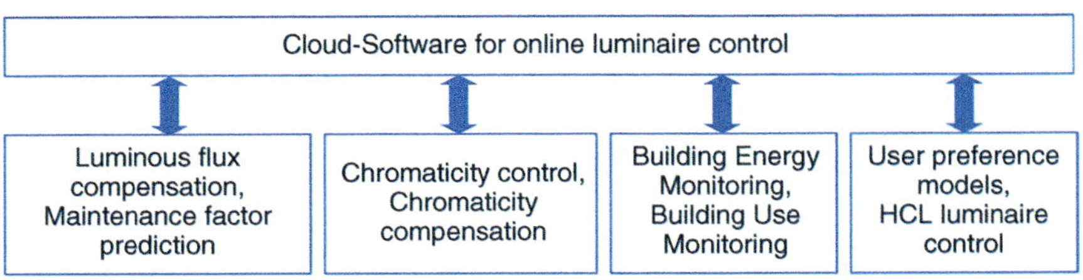

Figure 11.4 Structure and content of the cloud programmes for controlling luminaires. Source: TU Darmstadt.

industry, although the degree of connection to lighting technology/lighting engineering is only moderate.

2. *Physical conditions of the luminaires and luminaire systems*: The luminaires in the building provide light with a (more or less) precisely defined amount of light (luminous flux), colour temperature, spectrum, and direction of light, in conjunction with daylight, which in turn depends on the time of day and year, as well as on the weather. Since the components of LED luminaires (LED light sources, printed circuit boards, connecting elements, optics, and electronics) are subject to a degradation process, which in turn depends on service life, temperature, air quality, humidity (e.g. in a chemical plant, in a food industry factory), and solar radiation in the building, recording or calculating the degradation of the photometric and spectral parameters in the building is a complicated task that cannot be solved by today's *tables of maintenance factors*. This is a challenge for the future, which will become even more difficult because the future maintenance factor will no longer be based on luminous flux loss alone, but also on the *chromaticity shift* of a multi-channel LED luminaire with variable colour temperature and high colour rendering.
3. *User preference and HCL*: The users in any building have different ages, habits, occupations, professional contents, and expectations. They work at differently arranged workplaces (one-man office, open-plan office) with proximity to windows or further away from windows. Buildings have different orientations and compositions. Current lighting technology standards prescribe minimum values for lighting parameters. Research work in today's modern lighting technology and sleep research determines preferred values for colour temperatures, spectra, luminance, or illuminance distributions as mean values in test trials with a manageable number of test persons (approximately between 20 and 40) from a specific cultural group, at a specific geographical location, over a relatively short period of time (a few days or weeks). The automatic collection of individual lighting usage profiles over a longer period of time (over months) and the processing of this large amount of data with a future cloud programme will then lead in the future to user preference models and algorithms that can best be developed with artificial intelligence (AI) methods, e.g. machine learning (ML) and deep learning. The results of these computations will be used online for the instant regulation of the luminaires at any location for different weather conditions, seasons, and times of day.

While the parts of the Cloud Software (Figure 11.4) on luminous flux degradation, chromaticity coordinates, or the monitoring of light energy are more of technological and physical nature, the '*user preference*' part of the programme represents the entire content of current and future HCL technologies, which are therefore described in more detail in Chapters 4–9 of this book. In Section 11.2, the technical principles and concepts of *Connected Lighting* will be presented. Section 11.3 will discuss the structure of *cloud software* and application cases (so-called *use cases*).

11.2 Technical Principles of *Smart Lighting*

Until the beginning of the 2000s, there was a fixed assignment of switching elements to luminaires, so changes of assignment required a modification of the installation, and multiple assignments led to complex systems [3]. Flexible and adaptive lighting scenes were therefore difficult to realise. Since the beginning of the 2000s, additional *bus integrations* of simple sensors and presence sensors were introduced, which primarily served to control the direct lighting situation, e.g. the absence of room users. In the meantime, control via apps and the first approaches to connecting to cloud services have been developed. It can be assumed that sensor and user data will be processed in the cloud from 2023–2025, user preferences will be determined by means of ML algorithms, and HCL systems and solutions can be disseminated as a result.

Such component solutions must be considered as part of smart concepts for buildings, which in turn can be classified as an integrative part of smart urban development, see Figure 11.5.

The development of networking technology (so-called 'Connecting Technology') goes through several development stages. In Development Stage #1 (see Figure 11.6),

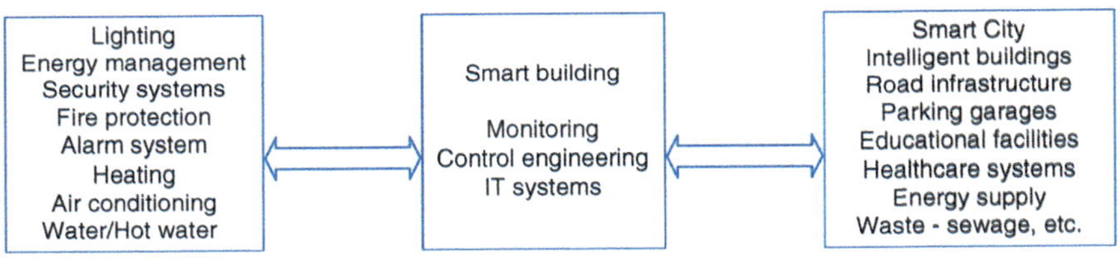

Figure 11.5 Path from building solution components via 'Smart Building' to 'Smart City'. Source: TU Darmstadt.

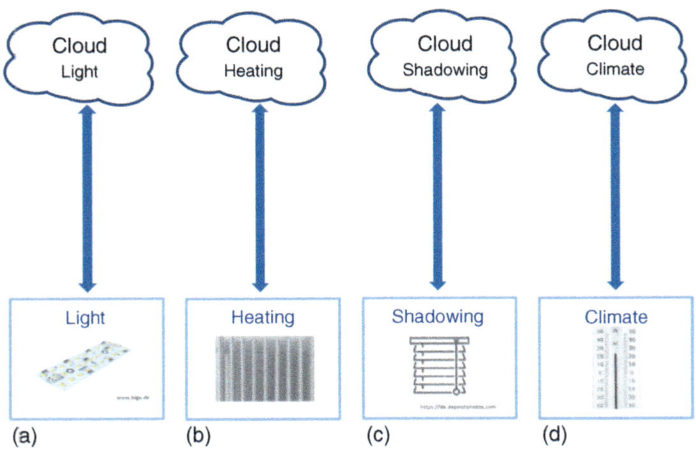

Figure 11.6 Building networking technology in Development Stage #1. (a) Light. Source: bigE. (b) Heating. Source: TU Darmstadt. (c) Shadowing. Source: FishCoolish/Adobe Stock. (d) Climate. Source: TU Darmstadt.

each service system of the building (lighting, heating, and window systems) has different types of control with different IT protocols, e.g. DALI, KNX, and LON. There is no connection between the data of the individual systems in the building and each other, and thus no overarching communication. The results of this are duplicate sensors and additional work for the user and building management.

The above-mentioned disadvantages are to be overcome – probably in the years 2022–2025 and later – in Development Stage #2. In this stage, all systems of the building communicate via the common gateway platform and with a common and well-defined semantics, e.g. via MQTT-formats, with one or more cloud servers, which are in turn organised by a common management software (see Figure 11.7). Thus, interactions between the individual building systems, sensor signal sharing, sensor fusion, as well as cross-system communications and evaluations of the data, will become possible. In this way, networking technology will enable the lighting industry to come much closer to the goal of '*Light and Building as a Service for People*' by generating application cases (*use cases*) and setting up practical business models through data evaluation and behavioural analysis of the processes in the building.

If we move from an organisational platform level as shown in Figure 11.7 to the data structure and processing level, the chain of signal and data structures (sensors, signal transmitters, nodes, processing locations, and data extraction stages) of modern luminaires can be illustrated as in Figure 11.8.

If we re-interpret the structure of Figure 11.8 from the physical locations and types of sensor signals to the spatial, social, user-specific, and temporal locations of these signals, then we will arrive at Figure 11.9, which better describes the data structure and evaluation of the data from the user's point of view.

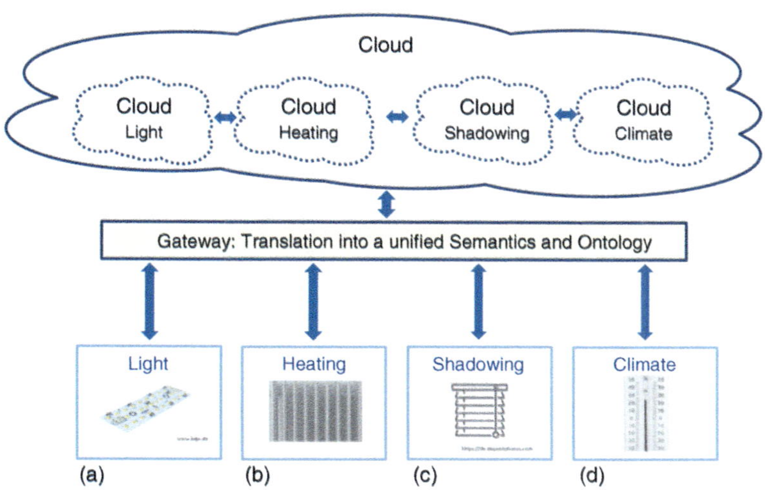

Figure 11.7 Building networking technology in Development Stage #2. (a) Light. Source: bigE. (b) Heating. Source: TU Darmstadt. (c) Shadowing. Source: FishCoolish/Adobe Stock. (d) Climate. Source: TU Darmstadt.

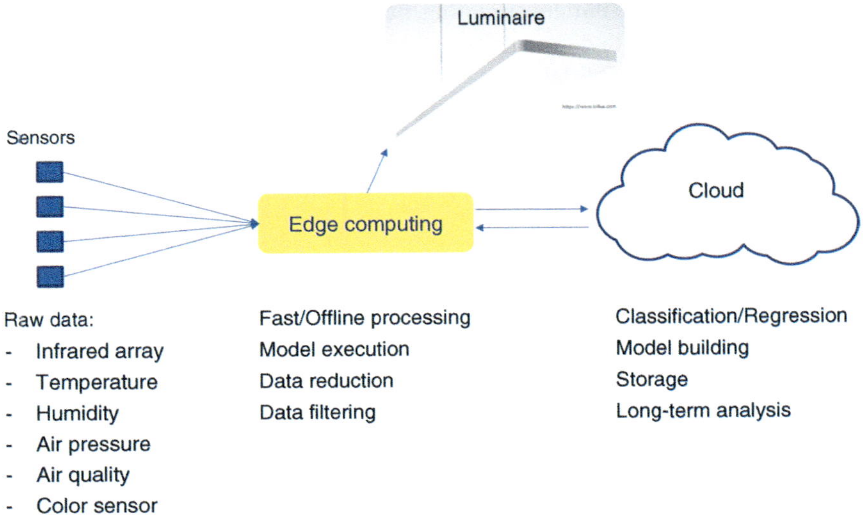

Figure 11.8 Scheme of an Intelligent Lighting System. Source: Technische Universität Darmstadt.

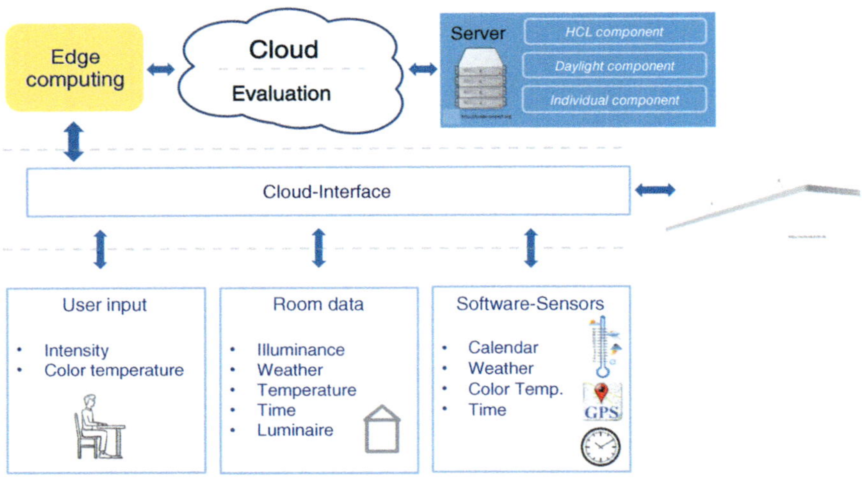

Figure 11.9 Data structure and evaluation of the data from the user's perspective. Source: Technische Universität Darmstadt.

Table 11.2 explains the terms, the meanings, and the applicability of the components of the above-described networking technique.

With the introduction of Smart Lighting and Smart Building concepts, as well as during the realisation of Industry 4.0 processes, the importance of sensors is increasing strongly. In the context of creating user profiles in the room, the following physical sensor types are important:

- *Air quality*: Is improving air quality crucial for concentrated tasks?
- *Pressure*: Are the windows/doors open or closed?

Table 11.2 Terms, meanings, and applicability of the components of the networking technology described in the text.

Term	Explanation
Cloud	Provides resources and services such as storage space and computing power in data centres. *Definition*: • A network of multiple devices, such as computers and servers, that are connected to each other Three main groups of services that a cloud platform can provide: • *IaaS (Infrastructure as a Service)*: A remote data centre with resources such as storage space, computing power and network components • *PaaS (Platform as a Service)*: A development platform with tools and components for building, testing, and launching applications • *SaaS (Software as a Service)*: A ready-made software for various business cases
Cloud IaaS	*Advantages*: • Large amount of storage space available (> terabyte) • Large number of computing nodes with great performance of RAM, CPU, and GPU • No downtime within the cloud • Resources can be scaled as required within a cloud *Disadvantages*: • High latency due to the path to the central computing cluster • Failures of the connection to the cloud • Security and privacy through central storage and transfer of data
Fog	Term originally coined by Cisco, a cloud concept that moves computing power and thus intelligence to the edge of the cloud. • Decentralisation of data processing = scalable • Not all data needs to be routed to the main computing unit • Latency and processing time decrease • Processes become real-time capable • Use of smaller, more local server farms (Fog-Nodes) • Intelligent switching and computing nodes. *Main objective*: Shortening of communication paths and reduction of data transmission volume; simple intelligent evaluation

(*continued*)

Table 11.2 (Continued)

Term	Explanation
Edge	Corresponds to a further decentralisation of 'Fog' • Process the data directly in the terminal or in the same room/building • Provides the collected data from the sensors • Very simple, fast processing without complex intelligence • Low computing power and memory *Tasks*: • Basic processing • Communication to other terminals or edges is possible
Gateway	To establish a connection between two different systems • Forwarding data between different protocols and transmission paths • Execution as software or as a physical device • Inserting a security layer for simple field protocols • Simple field protocols are used to minimise the computing power and thus the power consumption in the end devices.
Terminals Sensors Luminaires	*Sensors*: Collecting raw data • Calibration and error estimation are important • Mounting the sensors in the room • Observe privacy and security when collecting data *Luminaire*: Output device for Smart Lighting • Multiple colours (light channels) for dynamic lighting • Currently: Warm white and cool white LEDs • Near future: Three to six LED channels (e.g. RGB-LEDs, warm white, and cold white LEDs)

Source: Adapted from Khanh et al. [3].

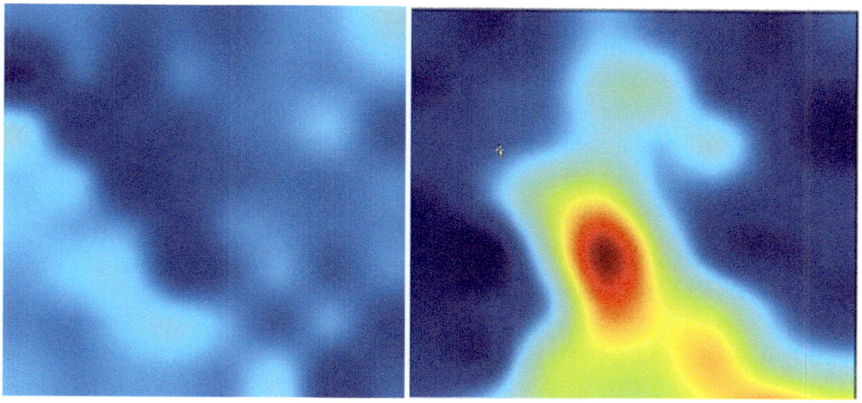

Figure 11.10 Signal representation of an IR array; left: the room is empty; right: position of a detected person. Source: Technische Universität Darmstadt.

- *Colour sensor*: extraneous light, chromaticity coordinates, illuminance/luminance, and colour temperature to control light sources, with or without consideration of daylight (see Section 11.5).
- *Temperature/humidity*: dependence of the user's well-being on environmental data at the time of the current lighting situation.
- *Array of thermal sensors (IR-array)*: Count the number of room users to estimate which tasks are performed in which room areas; determine the length of stay, see Figure 11.10.

These datasets, which are physically measured at the location, as well as other spatial and local data, such as GPS data, BIM data of the building, or weather data of the nearest weather stations (retrievable via the Internet), are read out and transmitted at a certain time interval. The following aspects regarding data pre-processing must be considered:

(a) *Raw data and its usability*: Different sensors differ in their ability to:
   - Data volume.
   - Readout speed.
   - Interpretation of the data.

Example 1: Temperature sensor:
- Data quantity: 1 × float.
- Readout speed = 5 minutes.
- Interpretation: Temperature in °C (direct).

Example 2: Infrared array:
- Amount of data: 64 pixels = 64 × float.
- Readout speed = 0.1 seconds.
- Interpretation: Temperature in °C.

(b) *Data reduction and storage: the following aspects should be considered*
- Large number of sensors.
- Different speeds of evaluation are required.
- Different time of storage (*Valid Time*).
- Data that are evaluated immediately do not have to be stored.
- Stored data have a validity period, after which they are no longer relevant.
- Lossy vs. non-lossy compression of sensor data.

Example 3: Infrared array:
- Amount of data: 64 pixels = 64 × float.
- Readout speed = 0.1 seconds.
- Daily data volume: = 64 float × 86 400 seconds/day × 10/seconds = 55 296 000 float/day/sensor.

Table 11.3 lists the data formats and database types for storage.

As soon as the data from sensors, room and building data, data from the virtual platform, and personal input data from individual users from different times are available, the data evaluation begins with the methods of *ML*; see Figure 11.11. This is a great advantage for future lighting research and the development of lighting systems. Previous lighting research findings on *user preference* regarding room usage and light settings (luminance distribution of ceilings and walls, illuminance on the work desk, colour temperature setting/light colour, interactions between lighting and architecture) were based on a number of visual and non-visual experiments with a limited duration (rather a maximum of several weeks, within one to two seasons), with a limited number of test subjects who tended to be younger, in a limited set of weather conditions, or in windowless rooms at different times of the day.

In [4], the following thoughts were formulated concerning the term '*ML*': '*Machine learning is a sub-field of artificial intelligence and describes dynamic algorithms that are capable of learning independently to improve results or*

Table 11.3 Data formats and database types for storage.

Data formats	JSON (RFC7159) • Very compact • Smaller data size • For structuring data	XML (RFC4825) • Metadata can be transferred • For outlining and formatting text and data
Database	SQL: Relational Database	NoSQL: Distributed Database
	Table-based:	Document-based:
	Rows and columns	Key-Value, document, graph
	Structured data	No standard scheme
	Vertically scaling	Unstructured data
	More power = hardware upgrade	Horizontal scaling
	Preferred for complex and frequent requests	More power = More servers in the pool
		Preferred for hierarchy data and large data volumes

Source: Adapted from Khanh et al. [3].

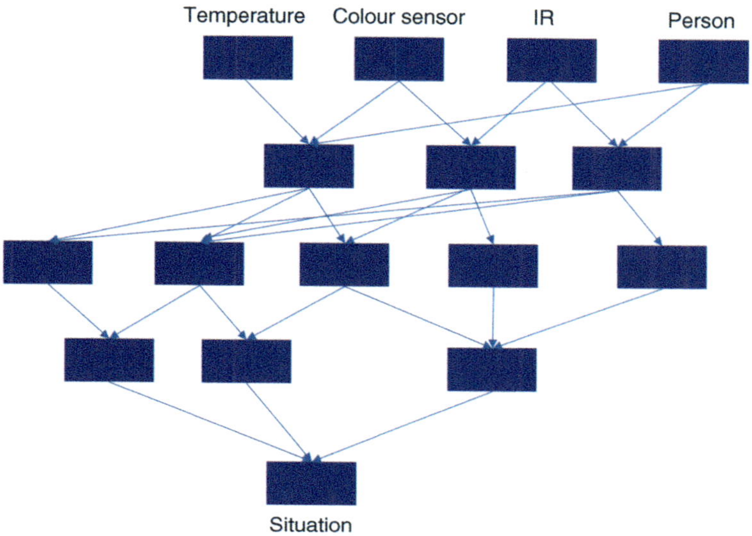

Figure 11.11 Machine Learning (ML)-based modelling and situation analysis using sensor data. Source: TU Darmstadt.

performance... According to the common definition by the computer scientist Tom Mitchell, the basic idea of machine learning is that a computer programme automatically improves its performance through new experiences (data) in a certain area... The term intelligence does not quite apply to machine learning, as it is less about intelligence and more about patterns and accuracy that can be recognised by machines or computers. Machine learning deals with the automated development of algorithms based on empirical data or training data. The focus is on optimising the results or improving the predictions based on 'learning processes".

Using the methods of ML, large amounts of data from different sensors under different conditions of use can be collected, processed, and modelled to derive rules, usage profiles, user models, or even regularities of the use of light and lighting by individual persons or groups of persons. Such models are stored in the cloud server, in the local building server, or in the central unit of the individual luminaire or luminaire groups of a room or a floor, constantly updated by new data from new room users as usage conditions and expectations change. These models are used to control the current luminaires under a specific condition, e.g. generation of a specific spectrum, colour temperature, luminance, and spatial light distribution; see Figure 11.12.

The generation and application of *ML* – based user models in lighting technology are characterised by the following aspects (see Figure 11.13):

- Input from persons and sensor data.
- Consideration of the time dependency of the data.
- Automatic adaptation to changing situations.
- Splitting training data and current application data.
- Predicting user profiles for new lighting situations.

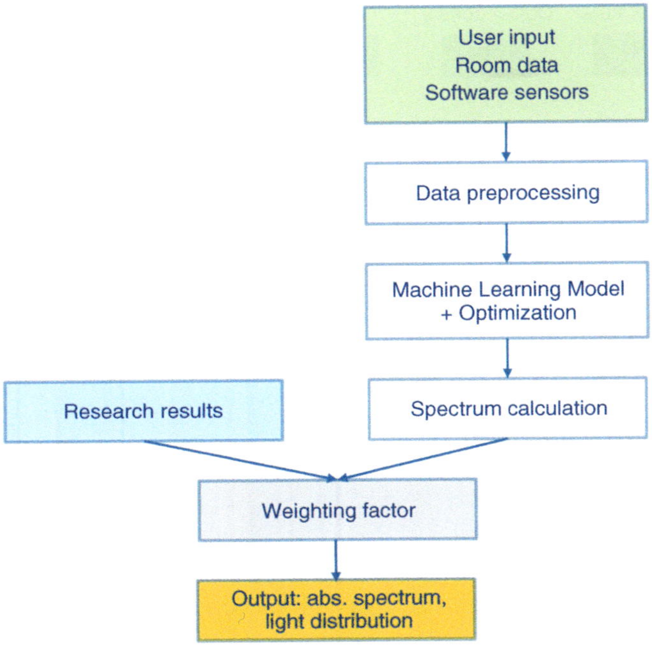

Figure 11.12 Sensor fusion and modelling using machine learning. Source: TU Darmstadt.

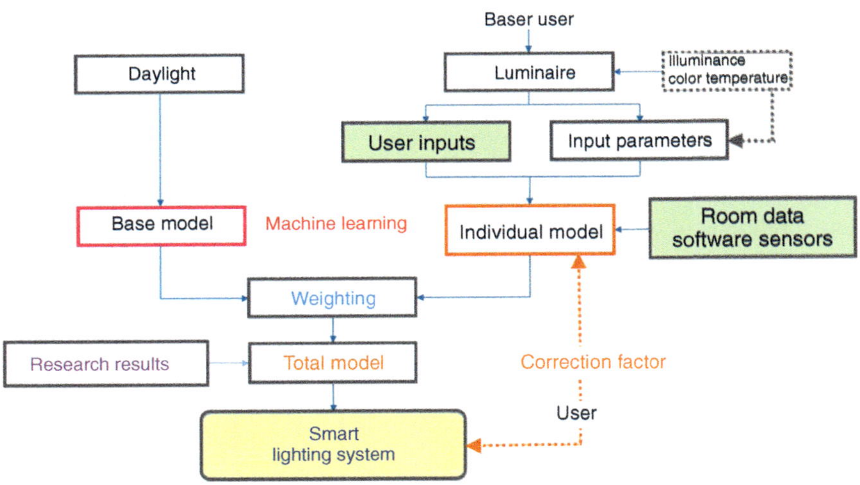

Figure 11.13 Generation and application of Machine Learning (ML)-based user models in lighting technology. Source: TU Darmstadt.

- Validate the prediction through subject studies.
- Limiting the selection range by existing experience or previous research results (e.g. limiting the colour temperature to a maximum of 6500 K, or, rather only to 6000 K).

- Output: CCT, illuminance values, circadian-effective values, ratio of indirect light to total light, colour quality parameters CIE CRI R_a, or TM-30-15 R_f.

Despite the complexity of generating user models, user preference is only one part of the cloud software and only one subtask of the '*Connected Lighting*' concept. Section 11.3 explains the structure of the cloud software and its *use cases*.

11.3 Cloud Software Structure and *Use Cases*

In principle, all essential subtasks for the operation of the lighting systems in connection with the entire infrastructure in a building are monitored and controlled by a cloud software. All monitoring and control processes take place in *real time* or near real time, whereby the monitoring intervals can also be selected selectively beforehand, according to a fixed scheme, or depending on the density of events. The software contains modules that can be processed independently or in groups. The most important lighting-specific modules are illustrated in Figure 11.14 by the structure of the TU Darmstadt's Cloud software for Lighting Technology.

With the help of sensor data such as daylight sensors, colour sensors, time of day sensors, degree of cloudiness sensors, and temperature sensors ON the LED-circuit boards in the room, the amount of daylight is varied by window and skylight systems, the illuminance or luminance distribution is changed depending on the position and number of room users, the colour quality of the lighting in the room is adjusted according to user preference, time of day (morning or afternoon), context of events in the room (a team meeting, a lecture in an auditorium, reading documents), and the energy consumption is balanced.

The technical systems only work reliably if the sensor data are relevant and reliable and the physical conditions are accurately reproduced. Therefore, the sensor systems must be measured or calibrated at a defined interval. It is a great challenge to qualify inexpensive sensor units to such an extent that accurate data can be generated with

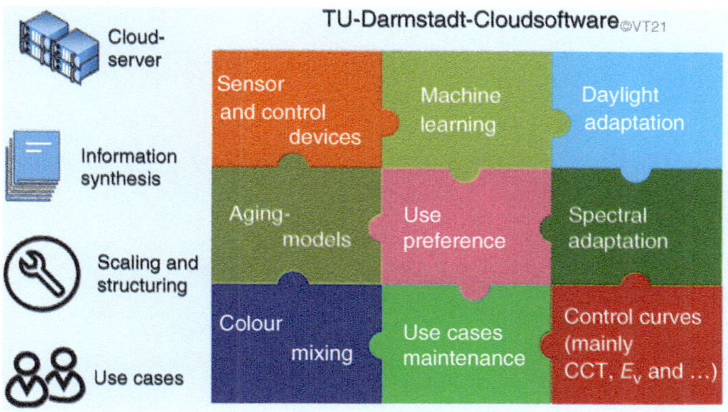

Figure 11.14 Structure of the lighting-specific modules in the cloud software. Example: the TU Darmstadt–Cloud software for Lighting Technology. Source: TU Darmstadt.

them. If the electrical data of the luminaires and other consumers in the building are to be used later to determine energy costs, it is important to also measure such electrical data accurately. Likewise, it is important to know the semiconductor-based colour and daylight sensors precisely to be able to adjust the colour and light settings as accurately as desired for the high perceptual requirements of the room users.

On the lighting design side (light generation of the luminaires, degrees of reflection of the room's boundary surfaces, and furnishings), the degradations and physical changes in material properties must be measured at regular intervals and considered in the cloud software. On the one hand, knowledge about these degradations is important to be able to precisely realise the colorimetric and photometric settings. On the other hand, maintenance factors, maintenance intervals, and replacement and repair measures must be reliably predicted and planned from the degradation data. As a rule, for economic and practical reasons, lighting installations cannot be re-measured too often. The ageing rate and maintenance intervals of LED luminaires are therefore to be calculated and verified by measuring the temperature at a specific location on the LED circuit board, by measuring the change in electrical data (LED voltage and current), and by processing this dataset in the cloud software, by the use of models, see Figure 11.15.

In accordance with the above considerations, an overview of the structure of the entire cloud software is shown in Figure 11.16.

Within the so-called '*Bottom – Up – Hierarchy*' of the HCL of '*Smart Buildings*', the structure in Figure 11.16 is intended to lead holistically through new intelligent technologies (sensors, data processing, communication systems, and internet platform) to new applications (so-called *use cases*). These applications have human light users at the centre, coupled with the concept of 'Light and Building as a

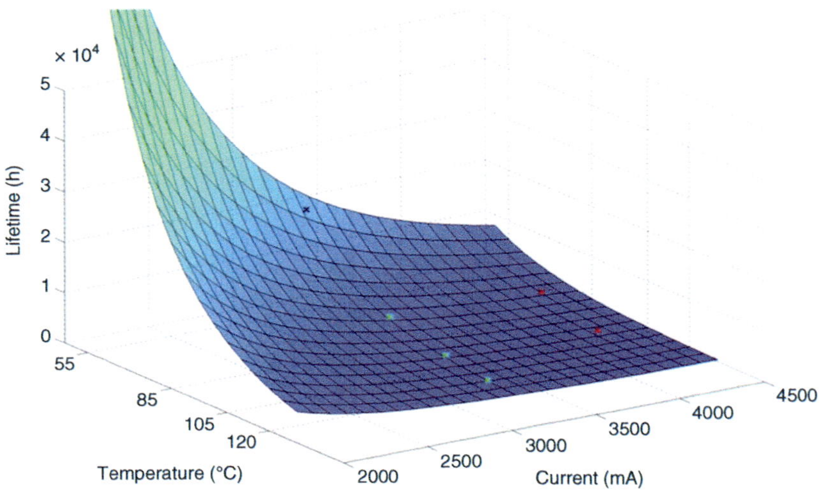

Figure 11.15 LED ageing model of a specific LED type as a function of current and PCB temperature. Source: Wagner [5]/TU Darmstadt/CC BY SA 4.0.

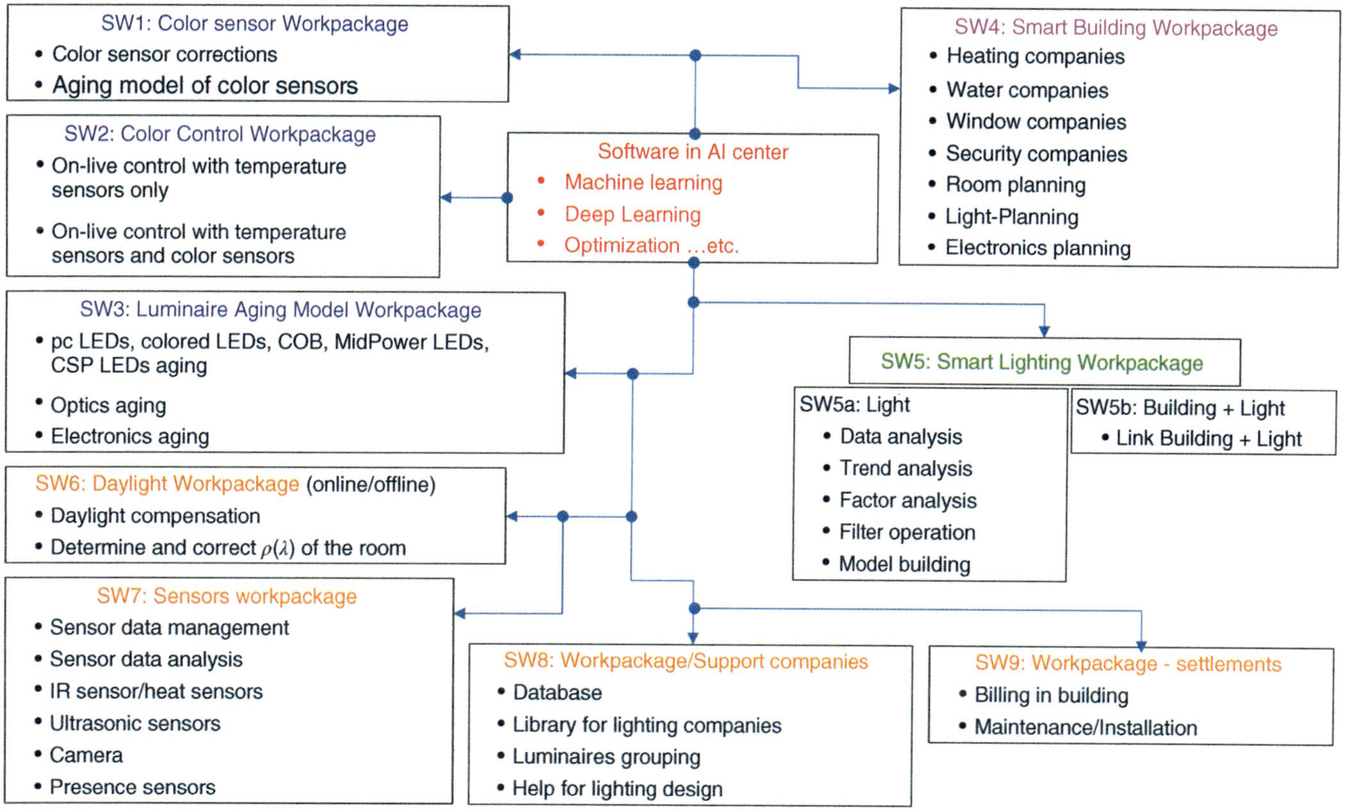

Figure 11.16 Structure of the entire cloud software. Source: TU Darmstadt.

Table 11.4 Overview of potential 'use cases'.

Energy management	Energy balancing, Use of dimming (situation-specific)
Maintenance	Online recording of luminaire data (I, U, T, H%), recording of service life Service contracts
Room management	Room occupancy-cleaning schedule Workplace rental
Safety	Emergency lighting Escape route marking
Navigation	Navigation for customers in a shop (Where are the individual product groups?)
Light quality	Control and regulation for high-quality, healthy light

Source: TU Darmstadt [3].

Service' and are oriented, among other things, towards economic aspects. These *use cases* cannot be fully elaborated at the time of publication of this book, as several years of experience with light and building services on a large scale are still missing. Table 11.4 outlines the initial considerations of such applications that are currently being pursued and worked on within the German lighting industry.

If we consider such *use cases* (Table 11.4) for a building that is currently being worked on, the following lighting-oriented concepts and operations are in the foreground and will be described in detail:

- Adaptive regulation and control of lighting systems for high-quality light that is beneficial to health. In this sense, the colour quality parameters (colour saturation, colour gamut, colour rendering, chromaticities), the circadian-effective parameters (MEDI, CS 2021, a_{mel}), as well as the photometric parameters (illuminance, luminance of walls and ceilings, cylindrical illuminance) are to be varied through targeted spectral optimisation and modification of the luminous intensity distribution of direct and indirect lighting by dimming the individual LED or OLED light channels. Changing the luminous intensity distribution is currently not yet the focus topic of the lighting industry. In lighting research, some relevant studies are known about the effectiveness of changing the direct-to-indirect ratios as a function of weather conditions, time point within the day, and light colour, e.g. in [6] and [7].
- The concept of '*Smart Lighting*' can only be successfully implemented if inexpensive colour sensors are qualified to determine the colorimetric, photometric and circadian parameters with sufficient accuracy to accurately describe the current lighting situation in the room and to control the lighting systems.

11.4 Light Control and Spectral Optimisation for High-Quality and Healthy Light

11.4.1 Stages of the Realisation Possibilities of the Luminaires for HCL Lighting Technology

Some aspects of high-quality and healthy light, especially for daytime working hours, can be described as follows:

- For the required visual and work performance at the 95% level (see Chapter 4), horizontal illuminance values of about 900–1000 lx, up to 1500 lx, with a ratio of $E_z : E_h = 1 : 2$, are desirable. However, this also means that a cylindrical vertical illuminance of about 450–750 lx at the eye should be achieved for good object recognition and visual communication.
- For psychologically and emotionally positive lighting, an illuminance level from 1200 lx on is also desirable for a good room assessment (or scene preference, see Chapter 6) with a colour temperature throughout the day between 4200 and 5500 K (maximum 6000–6500 K). Thus, vertical illuminance should also be greater than 600 lx.
- For '*good*' semantic judgements regarding colour quality and colour rendering, the general colour rendering index CIE CRI R_a should be at least 86, with saturation ΔC^* in the range of 1.0–2.0.
- Translating the above requirements with a vertical illuminance of 450 lx (or 600 lx) and a colour temperature of 4500–5500 K as the centre of gravity into the circadian-effective parameter CS *(model version 2018 of Rea et al., see Section 10.3 of this book)* and applying Eq. (10.9), we arrive at a CS value range of CS = 0.36 (for $E_v = 450$ lx and CCT = 4500 K) and CS = 0.49 (for $E_v = 750$ lx and CCT = 5500 K). This corresponds to the statement by Acosta et al. [8] that a CS value greater than 0.35 should be used for circadian-effective lighting. A vertical illuminance of 425 lx at the eye of the room user at a mostly preferred colour temperature of CCT = 4800 K corresponds to a MEDI-value of 336 lx.

It is therefore clear that, for high-quality, realistic, and immediately realisable HCL, several possible variants of the lighting system may come into question. What is important here is a holistic view of all possibilities. HCL lighting is much more variable than '*tunable white light*' (light with variable colour temperatures). For a professionally, financially, and technologically realistic HCL, there are therefore six levels of implementation options, as shown in Figure 11.17.

11.4.2 Levels 1 and 2 with Constant Colour Temperature

11.4.2.1 Basic Data of Circadian Effectiveness

Levels 1 and 2 (see Figure 11.17) enable room users and luminaire manufacturers to make an unproblematic transition to practical HCL. With a currently easily

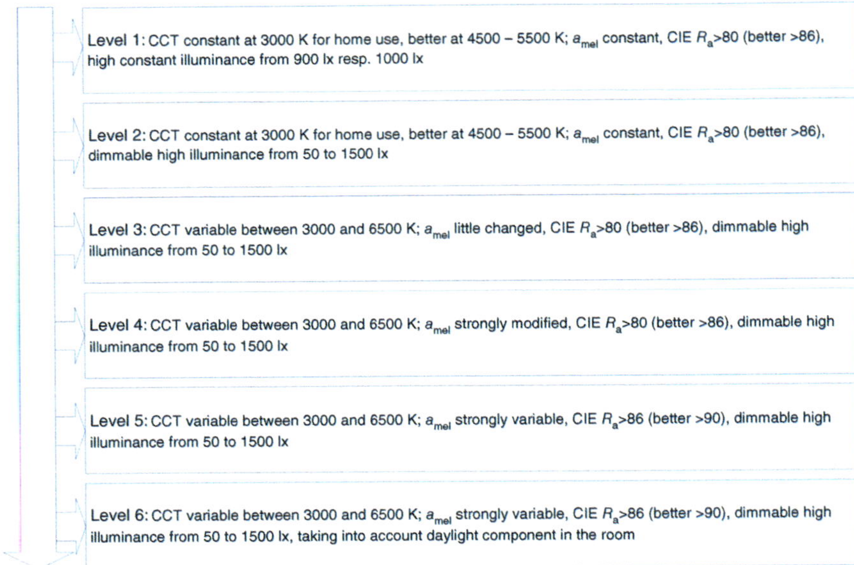

Figure 11.17 Levels of implementation options for practical Human Centric Lighting. Source: TU Darmstadt.

achievable LED phosphor with good (or very good) colour rendering for a constant colour temperature of around 3000 K for the home and between 4500 K (or 4000 K) and 5500 K, all that remains is to ensure that the LED luminaires are configured to be dimmable and that an acceptable dimming curve is programmed in over the day as a function of the phases of the day. Development levels 1 and 2 can be realised nowadays with the available technical possibilities. These white LEDs with good colour quality and their phosphors are reported on in Section 11.4.2.2. The advantage of stages 1 and 2 is that a single type of LED is sufficient, and a complicated and (sometimes highly) lossy *light mixing optic* does not have to be used. Table 11.5 shows CS values *(model version 2018 according to Rea et al., see Section* 10.3 *of this book)*, calculated according to Eq. (10.9), for four constant colour temperatures and four illuminance levels.

Table 11.5 CS values for four constant colour temperatures and four illuminances.

E_v | CCT (lx)	4000 K	4500 K	5500 K	6500 K
250	0.238	0.261	0.302	0.335
450	0.336	0.359	0.399	0.433
550	0.369	0.393	0.433	0.466
750	0.421	0.444	0.484	0.518

Source: TU Darmstadt.

Table 11.5 implies the following:

(a) For '*Level* 1' (see Figure 11.17) with a constant colour temperature of 4000–4500 K for professional lighting and a temporally constant vertical illuminance E_v of 450 lx (this corresponds to a horizontal illuminance E_h of about 900 lx), the requirement of Acosta et al. [8] with CS ≥ 0.35 can be almost or completely fulfilled. White LEDs with a colour temperature of more than 4000 K generally already have a high luminous efficacy, so their use for high illuminance in the room does not pose a major problem in terms of energy and costs.
(b) For '*Level* 2' with a constant colour temperature between 4000 and 5500 K and with dimmable horizontal illuminance up to 1500 lx ($E_{vertical}$ = 750 lx), the value of CS can be varied between 0.238 (for very late afternoon) and 0.484 (more for the most phases of the daytime in the office, in the school, or at the industrial workplaces). In both cases, a positive circadian effect with good visual performance can be expected. For '*Level* 2', a dimming curve (so-called *control curve*) for the variation of illuminance levels throughout the day should be determined depending on the situation. This is a question of lighting design and control programming of the system installation (see Chapter 12).

11.4.2.2 Previous Technologies for Generating White LED Light

In principle, the generation of white LED light has so far been based on the use of a blue LED, whose quasi-monochromatic radiation penetrates a phosphor layer and is partly absorbed and partly transmitted [9]. The absorption in the phosphor partially leads to the generation of longer-wave light (green, yellow, or red light), which is mixed with the transmitted blue LED light to form white light with a specific colour temperature and spectrum. The first technologically used phosphor was a YAG phosphor with a maximum wavelength in the range of around 560 nm. Later, phosphors based on LuAG (lutetium-aluminium-garnet, see [10–12]) with a maximum wavelength in the wavelength range between about 500 and 550 nm (see Figure 11.18) as well as nitride-based phosphors (e.g. Eu^{2+}-doped nitride material – $Sr[Li_2Al_2O_2N_2]:Eu^{2+}$ in [13]) with a maximum wavelength between about 600 and 670 nm (see Figure 11.19) [14] were used to enable high colour quality for red surface objects (e.g. red textiles, tomatoes, strawberries). These phosphors have in common that their excitation spectra have a maximum around 440–460 nm, so that a blue LED with a peak wavelength in this range leads to maximal radiation conversion.

Figure 11.20 shows the spectra of different white LED types. One can see the excitation wavelengths of the blue LEDs around 440–460 nm as well as different fluorescent spectra with maxima around 530, 580, 605, and about 637 nm.

The colour quality of the white LED types shown in Figure 11.20 with the colour temperature ranges around 3000 K (living area) and 4000–4100 K (office area) is shown in Table 11.6. The spectrum of the white LED WW56/80/600 in Figure 11.20, with R_a = 81, has a large gap of around 480 nm and a maximum of around 620 nm. In contrast, the spectrum WW307/97 has a maximum at 637 nm and a R_a-value of 97. However, the gap around 480 nm can also be seen here. This is roughly the state of

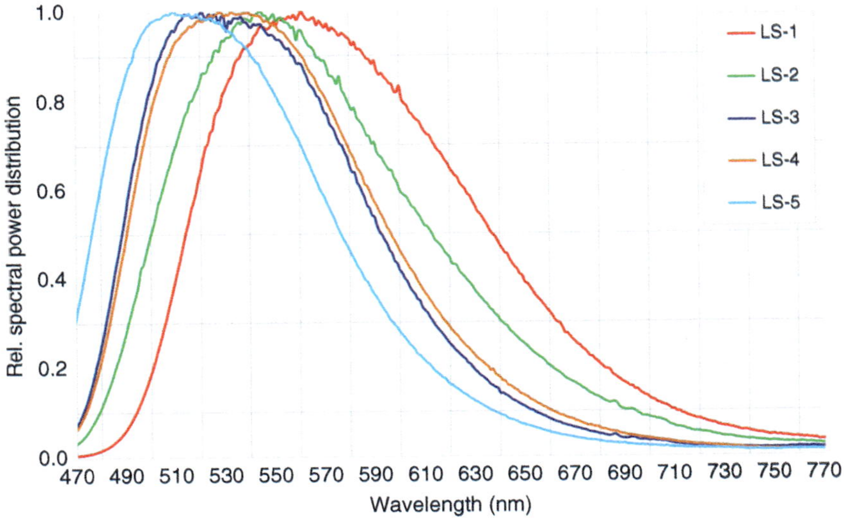

Figure 11.18 Emission spectra of phosphors based on YAG, LuAG. Source: TU Darmstadt.

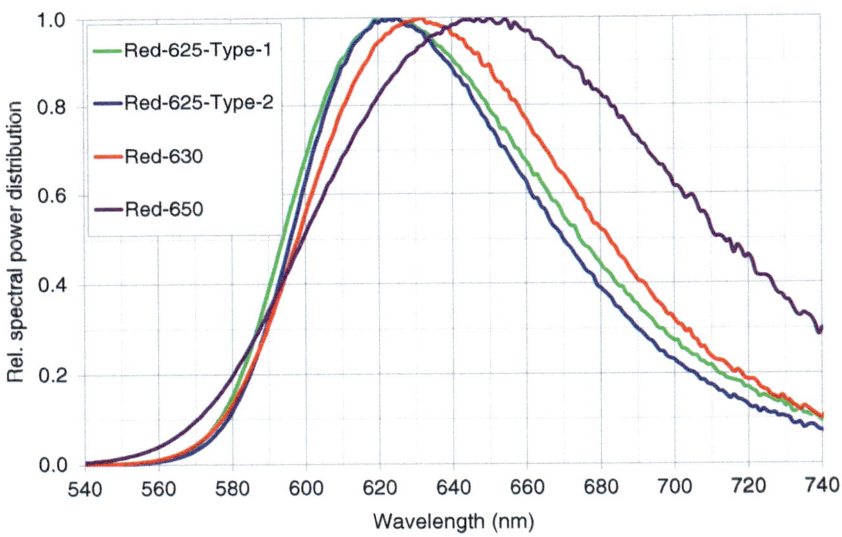

Figure 11.19 Emission spectra of red phosphors. Source : TU Darmstadt.

white LED light technology until around 2015, with the aim of increasing the colour quality of white LED light.

11.4.2.3 Newer Technologies for the Generation of White LED Light with Only One Colour Temperature

In the production of white LED light with a fixed colour temperature, there is currently a tendency to shift the maximum wavelength of the exciting LED from the

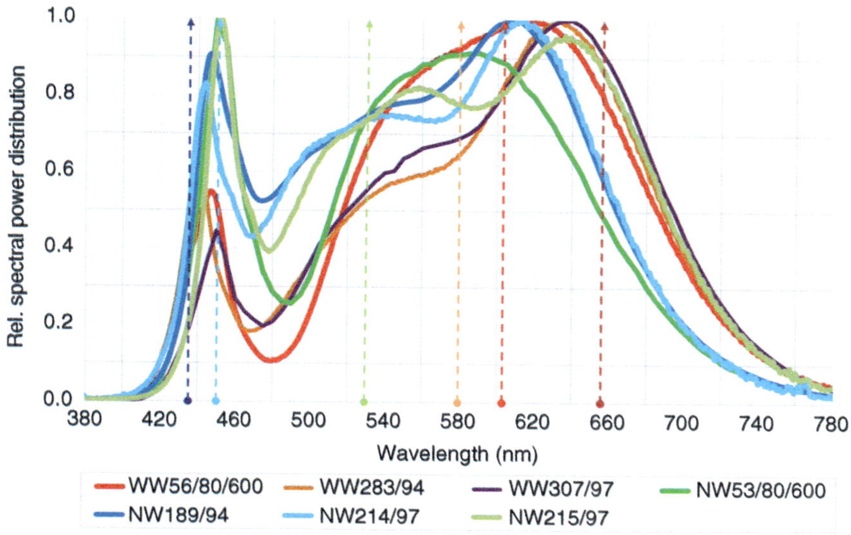

Figure 11.20 Spectra of different white LED types with current technologies. Source: TU Darmstadt.

Table 11.6 Correlated colour temperature (CCT) and colour rendering indices (CIE CRI) of the white LED types shown in Figure 11.20. CIE $R_{1,14}$: mean of the 14 special colour rendering indices.

LED	WW56/ 80/600	WW283/ 94	WW307/ 97	NW53/ 80/600	NW189/ 94	NW214/ 97	NW215/ 97
CCT in K	3091	3000	3027	4100	4129	4024	4043
CIE R_9	35	98	94	15	61	75	98
CIE R_a	81	94	97	81	95	97	97
CIE $R_{1,14}$	74	94	95	73	92	96	95

Source: TU Darmstadt.

blue range around 440–460 nm to the violet range around 410–420 nm, thus extending the light spectrum by the range 400–440 nm. The corresponding phosphor mixtures have multi-components with a phosphor with a secondary emission maximum around 495–500 nm and with a second phosphor with a main maximum either at 600 nm (LED-S-SPD827 and LED-S-SPD830, dashed curves, see Figure 11.21) or at 640 nm (see the other dashed curves in Figure 11.21). Moving the maximum wavelengths from 600 to 640 nm brings a significant increase in colour rendering (see Table 11.6) from $R_a = 83–85$ up to $R_a = 94–98$. All spectra of this type have a gap of around 440 nm.

Shifting the excitation wavelength down to about 415 nm is interesting if the new phosphors have a broad spectral emission distribution with a pronounced secondary maximum at 490–500 nm, approximately near the maximum wavelength

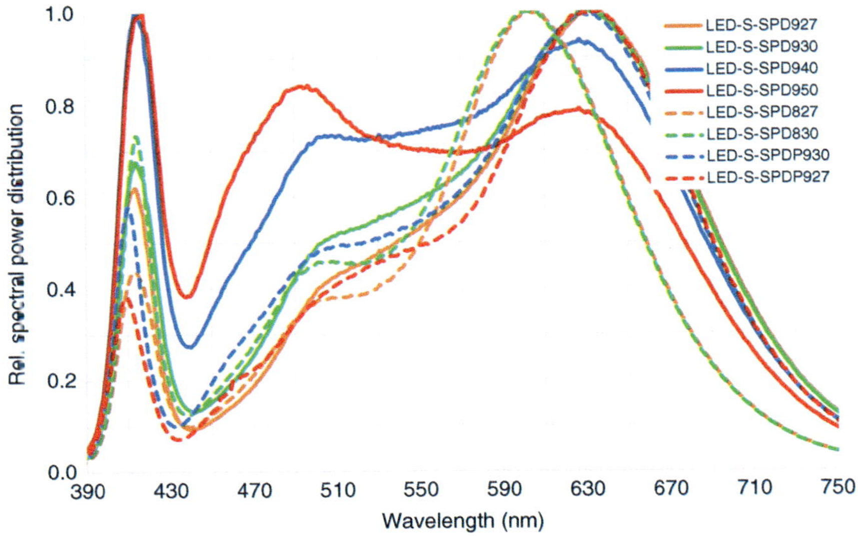

Figure 11.21 Spectra of different white LED types with more recent technologies. Source: TU Darmstadt.

Table 11.7 Correlated colour temperature (CCT), colour rendering indices (CIE CRI), and melanopic efficiency factor of the white LED types shown in Figure 11.21. CIE $R_{1,14}$: mean of the 14 special colour rendering indices.

LED	LED-S-SPD927	LED-S-SPD930	LED-S-SPD940	LED-S-SPD950	LED-S-SPD827	LED-S-SPD830	LED-S-SPDP930	LED-S-SPDP927
CCT in K	2763	3050	4014	5052	2726	3005	2980	2694
CIE R_9	89	93	94	93	6	13	96	96
CIE R_a	96	96	97	94	83	85	94	94
CIE $R_{1,14}$	94	94	96	93	79	81	92	93
$a_{mel,v}$	0.43	0.49	0.65	0.82	0.40	0.46	0.51	0.45

Source: TU Darmstadt.

of the spectral sensitivity of the intrinsically photosensitive retinal ganglion cells (ipRGCs) and thus near the maximum melanopic efficacy (see Figure 11.21, LED types LED-S-SPD940, LED-S-SPD950). This is also clearly shown in Table 11.7 in which the $a_{mel,v}$ value for the melanopic efficacy of the two LED types has risen sharply. These two LED types are therefore interesting for the illumination of workplaces during the daytime hours to ensure concentration, alertness, and, consequently, good work performance during the day. The colour rendering indices of these two white LED types are very high, see Table 11.7.

Likewise, with a constant colour temperature, but with an intended increase in melanopic efficacy, one can make an additional selection for a second conception. As before, a blue LED with a maximum excitation wavelength of around 450 nm and a conventional phosphor with an emission maximum of around 620 nm are used. In addition, a blue semiconductor LED component with a maximum wavelength of around 484 nm and a half-width of around 25 nm is integrated into the light source unit so that the entire spectrum receives a radiation increase of around 484 nm and a relatively high melanopic efficacy factor, see '*LED-B-3500 K*' and '*LED-B-4000 K*' in Figure 11.22 and in Table 11.8. The general colour rendering index CIE CRI R_a is in the moderate range between 82 and 84. These two LED types are therefore also interesting for the illumination of workplaces during the daytime hours.

For daytime workplaces such as offices, industrial halls, schools, and home offices with daytime working character, the colour rendering should achieve a minimum value of $R_a = 80$ with a high melanopic efficiency (see Figure 11.23, type 'LED-C-Day'). For the evening and night hours, the colour rendering shall also be good. The blue components in the radiation spectrum between 440 and 490 nm should nevertheless be reduced to enable melatonin production in the human body

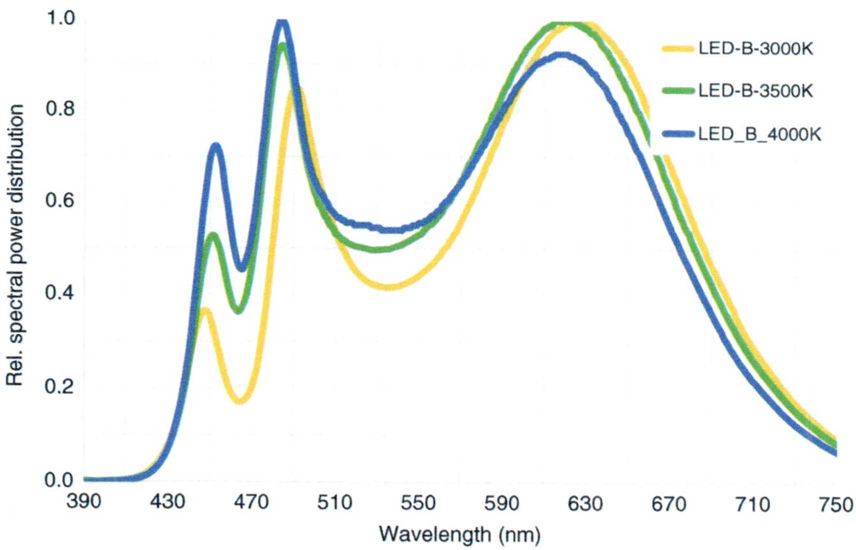

Figure 11.22 Spectra of white LEDs with increased melanopic effect factor. Source: TU Darmstadt.

Table 11.8 Correlated colour temperature, colour rendering index (CIE CRI), and melanopic effect factor of the white LED types shown in Figures 11.22 and 11.23.

LED	LED-B-3000K	LED-B-3500K	LED-B-4000K	LED-C-DAY	LED-C-NIGHT
CCT in K	2937	3312	3782	4004	3225
CIE R_a	82	83	84	83	81
$a_{mel,v}$	0.61	0.68	0.75	0.75	0.37

Source: TU Darmstadt.

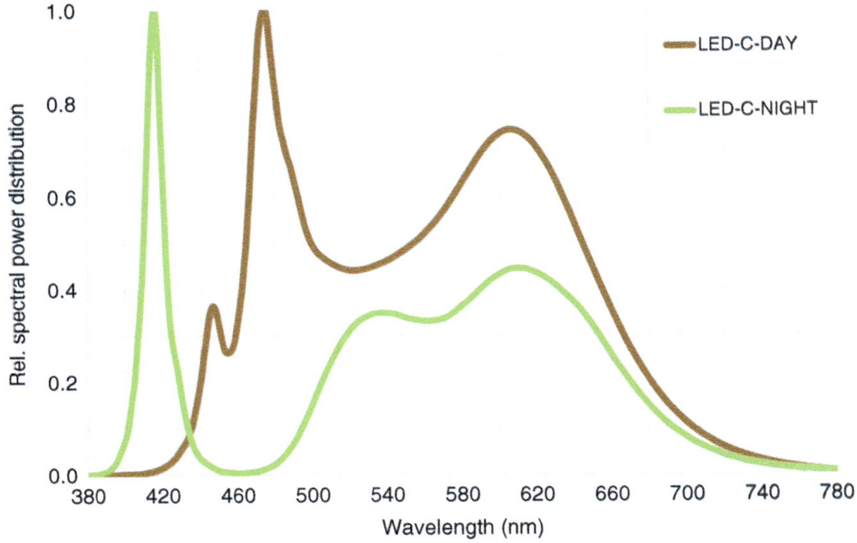

Figure 11.23 Spectra of white LEDs with increased (LED-C-DAY) or strongly reduced melanopic effect factor (LED-C-Night). Source: TU Darmstadt.

and thus a good start to sleep (see Figure 11.23, type 'LED-C-Night'). This is possible if one takes a closer look at the two spectra in Figure 11.23. The strong change in the melanopic effect factors of the two spectra with almost the same colour rendering (0.75 vs. 0.37) is shown in Table 11.8.

In recent years, there has been a discussion about certain special LED spectra, which are called '*full LED spectra*' or '*daylight-like spectra*'. Such a spectrum ('*7-LED-SuS-4K*') is shown in Figure 11.24 together with some reference spectra (Planckian radiator at 4000 K, CIE daylight phases for 5000, 5500, and 5800 K) and two previously known LED spectra with CCT = 4000 K and with different colour rendering properties ('*SPD-940*' and '*SPD-840*'). The red curve in Figure 11.24 that represents the '*full LED spectrum*' has a broad shape with a violet LED for excitation with a peak wavelength at 420 nm and with a multi-component phosphor mixture with a maximum wavelength of around 630 nm. The colour rendering indices as well as the melanopic effect factors of these LED spectra are listed in Table 11.9.

The '*full LED spectrum*' (7-LED-SuS-4K) with CCT = 3965 K has a slightly lower a_{mel}-*value* (0.61–0.64, Table 11.9) compared to the Planck emitter with 4000 K, i.e. a slightly lower melanopic efficacy. This can be partially explained by a gap in the spectrum around 480 nm. The development of this '*full LED spectrum*' apparently focused on increasing the colour quality of the illumination, which was very successful with high values (98) of the general colour rendering index CIE CRI R_a and the special colour rendering index CIE R_9 (for saturated red colours).

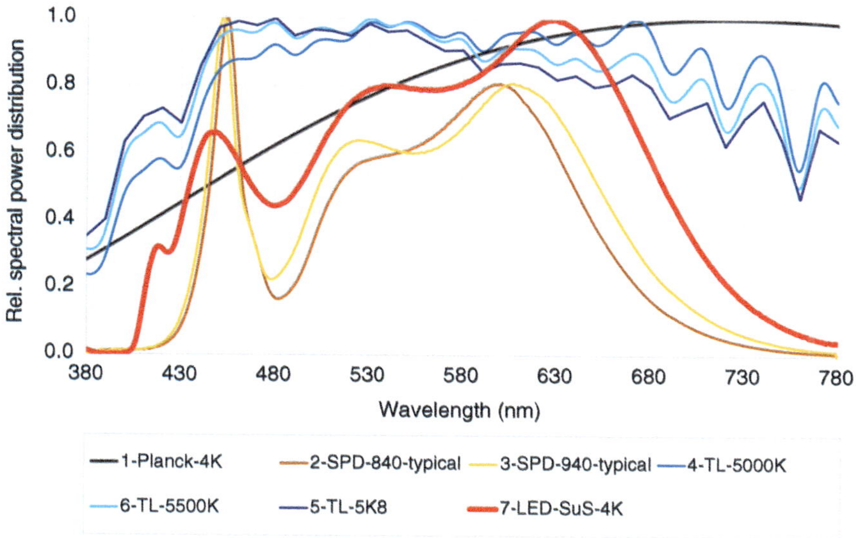

Figure 11.24 Spectra of some white LED types including a 'full LED spectrum' or 'daylight-like spectrum' (7-LED-SuS-4K) together with CIE daylight spectra and with Planck's radiator at 4000 K. Source: TU Darmstadt.

Table 11.9 Correlated colour temperature (CCT), colour rendering indices (CIE CRI), and melanopic effect factor of the white LED types shown in Figure 11.24, and the daylight phases (4-TL-5000K, 5-TL-5500K, 6-DL-5K8) as well as the Planck radiator (1-Planck-4K) at 4000 K. CIE $R_{1,14}$: mean of the 14 special colour rendering indices.

LED	1-Planck-4K	2-SPD-840 typical	3-SPD-940 typical	4-TL-5000K	5-TL-5500K	6-DL-5K8	7-LED-SuS-4K
CCT in K	4000	3933	4018	5000	5500	5800	3965
CIE R_9	100	12	58	100	100	100	98
CIE R_a	100	84	94	100	100	100	97
CIE $R_{1,14}$	100	78	90	100	100	100	96
$a_{mel,v}$	0.64	0.55	0.61	0.77	0.82	0.85	0.61

Source: TU Darmstadt.

11.4.3 Levels 3 and 4 (Figure 11.17) with Variable Colour Temperature and Variable Illuminance

In the context of HCL lighting, dynamic lighting plays an increasingly important role throughout the day, depending on the weather, the time of day, the context of the room, and the individual lighting needs of the room users. This requires the use of LED luminaires with variable colour temperatures from about 3000 to about 6000 K, with dimmable luminous flux, variable melanopic effect factor, and variable light distributions concerning the direct light component (directly onto the work surface) and the indirect component (indirectly, first onto the ceiling and onto the upper wall surfaces), whereby the indirect component is taking on an increasingly stronger role in today's lighting technology. In this context, white LED types with high colour quality and with a high but fixed melanopic effect at a *constant* colour temperature are desirable. But, in the long term, these LED types still only represent an interim solution.

Previous LED luminaires from leading luminaire manufacturers with *dynamic* light and *variable* colour temperatures, consisting of warm white and cool white LEDs, nevertheless have a relatively low melanopic efficacy, as the mixtures of these two LED types have a large gap at 480–484 nm in their spectra (see Figure 11.20). To achieve high and variable melanopic efficacy, blue LEDs with a peak wavelength of around 470–485 nm must be added to the warm white and cool or neutral white LEDs, whereby the entire light and colour mixing of these three LED components should be carried out carefully so that a colour locus very close to the curve of the Planckian radiators and the daylight phases and *good-very good* colour rendering can be guaranteed (in the sense of the international standards for interior lighting).

Such a solution was theoretically conceived by the authors of the present book and has been implemented in a practical project. Figure 11.25 shows the spectra of these LED colour mixtures between 3100 and 6000 K. One can see the high spectral components around 475–480 nm. From Table 11.10, it is clear that their CIE CRI R_a colour rendering indices are greater than 80.

To illustrate the advantage of the increased melanopic efficacy with this solution, Figure 11.26 shows a graphical comparison of the a_{mel} values of the colour mixtures (coloured crosses) and the a_{mel} values of the corresponding reference

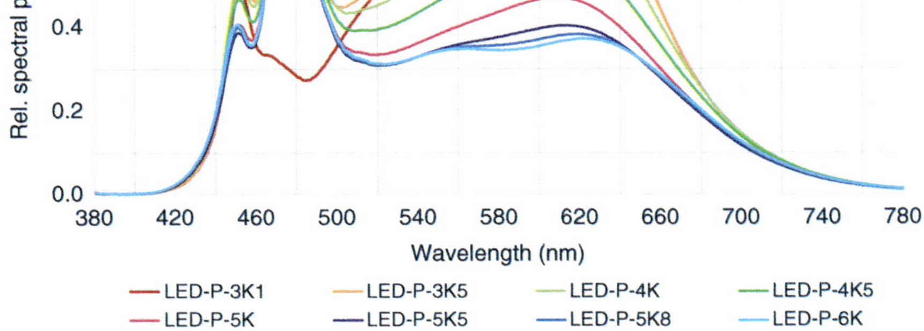

Figure 11.25 Spectra of LED-colour mixtures between 3100 and 6000 K, to achieve high and variable melanopic efficacy. Source: TU Darmstadt.

Table 11.10 Correlated colour temperature, colour rendering index (CIE CRI R_a), and melanopic efficiency factor of the white LED spectra shown in Figure 11.25.

LED	LED-P-3K1	LED-P-3K5	LED-P-4K	LED-P-4K5	LED-P-5K	LED-P-5K5	LED-P-5K8	LED-P-6K
CCT in K	3116	3500	4000	4500	5001	5501	5801	6001
CIE R_a	83	84	85	85	82	80	80	80
$a_{mel,v}$	0.44	0.61	0.72	0.82	0.91	0.98	1.01	1.03

Source: TU Darmstadt.

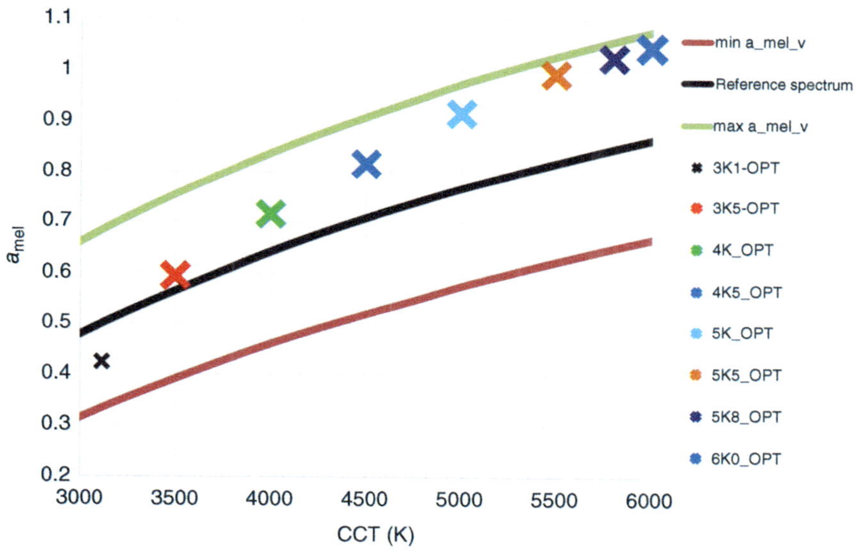

Figure 11.26 $a_{mel,v}$ values of the colour mixtures shown in Figure 11.25 (coloured crosses) and the corresponding reference spectra (Planckian radiators and daylight phases, black curve) at different colour temperatures from 3100 to 6000 K (according to [15]). Source: TU Darmstadt.

spectra (Planckian radiators and daylight phases, black curve) at different colour temperatures from 3100 to 6000 K. The crosses lie above the black curve in the diagram. This means a higher melanopic efficiency.

While the light mixing optics for a luminaire consisting of warm white and cool white LEDs is still relatively simple, with an optimised arrangement of these two LED types in relation to each other on LED circuit boards and a diffusing plate as luminaire cover optics, this mixing optics for three LED types (warm white, cool white, and semiconductor LEDs around 475 nm) is much more complex and lossy, as many more diffusion processes are forced by the optics in order to ensure a homogeneity of the light emission that is acceptable to the users.

11.4.4 Level 5 (Figure 11.17) with Variable Colour Temperature, Variable Illuminance, and High Colour Quality

A disadvantage of Levels 3 and 4 is that the mixture only has a general colour rendering index of a maximum of 85 (see Table 11.10), besides the great advantage that the circadian effectiveness, represented by the a_{mel}-values, can be very high (see Figure 11.26). These levels are already very good and will accompany HCL lighting, worldwide, for a long time. If, in the long term, lighting technology also wants to demonstrate and implement high colour rendering, this will also be possible: starting from the spectra of Level 4 (see Figure 11.25) with a warm white, and a neutral white LED type, and with blue semiconductor LEDs with a peak wavelength around 475 nm, a *red* semiconductor LED type with a peak wavelength between 625 and 660 nm can be added, and the resulting LED channels can be colorimetrically optimised again.

Such a LED combination was generated by the authors of the present book. Between 625 and 660 nm, the general CIE R_a colour rendering indices for the first eight CIE test colours hardly change. Nevertheless, the CIE R_9 colour rendering index (of the saturated red CIE test colour) improves rapidly with longer peak wavelengths of the red LED and has a maximum at about 650–660 nm. These colour mixing results for Level 5 are shown in Figure 11.27 and in Table 11.11. The general colour rendering indices CIE R_a are higher than 90 in the important colour temperature ranges of 3000–5000 K, whereby the a_{mel} value in this case slightly decreases compared to the values in Table 11.10 (Level 4). From a manufacturing point of view, the light mixing optics for the mixing of the four LED types (warm white, cool white, blue LED, and red LED) are much more complicated and lossy.

The result of this analytical and metrological execution states that the technology for generating white light now brings a high luminous efficacy of the LED products

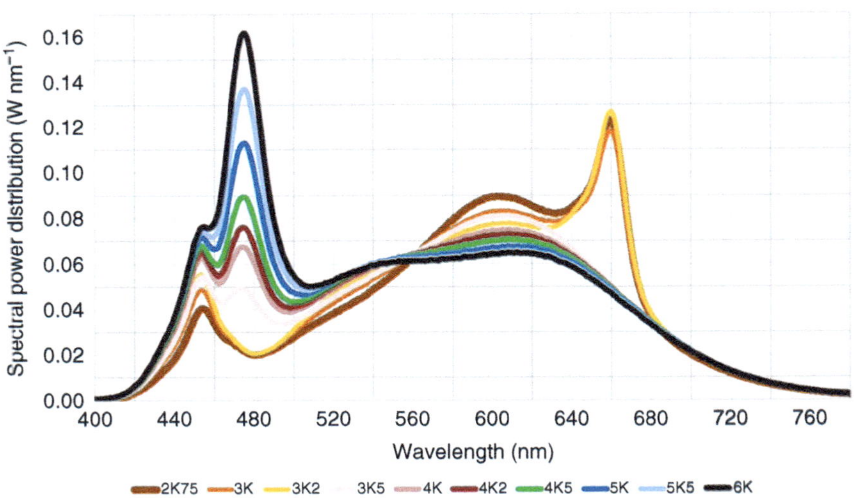

Figure 11.27 Spectra of LED colour mixtures in Level 5 (Figure 11.17) between 2750 and 6000 K. Source: TU Darmstadt.

Table 11.11 Correlated colour temperature (CCT), colour rendering index (CIE CRI R_a), and melanopic effect factor ($a_{mel,v}$) of selected LED spectra from Figure 11.27.

LED	2K75	3K	4K	4K5	5K	5K5	6K
CCT	2750 K	3000 K	4000 K	4500 K	5000 K	5500 K	6000 K
CIE R_a	89	91	92	92	90	86	82
$a_{mel,v}$	0.40	0.44	0.65	0.75	0.84	0.92	1.01

with high colour quality and melanopic effect. Dynamic light with a high variable melanopic effect and good colour quality in the sense of the standards for interior lighting is also technologically possible.

11.4.5 Level 6 with Variable Colour Temperature, Variable Illuminance, and Daylight Consideration

11.4.5.1 Introduction

With current and future luminaire systems of Levels 1–5 (see Figure 11.17), it is theoretically and practically possible to vary the spectrum, the luminous flux, and the light distribution curves for the direct and indirect components of the luminaires. Thus, certain *control curves* for the control of dynamic light pose no problem electrically and in terms of programming, which can be implemented in a microcomputer in the luminaires or in cloud software at any time. The decisive questions in the context of intelligent, energetically, and photometrically sensible dynamic lighting are how to integrate the valuable daylight into the '*dynamic lighting*' philosophy. Many concepts of *dynamic control curves* for the fixed setting of illuminance and colour temperature as a function of time and season have not yet been completed in this respect, for the following reasons:

- Influencing factors such as time of day, season, and therefore, the daily phases of the biological rhythm also include the amount of daylight, which can change dynamically due to the direction of incidence, amount of light, spectra, and weather conditions.
- Thus, there is a change in lighting resulting from the control curve and from the daylight because the daylight spectra add to the dynamically changed artificial light at any time, and this decisively influences the colour temperature, the lighting level, the colour quality, and thus the room atmosphere.
- Since daylight is much more available at the room zones close to the window than at the zones in the depth of the room, the question arises whether the control curve of the dynamic artificial light should rather be programmed for the average room or for different room zones (especially in open-plan offices)?
- In the context of the '*smart lighting*' concept with naturally variable daylight and dynamically programmable artificial light, two questions arise that are more directly related to the psychology of perception:

(a) Should the current daylight outside the building be considered as a reference and as the target of light mixing and colour mixing for the lighting in the room in the sense of the principle '*same colour temperature for inside and outside*'? The illuminance in the room is based on the specifications of the standardisation or on a fixed setting (e.g. 1000–1200 lx on the work desk at 10 o'clock in the morning).
(b) Should the room users or lighting designers instead decide on one or more control curves for colour temperatures and illuminance levels as a function of time? The time-dependent sum of the two lighting components (artificial light and daylight) should result in these control curves.

In any case, it is almost illusory or very costly to adapt the spectrum in the room to the spectrum of daylight with the help of multi-channel LED luminaires. Rather, for the observers and room users in the room, an adjustment of the lighting situation should be sufficient in such a way that a defined lighting level with a certain circadian effectiveness (*CS* or a_{mel} value) appropriate for the time of day, paired with good or excellent colour quality (see Chapters 6 and 7), should be achieved at any time at a desired colour temperature (or desired chromaticity).

To be able to realise these basic questions of the '*Smart Lighting*' concept, one must pursue at least three tasks:

(a) How can the colour and light properties of the room and of the daylight outside the building be determined with acceptable accuracy and with inexpensive but reliable measuring equipment at any time to achieve dynamic light control in the building? This question was answered in this chapter.
(b) As light researchers and light developers, do we have an experimental idea of how daylight changes dynamically? This question was partly analysed in Chapter 10 and will be further discussed in Section 11.4.5.2.
(c) How can the control curves for dynamic lighting be provided in the lighting design phase? This question will be answered in Chapter 12.

11.4.5.2 Variation of Daylight and Consequences for Indoor Lighting – Result of a Measurement

The following thought experiments are based on a semi-realistic consideration that in a room with windows as well as with diffuse, spectrally aselective (i.e. wavelength-independent) reflectance of room boundary surfaces, both in the early morning before sunrise and after sunset, the artificial light sources (in this specific case the white LEDs, CIE $R_a = 84$ and CCT = 3933 K) should provide a minimum illuminance of 500 lx [16].

On a cloudy day in Autumn 2019, in an office room facing north at the University of Darmstadt, the spectra prevailing in the room were measured, see Table 11.12 and Figure 11.28. In Table 11.12, the '*pure LED*' spectrum is illustrated for the case without daylight. The spectral gaps of the white LEDs (see 'pure LED' in Figure 11.28) around 480 nm are noticeably filled by daylight components. This makes it theoretically possible to improve the circadian effects during the day. Table 11.12 shows that the colorimetric data such as the colour rendering index

Table 11.12 Characteristics of light quality and colour quality as well as circadian characteristics for the spectra of white LEDs and of mixtures of LED and daylight ('LEDs + TL') from 9 a.m. to 5 p.m. in an office on an autumn day of 2019.

Light source	'Pure LED'	(LEDs + TL)-9 a.m.	(LEDs + TL)-11 a.m.	(LEDs + TL)-11 p.m.	(LEDs + TL)-3 p.m.	(LEDs + TL)-5 p.m.	(LEDs + TL)-6 p.m.
CCT in K	3933	4294	4409	4201	4192	3969	3956
Duv	4.94E-04	8.28E-04	8.04E-04	7.39E-04	8.57E-04	5.73E-04	5.85E-04
E_v in lx	500	625	700	600	590	510	505
CIE CRI R_9	11.8	29.6	35.7	25.3	24.4	13.6	12.8
CIE CRI R_a	84.3	88	89.1	87.0	86.9	84.6	84.5
IES TM30-15 R_f	84.0	87.9	89.3	87.1	86.9	84.4	84.2
IES TM30-15 R_g	96.4	97.1	97.3	96.9	96.9	96.3	96.3
a_{mel}	0.55	0.62	0.64	0.61	0.60	0.56	0.56
$L_{v,eq}$	384.7	500.6	567.0	476.3	468.1	394.3	389.8
CS_{2018}	0.33	0.42	0.45	0.40	0.40	0.34	0.33
$E_{v,mel,D65}$	306.4	431.1	499.7	402.9	395.0	316.6	312.0

Source: TU Darmstadt.

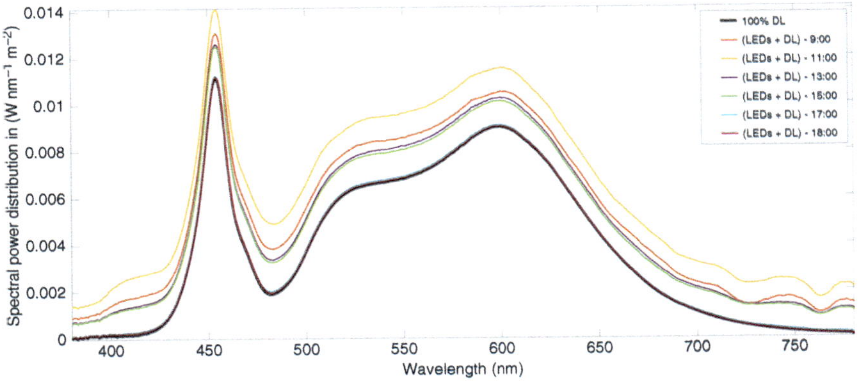

Figure 11.28 Irradiance spectra of white LEDs and of the mixtures of these LEDs with daylight over different phases of the day on a day in Autumn 2019. Source: TU Darmstadt.

CIE CRI R_a, the colour gamut measure IES TM-30-15 R_g, the brightness correlate $L_{v,eq}$ (Eq. (7.2)) with the quantity *luminance* L_v (instead of *illuminance* E_v), and the circadian-effective parameters a_{mel} and *CS* increase from 9 a.m. to 11 a.m. They then decrease again until 5 p.m. in the afternoon. The contribution of daylight to the quality of light and colour as well as to an increase in non-visual effects is quantitatively well recognised in terms of these quantities.

In another measurement, on a partly cloudy day in Summer 2019, the radiation spectra were measured from 9 a.m. to 6 p.m. outdoors. The results are shown in Figure 11.29. It can be seen, that the relative spectra change considerably during the course of the day, resulting in different colour temperatures, see also Table 11.13. If the lighting users in a modern building would like the spectra and colour temperatures of the indoor light to be adapted to the daylight spectra or daylight colour temperatures at any time during the day or to be oriented as closely as possible to them, then the spectra and the colour temperatures of the indoor light will have to be measured online and, if there is a certain deviation from the outdoor daylight, adjusted again with the help of modern online-controlled LED luminaires with hybrid multi-channel LED modules.

11.4.5.3 Approaches to Considering Daylight Components in Interior Space

To implement the above findings as well as other aspects of modern lighting technology (HCL, smart lighting) indoors, some further considerations are important from the lighting engineer's point of view:

(a) How do the photometric properties of light in an office room change as a function of the room's geometry, weather, time of day, and the height of use in the room? (e.g. usable area around 0.7–0.8 m; sitting height around 1.2–1.25 m; and eye height around 1.73 m with a person standing in the room).
(b) How can the missing contribution of daylight be compensated for by artificial light, generally depending on the time of day and the weather, if a default or a *user demand* is defined for the colour temperature and illuminance

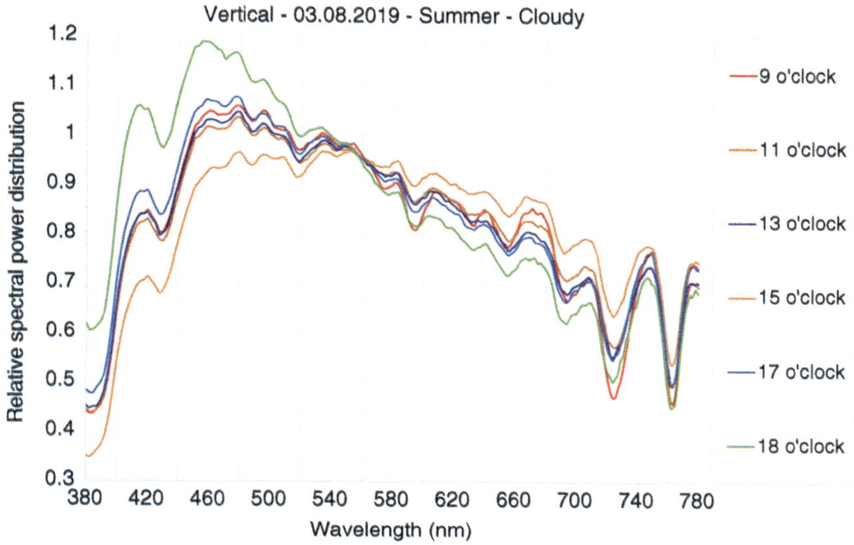

Figure 11.29 Irradiance spectra from 9 a.m. to 6 p.m. on a cloudy summer day. Source: TU Darmstadt.

Table 11.13 Colour temperatures and dominant wavelengths of the measured spectra on a cloudy summer day.

Spectrum	9 a.m.	11 a.m.	1 p.m.	3 p.m.	5 p.m.	6 p.m.
CCT (K)	6117	5879	5989	5361	6243	7102
λ_d (nm)	491.3	491.3	491.3	491.5	491.2	491.2

Source: TU Darmstadt.

distribution in the room, e.g. CCT (reference) = 5000 K and E_v (reference, in the room) = 800 lx? In principle, it is important to keep the main ideas for the compensation method in mind.

Today's practice of lighting design with daylight and artificial light, based on work experience and currently used by renowned lighting design companies, consists of dividing the work areas to be illuminated in modern offices into *three zones*. The so-called *zone near the window* or the façade is planned with 500 lx, as this is where the most daylight occurs. The so-called *interior zones* are planned with 1000 lx and the *intermediate zones*, where there is less daylight, with 750 lx. After [17], horizontal illuminance levels in the three zones of at least 1000 lx for at least three hours a day were achieved in an actual lighting design [17]. This is an empirical lighting design worthy of imitation for the first steps towards HCL design and practice.

In an office at the TU Darmstadt with two windows in two directions (north-west and north-east), the horizontal spectral irradiances were measured on a sunny day and on a rainy day in Summer 2019 using a close-meshed network of measuring points at three working heights (0.73, 1.25, and 1.73 m) at 9 a.m., 1 p.m., and 5 p.m. on a horizontal plane, respectively. Measurements were taken with an absolutely calibrated spectroradiometer at 9 a.m., 12 o'clock, 1 p.m., and 5 p.m., from which the horizontal illuminances (in lx), the colour temperatures, and the CS values (model version of M. Rea et al. see Section 10.3 of this book) were calculated as a function of the working heights, the time, the weather, and the measurement location in the room. The results for the *CS* value distributions are shown in Figure 11.30. It is clearly visible that these circadian parameters depend on the time of day, the weather, and, especially, on the measuring positions in the room. As expected, the measured values at the two windows have the highest levels, which are around CS = 0.5 and 0.6.

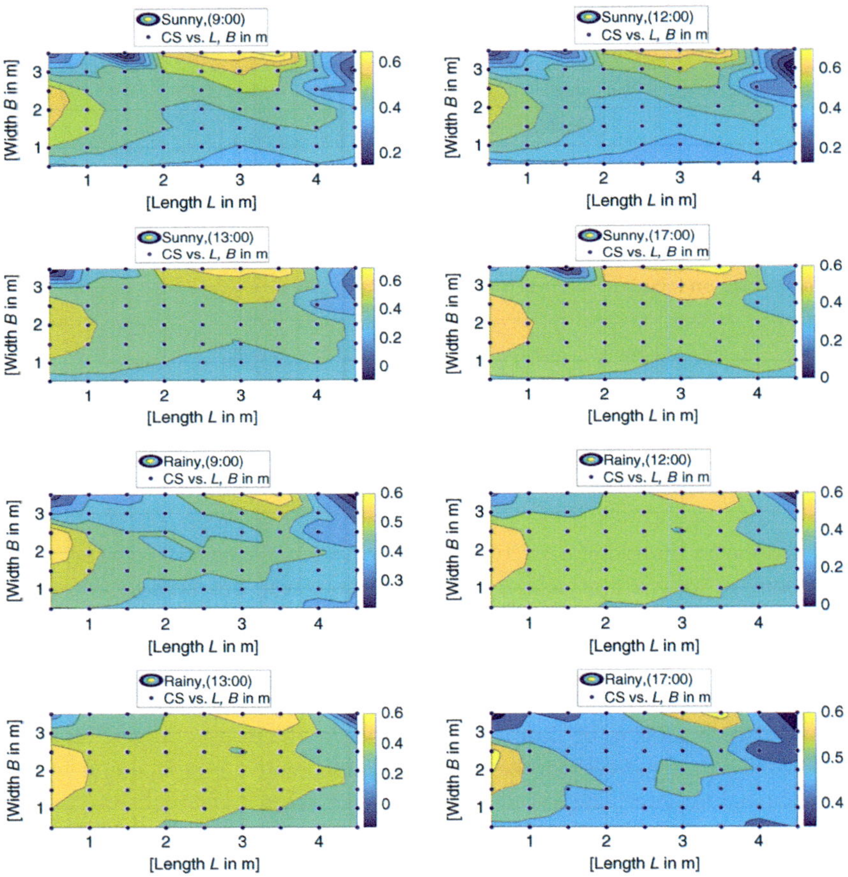

Figure 11.30 Distributions of the CS value (model version of 2018) on a sunny (top picture) and rainy day (bottom picture), at the working height of 1.25 m at four different times: 9 a.m., 12 o'clock, 1 p.m., and 5 p.m. Source: TU Darmstadt.

These CS value distributions (Figure 11.30) as well as the corresponding local distributions of horizontal illuminance levels and colour temperatures could only be determined with manual individual measurements for research purposes. If in future buildings, in addition to the modern LED luminaires with variable colour temperature and luminous flux adjustment options, a grid of light and colour sensors is also provided for the local determination of the above-mentioned photometric parameters, the individual luminaires at different locations in the room can be given different dimming levels depending on the weather, time of day, and the demands of the light users, in order to enable the distribution of illuminance and colour temperature in the room to be homogeneous with a certain average desired value. In an example in Figure 11.31, the amount of colour temperature to be compensated towards the desired value of CCT = 5000 K is shown as a function of the location in the room.

The consideration of variable daylight input (see the example in Figure 11.31) requires the following technological measures:

- Use of RGB colour sensors (optionally RGB colour cameras if permitted by law) to provide the colorimetric and photometric data for luminaire control outdoors and indoors.
- Use of dimmable or intelligent LED luminaires, the description of which is contained in the previous sections (see Levels 1–5 in Figure 11.17), and which are to be divided into zones in the actual room in the lighting design phase and addressed by the central computer (building computer, cloud software).
- Optional integration of position sensors (e.g. thermal IR matrix sensors with 88 pixels) to not only perform daylight compensation in the room roughly zonally but also to specifically change the light intensity to the current, narrow zone, depending on the specific position of room users and their possible viewing direction.

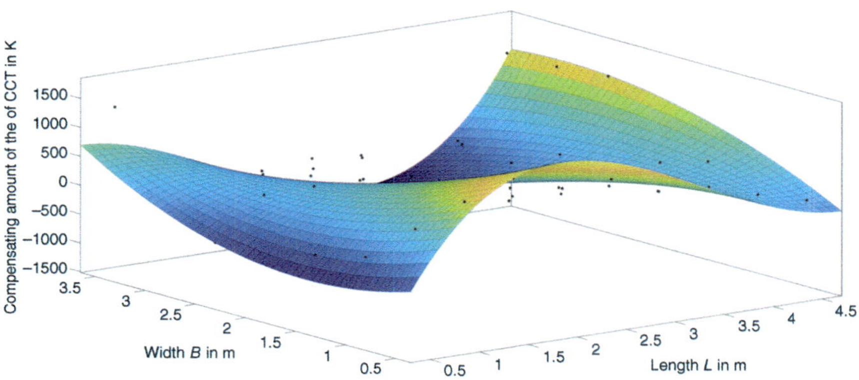

Figure 11.31 Example of the amount of colour temperature to be compensated towards the desired value of CCT = 5000 K, depending on the location in the room (W × H = 3.5 m × 4.5 m). Source: TU Darmstadt.

11.5 Measurement of Melanopic-Equivalent Daylight Illuminance (MEDI) with RGB Colour Sensors

11.5.1 Introduction Into the Context

In modern indoor and outdoor lighting applications, controlling or regulating adaptive semiconductor-based luminaires (LED-OLED) and window systems (daylight systems) with the aid of sensors in the room (e.g. presence sensors, position sensors, light and colour sensors) will help to reduce the energy consumption and to increase the health and emotion aspects of room users. To achieve a pre-defined value of the non-visual parameters such as *melanopic-equivalent daylight illuminance* (MEDI) at a specific location in the room (e.g. at the centre of the room or at locations further away from the windows) considering dynamically changing weather conditions and whereabouts of room users, in practice relatively inexpensive but sufficiently accurate optical sensors (RGB sensors) are needed. The aim is not only to obtain the target photometric and colorimetric parameters such as illuminance E_v (in lx), chromaticity coordinates (x, y, z) or CCT in K, but also to obtain the non-visual parameters MEDI and MDER, which are defined in Chapter 2 and in Section 10.7 of the present book. The principle of this Smart Lighting concept using RGB colour sensors is illustrated in Figure 11.32. The methods for measuring non-visual parameters with inexpensive but well-qualified RGB colour sensors form the content of this section, which has been published by the authors of this book [18].

11.5.2 *RGB* Colour Sensors: Characterisation and Signal Transformation

11.5.2.1 Characterisation of *RGB* Colour Sensors

Colour sensors (see Figure 11.33) are arrays of individual sensors or of a group of sensors based on semiconductors with silicon as the semiconductor material, which is mostly used for the visible spectral range between 380 and 780 nm.

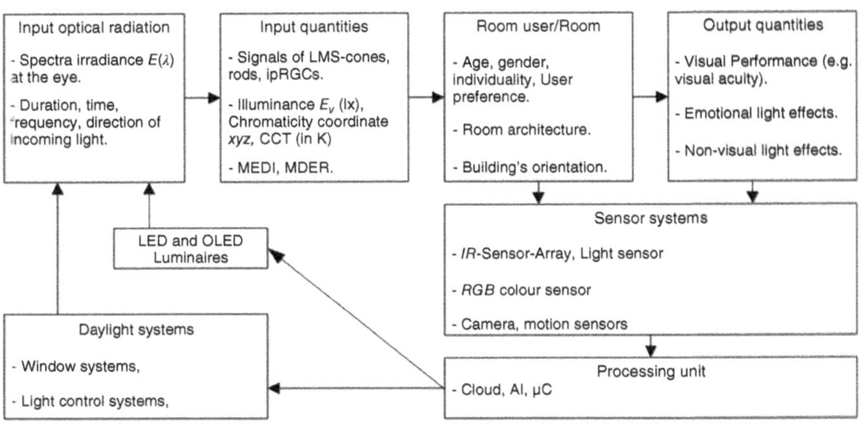

Figure 11.32 Principle of a Smart Lighting concept with *RGB* colour sensors. Source: TU Darmstadt.

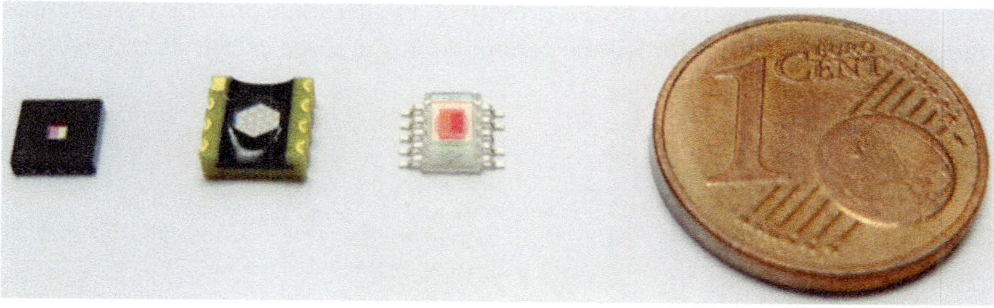

Figure 11.33 Examples of *RGB* colour sensors. Source: Technische Universität Darmstadt.

To generate the appropriate spectral sensitivity curves of the individual colour channels in the red, green, and blue range (so-called RGB-channels), optical colour filters based on thin-film technology (interference filters), absorption glass filters, or colour lacquers are micro-structured and applied to the respective silicon sensor. For the correct spatial evaluation of the optical radiation according to the cosine law, a so-called cos prefix (usually a small diffuser plate made of optical scattering materials) is applied to the RGB sensor. When the optical visible radiation is incident, the photons of individual radiation wavelengths are absorbed in the respective RGB sensor, and in this way, photon currents are generated, which are then converted to a voltage in an amplifier circuit. Subsequently, these voltages are digitised by an A/D converter (analogue-digital) in different bit depths (8-bit, 12-bit, and 16-bit). The spectral sensitivity of the respective colour channel *R-G-B* is therefore made up of the components shown in Eqs. (11.1)–(11.3).

$$R(\lambda) = S(\text{silicon}, \lambda) \cdot \tau(\text{diffusor}, \lambda) \cdot \tau(\text{colour lacquer}_R, \lambda) \cdot K_R = \frac{\text{CV}_R(\lambda)}{E(\lambda)} \tag{11.1}$$

$$G(\lambda) = S(\text{silicon}, \lambda) \cdot \tau(\text{diffusor}, \lambda) \cdot \tau(\text{colour lacquer}_G, \lambda) \cdot K_G = \frac{\text{CV}_G(\lambda)}{E(\lambda)} \tag{11.2}$$

$$B(\lambda) = S(\text{silicon}, \lambda) \cdot \tau(\text{diffusor}, \lambda) \cdot \tau(\text{colour lacquer}_B, \lambda) \cdot K_B = \frac{\text{CV}_B(\lambda)}{E(\lambda)} \tag{11.3}$$

with,

- S(silicon, λ): spectral sensitivity of the silicon sensor;
- τ(diffusor, λ): spectral transmittance of the cos-diffusor;
- τ (colour lacquer$_i$, λ): spectral transmittance of the colour filter layer for the respective colour channel $i = R, G, B$.
- K_R, K_G, K_B: Absolute factors to account for current-voltage conversion and voltage digitisation.
- CV_i (λ): Output signals (analogue or digital) of the respective *R-G-B* colour sensor at wavelength λ.

If a certain spectral irradiance $E(\lambda)$ is present on the *RGB* sensor, three corresponding output signals, *R*, *G*, and *B* will be generated in the three colour channels *R*, *G*, and *B*, see Eqs. (11.4)–(11.6).

$$R = \int E(\lambda) \cdot R(\lambda) \cdot \mathrm{d}\lambda \tag{11.4}$$

$$G = \int E(\lambda) \cdot G(\lambda) \cdot \mathrm{d}\lambda \tag{11.5}$$

$$B = \int E(\lambda) \cdot B(\lambda) \cdot \mathrm{d}\lambda \tag{11.6}$$

For the calculation of the *RGB* channel signals, it is therefore necessary to know the spectral sensitivity function of each individual colour channel *R*, *G*, and *B* or to determine it by laboratory measurements. The spectral apparatus for determining the spectral sensitivity of the semiconductor sensors in the authors' light laboratory (see Figure 11.34) consists of a high-intensity xenon ultrahigh-pressure lamp, a grating monochromator (spectral half-width $\Delta\lambda = 2$ nm, spectral measurement steps for the entire spectrum between 380 and 780 nm, $\Delta\lambda = 2$ nm), and an integrating sphere for homogenising the quasi-monochromatic radiation emerging from the monochromator. The *RGB* colour sensor and a known calibrated reference sensor are located in two different places on the inner wall of the sphere behind a shutter.

During the spectral measurement of the *RGB* sensor, one can determine the spectral sensitivity for the red colour channel, for example, according to Eq. (11.7).

$$R(\lambda) = \frac{\mathrm{CV}_R(\lambda)}{E(\text{sphere}, \lambda)} \tag{11.7}$$

with $\mathrm{CV}_R(\lambda)$ as the output signal of the red colour channel and $E(\text{sphere}, \lambda)$, the spectral irradiance on the inner wall of the sphere when the monochromator is set to wavelength λ, which in turn can be determined by the measured output photocurrent $i(\text{reference}, \lambda)$ and the known spectral sensitivity of the reference sensor $S(\text{reference}, \lambda)$, we can write Eq. (11.8).

$$E(\text{sphere}, \lambda) = \frac{i(\text{reference}, \lambda)}{S(\text{reference}, \lambda)} \tag{11.8}$$

The absolute spectral sensitivity of the respective colour channel *R*, *G*, and *B* can thus be determined from Eqs. (11.7) and (11.8).

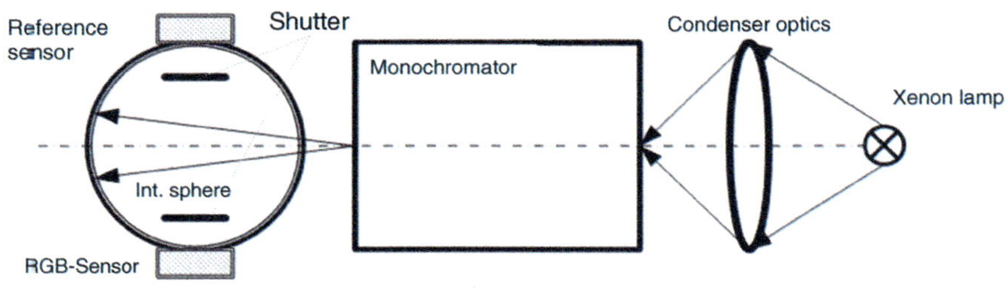

Figure 11.34 Schematic arrangement for measuring the spectral sensitivity of RGB sensors. Source: TU Darmstadt.

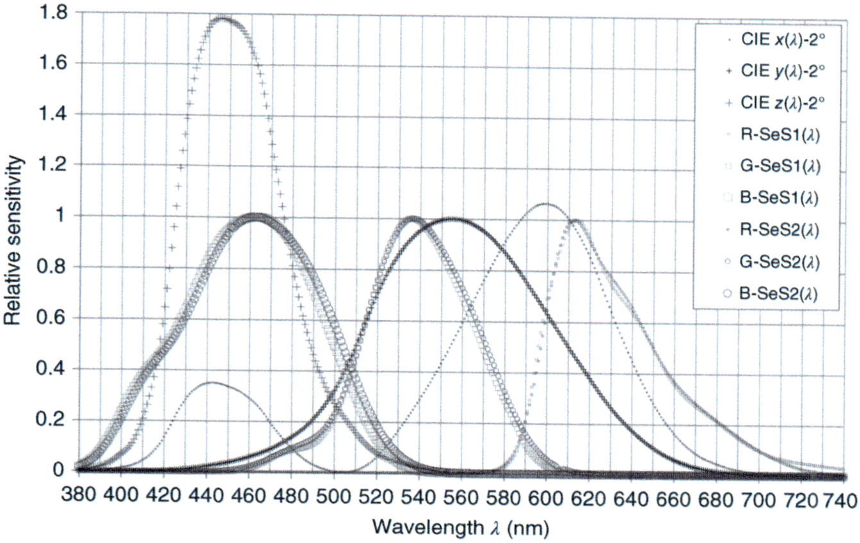

Figure 11.35 Examples of the relative spectral sensitivity curves of some *RGB* colour sensors. Source: TU Darmstadt.

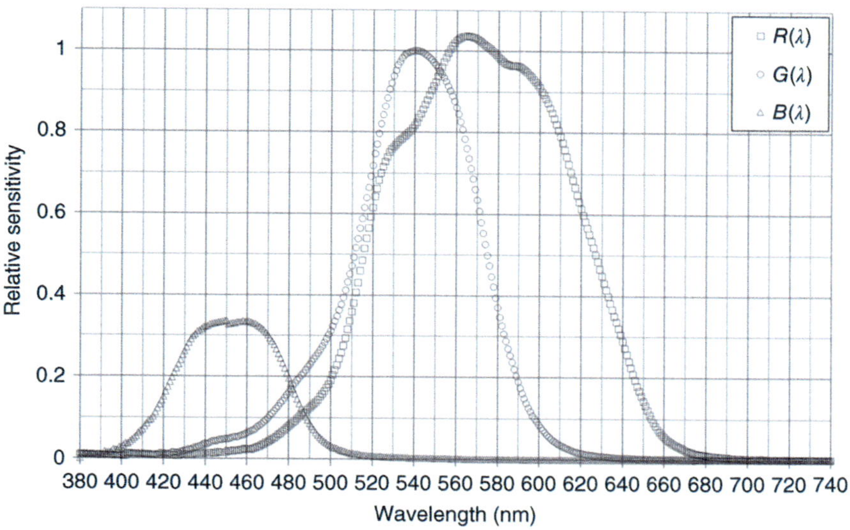

Figure 11.36 Relative spectral sensitivity curves of another *RGB* colour sensor type. Source: TU Darmstadt.

Figures 11.35 and 11.36 show examples of the spectral sensitivity curves of some RGB colour sensors determined in the present authors' lighting laboratory. In Figure 11.35, the two data sets of the same type, SeS1 and SeS2, differ by the different peak heights of the *R*-colour channel and by the different slopes of the *G*- and *B*-curves, because the two sensors SeS1 and SeS2 come from different production sets. The spectral sensitivity curves SeS1 and SeS2 in Figure 11.35 differ fundamentally from the spectral *RGB* sensitivity curves of the colour sensor type in Figure 11.36.

11.5.3 Method of Signal Transformation from *RGB* to *XYZ*

Figures 11.35 and 11.36 show that the *RGB* curves of the real colour sensors differ – more or less strongly – from the *xyz* curves of the CIE colour matching functions for a field of view of 2° [19]. To obtain the tristimulus values (*XYZ*), the generated RGB signals must be transformed into digital form with the aid of matrixing, see Eq. (11.9).

$$\begin{bmatrix} X_{\mathrm{SeS},i} \\ Y_{\mathrm{SeS},i} \\ Z_{\mathrm{SeS},i} \end{bmatrix} = \begin{bmatrix} m_{1,1} & \cdots & m_{1,n} \\ \vdots & \ddots & \vdots \\ m_{3,1} & \cdots & m_{3,n} \end{bmatrix} \cdot \begin{bmatrix} R_i \\ G_i \\ B_i \end{bmatrix} \tag{11.9}$$

Since the sensor technology, the amplification electronics as well as the A/D conversion also often have a, more or less pronounced, non-linear behaviour, one may have to use different matrixing types from 3 × 3 to 3 × 22, see Table 11.14.

If the sensor electronics are linear and the spectral response curves have a similar relative shape to the standard colour matching functions (*xyz*) or the sensitivity curves of the retinal photoreceptors (LMS), 3 × 3 to 3 × 8 matrices will yield contributions from RGB signals as a function of the first term. The more the sensor electronics deviate from a linear behaviour and the shape of the spectral curves of the real sensors differs from the *xyz* colour matching functions, matrices with RGB contributions in quadratic or cubic functions (see Table 11.14) should be set up. The procedure for finding the optimal matrix based on the chromaticity difference $\Delta u'v'$ is shown in Figure 11.37.

Table 11.14 Different matrix types for colour sensing [20].

No.	Size	Content
1	3 × 3	$[R\ G\ B]$
2	3 × 5	$[R\ G\ B\ RGB\ 1]$
3	3 × 7	$[R\ G\ B\ RG\ RB\ GB\ 1]$
4	3 × 8	$[R\ G\ B\ RG\ RB\ GB\ RGB\ 1]$
5	3 × 10	$[R\ G\ B\ RG\ RB\ GB\ R^2\ G^2\ B^2\ 1]$
6	3 × 11	$[R\ G\ B\ RG\ RB\ GB\ R^2\ G^2\ B^2\ RGB\ 1]$
7	3 × 14	$[R\ G\ B\ RG\ RB\ GB\ R^2\ G^2\ B^2\ RGB\ R^3\ G^3\ B^3\ 1]$
8	3 × 16	$[R\ G\ B\ RG\ RB\ GB\ R^2\ G^2\ B^2\ RGB\ R^2G\ G^2B\ B^2R\ R^3\ G^3\ B^3]$
9	3 × 17	$[R\ G\ B\ RG\ RB\ GB\ R^2\ G^2\ B^2\ RGB\ R^2G\ G^2B\ B^2R\ R^3\ G^3\ B^3\ 1]$
10	3 × 19	$[R\ G\ B\ RG\ RB\ GB\ R^2\ G^2\ B^2\ RGB\ R^2G\ G^2B\ B^2R\ R^2B\ G^2R\ B^2G\ R^3\ G^3\ B^3]$
11	3 × 20	$[R\ G\ B\ RG\ RB\ GB\ R^2\ G^2\ B^2\ RGB\ R^2G\ G^2B\ B^2R\ R^2B\ G^2R\ B^2G\ R^3\ G^3\ B^3\ 1]$
12	3 × 22	$[R\ G\ B\ RG\ RB\ GB\ R^2\ G^2\ B^2\ RGB\ R^2G\ G^2B\ B^2R\ R^2B\ G^2R\ B^2G\ R^3\ G^3\ B^3\ R^2GB\ RG^2B\ RGB^2]$

Source: TU Darmstadt.

11.5.4 Matrix Transformation in Practice, Verification with an Actual *RGB* Colour Sensor

With the colour sensor type, whose spectral sensitivity curves are shown in Figure 11.36, the optimisation processes are carried out according to the scheme in Figure 11.37 with a series of matrix types from 3×3 to 3×22 with nine different lamp spectra (as a training set). These nine lamp spectra are described in tabular form in Table 11.15 and shown in Figure 11.38. The illuminance was selected for the optimisation of the matrix to a value of $E_v = 750$ lx. In this training set, following lamp types were selected to reflect the variety of types used in practice today: fluorescent lamps between 2640 and 4423 K with many spectral lines (see Figure 11.38), one halogen incandescent lamp, two phosphor-converted white LEDs, and a combination of RGB LEDs and warm white LEDs (RGBWW4500K).

The calculation with different matrix types and with individual lamp spectra yields chromaticity differences $\Delta u'v'$ of varying magnitude, with the 3×3 matrix form yielding the smallest colour difference (see Table 11.15). Optimisation with more lamp types did not bring any significant improvement in the present case.

The final 3×3 optimised matrix for the transformation from *RGB* to *XYZ* as well as the formula for the calculation of the illuminance from the *RGB* signals are shown

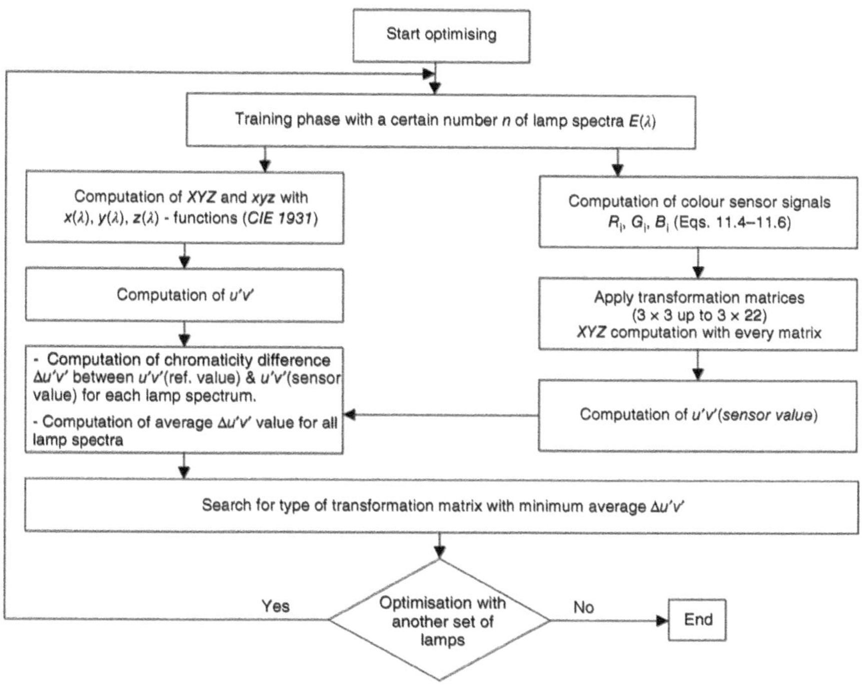

Figure 11.37 Flowchart of optimising a transformation matrix from *RGB* to *XYZ* based on the chromaticity difference $\Delta u'v'$. Source: TU Darmstadt.

Table 11.15 Properties of the lamp types for matrix optimisation.

Lamp type	Tungsten halogen	Xenon-2	CFL 3000K	CFL 5000 K	FL 627	FL 645	LED C3L	LED c3N	LED RGBWW4500
CCT (in K)	2762	4100	2640	4423	2785	4423	2640	4580	4500
E_v (in lx)	750	750	750	750	750	750	750	750	750

Source: TU Darmstadt.

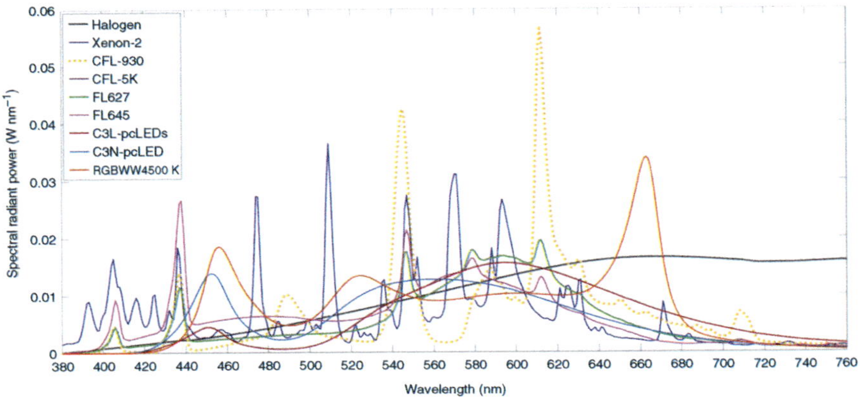

Figure 11.38 Spectra of the lamp types for matrix optimisation. Source: TU Darmstadt.

in Table 11.17. To verify the prediction quality (see Table 11.16) of the *selected prediction* formula (see Table 11.17), 13 spectra were selected in the next step. These spectra included white phosphor-converted LEDs (pc-LEDs) and mixtures of daylight spectra with white phosphor-converted LEDs (TL-pc-LEDs), as is very often the case in practical indoor lighting technology (see Figure 11.39). The illuminance was again set to 750 lx. The result of the verification with these thirteen spectra is shown in Table 11.18.

The deviation of the illuminances, once calculated directly from the lamp spectra and once via the RGB colour sensor signals and via the formula in Table 11.17, was below 0.65%, see Table 11.18. Table 11.18 also shows that, for most of the pc-LEDs, and for the combinations of daylight and white LEDs, the chromaticity difference $\Delta u'v'$ is in a small or moderately small range from the point of view of practical lighting technology. An exception is the spectrum R12GB12-5000 K (thick green curve in Figure 11.39) with the three distinct peaks of the three RGB LEDs (*B* around 455 nm, *G* around 525 nm, and *R* around 660 nm). There, the chromaticity difference equalled $\Delta u'v' = 1.01 \times 10^{-2}$.

11.5.5 Measurement of the Non-visual Quantities MEDI and MDER

The fitting formula for MDER in terms of the chromaticity coordinate z is shown in Eq. (11.10), see also Section 10.7.2.

$$\text{MDER} = a \cdot e^{b \cdot z} + c \cdot e^{d \cdot z} \tag{11.10}$$

Table 11.16 Chromaticity difference $\Delta u'v'$ for different matrix types for the lamp types of the training set.

Name	Xenon-2	CFL-930	CFL-5K	FL627	FL645	C3L-pcLEDs	C3N-pcLED	RGBWW 4500 K	Max.
$\Delta u'v'_{3\times3}$	7.4×10^{-3}	7.7×10^{-3}	1.5×10^{-3}	4.5×10^{-3}	1.5×10^{-3}	4.2×10^{-4}	2.7×10^{-3}	8.3×10^{-3}	8.3×10^{-3}
$\Delta u'v'_{3\times5}$	4.8×10^{-3}	4.8×10^{-2}	1.2×10^{-2}	4.1×10^{-2}	1.2×10^{-2}	5.1×10^{-2}	1.6×10^{-2}	1.6×10^{-2}	5.1×10^{-2}
$\Delta u'v'_{3\times7}$	8.5×10^{-3}	5.3×10^{-3}	3.1×10^{-7}	1.6×10^{-3}	1.8×10^{-6}	7.0×10^{-3}	2.0×10^{-2}	6.4×10^{-3}	2.0×10^{-2}
$\Delta u'v'_{3\times8}$	6.9×10^{-3}	3.2×10^{-2}	1.1×10^{-2}	4.0×10^{-2}	1.1×10^{-2}	4.8×10^{-2}	7.0×10^{-3}	1.6×10^{-2}	4.8×10^{-2}
$\Delta u'v'_{3\times10}$	2.9×10^{-3}	4.5×10^{-3}	4.4×10^{-7}	9.2×10^{-4}	8.1×10^{-7}	1.2×10^{-2}	2.8×10^{-8}	3.8×10^{-3}	1.2×10^{-2}
$\Delta u'v'_{3\times11}$	7.9×10^{-3}	2.8×10^{-10}	2.2×10^{-3}	3.7×10^{-2}	2.2×10^{-3}	4.2×10^{-2}	1.2×10^{-2}	8.1×10^{-3}	4.3×10^{-2}
$\Delta u'v'_{3\times14}$	7.1×10^{-3}	1.0×10^{-2}	9.8×10^{-7}	7.8×10^{-3}	3.5×10^{-6}	5.4×10^{-8}	1.7×10^{-3}	6.7×10^{-3}	1.0×10^{-2}
$\Delta u'v'_{3\times16}$	8.4×10^{-3}	5.7×10^{-3}	5.9×10^{-9}	3.0×10^{-3}	4.1×10^{-6}	2.5×10^{-3}	1.9×10^{-3}	1.0×10^{-2}	1.0×10^{-2}
$\Delta u'v'_{3\times17}$	8.4×10^{-3}	$5.7\times \times10^{-3}$	5.9×10^{-9}	3.0×10^{-3}	4.1×10^{-6}	2.5×10^{-3}	1.9×10^{-3}	1.0×10^{-2}	1.0×10^{-2}
$\Delta u'v'_{3\times19}$	7.2×10^{-3}	9.1×10^{-3}	3.6×10^{-5}	4.3×10^{-3}	3.6×10^{-5}	9.3×10^{-9}	5.7×10^{-3}	7.5×10^{-3}	9.1×10^{-3}
$\Delta u'v'_{3\times20}$	7.2×10^{-3}	9.1×10^{-3}	3.6×10^{-5}	4.3×10^{-3}	3.6×10^{-5}	9.3×10^{-9}	5.7×10^{-3}	7.5×10^{-3}	9.1×10^{-3}
$\Delta u'v'_{3\times22}$	6.6×10^{-3}	8.9×10^{-3}	4.8×10^{-7}	2.2×10^{-3}	1.8×10^{-6}	2.3×10^{-7}	3.2×10^{-3}	7.2×10^{-3}	8.9×10^{-3}

Source: TU Darmstadt.

Table 11.17 Optimum matrix (3 × 3) for the transformation *RGB* to *XYZ*. Formula for the calculation of the illuminance E_v from the *RGB* signals.

E_v (lx) from *R*, *G*, *B*	$E_v = 683 \cdot (1.0001 \cdot E_{v,RGB} - 0.0066)$ $E_{v,RGB} = 0.6802 \cdot R + 0.3651 \cdot G + 0.1751 \cdot B$ $R^2 = 1.0$; RMSE = 0.15
Matrix 3 × 3 in case of the nine light sources of training set	$\begin{bmatrix} 5.73 \cdot 10^5 & -4.17 \cdot 10^5 & -2.27 \cdot 10^5 \\ 3.64 \cdot 10^5 & -4.50 \cdot 10^4 & -3.26 \cdot 10^5 \\ -6.61 \cdot 10^4 & 2.04 \cdot 10^5 & 1.13 \cdot 10^6 \end{bmatrix}$

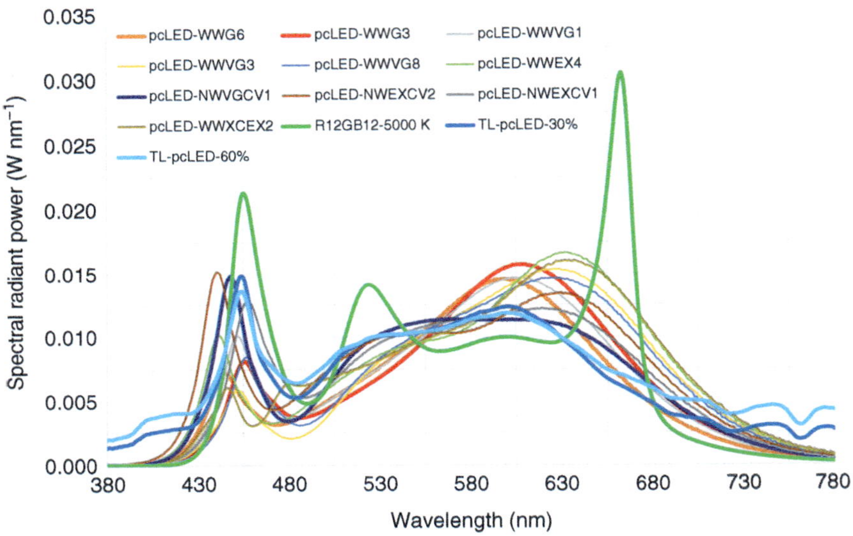

Figure 11.39 Spectra of the 13 light source types for the validation of the 3 × 3 matrix in Table 11.17. Source: TU Darmstadt.

The parameters *a*, *b*, *c*, and *d* for the fitting function and the correlation coefficient r^2 are shown in Table 11.19. The correlation between MDER (ordinate) and the chromaticity coordinate *z* (abscissa) is shown in Figure 11.40.

To verify the quality of the characterisation of the RGB colour sensor under test (see Figure 11.36), the quality of the '*matrixing*' in the transformation from *RGB* to *XYZ* and E_v (see Table 11.17) and the quality of the transformation from *z* to MDER (see Eq. (11.10)) and then, also to MEDI, we set each one of the 13 light source spectra in Figure 11.39 to 750 lx. From these 13 spectra, the chromaticity coordinates *x* and *y* as well as the quantities CCT, CRI R_a, E_v, $\text{MDER}_{\text{original}}$ and $\text{MEDI}_{\text{original}}$ were calculated. They are listed in Table 11.20. For these 13 spectra, the chromaticity coordinates x_{new} and y_{new}, the value of $E_{v,\text{new}}$ and the values MDER_{new} and MEDI_{new} were also computed from the RGB colour sensor values, via matrixing and the transformation from *z* to MDER or to MEDI. The maximum relative

Table 11.18 Verification of the optimum matrix (3 × 3) in Table 11.17 with 13 LED and mixed daylight spectra.

Name	pcLED-WWG6	pcLED-WWG3	pcLED-WWVG1	pcLED-WWVG3	pcLED-WWVG8	pcLED-WWEX4	pcLED-NWVG1	pcLED-NWEXCV2	pcLED-CWVG1	pcLED-CWEX2	R12GB12-5000 K	TL-pcLED-30%	TL-pcLED-60%
CCT (K)	3134	2801	3105	2782	2998	2969	4614	3942	5112	5059	5001	4391	4571
CRI R_a	80.27	84.06	85.63	88.10	90.56	94.41	90.91	93.09	90.06	95.99	89.73	89.85	92.48
x	0.4285	0.4434	0.4217	0.4557	0.4352	0.4252	0.3569	0.3764	0.3424	0.3439	0.3449	0.3644	0.3581
y	0.4025	0.3929	0.3841	0.4135	0.4003	0.3765	0.3608	0.3550	0.3537	0.3557	0.3495	0.3650	0.3605
E_v (lx)	750	750	750	750	750	750	750	750	750	750	750	750	750
$E_{v,new}$	750.35	752.29	752.54	750.90	751.14	754.17	751.40	754.79	751.31	750.01	747.60	751.94	751.91
x_{new}	0.4285	0.4434	0.4217	0.4557	0.4352	0.4252	0.3546	0.3779	0.3361	0.3368	0.3319	0.3665	0.3572
y_{new}	0.4025	0.3929	0.3841	0.4135	0.4003	0.3765	0.3643	0.3698	0.3545	0.3571	0.3529	0.3696	0.3640
$\Delta u'v'$	1.09×10^{-10}	1.36×10^{-10}	1.20×10^{-10}	1.66×10^{-10}	1.53×10^{-10}	1.68×10^{-10}	3.15×10^{-3}	8.76×10^{-3}	4.50×10^{-3}	5.32×10^{-3}	1.01×10^{-2}	2.42×10^{-3}	2.51×10^{-3}
ΔE_v in %	0.05	0.31	0.34	0.12	0.15	0.56	0.19	0.64	0.17	0.00	-0.32	0.26	0.25

Source: TU Darmstadt.

Table 11.19 Parameter *a*, *b*, *c*, and *d* for the fit function in Eq. (11.10); correlation coefficient r^2.

Parameter	*a*	*b*	*c*	*d*	r^2
Value	0.53806	2.3212	−0.37585	−0.85808	0.968

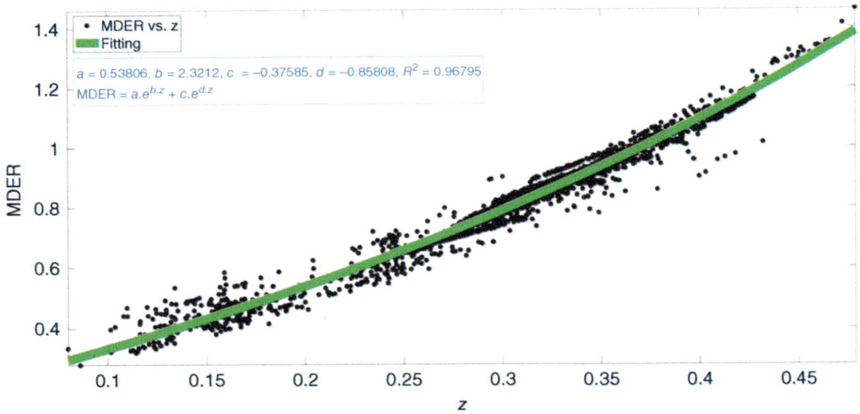

Figure 11.40 Correlation between MDER (ordinate) and chromaticity coordinate *z* (abscissa). Fitting function (Eq. (11.10)): green curve. Source: TU Darmstadt.

deviations ΔMDER (in %) and ΔMEDI (in %) were found to be in the range of ±3.3%, see Table 11.20.

11.5.6 Summary

In indoor lighting technology, the daily task is to assess the lighting systems after completion of the building or after reconstruction, to assess the photometric and colorimetric quality of the newly developed interior luminaires, and, in the context of intelligent lighting (smart lighting, HCL lighting, integrative lighting), to adaptively control lighting systems to provide the room users with the best living and always working conditions and room atmosphere. To this end, non-visual input parameters such as MEDI and MDER need to be recorded and processed with sufficient accuracy, reliability, and reasonable effort.

Previous illuminance metres, luminance metres, colour and luminance cameras, and small portable colour metres are currently only capable of determining photometric quantities such as illuminance E_v, luminance L_v, and colorimetric parameters such as chromaticity coordinates *xyz* and CCT. To measure the non-visual parameters, a transformation is needed to convert the tristimulus values to the non-visual parameters, as well as a sensory platform to get from RGB sensor signals to the tristimulus values *XYZ* and E_v (in lx).

Table 11.20 Verification of the MDER and MEDI values with the colour sensor under test on the basis of the 13 spectra in Figure 11.36 set to 750 lx.

Name	pcLED-WWG6	pcLED-WWG3	pcLED-WWVG1	pcLED-WWVG3	pcLED-WWVG8	pcLED-WWEX4	pcLED-NWVG1	pcLED-NWEXCV2	pcLED-CWVG1	pcLED-CWEX2	R12GB12-5000 K	TL-pcLED-30%	TL-pcLED-60%
CCT (K)	3134	2801	3105	2782	2998	2969	4614	3942	5112	5059	5001	4391	4571
CIE R_a	80.27	84.06	85.63	88.10	90.56	94.41	90.91	93.09	90.06	95.99	89.73	89.85	92.48
x	0.4285	0.4434	0.4217	0.4557	0.4352	0.4252	0.3569	0.3764	0.3424	0.3439	0.3449	0.3644	0.3581
y	0.4025	0.3929	0.3841	0.4135	0.4003	0.3765	0.3608	0.3550	0.3537	0.3557	0.3495	0.3650	0.3605
E_v (lx)	750	750	750	750	750	750	750	750	750	750	750	750	750
$MDER_{original}$	0.47	0.46	0.51	0.40	0.48	0.53	0.75	0.67	0.81	0.83	0.83	0.71	0.75
$MEDI_{original}$	349.06	347.67	383.55	300.60	358.27	395.93	562.91	500.06	610.69	625.10	625.79	532.40	563.22
$E_{v,new}$ (lx)	750.35	752.29	752.54	750.90	751.14	754.17	751.40	754.79	751.31	750.01	747.60	751.94	751.91
x_{new}	0.4285	0.4434	0.4217	0.4557	0.4352	0.4252	0.3546	0.3779	0.3361	0.3368	0.3319	0.3665	0.3572
y_{new}	0.4025	0.3929	0.3841	0.4135	0.4003	0.3765	0.3643	0.3698	0.3545	0.3571	0.3529	0.3696	0.3640
$MEDR_{new}$	0.47	0.46	0.53	0.39	0.46	0.54	0.74	0.66	0.82	0.81	0.83	0.69	0.73
$MEDI_{new}$	353.75	346.28	396.00	295.19	346.92	403.79	554.58	500.85	612.52	604.47	621.78	521.19	550.29
$\Delta u'v'$	1.09×10^{-10}	1.36×10^{-10}	1.20×10^{-10}	1.66×10^{-10}	1.53×10^{-10}	1.68×10^{-10}	3.15×10^{-3}	8.76×10^{-3}	4.50×10^{-3}	5.32×10^{-3}	1.01×10^{-2}	2.42×10^{-3}	2.51×10^{-3}
ΔE_v in %	0.05	0.31	0.34	0.12	0.15	0.56	0.19	0.64	0.17	0.00	-0.32	0.26	0.25
ΔMEDR in %	1.3	−0.7	2.9	−1.9	−3.3	1.4	−1.7	−0.5	0.1	−3.3	−0.3	−2.4	−2.5
ΔMEDI in %	1.35	0.40	3.25	1.80	3.17	1.99	1.48	0.16	0.30	3.30	0.64	2.10	2.30

Source: TU Darmstadt.

The section of this book characterises some exemplary RGB colour sensors based on laboratory measurements, calculations, and optimisations, in creating one or more matrices for an optimal transformation of RGB sensor signals to the tristimulus values and then, based on extensive calculations, achieving a transformation from the chromaticity coordinate z to MDER (see Section 10.7 of this book) and, via E_v, also to MEDI with good accuracy. This work and the methodology described therein enable an accurate and financially justifiable assessment of lighting installations according to non-visual criteria.

References

1 Khanh, T.Q., Bodrogi, P., and Vinh, Q.T. (2018). Intelligente Beleuchtung und HCL-Lichttechnik- Grundsätze und Zusammenhänge (Intelligent lighting and HCL lighting technology- principles and contexts). *Z. Licht* 9-2018: 72–79. ISSN: 0024-2861.

2 Statistisches Bundesamt (2020). Pflegeheime und ambulante Pflegedienste (Federal Statistical Office. Nursing Homes and Outpatient Care Services). https://www.destatis.de.

3 Khanh, T.Q., Klir, S., and Benkner, S. (2019). Lecture series on "Smart Lighting," Lectures Summer Semester 2019. Department of Lighting Technology, TU Darmstadt.

4 https://www.sem-deutschland.de/inbound-marketing-agentur/online-marketing-glossar/was-ist-maschinen-learning-definition-funktionsweise-bedeutung (retrieved 17 April 2023).

5 Wagner, M. (2018). Einfluss von Temperatur und Strom auf die Zuverlässigkeit und das lichttechnische Verhalten von LED-Systemen (Influence of temperature and current on the reliability and photometric behaviour of LED systems). Dissertation, TU Darmstadt.

6 Fleischer, S.E. (2001). Die psychologische Wirkung veränderlicher Kunstlichtsituationen auf den Menschen (The psychological effect of variable artificial light situations on humans). Dissertation. ETH No. 14033, Eidgenössische TH Zürich, ETH Library. https://doi.org/10.3929/ethz-a-004122647.

7 Houser, K.W. (2002). The subjective response to linear fluorescent, direct/ indirect lighting systems. *Light. Res. Technol.* 34 (3): 243–264.

8 Acosta, I., Leslie, R.P., and Figueiro, M.G. (2017). Analysis of circadian stimulus allowed by daylighting in hospital rooms. *Light. Res. Technol.* 49: 49–61.

9 Vinh, Q.T. and Khanh, T.Q. (2020). Weißes LED-Licht – Erzeugung des weißen LED-Lichts und technische Fortschritte. (White LED light – White LED light generation and technical advances). *Z. Licht* 7-2020: 62–67. ISSN: 0024-2861.

10 https://www.merckgroup.com/en/news/full-spectrum-phosphors-06-04-2017.html (retrieved 6 June 2023).

11 Winkler, H., Benker, A., Petry, R., and Vosgroene, T. (2013). United States Patent Publication Application No. US2013/0063926 A1. Merck KGaA, 14.03.2013.

12 Benker, A., Petry, R., Vosgroene, T., and Winkler, H. (2011). Patent No. DE102010021341 A1, Merck KGaA, 24.11.2011.

13 Hoerder, G.J., Seibald, M., Baumann, D. et al. (2019). $Sr[Li_2Al_2O_2N_2]:Eu^{2+}$ – A high performance red phosphor to brighten the future. *Nat. Commun.* 10: 1824. https://doi.org/10.1038/s41467-019-09632-w.

14 Pust, P., Weiler, V., Hecht, C. et al. (2014). Narrow-band red-emitting $Sr[LiAl_3N_4]:Eu^{2+}$ as a next-generation LED-phosphor material. *Nat. Mater.* 13: 891–896. https://doi.org/10.1038/nmat4012.

15 Dr. Sebastian Knoche (Trilux-Arnsberg), Personal Communication.

16 Khanh, T.Q. and Vinh, Q.T. (2020). Tageslicht in der Beleuchtungstechnik – eine quantitative Charakterisierung (Daylight in lighting technology – A quantitative characterisation). *Z. Licht* 4-2020: 60–67. ISSN: 0024-2861.

17 Editorial of the magazine Licht. Interview with Peter Andres - No “New Work” without “New Light” - Performance-enhancing light in the workplace. (Editorial office of the Z. Licht. Interview with Peter Andres - No “New Work” without “New Light” - Performance-enhancing light in the workplace). Z. Licht 2: 64-66 (Further information: Peter Andres Unabhängige Lichtplanung und -beratung GbR, Hamburg, www.andres-lichtplanung.de). www.andres-lichtplanung.de.

18 Vinh, Q.T., Bodrogi, P., and Khanh, T.Q. (2023). Determination and Measurement of Melanopic Equivalent Daylight (D65) Illuminance (mEDI) in the Context of Smart and Integrative Lighting. *Sensors* 23 (11): 5000. https://doi.org/10.3390/s23115000.

19 Commission Internationale de l’Éclairage (2004). *Colorimetry – CIE Technical Report 15:2004*, 3e. Vienna, Austria: CIE.

20 Westland, S., Ripamonti, C., and Cheung, V. (2012). *Computational Colour Science Using MATLAB*. Wiley.

12

HCL-Oriented Lighting Design: Basic Aspects and Implementation

In this chapter, the basic concepts, important aspects, influencing parameters and the process of Human-Centric Lighting (HCL) oriented lighting design are discussed. On this basis, the processes of daylight planning, dynamic lighting, and the planning of lighting for elderly room users (e.g. in nursing homes or in homes for the elderly), which are decisive for the planning of HCL lighting, are then described in the further course of the chapter. The focus of this chapter is on establishing and justifying a design for the specification of a lighting installation with good lighting quality for daytime and night-time.

12.1 Classification of HCL-Oriented Lighting Quality Concepts

12.1.1 Conceptions and Thought Processes on Lighting Quality Until 2002

In its historical development since the invention of the incandescent lamp by Edison, lighting technology in general can be divided into four main phases with somewhat blurred temporal boundaries [1]. In the first phase from the end of the nineteenth century or the beginning of the twentieth century to the beginning of the 1990s, there were several stages of the industrial revolution in mechanical engineering and electrical engineering with thermal radiators and discharge lamps as light sources. The lighting technology of the time was essentially about visual performance, avoiding accidents at work and saving energy after the oil crisis in the 1970s.

There were many fundamental studies and proposals from the 1930s to the mid-1990s – the first main period of classical lighting research – to define the thresholds of illuminance and luminance for adequate speed and accuracy in processing visual information. It began with the work of Luckiesh and Moss ([2], 1934), and Blackwell ([3], 1946), which led to the definition of the standard values for indoor lighting in the USA and England, and were based on the perception of contrast at a certain ambient luminance. This work was completed by the so-called RVP (Relative Visual Performance) model by Rea and Ouelette ([4], 1991, see Chapter 4). According to this RVP model, the RVP of observers enters a saturation

Human Centric Integrative Lighting: Technology, Perception, Non-Visual Effects, First Edition.
Tran Quoc Khanh, Peter Bodrogi, and Trinh Quang Vinh.

range at a visual condition with a contrast of approximately 0.5 and at an ambient *luminance* of more than 35 cd m^{-2} (which often occurs in a contemporary office space). Visual performance (and thus the physiological prerequisite for easy and effective working) is thus mostly guaranteed in an average office with a horizontal illuminance of around 350–600 lx. Therefore, classical lighting design reached the limit of its performance and lost its importance toward the end of the 1990s concerning the analysis and optimisation of the user's well-being, performance, and experiencing an interesting and pleasing lighting situation.

With the development of fluorescent lamps based on the three-band concept with RGB-fluorescent material combination with relatively high colour-rendering and good luminous efficacy, lighting research was intensified in 1971 by Koedam and Opstelten [5] on the psychological–sociological and emotional effects of lighting in interiors. In this process, lighting research applied new principles with roots in environmental psychology, occupational psychology, sociology, and lighting architecture in the 1970s and 1980s toward a new and exciting field of works investigating the psychological behavioural processes underlying the quality of light and lighting. The foundation stone for this was laid by the information processing model for environmental detection and assessment by Kaplan [6], according to which the amount of information presented in a scene is essential for the observer's perception of the environment.

The four psychological dimensions of environmental appraisal in this model range from the coherence and visibility of the information presented to the perception of the so-called '*mystery*' of the scene (the scene continues to maintain the observer's interest) and its complexity, which requires a detailed interpretation of the environment and the scene. Due to the fact that the spectral distribution of light sources (halogen incandescent lamps, fluorescent lamps) and thus the colour temperature, saturation and chromaticity could not be varied on a large scale in the years 1970–2000, lighting architecture and lighting psychology were primarily concerned with the luminance distribution and with the absolute illuminance/luminance level within the room. Of many research groups, three representatives should be mentioned here who decisively shaped the research direction until around 2002.

12.1.1.1 Flynn et al.

When viewing and perceiving an illuminated scene in a specific context – with a variety of lighting options – there are several adjectives, attributes, and descriptors that describe different aspects of the viewer's emotions and impressions. Toward the beginning of the 1970s, lighting research began to cover the interpretative dimensions of the scene and the spatial impression. The first work in this field was done by Flynn et al. in 1973 [7] in a *meeting room* with 6 different lighting configurations, where 96 test persons in 12 groups were given a questionnaire with 34 different scales to assess these lighting configurations, such as '*clear–hazy*', '*pleasant–unpleasant*', '*bright–dim*', '*stimulating–boring*'. The analysis of these 34 scales led to the determination of the relevant categories of the impression of a scene according to Flynn et al. [8], see Table 12.1.

Table 12.1 Categories for describing the impression of a scene according to Flynn et al. [8].

Perceptual categories	Impression of visual clarity Impression of space Impression of spatial complexity Impression of shades Impression of glare
Behavioural attitude	Impression of public vs. private spaces Impression of relaxation vs. tension
Overall preference	Impression of preference (like, dislike) Impression of friendliness and convenience

Source: Adapted from Flynn et al. [8].

Flynn et al. [8] conducted a multidimensional analysis of visual scaling, innovative for the time, to identify the lighting properties and (psychological) dimensions of illumination or light. These dimensions resulted in the subjective impressions in Table 12.1 in the context of scene evaluation. The following four basic characteristics were found:

1. homogeneity (even/uneven);
2. brightness (light/dark);
3. '*Light comes from above – light comes peripherally*';
4. *visually warm/visually cold.*

12.1.1.2 Rowlands and Loe

In the mid-1980s, Rowlands and Loe's research group conducted extensive experiments with 18 lighting situations in an office [9] and determined two dimensions for the perception of the room, '*brightness*' and '*degree of interest*'. The latter was related to the presence of inhomogeneities of light distributions in the room. Thus, the two dimensions of lighting in the analysis of Rowlands and Loe fit into the four light dimensions of Flynn et al.

12.1.1.3 Veitch and Newsham

In 1998, Veitch and Newsham conducted experiments to assess room appearance with 292 test subjects and with 9 lighting systems using 27 semantic scales [10]. The most important characteristics for describing the impression of room appearance turned out to be '*visual attraction and complexity*' and '*brightness*'. These results by *Veitch and Newsham* do not necessarily agree with the results of Flynn et al. or *Rowlands and Loe*. They point to a great difficulty in lighting research and lighting architecture to find out which concise terms can describe the variety of aspects of room perception. The characteristics '*visual attraction and complexity*' of *Veitch and Newsham* mean a higher level of interpretation with more cognitive and emotional depth for lighting research compared to the characteristics '*brightness and homogeneity*' of Flynn et al. or *Rowlands and Loe*. They include with a high probability the lit scene of the room, and the architecture as well as the mutual effect between room objects and lighting.

Table 12.2 Room appearance characteristics and the associated photometric and colorimetric descriptor parameters.

Room appearance features	Photometric parameters	Colorimetric parameters
• Homogeneity • Brightness • 'From above'/peripherally • Interest (inhomogeneity) • Complexity • Visual attraction • Visual clarity	• Horizontal or cylindrical illuminance • Luminance of ceiling and wall • Luminance distribution in the field of view	• Colour temperature • Colour rendering • Chromaticity of light colour or white tone (white point) • Colour saturation, hue • Further parameters from the light spectrum

Source: TU Darmstadt.

Table 12.2 shows the photometric and colorimetric parameters that quantitatively characterise the above-mentioned subjective descriptors of scene appearance according to the work of the three research groups mentioned above, from the point of view of the authors of this book.

In parallel with these three fundamental investigations, the previous German Standard DIN 5035 [11] mentioned the following quality features of lighting systems (according to Dehoff [12] and [13]):

- Illuminance level
- Brightness distribution
- Glare limitation
- Colour rendering
- Light colour
- Light direction
- Shadiness

The above-mentioned quality features were characterised in the IESNA Lighting Handbook [14] by the following three terms:

- Ergonomics of lighting systems
- Subjective criteria: room harmony, lighting mood, and lighting design
- Economic efficiency

While the quality features according to DIN 5035 of 1990 rather describe the eye-physiological features, the supplementary subjective criteria such as room harmony, lighting mood, and lighting design are clearly related to the cognitive psychological level. These criteria have a similar significance to the features '*visual attraction and complexity*' as found by Veitch and Newsham in [10].

Lighting quality has been summarised in the IESNA Lighting Handbook [14] as an integration of '*Human needs*', '*Architecture*', and '*Economics and Environment*', see Figure 12.1.

Figure 12.1 shows a holistic approach and considers lighting quality as the result of an integrative process, with the term '*Human Needs*' rather encompassing the core

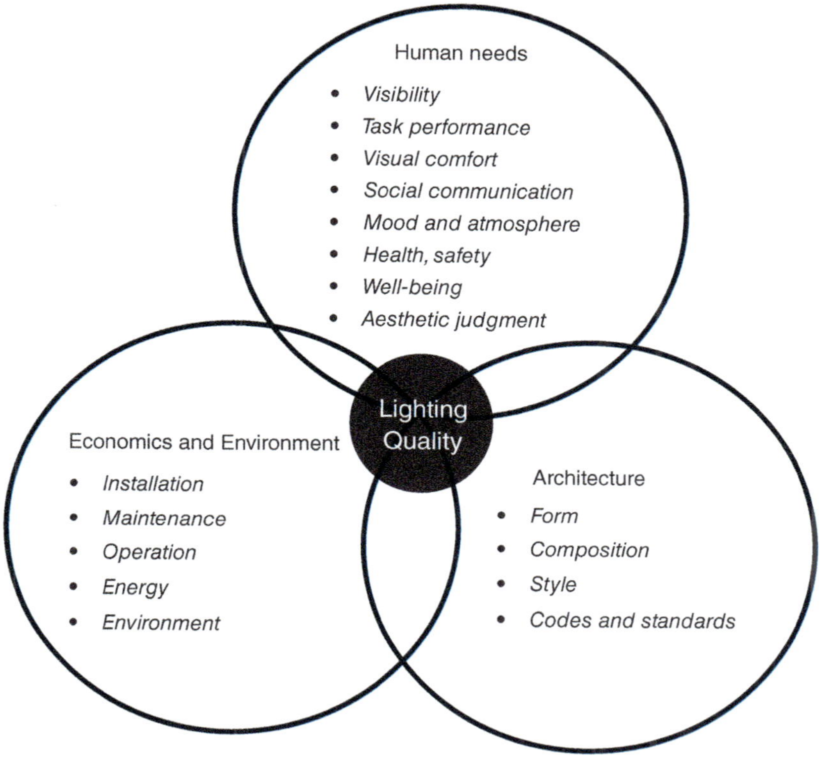

Figure 12.1 Lighting quality is the integration of 'Human needs', 'Architecture' and 'Economics, and Environment' according to the American Lighting Manual (after Dehoff [12] and IESNA [14]). Source: Reproduced with permission from IESNA [14].

of today's HCL aspects with visual performance, psychological–emotional aspects, and non-visual aspects, such as '*mood, health, well-being, and visual comfort*'. This is also an impressive proof that the HCL idea (that humans should be in the focus of lighting) is not new: this idea has a long history. The difference from today's HCL approach is the discovery of the ipRGCs at the beginning of the twenty-first century which accelerated research on non-visual light effects such as sleep quality, stabilisation of the circadian rhythm (or wake–sleep-rhythm) as well as the measurement of light effects not only by subjective attributes but also by vital parameters such as melatonin, cortisol, EEG, or ECG. Thus, the focus of the discussion has shifted in recent years to the non-visual and health aspects. The overall holistic approach of Figure 12.1 still has the same validity.

Answering the *important question of* whether the research results on visual performance and psychological–emotional lighting effects up to the beginning of the twenty-first century, continued until today, as well as the results of interdisciplinary research rather on non-visual effects mainly since about 2001, *lead to the same orders of magnitude* of lighting parameters and they complement each other, is the main task of this chapter.

The IESNA approach (Figure 12.1) goes further and includes architecture (room layout, form, and materials) in the description and modelling of lighting quality. The interaction between the users of the space, the lighting and the architecture for a harmonious atmosphere and working climate is thus conducive to health and work productivity.

William Lam, quoted in [15] and in [12], categorised the user's need concerning lighting into the following three components:

- '*Activity Needs*' (functional): lighting to perform a visual task;
- '*Biological Needs*' (emotional): for the unconscious perception of the environment;
- '*Architectural Needs*' (aesthetic): perception of architectural lighting design.

William Lam formulated: '*To design good lighting, the designer must clearly understand the principles and processes of visual perception and the human needs for visual information*'. Kramer, quoted in [12], pursued the same ideas as Lam, albeit with slightly different emphases: '*We need a definition of lighting quality that is equally valid for 'aesthetic' light and 'working' light, for natural and artificial light, and that takes into account not only people's bodies but also their psyches and sensibilities, i.e. the whole person*' [15, 16].

In the room to be illuminated, people move around and do their work. Conscious and unconscious demands are placed on this room as a whole, and also, on the individual room zones. Ganslandt and Hofmann wrote about this [17]:

> Architectural requirements for a lighting concept result from the structures of the architecture to be illuminated. The task of lighting in this context is to clarify the structure of the space, its forms, rhythms, and modules, to emphasise special architectural features and to support the planned atmosphere of the building. Both the arrangement of the luminaires and their lighting effects are thus intended to support the architecture, but also to actively change its effect, if necessary.

In these considerations, the quality of lighting is the goal, and the planning or design of lighting is the process or the operation. The path to the goal is roughly the same by IESNA, W. Lam, H. Kramer, and P. Dehoff, namely that people as room users should be focused and centred in the design. In this context, it is important, after the qualitative formulation of the goals and ways of lighting quality and lighting design, to concretely analyse and document the metrics of lighting, the input and output parameters, the influencing parameters, and the concepts for the implementation. In the context of this book, the terms '*light quality*' as used by German-speaking lighting designers and lighting engineers and '*lighting quality*' as applied in American and English literature are the same. Strictly speaking, however, these two terms are not always identical.

At the end of the twentieth century and in the first years around 2000–2001, publications by Veitch and Newsham appeared, with guiding ideas about the definition of lighting quality and relationships between lighting conditions, individual processes and individual sensation outcomes [18–20]. In [19] Veitch formulated:

> Lighting quality has been defined as the degree to which a lighting installation fulfils the requirements of human needs and architectural integration within its unique constraints of economics, energy consumption, maintenance and the like. Lighting designers, and some lighting researchers, hold the implicit theory that occupant preference for luminous conditions can identify the conditions that will satisfy the human needs in the equation. If this is true, then recommended practice documents should present guidelines based on knowledge about the preferred luminous conditions in a variety of settings. Information on preferred luminous conditions might also address the fear that lighting quality will inevitably decline as energy efficiency regulations limit lighting power densities. Such information would also prove useful to researchers wishing to test the theory that certain, preferred, luminous conditions create a state of positive affect that leads to improved work performance, creativity, and social behaviour.

With this formulation, the *lighting quality* of a lighting installation represents the degree to which it meets the requirements of human needs and architectural integration, with factors such as economy, energy consumption, maintenance, etc., also playing a role. The output variables and outcomes of this process are improved work performance, creativity, and better social behaviour. The relationship between lighting conditions, individual processes, and individual sensory outcomes is shown in Figure 12.2 according to Veitch [19]. The processes of the room users are divided into:

Psychological–biological processes: visibility, photobiology, arousal (stimulation), stress; and

Psychological processes: perceived control (e.g. via the adjustment options of lighting systems), attention due to high contrast or spotlighting or light distribution on the work plane, an environmental assessment (see Table 12.2) and 'influence' (so-called affect).

These two types of processes are in turn influenced by workplace conditions such as lighting, social, and organisational processes and by individual concerns. The processing of these influencing factors in these two processes provides individual (i.e. person-specific) and organisational output results (so-called *outcomes*). The formulations and conceptions in the two figures of IESNA (Figure 12.1) and Veitch (Figure 12.2) correspond to each other to a high degree. The second main phase of lighting research came to an end around 2001–2002.

12.1.2 Literature Analysis and New Thoughts on Lighting Quality

With the development of white high-power LEDs as light sources based on semiconductors in the years between about 1999 to about 2019 – in the third main phase of the development of lighting science and technology – the focus on the technological side was on the development of energy-efficient and reliable LED – light systems with a long service life at a defined colour temperature with a constant light spectrum. The first solution approaches of HCL-oriented lighting systems have

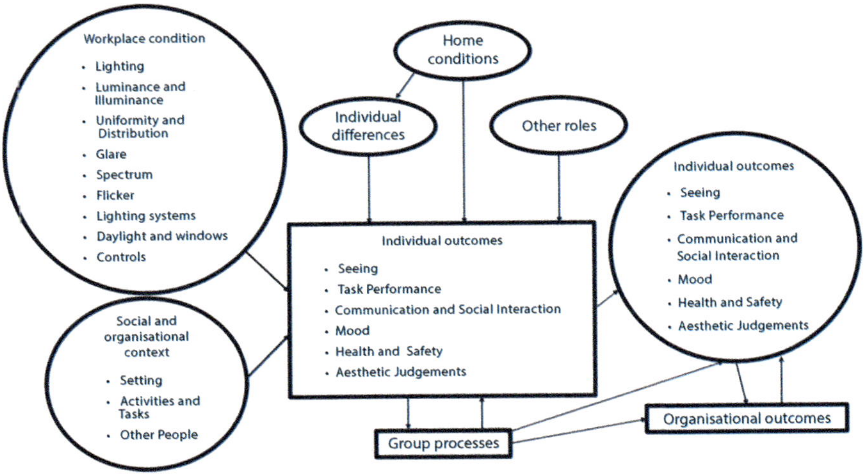

Figure 12.2 Relationship between lighting conditions, individual processes, and individual sensation outcomes according to Veitch [19]. Source: Reproduced from Veitch [19] with permission from Lighting Research and Technology.

been conceived since about 2012. In lighting research, a number of new colorimetric and non-visual (i.e. melanopically effective) parameters have been defined, coupled with a sound and extensive analysis of the psychological-emotional, non-visual short-term effects of lighting systems. Literature items up to 2019 on different lighting effects such as wakefulness, sleep quality, and circadian rhythm were analysed and reviewed in 2018–2019. These results form an important part of the contents of Chapter 9 of this book.

The relatively recent research results in lighting technology up to 2020 were comprehensively analysed by SHC (Solar Heating Cooling Programme/International Energy Agency) in a literature review in 2020 entitled '*Literature review of user needs, toward user requirements*' [21]. The objectives of this literature study were to formulate user requirements regarding daylight and artificial lighting through an extensive literature review. The starting point for the discussion was the following definition of lighting quality by Kruisselbrink et al. from 2018 [22]: '*Lighting quality is a concept that allows excellent vision while providing high comfort*'. These authors came to the conclusion that in this above-mentioned definition as well as in the publication of these three authors, only direct and measurable metrics for describing lighting quality such as intensity, glare, spectrum, light distribution, light direction, and dynamics are taken into account. The factors that have an indirect but substantial impact on the health and well-being of room users, such as non-visual effects, are not included in this definition. This literature study considers the following four aspects of lighting quality:

- Visual perception of light;
- Visual comfort;
- Psychological aspects of light (e.g. quality of room perception, privacy);
- Non-visual aspects of light (action spectrum of ipRGC-cells, hormone levels).

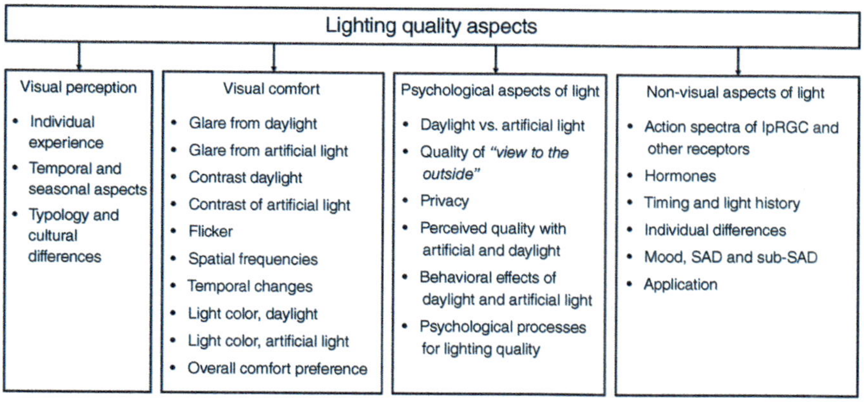

Figure 12.3 Research topics for lighting quality according to the SHC literature study [21]. Source: Adapted from SHC [21].

For a discussion of the process of lighting design, it is necessary at this point to show the topics of these four aspects schematically, see Figure 12.3.

The new aspects in the SHC study compared to previous concepts of lighting quality are the non-visual effects of light, the integration of daylight, the emphasis on light colour, cultural differences, the consideration of the individuality of room users as well as temporal aspects (so-called *timing*, i.e. dynamic light).

At the end of the historical and methodological description of the characteristics, components, and interrelationships within *lighting quality*, one should formulate a final definition for it, to give a meaning and thus, a well-defined task to lighting design. A sustainable and comprehensive definition was given by Dehoff and Tralau in 2017 in [23] as follows: '*Lighting quality describes the extent to which light is experienced in a space or in an environment. It presents itself as dynamically variable. Lighting quality depends on the functional, biological, psychological, and architectural requirements made by the user. Feedback processes and methods are necessary to determine lighting quality. The measure of lighting quality results from the comparison between the user's requirements for a lighting solution and the evaluation of the implemented lighting solution*'.

For the authors of this book, the following important task was derived from this, for lighting research and lighting application: the lighting designer must know the functional, biological, psychological, and architectural requirements of the user. The specifications of photometric values in the international standards for interior lighting are only minimum values. In the historical course of lighting research as well as in the future, the following solutions exist for the above task:

(a) The light researchers conduct experiments in the laboratory or in real rooms over a defined test duration, with regard to certain visual and non-visual criteria and determine mean values and confidence intervals for the collected subjective data. One must take into account the subject size, light history, time, duration, season, daylight percentage, test environment, gender, age, culture of the test subjects and the limitations of the experiments. The mean values are then the control variables to be set for the lighting systems. This approach is a general scientific method and is often used in lighting research to date. The results of such experiments are described in Chapters 4–11, among others, as well as in lighting technology reference books and technical articles and represent the mean values of a group of people and not a person-specific requirement or preference.

(b) One conducts experiments and interviews in concrete lighting projects with the people whose stay is planned in the rooms to be built later. In principle, the methodology of the investigation is similar to that of (a) above. The effort is higher and afterwards one is able to formulate person-specific requirements in the concrete case for the project planning, in contrast to the (a) method.

(c) In the age of technologically advanced smart lighting systems, the intelligent luminaires connected to cloud servers (see Chapter 11) are initially pre-programmed with the average values of the preferred lighting parameters in case (a). By means of sensors in the lighting systems, GPS data, weather data and by means of data evaluation regarding movement sequences of people in the room, weather conditions, and preference of people in the specific room over a longer period of time (weeks, up to two months), the intelligent luminaire at the workplace in question will determine the person-dependent, favoured lighting parameters with the help of artificial intelligence (AI) methods in the cloud software and adjust itself according to the situation. As soon as the weather conditions, the working conditions and the people in the room change, new values to be determined with AI methods will be calculated and set. This method has the advantage that in the worst case, the average values in case (a) are set, which is already a great step forward compared to current practice.

12.1.3 Summary of the Concepts on HCL and Lighting Quality – A Draft Overall Concept

The lighting technology process in the fourth main stage of the development of lighting research and technology, which began in 2018 and will continue for many years, does not involve any fundamentally new aspects compared to the third main stage from a purely lighting technological point of view. In the opinion of the authors of this book, the design elements in the lighting quality conceptions in Figures 12.1–12.3 are essentially the same, with different shifts in emphasis and with a current deepening into colour quality as well as into the non-visual aspects that entail individuality, daylight, the timing of light switch-on as well as changes in lighting parameters over time.

Nevertheless, the social framework conditions have changed in the years since 1990 until today and the following aspects should be mentioned:

- The dynamic development of the Internet platform (Internet of Things [IOTs]), sensor technology, computing power and the networking of the most diverse digital systems. This leads to the development of *smart buildings* or *smart lighting*, which was discussed in detail in Chapter 11.
- Increasing research activities in the field of non-visual effects and colour quality (see Chapters 5–9 of this book).

- The development of LED or OLED lighting technology with a variety of spectra, wavelengths (see Chapter 11), dimming options as well as increased luminous efficacy and reliability of the components and lighting systems.
- The urbanisation of living spaces, the strong change in the character of industrial production (through automation, robotics) as well as the character of office spaces (more individuality, more *home office*, and more flexibility of office occupation).
- The demographic development with the consequence that more people remain in working life until old age and the number of nursing homes as well as health care facilities is growing rapidly. From Chapter 9, the reader learns that the non-visual effects are highest among those population groups that need light the most, e.g. the elderly in working life, residents, and medical staff in nursing homes as well as in clinics.

In this main stage of the development of lighting research, lighting technology and lighting design, it is mainly a matter of putting the research results into practice in a meaningful and professional way with qualified lighting design. Basically, there are already several available development stages of luminaire systems for HCL-oriented lighting (see Chapter 11). It is not so much about the '*hardware*' of the luminaire systems, but about a creative and intelligent application of '*light*' and lighting concepts to achieve targeted human-centred lighting. The realisation of such lighting concepts currently, and certainly increasing in the future, takes place via software solutions that can be updated and improved from time to time. Accordingly, Figure 12.4 shows schematically the significance of intelligent lighting systems in connection with daylight systems in the concept of *Smart Lighting* in an adaptive process with light users and room scenes with regard to the consideration of lighting quality and the classification of lighting design.

In the context of the realisation of lighting quality, as well as with the technological prerequisite of modern luminaire systems and Smart Lighting (see Figure 12.4), daylight, temporal variation (dynamic lighting), and lighting for the population group with the highest lighting needs (lighting for the elderly) will play a major role. For the practical implementation of lighting design (lighting planning), the specification of the lighting parameters preferred by people and conducive to health (photometric, colorimetric, and non-visual parameters) as well as the concrete methodology of the approach are of high importance and are therefore discussed in detail in the next sections.

12.2 Lighting Design: The Process and the Influencing Factors to Achieve Lighting Quality

12.2.1 Goals and Classification of HCL-Oriented Lighting Design

The tasks of lighting intelligent lighting systems in connection with daylight systems design were already outlined in 1996–2000. In 2000, Dehoff formulated [12]: '*Visual comfort, visual performance and room ambience are listed as priorities in the quality consideration of lighting design*'. In the course of the development of lighting

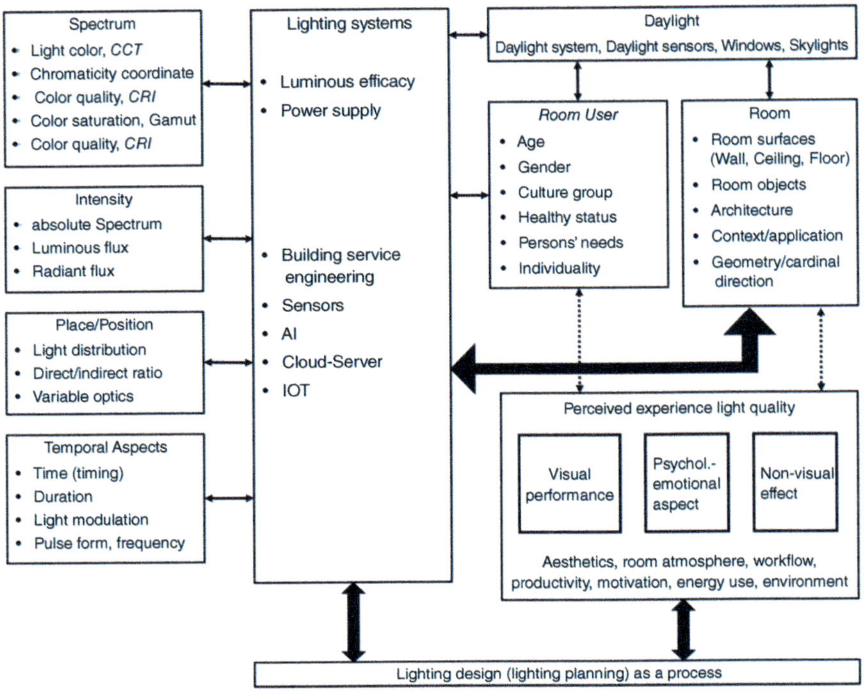

Figure 12.4 Intelligent lighting systems, daylight in an adaptive process with light users and room scenes in the context of the consideration of lighting quality and the classification of lighting design. Source: TU Darmstadt.

technology and because of the greater awareness of lighting quality, fundamental yet application-oriented literature has been published in recent years on the processes involved in implementing human-centred lighting design. The following is based on the two publications ([23], Dehoff and Tralau) and [24]. The lighting system design process is, from an operational point of view, one of the steps of an entire lighting project [24], see Figure 12.5.

The lighting design is commissioned by the client, who has a specific idea and formulates the user requirements. The lighting users in the space to be built

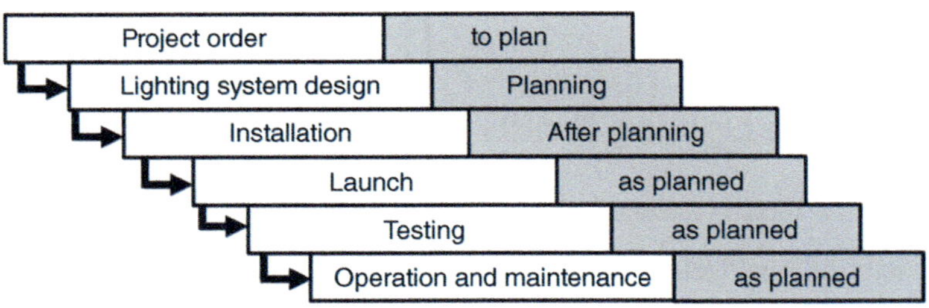

Figure 12.5 Steps of a lighting project. Source: Adapted from Fördergemeinschaft Gutes Licht [24].

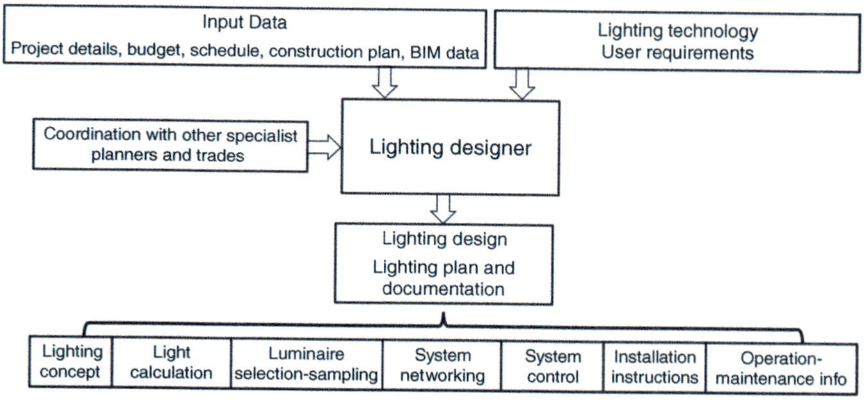

Figure 12.6 Role, classification of work, and work results of a lighting designer in the overall lighting design process according to [2]. Source: TU Darmstadt.

later are not always the ones who have these requirements. An efficient lighting design should focus on the four components visual performance – psychological effects – non-visual effects – architectural and spatial aspects from the very beginning and be designed for long-term operation of the room, late also including maintenance. The lighting design process should not be carried out in isolation but should be coordinated holistically and in an interdisciplinary way with all relevant services (heating, air conditioning, water supply, power supply, window technology, and sanitary engineering), with the experts involved (e.g. construction managers, architects, and power planners) and the users.

The role, the classification of the work, and the work results of a lighting designer in the overall lighting design process are shown in Figure 12.6 (after [24]). The formulation and establishment of user requirements, the weighting of the individual requirements in the specific application, the determination of design means to fulfil the requirements, as well as the methodology, the way in which the lighting results are first presented to the client and later tested in the constructed interior space and compared with the user requirements, are of great importance and are described in Section 12.2.2.

12.2.2 Process Steps of HCL-Oriented Lighting Design

According to [23], the process of achieving high-quality lighting can be divided into three steps as follows.

Step 1: *Determine the project-specific (individual) user requirements*

In a concrete project for a specific application, the individual requirements are weighted and classified separately. The weighting factors are derived either from literature values, from the assessment of the client and user and/or from discussions between lighting designers and users. These weighting factors should consider the relevance of the individual requirements between '*not important*' and '*very*

Table 12.3 Semantic scaling for weighting user requirements [23].

Scale	Semantic meaning
0	Not applicable
1	Weakly relevant
2	Relevant
3	Very relevant
4–5	Scope for increasing the requirement
5	Maximum

Source: Adapted from Dehoff and Tralau [23].

important' and can be on scales between 1 and 5 or 1 and 7. Requirements with a high weighting must be considered with great attention and effort, see Table 12.3.

The entire catalogue of requirements can be divided into four groups:

- Functional,
- Biological,
- Psychological, and
- Architectural.

Step 2: *Lighting design using design tools and measures to meet requirements*

The lighting design can be carried out in this second step with known requirements and weightings of the individual requirements, whereby the design means relevant for the lighting design are selected by lighting concepts, lighting calculations, applications of table values for the specification of HCL-oriented lighting systems (see Section 12.5), luminaire selection, system interfaces, lighting simulation, and testing to meet the requirements. There are a number of design tools, the most important of which are listed below:

- *Luminous flux of the light source (adjustable by dimming, luminous flux of the direct and indirect components);*
- *Relative size of the light-emitting surface;*
- *Number, arrangement and position of the light source(s) (as a result of the light calculation, and light simulation);*
- *Light colour (CCT, colour location, selection by choice of LED – combination, luminaire selection);*
- *Light distribution (luminous intensity distribution curves);*
- *Spectrum of the light source (through single or multichannel LED luminaires, mixable via software);*
- *Control and controllability of the lighting;*
- *Operation of the lighting.*

Table 12.4 shows the requirements according to Dehoff and Tralau [23] in four groups in the vertical direction and the design tools in the horizontal direction. This also means that each requirement may need one or more design tools. However, a

Table 12.4 List of user requirements in four different groups according to Dehoff and Tralau [23].

Requirements	Design tool							
	Luminous flux of the light source	**Relative size of the light-emitting surface**	**Number, requirement, and position of the light source**	**Light colour**	**Light distribution**	**Light source spectrum**	**Adjustable or controllable lighting**	**Operating the lighting**
Functional requirements								
F1 Seeing – Identifying details								
F2 Seeing – Identifying shapes								
F3 Seeing – Identifying colours								
F4 Speed of vision and identification								
F5 Seeing and identifying overtime/visual performance								
F6 Attention guidance								
F7 Order/inconspicuousness								
F8 Physical security								
F9 Object protection								
B1 Activation								
B2 Recreation								
B3 Circadian rhythm								
B4 Radiation protection								
B5 Radiation physiological effect								

Table 12.4 (Continued)

Requirements	Design tool							
	Luminous flux of the light source	Relative size of the light-emitting surface	Number, requirement, and position of the light source	Light colour	Light distribution	Light source spectrum	Adjustable or controllable lighting	Operating the lighting
P1 Spatial orientation								
P2 Temporal orientation								
P3 Orientation about what is happening								
P4 Privacy								
P5 Personal territory								
P6 Self-expression/representation								
P7 Safety feeling								
P8 Own control								
P9 Mental activation								
P10 Mental recovery								
P11 Familiarity								
A1 Structure of the room according to form								
A2 Arrangement of the room according to rhythms								
A3 Structuring the room according to zones								
A4 Support the character of the architecture								
A5 Architectural (design features emphasised)								

Source: Adapted from Dehoff and Tralau [23].

particular design tool can also be used to fulfil several requirements. In the case of a requirement with a high weighting, the applicable design tool must be planned, conceptualised and executed with care.

Step 3: *Evaluation of the lighting solution and comparison with user requirements*

The process is completed and accepted as a project result when the planned and implemented lighting solution has been assessed with regard to the specified requirements. The assessment is carried out either by surveying the lighting users after a defined period of use or by measuring the installed lighting system at individual workplaces and on the walls and ceilings (with and without daylight) or by combining the two variants. According to Dehoff and Tralau [23], the following evaluation variables can be considered when assessing lighting quality:

- *Illuminance in the visual task area* (E_v);
- *Uniformity of illuminance* (U_o);
- *Colour contrast (FK);*
- *Brightness contrast (creative) (HK);*
- *Discomfort glare* (B_{psy});
- *Disability glare* (B_{phy});
- *Reflex glare* (B_{Re});
- *Correlated colour temperature* (CCT);
- *Colour-rendering index* (CIE CRI R_a);
- *Contrast Fidelity* (CRF);
- *Drop Shadow* (SS);
- *(Balanced) luminance distribution* (B_{al});
- *Modelling (Mod);*
- *Absence of flicker (Fl);*
- *Melanopic effect factor* ($a_{mel,v}$);
- *Damage potential* (H_{dm})*; and other*
- *Qualitative factors (Q).*

However, other evaluation variables can also be considered:

- *Colour-rendering index* R_f *and colour gamut index* R_g *according to* IES [25];
- *Colour-rendering index* R_f *according to CIE 2017* [26];
- *Circadian Stimulus CS after* [27] in the version of the year 2018 and version 2021 (CS2021) in Chapter 13 of this book;
- *'Colour preference' CP according to Section* 7.4.3 *of this book;*
- *'Scene preference' SP according to Section* 7.4.4 *of this book.*

For a final consideration of the subject of lighting design, reference is made to the following formulation by Dehoff and Tralau [23]: '*...good lighting quality represents the correspondence between the user's requirements on the one hand and the assessment of the lighting solution on the other. Quantitative variables of lighting technology were listed, which enable an evaluation of the lighting situation. The combination of requirements and assessment, as well as the degree of fulfilment of the requirements, represents the measure of lighting quality*'.

12.3 Daylight and Daylight Planning

In this section, the characteristic values of daylighting standards in Germany and Europe are summarised. Some new research results on the effect of daylight on acceptance, on work productivity, and on absenteeism are presented. In this context, the effects of the dimensioning of the percentage window shares and the degrees of reflection of the room boundary surfaces are highlighted.

12.3.1 Introduction

Daylight includes direct sunlight and skylight scattered by the atmosphere and clouds. Its intensity and spectral composition depend on the time of day, season, weather, and geographical location, see Figure 12.7.

Sunlight, the sky and the clouds with all their properties have shaped the biorhythms and evolutionary development of plants, animals, and humans for hundreds of thousands of years and are described in works of art, buildings, churches, in historical, philosophical, cultural, scientific, and religious collections. Until the end of the nineteenth century, the diurnal cycle of outdoor light shaped the way of working and the timing of production, mainly in agriculture, fishing, handicrafts, and events in social institutions. At that time, knowledge about culture, civilisation, and technology was closely linked to knowledge about daylight, the position of the sun and its dynamics.

At the end of the nineteenth century, the first industrial revolution began with electrification, the introduction of the incandescent lamp as well as the development of discharge lamps in the 1930s, which allowed the place of value creation to be increasingly indoors. Lighting research on visual performance and colour quality and lighting technology in the twentieth century was mostly based on laboratory work with electrically powered light sources such as incandescent lamps, halogen lamps, and discharge lamps. This has changed markedly since the beginning of the twenty-first century mainly for three reasons:

Figure 12.7 Three sky conditions on a summer day (19.08.2020) in Darmstadt (Germany). Source: Technische Universität Darmstadt.

(a) through the development of LED and OLED technology, with the current high luminous efficacy and with spectral compositions that can be varied to a great extent. The spectrum of the semiconductor light sources can be adapted very closely to the daylight spectrum;
(b) through the discovery of intrinsically photosensitive ganglion cells (ipRGCs) and the qualitative justification of the effect of light on the circadian rhythm, wake–sleep behaviour, phase shift, metabolism, alertness, mood, and health. Daylight with high intensity during the day is conducive to the positive control of the circadian process;
(c) through the discussion on climate change and energy conservation. Solar energy as renewable energy and daylight are often abundant, depending on geography and weather, and can supplement electric light to a high degree.

Due to these three motivating factors, as well as social and technological circumstances such as digitalisation, the IOTs, demographic, and health developments, the increasingly critical food supply problem resulting from the rapid growth of the world's population on an international scale, and the 24/7 rhythm of life and work, the fundamental knowledge about the effects of light on humans, animals, and plants must be renewed, deepened, and supplemented. Through the introduction of smart lighting with all the regulation and control methods, the research results should not only be carried out in the laboratory. We should not only formulate the results qualitatively. The lighting technology measures are to be planned and programmed quantitatively for the sensor-based lighting controls with digital systems in the interiors. This requires a new strategy, a new approach and a practical approach to lighting research and lighting technology, which now have a long way to go quickly.

The aim of this section is to outline such topics of lighting design and some recent research results directly related to them, without claiming to be exhaustive. Part of this work by the authors of this book deals with the metrological determination of the spectral, colorimetric, and photometric properties of daylight on certain exemplary days and with the interpretation of these measurement data with regard to Smart Lighting. The results of the daylight measurements have already been described to a large extent in Chapter 11.

12.3.2 Daylight from a Lighting Design Perspective – Daylight Design in the Context of Standardisation

The dimensioning of daylight and the question of how to design daylight in buildings belong to the lighting design area, for which the German DIN standard series 5034 [28] in part and, since March 2019, the European standard DIN EN 17037 [29] can be used. By using local weather data, a simulation for daylight supply can be determined for three levels, 'Low', 'Medium', and 'High', in the interior. A certain level is reached when for 50% of the room area for 50% of the daylight hours (24 hours × 365 days/2 = 4380 hours) a target value such as 100, 300, or 500 lx [30] is met. Table 12.5 shows recommended values [30] for the supply of daylight through daylight openings in vertical and inclined surfaces.

Table 12.5 Recommendation for the supply of daylight through daylight openings in vertical and inclined surfaces.

Recommended level	Target illuminance (lx)	Room share for the target value (%)	Minimum target illuminance (lx)	Room share for the minimum target value (%)	Proportion of daytime hours (%)
Low	300	50	100	95	50
Medium	500	50	300	95	50
High	750	50	500	95	50

Source: Adapted from Jakobiak et al. [30] https://www.initiative-tageslicht.de/last accessed January 19, 2023.

The illuminance values in Table 12.5 are values that can be achieved by providing daylight alone. The lighting preferences of office users were extensively investigated in a field study by Moosmann [31] in nine office buildings in Germany. The results of the evaluation of almost 1000 questionnaires are summarised below:

- 67% of the respondents stated that they would not like any change in the lighting level at the mean value of the horizontal illuminance at the workplace of 966 lx;
- The desired horizontal illuminance depends on the season and is higher in spring (1046 lx) than in summer (927 lx) and autumn (875 lx), and winter (787 lx);

The difference between the above-mentioned preferred illuminance levels and the target illuminance levels shown in Table 12.5 can be interpreted as the necessary contribution of electric light, which is to be recorded in a time-dynamic manner and implemented by intelligent dimmable luminaires. According to the above-mentioned empirical study [31], a good supply of daylight from the point of view of office users is guaranteed for workplaces near windows with a window area proportion of around 60 %. In [32] some calculations were made with different sample rooms with the result that the daylight standard values in Table 12.5 can be met by a window area of 20% to 25% of the floor area of the room.

In a field study in four office buildings conducted by Gramm et al. in 2014 [33], the test subjects only rated a daylight coefficient D of 2% at the workplace as '*sufficiently bright*'. Therefore, the arrangement of workplaces and communal desks in the area near the window is important. In Table 12.6, values of the daylight coefficient between 1.8 % and 2.2% (i.e. close to 2%) are shown for daylight openings with an illuminance of more than 300 lx for a daylight hour share of 50% for the city of Berlin (taken from [30]). From this context, one can see the potential of windows and skylights to provide sufficiently bright *room lighting* at $D = 2\%$ for a target illuminance of 300 lx by providing daylight for 50% of the daylight hours.

In a study by Giuliani et al. [34] with 561 students from 8 European universities of architecture in different rooms with different proportions of windows, sky conditions (rainy day, degree of cloudiness, clear sky), and room geometry, a series of

Table 12.6 Values (taken from [30]) of the daylight coefficient between 1.8% and 2.2% for daylight openings with an illuminance of more than 300 lx for a daylight hour proportion of 50% for the city of Berlin.

	Mean external diffuse illuminance	*D* (Reference 300 lx)
Windows	13 900 lx	2.2%
Roof skylights	17 100 lx	1.8%

Source: Adapted from Jakobiak et al. [30] https://www.initiative-tageslicht.de/last accessed January 19, 2023.

questions were asked regarding subjective assessments of perception, mood, amount of light, and pleasantness. The statistical evaluations of the answers showed a significant relation between the estimation of the amount of incident daylight and the daylight coefficient *D*. In the rooms with *D* greater than 3%, the assessment of the perceived amount of daylight is between 3 (acceptable) and 4 (high). The assessment of the pleasantness of daylight also correlates with the daylight coefficient. In the rooms with *D* greater than 5%, no problems regarding comfort occurred. If one combines the research results of Gramm et al. [33] and Giuliani et al. [34], the daylight coefficient should be greater than 3% in order to achieve a sufficiently bright and acceptable or high perception for room users.

12.3.3 Daylight Planning for Non-visual Effects

Due to the findings that a good supply of daylight can lead to a stable circadian rhythm, associated with alertness and concentration during the day and with a comparatively good quality of sleep, daylight planning, and building architecture have for a few years now begun to include the level and duration of daylight exposure for stabilisation of circadian processes in room and window design.

The procedure for daylight planning in the context is described in [35] by Acosta et al. In 12 model rooms in the context of a hospital or a nursing home with depth = 6 m, height = 3 m and width = 3 m with calculated varying window sizes from 10% to 80% of the *façade area* and with two different, assumed groups of reflection degrees of the room boundary surfaces (see Table 12.7), daylight potentials were calculated with a daylight programme. The windows have a transmittance of 0.75 and are oriented toward the north so that these calculation results can be considered as '*worst case*'.

The calculations [35] were performed for London as a city with predominantly overcast skies and for Madrid as an example of a city with mostly clear skies. The calculation addresses the question of what percentage of all days in the year reaches or exceeds a Circadian Stimulus CS of 0.35 (CS model in the version of 2005 or 2018), considering several locations in the room for a person lying down (e.g. a patient in a hospital; horizontal plane, working height 0.6 m) and for a person sitting down (vertical plane, working height 1 m). It was assumed that the daylight in the room has a spectrum like a reference daylight type D65. A CS value of 0.35 for D65 means

Table 12.7 Reflectance of the room boundary surface of the assumed model rooms.

Group	Ceiling	Walls	Floor
Bright	0.8	0.8	0.6
Dark	0.6	0.4	0.2

Source: Adapted from Acosta et al. [35].

an illuminance at the eye of the person under consideration of 233 lx, which, in the opinion of the authors of [35], provides stable control of the circadian rhythm when administered for an hour or more in the early morning. The results (CS mean values between 8:00 a.m. and 12:00 p.m.) for London [35] can be summarised as follows:

- With a window size between 60% and 80% of the façade size, the CS values are similarly high.
- To achieve the value of CS = 0.35, the window size should be larger than 40% for the room with high reflection coefficients of the room boundary surfaces.
- For the period in the morning between 8:00 a.m. and 12:00 p.m., daylight provides higher illuminance levels for seated persons than for reclining persons. For a window size of 20% in the room with higher degrees of reflection, the value of CS = 0.35 is only achieved for lying persons up to the room depth of 3 m, and for seated persons up to 5 m.
- For the model room with lower reflectance values (see Table 12.7), even with the window size of 80%, the daylight is not able to produce the CS-value of 0.35 for lying persons for all locations in the room. The calculation of the potential of daylight for a room is therefore not only related to the window size but also to the reflectance of the room's boundary surfaces such as walls and ceiling.
- For a lying patient in the area 0–2 m next to the window, the value CS = 0.35 is achieved for 75% of all days in the year if the window size is 30% of the façade size and high reflection coefficients of the room boundary surfaces are included in the calculation. For the middle room area between 2 and 4 m, this is only possible if the window size would be at least 60%. Thus, it tends to be desirable to select a window size of 60% of the façade size in order to positively shape the view to the outside, among other aspects.

In any case, daylight planning should focus on the non-visual effects, taking into account the positions of the room users and the spectrum of light transmitted through the window panes. The spectral transmission of the window panes plays a major role in this context.

12.3.4 Some Data on Daylighting Effects

Today's society in industrialised nations is undergoing a pronounced information technology process with digitalisation, global finance, and trade around the clock, with mature automation of processes in the industry (Industry 4.0), traffic and

Table 12.8 Illuminance and exposure data as median values for the time in the office and daytime exposure.

Data characteristics	For the time in the office	Daylighting
Exposure	2244 lxh (1199–3392 lxh)	7394 lxh (5234–11683 lxh)
Median illuminance	308 lx (202–572 lx)	183 lx (118–249 lx)
Duration over 100 lx	240 minutes (204–283 minutes)	534 minutes (436–626 minutes)
Duration over 1000 lx	16 minutes (2–48 minutes)	105 minutes (61–187 minutes)
Duration over 2500 lx	1 minute (0–8 minutes)	43 minutes (16–92 minutes)

Source: Adapted from Hubalek et al. [37].

transport, and with ever greater use of information within a short time. These processes mostly take place indoors [36]. According to Schweizer et al. from 2007 [36], Europeans spend an average of about 22.25 hours per day indoors. In [37] a long field study by Hubalek et al. on lighting conditions in offices was described. Part of these results can be found in Table 12.8 where the illuminance and exposure data are given as median values for the time in the office and the daytime exposure (in lux-hours, lxh, for the whole day). The time period during which the threshold values 100, 1000, and 2500 lx were exceeded by current measurement data is also given. It can be seen that the time duration for exceeding the illuminance of 1000 lx within office working hours is only 16 minutes.

Stefani [38] has analysed the literature on labour productivity. According to this, sleep disturbance associated with a lack of daylight use causes a mean production loss of 5.5% and insomnia causes a production loss of 6.1% [39]. That is basically a considerable balance. In addition, there are certain aspects such as safety at work and the development of illness. An increase in daylight, on the other hand, brings a gain in labour productivity. According to [38, 40] employee absenteeism at Lockhead Corporation, with 2700 employees, decreased by 15% and productivity increased by 15% at the same time when the move to a new building with more daylight took place.

Verifone (Los Angeles/USA) achieved a productivity increase of around 5 % when it moved into a building with more daylight (according to [38, 41]). Basically, a medium or large company will only achieve a 5% increase in productivity if the logistical internal flow and process coordination is improved at great expense and care. For this reason, improving the supply of daylight is an economically sensible and, in most cases, cost-effective solution.

12.4 Specification of HCL Lighting Systems for Daytime – Draft Recommendation

12.4.1 Introduction

With regard to the daily working time situation (daytime, night shift, and rotating shift), the Federal Institute for Occupational Safety and Health

Table 12.9 Location of working hours of employees ($n = 17\,934$).

Position of the working time	Percentage share-mean value across all economic sectors (%)	Percentage share in industry (%)	Percentage in the public sector (%)
Working time between 7 a.m. and 7 p.m.	80	74	81
Staggered working hours	8	7	6
Alternating shifts without night work	5	6	3
Alternating shifts with night work or continuous night work	7	13	8

Source: Adapted from BAuA [42].

(Dortmund/Germany) collected the results of an investigation based on the data of 17934 employees [42] in 2016, which are shown in Table 12.9.

According to this, 80% of employees work in a regular time range between 7 a.m. and 7 p.m., i.e. during the day. In industry as well as in the public sector, the percentage of night shift workers is high, 13%, and 8%, respectively. The purpose of this section is to formulate, justify and establish the specification on the design and assessment of indoor HCL lighting systems for daytime hours based on the previous Chapters 4–11 and other literature. An additional table of value specification of the latest SHC literature study [21] is also given and commented on in this section.

Based on the lighting design recommendation of Dehoff and Tralau [23]the studies by Flynn et al., and Veitch and Newsham, and the experience of the authors of this book, Table 12.13 will show an *HCL oriented* lighting system *with recommended (preferred) values*. The values in Table 12.13 may differ from the standard values and values in other guidelines and are intended as a contribution to the discussion on the way to designing sustainable HCL (integrative lighting) installations. We will explain the values to be presented later in Table 12.13.

12.4.2 Illumination Level, Circadian-Effective Illuminance Levels

On the basis of work performance (Chapter 4): In the years up to the end of the 1990s, when lighting research was concerned with *visual performance* criteria, the horizontal illuminance levels on the work table or work plane, which are mostly horizontally oriented, were determined on the basis of determining work performance at the 90% or 95% level. These results are described in detail in Chapter 4 and listed in Tables 4.9, 4.10, and 4.14. Their significance is also supported by the suggestions in Table 4.19. The ranges of illuminance on the horizontal working plane are thus around 1000–1200 lx (or 740–1400 lx) depending on the reflectance of the working tools and papers.

On the basis of subjective (psychophysical) studies: In numerous *psychological (psychophysical)* studies on the determination of preferred illuminance levels, a large individual variation was registered. This variation may depend on the application, the type of work, and the age and gender of the users. Most of the studies were conducted in rooms with office characters. From 1994 to 1996, a working group led by Begemann et al. [43] carried out a study in offices with windows, involving 96 test persons who had chosen an illuminance for artificial light of 800 lx on average in addition to daylight. The interpretation of these data by the authors of this book is that, according to the study by Seidl [44], a daylight ratio of 0.9%, and, according to [45], a global illuminance of about 6000 lx in the winter months and 70 000 lx in the summer months (except for the peak values) can be assumed. This means that the additional daylight component can be calculated to be approximately between 54 and 630 lx so that the total illuminance value can lie in the preferred range between 854 and 1430 lx. Begemann et al. did not register any *age influence* or *seasonal variation* in their experiments. Tops et al. [46] analysed the data of this same research group again a few years later and gave a mean preferred illuminance of around 1400 lx.

In addition, according to Moosmann's already mentioned results [31] (see also above), 67% of the respondents said they did not want any change in the lighting level at the mean value of horizontal illuminance at the workplace of 966 lx. The lighting situations in which the lighting users did not express any wish to change the lighting level were subject to the influence of age. The horizontal illuminance at the workplace where the light users did not make any change requests was 1168 lx for people under 30 years of age, 1006 lx between 30 and 39 years of age, 883 lx between 40 and 49 years of age and 834 lx for people over 50 years of age. These results were statistically significantly different. Also, the desired horizontal illuminance depended on the season and was higher in spring (1046 lx) than in summer (927 lx) or autumn (875 lx), and winter (787 lx).

The above results of Moosmann from 2014 in the context of field tests over a longer period of time in actively used office spaces are thus smaller than the preferred illuminance levels of Begemann under the condition of real laboratory spaces within a short period of time. The results of Begemann and Moosmann were obtained in office spaces (university buildings, bank buildings, research institute buildings) in Central Europe. There has been very little research on lighting in industrial workplaces in countries with a long winter. An exception to this is Juslen's doctoral thesis in 2007 in six industrial plants of an assembly nature in Holland, Finland, and Germany [47]. In an industrial plant in Finland, a field study was carried out over 2 years with 23 assembly workers (3 men, and 20 women with an average age around 35) and with fluorescent luminaires with a colour temperature of 4000 K. The desired illuminance levels were determined by the lighting system. The desired illuminance levels were set with an IR remote control. The illuminances set by the test subjects were between 1340 and 1370 lx.

Chapter 7 of this book described the experiments to set the preferred illuminance at several colour temperatures for a real scene in a real office-type room. The results in Figure 7.11 show that the illuminance for the criterion '*just fine*' (scale value 80) at 4000 K equals 1250 lx. The results of Tops, Moosmann, Juslen, and TU Darmstadt are summarised in Table 12.10.

Table 12.10 Preferred horizontal illuminance levels of four field studies.

Study	Result
Tops et al. [46]	1400 lx
Moosmann [31]	966 lx: depending on age (834 lx for ages over 50 and 1168 lx for people under 30) and depending on the time of year
Juslen [47]	1340–1370 lx
TU Darmstadt (Chapter 7, Figure 7.11)	1250 lx
Mean value (from four results in this table, without taking the number of the test persons into account)	1243 lx

Source: TU Darmstadt.

For the setting and dimensioning of indoor lighting solutions, an important range of horizontal illuminance levels for a possible satisfaction of light users around 1243 lx (rounded to 1250 lx) is thus conceivable, which is to be considered in the context of HCL lighting and the promotion of people's health. Office areas and industrial halls have thus also been examined in terms of lighting technology.

On the basis of the studies on non-visual effects: An extensive field measurement was carried out by Hubalek et al. [37, 48] in 2005. Using an electronic device attached to the subject's glasses and containing two sensors each for measuring the vertical illuminance (in lx) and the effective irradiance according to the spectral sensitivity curve for melatonin suppression by Gall [49], measurements were carried out over three weeks in April and June 2005 on seven consecutive days with 23 test persons in office rooms in the Zurich area (Switzerland), with the test persons spending between seven and nine hours per day in the office. The sensor data were managed with a control unit and processed with a mini-computer.

On each evening before sleep, a questionnaire was filled out by the respective test person regarding the mood dimensions '*pleasure*' and '*arousal*' (latter is equivalent to wakefulness or being active), as well as *sleep quality* of last night. Regarding sleep quality, two questions had to be answered: 'How did you sleep last night compared to an average night?' and 'How restless did you sleep?'. After scoring and transforming the results, the number −4 meant *low sleep quality* and +4 meant *high sleep quality*. According to the interpretation of the authors of this book, the number 0, therefore, means a night with normal sleep quality.

Table 12.8 showed the illuminance and exposure data as median values for the time in the office and the daytime exposure (in lxh, for the whole day). Some remarkable results [37, 48] can be summarised as follows:

- Sleep quality has a significant relation to daylight exposure (in lxh) and to the duration during which the (vertical) illuminance levels of 1000 and 2500 lx are exceeded. This is a scientific indication that *high daylight illuminance levels* (due to high illuminance in the office or a longer time spent outdoors) are of *great importance* for good sleep quality.

- In the office during office working hours, the duration during which the values of 1000 and 2500 lx are exceeded is too short (16 minutes and 1 minute, respectively). The room user should go to windows more often or more frequently during breaks, and as far as possible, hold team meetings outside, or in the area near the window in the building.
- For a sleep quality scale value of +1.03 and 2.35, daytime exposures of 7394 and 21815 lxh, respectively, were determined. Interpreting the numbers 1.03 and 2.35 as indications of '*relatively good*' and '*good*' sleep quality, respectively, the exposures, vertically on the person's face, are desirable between 7394 and 21 815 lxh.
- The daytime exposure of 7394 lxh applies – with high probability – to the measurement period from April to June. Thus, results for the winter are missing.

The results of Hubalek et al. [37, 48] are expressed in photometric parameters such as illuminance. No other colorimetric and spectral parameters were included in this study so that circadian-effective metrics such as CS according to Rea et al. (see Chapter 10, in the model version 2005 and 2018) or melanopically effective illuminance according to DIN SPEC 5031-100 (see Chapter 2) cannot be calculated. The interpretation of the results of Hubalek et al. will therefore be carried out here.

Table 12.8 showed (Hubalek et al. [37, 48]) a median vertical illuminance value in office rooms of test subjects for the time in the office as low as 308 lx. Since further information on spectra and colour temperatures is missing in the cited literature, the authors of this book assumed two options for the colour temperatures, 4800 and 5300 K, for the furnished office rooms with large windows (yellow floor coverings, slight yellowing of white walls and ceilings, coloured shelves and cabinets, intermittent daylight). The CS values for a vertical illuminance of 308 lx and these two colour temperatures are given in Table 12.11. The CS values of 0.309 and 0.329 fulfil the condition of Rea and Figuiero [52] that CS should be greater than 0.3 to have a significant health effect. These values are slightly below or equal to the CS value of 0.328, which corresponds to a vertical illuminance of 240 lx at CCT = 6500 K (see Table 12.11), as recommended in DIN SPEC 5031-100 [50]. In addition, it must be taken into account that the days for the field measurements of Hubalek et al. were in the period April and June, rather than spring and summer. The values for the winter period could probably be lower.

More realistic and of practical relevance is the following (theoretical) consideration for regions with a long winter and with a long dark phase on a winter day (e.g. in Canada, North America, the Scandinavian countries, and Russia). The daylight level on the way to work (to the office, to the company) as well as from work to home in the afternoon is low and the lighting in the living area is also rather low so that the required daylight exposure of 7394 lxh for a relatively good quality of sleep should rather be supplied by artificial lighting within working hours in the office or industrial hall. The exposure of 7394 lxh over eight hours of working time means a vertical illuminance at the eye of 924 lx. With the assumed colour temperatures of 4800 and 5300 K, the CS values equal 0.491 and 0.512, respectively. It can thus be stated that the circadian effective stimulus should be planned around the value of CS = 0.5 in office rooms in regions with a hard winter time. Assuming a ratio of $E_v : E_h = 1 : 2$,

Table 12.11 Interpretation of the results of Hubalek et al. [37, 48] with CS values (model version 2005 or 2028).

Data characteristics	For a time in the office in lx	For time spent in the office in CS
Median vertical illuminance in the office (Hubalek et al. data, Table 12.8)	308 lx	CS = 0.309 at CCT = 4800 K CS = 0.329 at CCT = 5300 K
Recommendation of DIN SPEC 5031-100 [50]	240 lx with D65	CS = 0.328
Recommendation of Brown et al. [51] in 2022	250 lx D65 equivalent	CS = Approx. 0.335
Considerations of the authors of this book For regions with a long, hard winter	7394 lxh: 8 h = 924 lx Over 8 hours constantly in the office	CS = 0.492 at CCT = 4800 K CS = 0.512 with CCT = 5300 K
Considerations of the authors of this book For regions with moderate winter	4394 lxh: 8 h = 550 lx Over 8 hours constantly in the office	CS = 0.406 at CCT = 4800 K CS = 0.429 at CCT = 5300 K

Source: TU Darmstadt.

the vertical illuminance value of 924 lx means a horizontal illuminance of 1848 lx, which is certainly relatively high. Therefore, in northern regions in winter, people should nevertheless be outside or at the windows more often during breaks in working hours.

For regions in Europe such as Frankfurt, Berlin or Paris with a moderate winter with cloudy skies throughout the day, another rough calculation can be made so that a person spends about 30 minutes each on the way to work and home in the afternoon outdoors and also half an hour outdoors during the lunch break. If we assume the measured data in *Frankfurt am Main* in winter in Figure 10.15 (in Chapter 10) with a vertical illuminance of only 2000 lx, this results in an illumination of 1.5 hours × 2000 lx = 3000 lxh. The remaining exposure that should lead to a relative quality of sleep is then $H = 7394 - 3000$ lxh = 4394 lxh. Since this exposure is to be taken within the eight-hour working period, the average vertical illuminance at the eye should be $E = 4394$ lxh : 8 h = 550 lx. This means a horizontal illuminance of 1100 lx.

This estimate, based on the measurement data of Hubalek et al. and on the interpretation that office workers need about 7394 lxh to maintain a relatively good quality of sleep, also leads to a horizontal illuminance around 1100 lx and to a CS value (model version 2005 or 2018) in the order of CS = 0.406–0.429, see Table 12.11. Further literature research and conversion of the information given by the authors of this literature for their experiments in illuminance and colour temperature into CS values according to Eq. (10.9) finally lead to Table 12.12.

Table 12.12 Photometric, colorimetric, and circadian-effect parameters (CS model in version 2005 and 2018) and their value ranges for different assessment purposes.

Literature or research group	Illuminance in lx	Colour temperature in K	CS	Effect on the test subjects in the test
TU Darmstadt	1200 lx horizontal 600 lx at the eye	4800	0.428	Preferred light quality
Smolders et al. [53]	1000 lx	4000	0.469	Significant improvement in subjective alertness and heart rate
Smolders et al. [53]	250 lx at the eye	2700 K	0.16	No effect
Smolders et al. [53]	250 lx at the eye	6000 K	0.32	Slight improvement
Figuiero et al. [54]	415 lx	6000 K	0.403	Significant improvement
Current minimum values in standards	500 lx horizontal 250 lx vertical at the eye	4000 K	0.240	With high probability no effect or little effect

Source: TU Darmstadt.

To find out sustainable values for lighting design with positive non-visual effects, the authors of the present book focused on the work of two research groups:

(a) The research group of Figuiero et al. (*USA*) [54] tested the effect of light during the day in various buildings in the USA and in American embassies abroad and found a CS value of $CS \geq 0.3$ (e.g. 415 lx at the eye at 6000 K), which led to significantly better alertness. In [54] Figuiero et al. found increased sleep duration and less depression with $CS \geq 0.3$ for a home for patients with dementia and Alzheimer's disease. In another study [55] – also in a home for people with dementia – a longer sleep duration and less depression were achieved with a light intervention with $CS \geq 0.6$.

(b) The research group around Smolders et al. [53] (The Netherlands) found, among other things, by comparing 200 and 1000 lx at the eye and at 4000 K, a significant positive effect at 1000 lx on subjective alertness, vitality, heart rate and heart rate variability. In another study with two light interventions, each with 500 lx on a table (presumably approx. 250 lx at the eye) and at 2700 and 6000 K, no effects were found at 2700 K and only a smaller improvement in performance and subjective vitality at 6000 K.

Thus, through the evaluation of the literature and the authors' own research in the psychophysical context, the following conclusions emerge for the daytime use of lighting systems:

- $CS < 0.16$: no effect according to Smolders et al.; Figuiero et al. also assume a small to negligible effect at $CS = 0.15$;

- 0.3 < CS < 0.4: slight to visible improvement of the non-visual lighting effect; this also corresponds to the recommendation of Figuiero et al. with CS ≥ 0.3;
- CS > 0.4: good quality of light and significant improvement of non-visual lighting effect.

Thus, the first hints were described above for lighting design and lighting product development. The minimum value in today's standards of 500 lx (e.g. European DIN EN 12464) at the workplace (250 lx at the eye) with a typical colour temperature in today's office at 4000 K only provides CS = 0.24. With this, it is highly unlikely to trigger effects such as alertness and vitality (see Table 12.12). Further investigations are necessary to substantiate the list of planning values.

Currently, there are some circadian-effective metrics for considering non-visual effects such as D65 – equivalent illuminance according to DIN 5031-100, melanopic illuminance according to CIE and CS (CS, *model 2005 or 2018*), and there are discussions about which metric would be the right one and how to apply and interpret it. Different research and lighting design groups use different metrics. Nevertheless, in order to be able to combine the meaning of the experimental results obtained from the different publications and research reports when different metrics are specified, the authors of this book calculated these three current metrics on the basis of 553 light source spectra (fluorescent lamps, high-discharge lamps, white LEDs) at 3500 K and determined a linear relationship between each of them, see Figure 12.8. With these linear functions, one metric can be transferred into the other with usable accuracy and the research results of the international working groups can be brought onto a common basis.

As already mentioned above, Table 12.13 accumulates the knowledge presented above and proposes a draft recommendation on how a preferred and

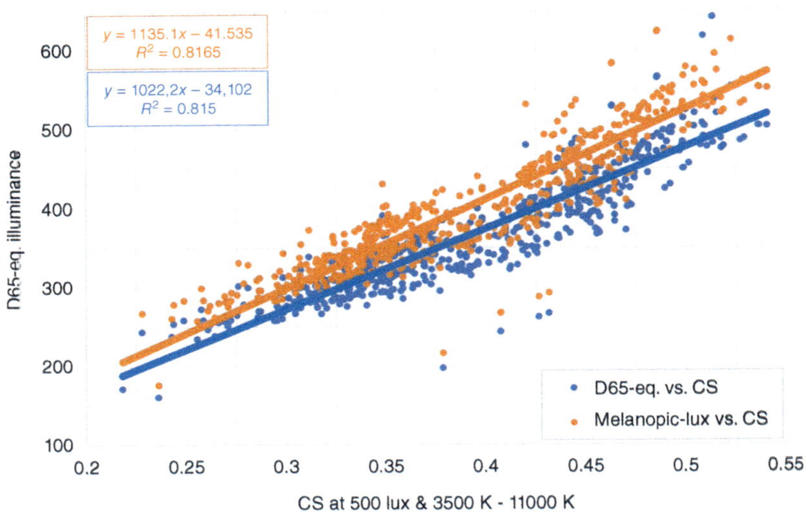

Figure 12.8 D65 equivalent illuminance or melanopic effective illuminance ([56, 57]) as a function of the circadian stimulus CS (version 2018). Source: TU Darmstadt.

Table 12.13 Draft recommendation for a preferred and circadian-effective HCL lighting system for daytime.

Assessment quantity	Recommendation of this book	Reason or comment Literature cited
Horizontal illuminance in the visual task area	1000–1250 lx	See explanations in Chapters 4, 6, 7, and 12
Vertical illuminance in the visual task area with a ratio $E_v : E_h = 1 : 2$	500–625 lx	
Circadian Stimulus CS (vertical, model version 2005 or 2018)	0.3 < CS < 0.4 for cases with acceptance; CS > 0.4 for good non-visual effects	
D65-equivalent illuminance (vertical)	250 lx < $E_{D65,eq}$ <375 lx for cases with acceptance; >375 lx for good non-visual effects	
Discomfort glare UGR	≤19	DIN EN 12464-1
Light colour, CCT	4000 < CCT < 5800 K 4300–5000 K preferred	Chapter 6
Distance from the Planckian locus or daylight locus, $\Delta u'v'$	−0.003	Below the curve
Colour-rendering index CIE CRI R_a	At least 80 Preferred >86	Chapter 6
Modelling	Between 0.3 and 0.6	
Luminance distribution within the 40° field of view	Max : Min = 20 : 1	Literature [16]
Luminance of the wall	>100 cd m^{-2}	Chapters 6 and 9
Ceiling luminance	>200 cd m^{-2}	Chapter 9
Uniformity of illuminance	$U_0 > 0.6$	DIN EN 12464-1
Indirect proportion of total horizontal illuminance on the work table	>60%	Chapter 9
Frequency of the PWM light	Minimum value: 400 Hz Preferred >1000 Hz	Chapter 9
Daylight coefficient	>3%	—
Window proportion, related to the façade area (for new construction or after renovation)	Good: >40% Very good: >60%	—

Source: TU Darmstadt.

circadian-effective HCL lighting system should be planned in terms of lighting technology. This knowledge was derived from the results in this book from Chapters 4–12 as well as from literature studies. Table 12.13 applies to workrooms and rather to younger (or middle-aged) room users, up to about 50 years of age.

As a supplement and as a comparison, recommendations of the SHC study of the International Energy Agency [21] for a lighting installation for good lighting quality are shown in Table 12.14 in its original form.

12.5 Dynamic Lighting, Control Curves

The previous lighting concept according to international or national standards (e.g. DIN EN 12464) with its focus on the most important lighting parameters such as illuminance at the working plane, glare (e.g. UGR model), homogeneity U_0, light colour, colour rendering and modelling has been extended in the course of the previous discussion on lighting quality to include the colour quality (colour gamut, a moderate increase in saturation, see Chapter 6), luminance distribution in the 40° field of vision, i.e. a moderate, conscious inhomogeneity in the scene to increase attraction and perceived brightness. Visual clarity (see Chapter 5) and the non-visual lighting quality component have also been optimised. In recent literature, non-visual components have been linked either to the recommendation of achieving the vertical D65- equivalent illuminance of 240 lx (or 250 lx after Brown et al. [51] in 2022), or, according to the recommendation of Figuiero et al. [54], to CS >0.3 or CS >0.35.

How high HCL lighting quality can be achieved is summarised as follows:

- With an increase in horizontal illuminance from a minimum value of 500 lx (as specified by the standards) to about 1000–1250 lx and the associated increase in vertical illuminance at the eye;
- By increasing the indirect luminous flux components rather on the ceiling and partly on the upper half of the walls to achieve:
  (a) a higher vertical illuminance at the eye;
  (b) increased wall luminance in the observer's field of view;
  (c) lower glare according to the UGR model due to increased ceiling luminance;
  (d) a better *modelling factor*;
  (e) more circadian-effective light at the lower half of the retina, where most ipRGCs are located (see Figure 10.3);
  (f) a better spatial room-architecture perception.
- With an increased awareness of the use of daylight with higher daylight coefficients (higher than 3%, preferred 5%), with a higher proportion of windows in the room, with the arrangement of workplaces near windows and a better view to the outside.

It is important to emphasise that an increase in vertical illuminance at the eye and an increase in luminance in the field of vision (e.g. on the wall), *without* synchronously increasing the *glare* of light-emitting luminaire surfaces, is only possible

Table 12.14 Recommendations of the SHC study [21] for a lighting installation for good lighting quality, in original form.

	Daylight		Electric light	
Parameter	**Measure**	**Standard value**	**Measure**	**Standard value**
Workplace illuminance General	Target illuminance of daylight provision from windows	≥300 *lx* on the working place level ≥50% of the yearly daylight hours ≥50% of the space area	Mean E_h on the desk	Together with daylight ≥500 *lx*
Workplace illuminance General	Spaces with skylights	as for windows but ≥95% of the space area	—	—
Workplace illuminance Visual demanding	daylight provision from windows	≥750 *lx* on the desk ≥50% of the yearly daylight hours	Mean E_h on the desk	1000 lx
Workplace illuminance homogeneity	Minimum Target illuminance of Daylight provision from windows	≥100 *lx* on the working level in room ≥50% of the yearly daylight hours ≥95% of the space area	Uniformity U_o ($E_{min} : E_{mean}$) on the desk	≥0.6
Workplace illuminance homogeneity Visual demanding	Minimum Target illuminance of Daylight provision from windows	≥200 *lx* on the working level in room ≥50% of the yearly daylight hours ≥95% of the space area	Uniformity U_o ($E_{min} : E_{mean}$) on the desk	≥0.7
Movement area illuminance	No measure	Low daylight illuminance is accepted	Mean E_h	200 lx
Movement area illuminance homogeneity	No measure	Low daylight illuminance homogeneity is accepted	U_o ($E_{min} : E_{mean}$)	0.4
Colour of light in general	No measure	Natural variation of colour of daylight is appreciated	CCT (K)	3000–5000 K
Colour of light Wards	No measure	blue and green tint of window glass should be avoided	CCT (K)	$4000 \leq CCT \leq 5000$
Workplace colour rendition	Colour rendering for window glass	hue shift ≤30 saturation shift ≤30	CIE CRI (additional R_{12})	≥80

Table 12.14 (Continued)

	Daylight		Electric light	
Parameter	**Measure**	**Standard value**	**Measure**	**Standard value**
Workplace colour rendition Colour quality demanding	Colour rendering for window glass	hue shift ≤10 saturation shift ≤20	CIE CRI (additional R_{12})	≥90
View out to the outside from the workplace	Width of the view	>14°		
Length of the view	>6*m*			
Number of view layers (ground, landscape, sky)	Minimum landscape layer is visible			
Glare	Daylight glare probability DGP e <5%	<0.45 besides 5% of the occupation time	UGR	≤19
Luminance in the visual field	Max. luminance of the window surface for workplaces in the neighbourhood of window	≤4000 cd m^{-2}	Max. luminance at gamma >60°	≤3000 cd m^{-2}
Repeating luminance contrasts	Spatial frequency on the window surface	Avoid strong luminance contrast in the medium frequencies (0.2–0.5)	Multi shadows from lamps Mean illuminance at the surface level	≥100 *lx*
Homogeneity of the light-emitting surface	$L = (L(90\%): L(10\%)/L\text{everage}$			
Non-visual effects of light (daytime work)	Daylight is recommended wherever possible	Circadian stimulus (CS)	≥0.3 (Practically ≥0.4) throughout the day	
Non-visual effects of light (shift work)			Circadian Stimulus (CS)	$CS \geq 0.3$ throughout the first part of the shift (until about 05:00 a.m.), and then shielding themselves from exposure to circadian-effective light until they are home and in their darkened bedroom for sleep

Source: Adapted from SHC [21].

by means of a targeted, homogeneous distribution of the indirect luminous flux component to the ceiling and to the upper half of the walls.

In addition to these setting options, with the physical criteria intensity (luminous flux, illuminance), spectra (light colour, colour quality) and light distribution (indirect/direct, diffuse, focussed), the *time factor* plays a *major role* in the context of HCL-oriented lighting, which currently determines the discussions in lighting research as well as the strategies of luminaire development. The following aspects belong to the concept of the '*time factor*':

(a) The time or '*timing*' when the illuminance and spectra (colour temperatures) should be adjusted. From the point of view of considering non-visual effects, the effect of light is highest in the early morning ('*morning light*') as well as in the late evening and at night (see Chapters 3 and 9). For the criteria visual performance/work performance, and the psychological factors, the planned values should be set and maintained, at all times during the work process.
(b) The *duration of* each light setting (illuminance and/or colour temperature) should be effective. This question can only be answered under the concrete working conditions of the light users, e.g. according to the actual duration of the breakfast break in the morning, the duration of lunch or the afternoon break, the duration of an event such as a meeting or a production check, or the end of working hours.
(c) The *temporal history* ('*pre-exposure*'): For the early shift workers who start their shift at about 5:30 or 6:00 a.m., or for the office workers with office hours starting at 7:00 or 7:30 a.m., the way to work in winter as well as after office hours after 5 p.m., for example, tends to be dark or with little light. In summer, the situation is different. Therefore, the lighting in the starting hour and at the end of work can be designed according to the season and specific to the workplace.

After a detailed evaluation of the measurement data in the outdoor area depending on the time of day, it can be assumed of natural light (daylight) that the colour temperature does not change much on a cloud-free summer day or on a very cloudy autumn or winter day, except for the period duration of sunrise and sunset. The dynamics of daylight are based to a considerable extent on the change in illuminance on a horizontal or vertical plane (see Figure 10.24 for a summer day and Figure 12.9 for a winter day). Changes in illuminance, therefore, have a greater and demonstrable influence on the non-visual conditions of light users, whereas changes in colour temperature have a great psychological-emotional significance, see Chapter 9.

In parallel with the analysis of previous research results in the literature on the effect of dynamic light on sleep behaviour, phase shift or wakefulness, which was analysed in Chapter 9, the present section discusses and comments on the current dynamic *light control curves* that have been applied or recommended by some lighting companies and research groups.

The publication [24] shows some *control curves* for work in offices, in industry as well as in schools. For each of these applications with different starting times and working durations, there are different control curves with some common features, see Figure 12.10.

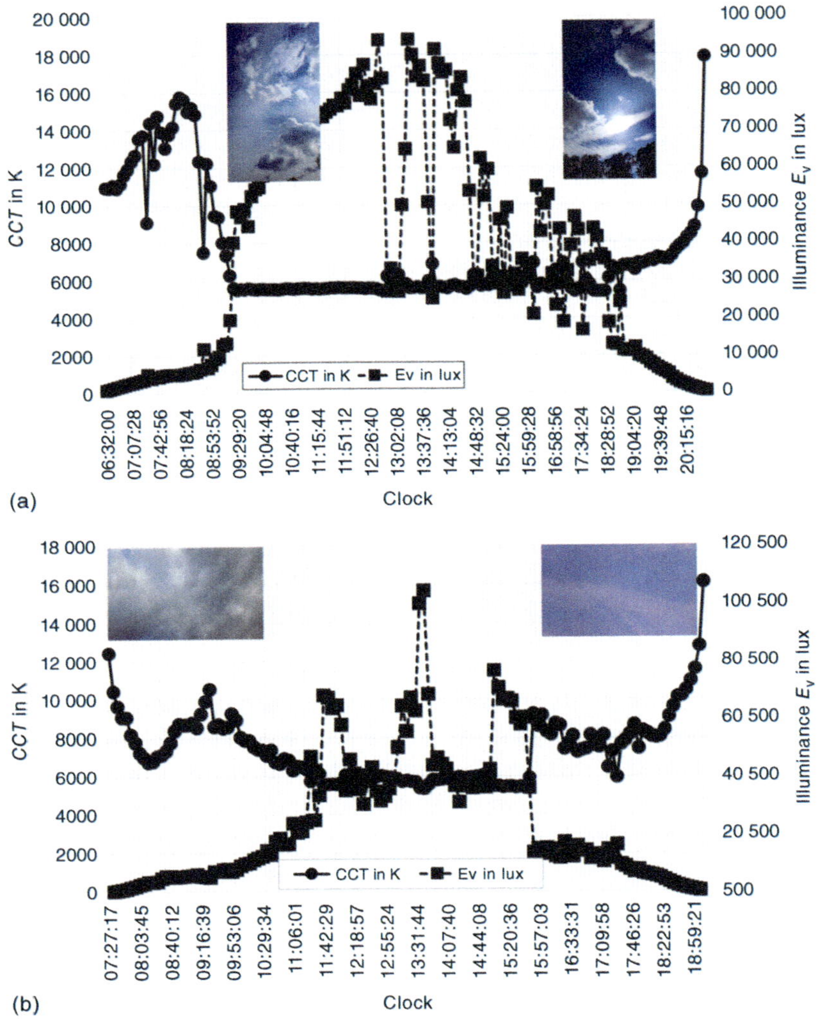

Figure 12.9 Daylight distribution of illuminance and colour temperature on a sunny winter day (left) and a cloudy winter day (right) in Darmstadt, Germany. Source: Technische Universität Darmstadt.

The following can be seen from Figure 12.10 based on the control curves shown for the variation of illuminance:

- For industry, the early shift begins at 5:30 or 6:00 a.m. with half an hour of preparation time, starting with an illuminance at about 40–50% of the maximum. After this half hour, the lighting system is set to 100% until 9 a.m. The core phase with 100% light lasts until 8 p.m. for industry with the late shift. After that, the light level drops linearly to 0% at 10:30 p.m. The end of the shift usually takes place 30 minutes before.

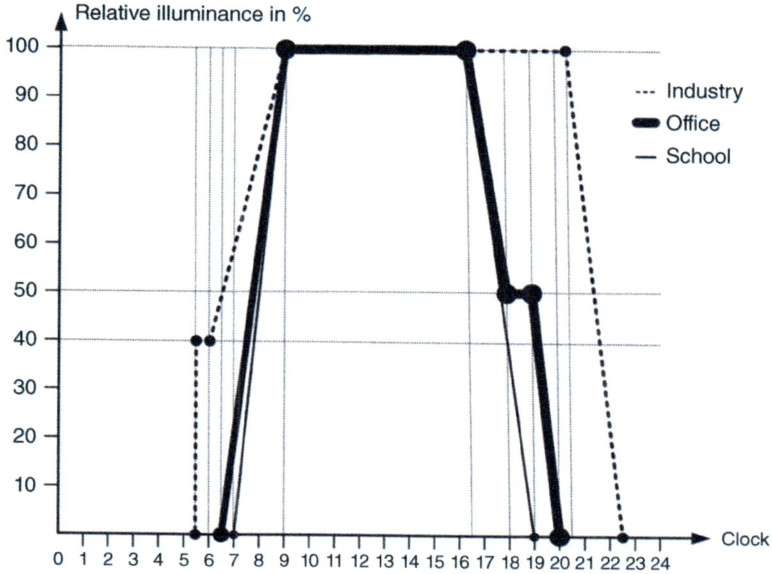

Figure 12.10 Exemplary light control curves [24] for applications in offices, schools, and industry. Source: Adapted from Fördergemeinschaft Gutes Licht [24].

- The control curves for school and office are similar. For these two applications, the 100% light level remains constant from 9:00 a.m. to 4:30 p.m. The end of this core phase can be set flexibly from office to office or from classroom to classroom depending on the workload during these hours at the end of the afternoon.
- The phases of light increase in the early morning and in the afternoon and tend to last longer over several hours. For the phase in the morning, it makes more sense to quickly reach the maximum light level. For this reason, the authors of the present book developed some suggestions for improvement and present them in Figure 12.11.

The difference between the control curves in Figure 12.11 and in Figure 12.10 is marked by the faster change in illuminance in Figure 12.11. The control curves in Figure 12.11 can be characterised as follows:

- The maximum (100% light) is at 1000–1250 lx horizontal illuminance, see Table 12.13. Half of this (in the 50%-light phase) is 500–625 lx according to the current standard for indoor lighting and this guarantees good visual performance in the preparation time.
- Despite the apparently rapid change in light, experience has shown that the setting speed for the lighting system is less than 1 lux per second [20] and less than 300 K per 90 seconds for colour temperature change [58] should be maintained. This means that the change in light is not detected by the light users and can therefore be readily accepted. For a change from 50% to 100% of the maximum, i.e. from 500 to 1000 lx with a difference of 500 lx, a light setting duration of about 20–25 minutes is optimal.

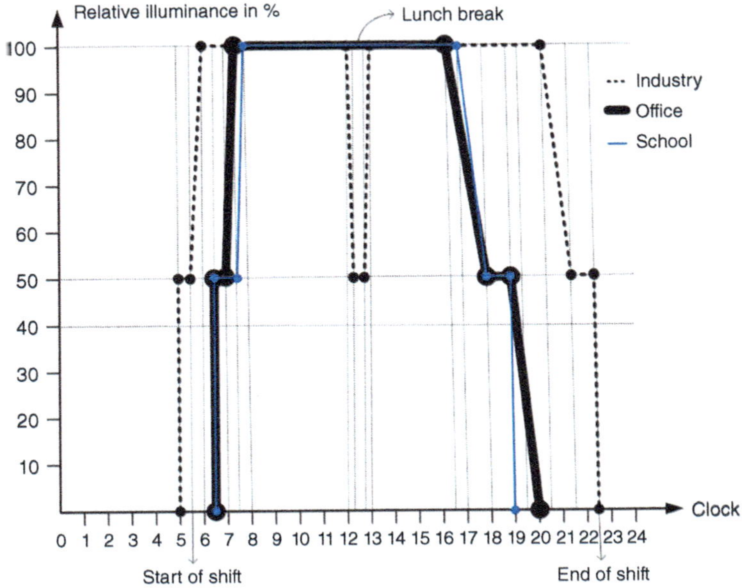

Figure 12.11 Control curves, proposed by the authors of the present book, for applications in offices, schools, and industry. Source: TU Darmstadt.

- For the lunch break of about one hour, the light level is reduced to 50% of the maximum to foster relaxation.
- For the offices and schools, the phases with 100% light level (1000–1250 lx) with a colour temperature approximately between 5000 and 5500 K and the phases with 50% light level with 4000 K were proposed. This results in Table 12.15 with the CS-values.
- For industrial applications, the values for illuminance and colour temperature up to 4:30 p.m. and 5:00 p.m., respectively, are roughly the same as for offices and schools. Therefore, Table 12.15 also applies. For the late shift around 5:00 p.m. to

Table 12.15 Circadian Stimulus (CS, model version 2018) values and D65 -equivalent illuminance at the eye when realising the control curves in Figure 12.11 for offices and schools.

100% light-level phase	Horizontal illuminance = 1000 lx. Vertical illuminance at the eye: approx. 500 lx; colour temperature CCT = 5250 K	CS = 0.408 $E_{D65,\,eq.}$ = 383 lx
50% light-level phase	Horizontal illuminance = 500 lx. Vertical illuminance at the eye: approx. 250 lx; colour temperature CCT = 4000 K	CS = 0.238 $E_{D65,\,eq.}$ = 209 lx

Source: TU Darmstadt.

Table 12.16 Circadian Stimulus (CS) values and D65-equivalent illuminance at the eye when realising the control curves in Figure 12.11 for industrial applications.

100% light-level phase until 5 p.m.	Horizontal illuminance = 1000 lx Vertical illuminance at the eye: approx. 500 lx Colour temperature CCT = 5250 K	CS = 0.408 $E_{D65,\ eq.} = 383$ lx
100% light-level phase 5 p.m. until 8:30 p.m.	Horizontal illuminance = 1000 lx Vertical illuminance at the eye: approx. 500 lx Colour temperature CCT = 4000 K	CS = 0.353 $E_{D65,\ eq.} = 326$ lx
50% light-level phase	Horizontal illuminance = 500 lx Vertical illuminance at the eye: approx. 250 lx Colour temperature CCT = 4000 K	CS = 0.238 $E_{D65,\ eq.} = 209$ lx

Source: TU Darmstadt.

8:30 p.m., colour temperature can be reduced to 4000 K. Table 12.16 shows these CS and D65-equivalent illuminance values.

Worth mentioning in Figure 12.11 is the phase from 8:30 p.m. to 10:30 p.m. with CCT = 4000 K and with a steadily decreasing illuminance, so that the CS values are reduced from 0.353 starting at 8:30 p.m. to 0.238 at 9:30 p.m. The value CS = 0.238 remains for the last hour before the end of the shift, on the one hand, to keep *sleep quality* relatively stable in the following night, and, on the other hand, to enable good *work performance* in the last two hours before the end of the shift.

Figure 12.12 shows a control curve already used by a lighting company for nursing homes and homes for the elderly. Here, too, the phase from 6:00 a.m. to 9:00 a.m. was realised with *cool white* light up to 70% of the maximum light level. The phase thereafter, until 12 noon, has neutral white mixed light. It is interesting to consider a relaxation phase from about 12:00–3 p.m. with *warm white* light and with a 50% light level between 1:00 and 2:00 p.m. The phase in the afternoon between 3 and 4 p.m. is designed with 100% light and with *neutral white* light. After dinner and after a body hygiene phase, the *warm white* light is steadily reduced from 7:00 p.m. This resting phase from 7 p.m. has several smaller stages, which can certainly be programmed individually after consultation with the residents.

12.6 Lighting for Users with Higher Lighting Requirements

Demographic development or the ageing of the population is progressing strongly in many countries, with two main consequences. On the one hand, the age of people

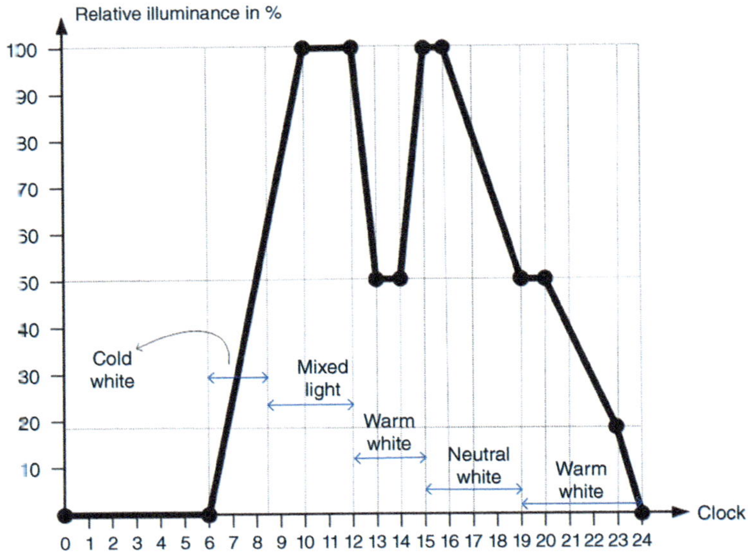

Figure 12.12 Control curve of a lighting system in a German nursing home. Source: TU Darmstadt, based on data from the company Waldmann (Germany), with permission from Waldmann.

who are still actively participating in working life is increasing strongly and leads to the fact that the working conditions for such groups of people should be designed in an age-appropriate way, whereby the lighting condition for receiving and processing the optical information from the working environment plays a major role. On the other hand, life expectancy is increasing, as is the number of people living as inpatients in elderly people's homes and nursing homes, as well as older people who need outpatient medical care in their own homes until old age. In both cases, the questions for human-centred lighting technology are as follows:

(a) How can the lighting conditions in workspaces for older professionals contribute to vitality, work productivity, well-being, sleep quality, and concentration?
(b) How can the lighting system in the living areas for elderly people as well as for nursing staff contribute to reducing sleep deficits, dementia, depression, risk of falls, consumption of medication for the residents, and stress, tension, and accidents at work for the medical staff as far as possible?

In this section, the visual problems in old age are presented, a new consideration for the partial compensation of spectral losses in old age by age-dependent colour mixtures in a multichannel LED luminaire system is described, and some aspects for the design of better visual conditions in a nursing home are outlined.

12.6.1 Vision in Old Age – Some Aspects

With age, several problems occur that are optical, neurobiological, and chemical, and result in visual losses. The main photometric features are explained below:

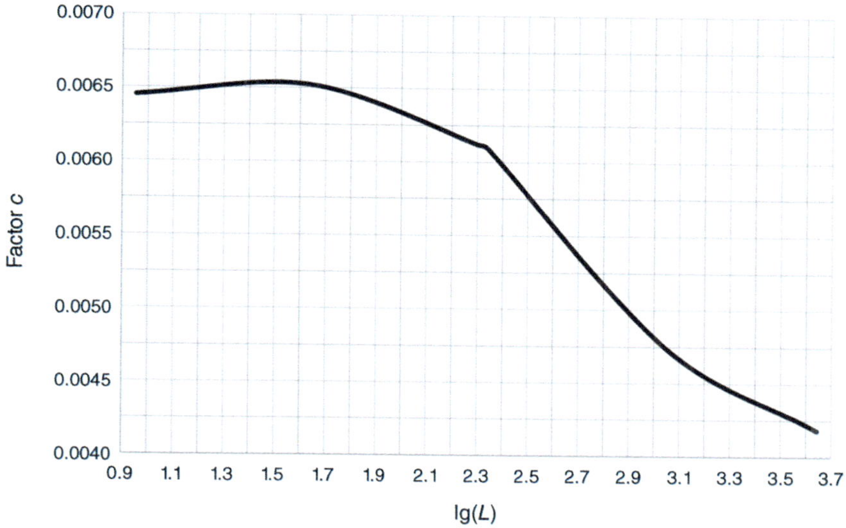

Figure 12.13 Luminance-dependent factor $c(L)$ for the eye ageing model for the luminance range 10–1000 cd m^{-2}. Source: Adapted from DIN SPEC [50].

Table 12.17 Factor $k_{\text{pupil}}(A)$ as a function of age.

Age	$A = 25$	$A = 32$	$A = 50$	$A = 75$	$A = 90$
$k_{\text{pupil}}(A)$	1.088	1.000	0.792	0.543	0.416

Source: Adapted from DIN SPEC [50].

- A smaller pupil area with age, which means a lower radiant flux or a lower luminous flux. In the European standard DIN SPEC 5031-100 [50], a correction factor for age-related pupil reduction is introduced. It is a ratio of the pupil area of persons with age A to that of 32-year-old persons, calculated with Eq. (12.1).

$$k_{\text{pupil}}(A) = [1 - c \cdot (A - 32)]^2 \tag{12.1}$$

  In Eq. (12.1), c is a luminance-dependent factor, see Figure 12.13. The symbol A means age in years. Table 12.17 lists the values of the factor $k_{\text{pupil}}(A)$ as a function of age. This shows that the active pupil area in the age range between 75 and 90 years is only about half the area of a 32-year-old person.
- A reduction in the spectral transmission of the eye lens. Figure 12.15 shows the spectral transmission curves of the eye, as a function of age [50]. It is clearly visible that the spectral transmission in the blue wavelength range approximately before 580 nm experiences a much greater loss than that in the wavelength range from 600 nm. It, therefore, leads to a yellowing of the eye lens and to a strong reduction of the circadian effective irradiance, because the spectral sensitivity of the blue cones, the rods and the ipRGCs is relatively high in the wavelength range up to 520 nm. After [50], the ratio of the spectrally weighted transmittance of

the ocular media of persons of age A to that of 32 years old persons is calculated with Eqs. (12.2)–(12.4). The optical density $D(\lambda, A)$ of the ocular media is shown in Figure 12.14 and the spectral transmittance of the ocular media is shown in Figure 12.15.

$$k_{\text{eye,trans}}(A) = \frac{\int_{\lambda_{380}}^{\lambda_{780}} S_\lambda \cdot \frac{\tau(\lambda,A)}{\tau(\lambda,A=32)} \cdot V(\lambda) \cdot d\lambda}{\int_{\lambda_{380}}^{\lambda_{780}} S_\lambda \cdot V(\lambda) \cdot d\lambda} \tag{12.2}$$

$$\tau(\lambda, A) = 10^{(-D(\lambda,A))} \tag{12.3}$$

$$\begin{aligned} D(\lambda, A) = {} & 0.11 + (0.225 + 0.000031 \cdot A^2)(400/\lambda)^4 \\ & + \exp(-0.003249 \cdot (\lambda - 273)^2)(151.5492) \\ & + \exp(-0.000841 \cdot (\lambda - 370)^2)(2.12574 - 0.00013419 \cdot A^2) \\ & + \exp(-0.000441 \cdot (\lambda - 325)^2)(0.70505 + 0.00222270 \cdot A^2) \\ & + \exp(-0.000064 \cdot (\lambda - 325)^2)(0.02288 - 0.00018876 \cdot A^2) \end{aligned} \tag{12.4}$$

Symbols have the following meanings in Eqs. (12.2)–(12.4):
- $\tau(\lambda, A)$: Spectral transmittance of the ocular media for age A, see Figure 12.15;
- S_λ: Spectrum of the light source;
- $V(\lambda)$: Luminous efficiency function for daytime vision and for a 2° field of view;
- $D(\lambda, A)$: Optical density of the ocular media, see Figure 12.14.

- Equations (12.2)–(12.4) result in the ageing factor $K_{\text{eye}}(A)$ for the correction of the radiation magnitude in the observation of people at age A, taking into account the age-dependent transmission change and pupil reduction, which is calculated

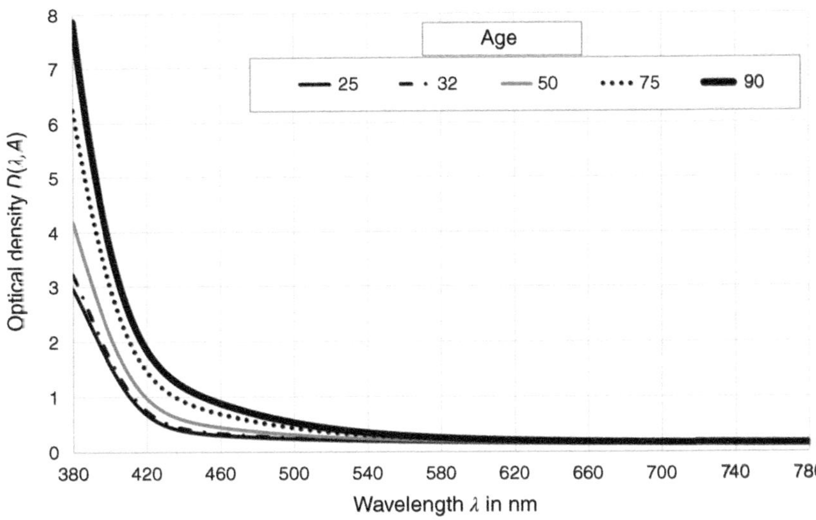

Figure 12.14 Optical density of ocular media $D(\lambda, A)$, as a function of wavelength λ (in nm) and age A (25–90, see legend). Source: Adapted from DIN SPEC [50].

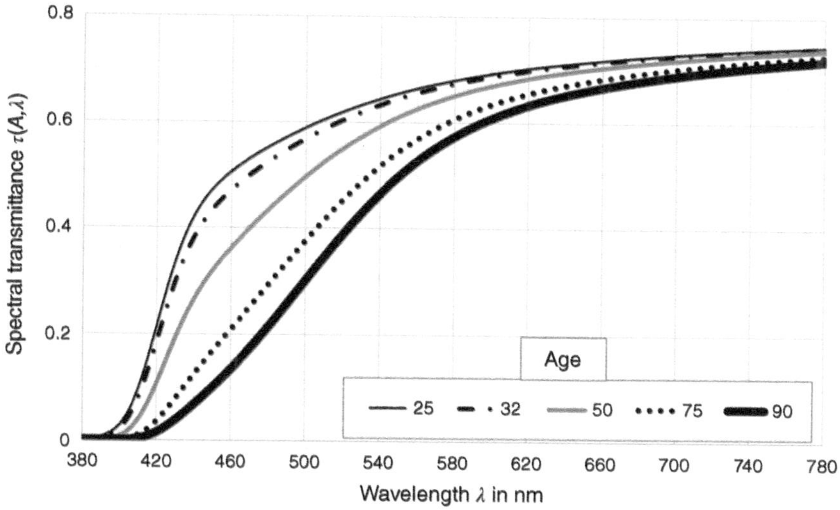

Figure 12.15 Spectral transmittance $\tau(\lambda, A)$ of ocular media as a function of age A (25–90, see legend). Source: Adapted from DIN SPEC [50].

according to Eq. (12.5), with $k_{\text{pupil}}(A)$ from Table 12.17.

$$K_{\text{eye}}(A) = k_{\text{eye,trans}}(A) \cdot k_{\text{pupil}}(A) \tag{12.5}$$

- The sharp decrease in *visual acuity* (*VA*) with age. The retinal receptors in the fovea are responsible for visual acuity or the recognition or identification of objects, e.g. reading documents, and recognising signs or shapes. The decrease (VA_{rel}) can be calculated, for the contrast value of e.g. $C = 0.9$ according to Adrian [59], see Eq. (12.6).

$$\text{VA}_{\text{rel}} = 0.556 \cdot \cos(0.0745 \cdot A^{1.606}) + 0.4597 \tag{12.6}$$

- An increase in disability glare with age, caused by the opacity and light scattering of the crystalline lens and vitreous humour as well as the change in structure on the retina. The increasing scattered light is distributed relatively homogeneously like a *veil layer* over the entire retina and thus reduces the object contrast and consequently the object detection. This so-called *equivalent veiling luminance* $L_{\text{S,eq}}$ is calculated according to the original formula of Holladay [60] and with some factor corrections, see Eq. (12.7).

$$L_{\text{S,eq}} = k \cdot \frac{E_{\text{glare}}}{\theta^n}; k = 9.05\left(1 + \left(\frac{\text{Age}}{66.4}\right)^4\right); n = 2 \tag{12.7}$$

In Eq. (12.7), the symbols have the following meaning:

- E_{glare}: Illuminance at the eye caused by the glare source;
- θ: angle at which the glare source is viewed;
- k: factor for the age dependence of the glare.

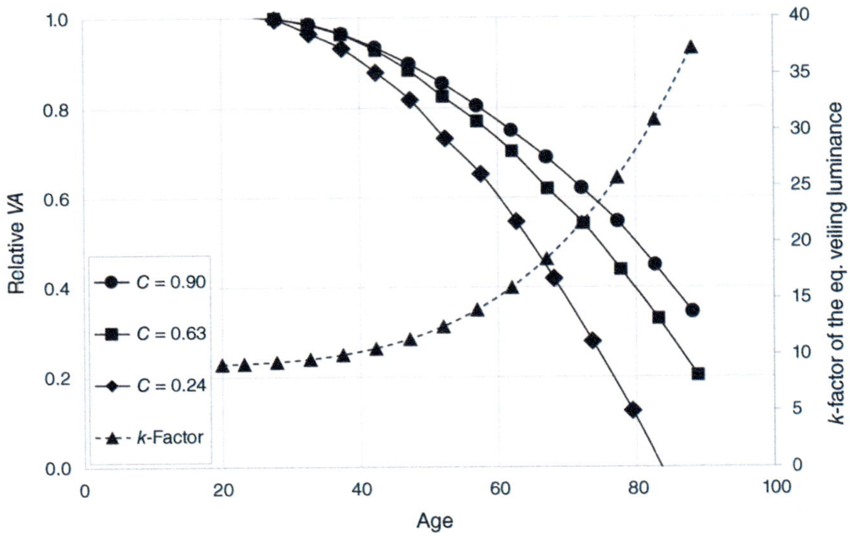

Figure 12.16 Relative visual acuity VA_{rel} (Eq. (12.6)) and ageing factor k for glare (Eq. (12.7)), as a function of age. Source: Holladay [60] and Schweitzer et al. [61].

Figure 12.16 shows three curves for the decrease in visual acuity caused by *glare* depending on age with three different contrast values C as a parameter. The factor k of the age dependence of glare is also shown in Figure 12.16. The simultaneous depiction of the decrease in visual acuity and the increase in disability glare with age in Figure 12.16 is intended to point out the following: If we look at these two dependencies and compare the relative change between 20 and 80 years of age, we learn that visual acuity is reduced by a factor of 2.5. People actually need *much more* light to compensate for the tendency to decrease. At the same time, however, glare increases by a factor of 3.1, so that the necessary increase in the level of light also leads to severe glare. Therefore, in addition to the purely physiological consideration of the age dependency of visual acuity, glare, pupil constriction, and the reduction of spectral transmission of the ocular media (predominantly in the blue range), it is also very important to record which light level the different age groups subjectively favour at their workplace over longer periods of use of the lighting systems.

According to Figure 12.16, for an 80-year-old person and a contrast value of $c = 0.63$, the relative visual acuity is only about 40% of the relative value that a 20-year-old person.

It should be emphasized that in the comprehensive field study by Moosmann [31] (already discussed above), the lighting situations where lighting users expressed no desire to change the lighting level were clearly influenced by *age*. The horizontal illuminance at the workplace where lighting users made no *change requests* was 1168 lx for people under 30 years of age, 1006 lx between 30 and 39 years of age, 883 lx between 40 and 49 years of age and 834 lx for people over 50 years of age. These results were statistically significantly different.

The above results, as well as the results of another study [61] with 80 test subjects, showed that *younger* test persons favour a higher illuminance compared to the older group (in [61] up to the age group around 80). Roughly calculated for both extensive studies ([31, 61]), the test groups over 50 years of age choose about 30% less illuminance than the group around 20 years of age, most likely in order to *reduce* glare. These results should be taken into account when dimensioning lighting systems in elderly people's homes and nursing homes or at the workplace for elderly professionals.

If one analyses the test results of Lindner's study described in Section 4.3.3 with 48 test persons in the age group 20–40 years and with 48 test persons with 50-years, one comes to the conclusion according to Figure 4.14 that the performance of the younger group is clearly higher than the performance of the older group, which caused more errors than the younger age group. Even with very high illuminance up to 2600 lx, one cannot eliminate these differences in performance. We can state that the older age group has a reduced performance under certain circumstances and favours a lower illuminance to avoid glare. An HCL lighting system for elderly people at the workplace or in a retirement home should therefore be designed with more planning care and more effort. Some research results as well as some planning aspects will therefore be described.

12.6.2 Lighting for Elderly People's Homes and People Suffering from Dementia

Due to the advancing demographic development on a global scale, lighting technology for elderly people's homes, nursing homes, and residential homes for elderly people in the outpatient sector as well as for people suffering from dementia is becoming increasingly important. Some main characteristics of these persons are:

- *No distinct sleep–wake rhythm* (frequent tiredness during the day, poor sleep at night);
- *The eye reacts more slowly* with much less optical information from the surroundings;
- Differences in brightness unsettle perception and movement safety;
- Shiny spots and surface areas, dark spots and shadows cause anxiety;
- Dream images and misinterpretations lead to wrong decisions and actions.

The results of this are:

- Risk of falling;
- No vitality;
- Physical inactivity;
- Increased care requirements for the medical staff, more medication consumption;
- More costs for care, treatment, and medication.

The first relevant research articles on the effect of light on elderly people were written at the end of the 1990s, after the knowledge of the authors of this present book. In [62] and in [63], experiments were conducted in the range of 3000–8000 lx on the eye for at least one hour *in the morning* over a period of more than two weeks for the stabilisation of nocturnal sleep and the sleep–wake rhythm for persons with Alzheimer's disease and dementia (**ADRD** = Alzheimer's Disease and Related Dementias).

At the same time, around 1997, Van Sommeren's research group [64] conducted experiments in a communal room where 22 elderly people with partial ADRD were illuminated with ceiling lights with fluorescent lamps and with high light intensity. The illuminance was 1136 ± 89 lx. The administration of the high amount of exposure resulted in an improvement of the rest–activity rhythm, especially in subjects with defective ocular apparatus. The result was expressed in the *consistency and improvement of harmonisation* (so-called '*clocking-in*') of one's rest-activity rhythm compared with *natural timing*.

In a study in 2014, the research group of Figuiero et al. [55] tested the administration of light with an illuminance of around 400 lx with a high colour temperature and a high blue light component and was able to increase the quality of sleep and reduce depression in the homes for ADRD patients. According to the calculation of the authors of this book, an illuminance of 400 lx at a colour temperature of around 6000 K leads to a CS value of about 0.4. In another study [54], Figuiero et al. experimented with a CS value greater than 0.3 in the daytime and less than 0.1 in the evening and found a reduction in depression symptoms and increased *sleep time* among ADRD patients. In [65], Figuiero et al. also found better sleep quality and mood in ADRD patients who experienced lighting with CS higher than 0.6.

For a new LED lighting system in a retirement home in Neubrandenburg (Germany), dynamic lighting was programmed according to Figure 12.12 for the variation for illuminance and to achieve dynamic colour temperatures throughout the day. The new system was installed in the living area (in the resident's room, bathroom, and toilet) and in the common rooms (fireplace room, lounge, dining room, dementia laboratory, and therapy room) with a pre-programmed lighting management system [66], see Figure 12.17. The study was scientifically accompanied in 2019 by an Institute of Psychology with the aim of evaluating the non-visual effects of the biodynamic light management system on persons with neurodegenerative diseases by collecting data over a period of three months in the test facility in Neubrandenburg compared to a control group without a light management system (see also [67]). The most important results [66] are shown in Table 12.18.

Two important results [66] could be quantified in the comparison between the test facility and a control group with comparable demographic characteristics in this typical retirement home:

- The test facility with light intervention delivered an average cortisol value of 17.7 $pg\,ml^{-1}$ compared to the control group with 5.7 $pg\,ml^{-1}$. Since the '*stress hormone*' cortisol is a characteristic value of the alertness and activity of the test participants, the use of light intervention in the described scheme makes sense.
- Through increased illuminance and the resulting visibility in the traffic zone, in the bathroom as well as in the living area, through the increased use of indirect diffuse light components (see Figure 12.17) to reduce shadows and inhomogeneity, the *fall rate* could be greatly reduced. The fall rate in the test facility was 1.5%.

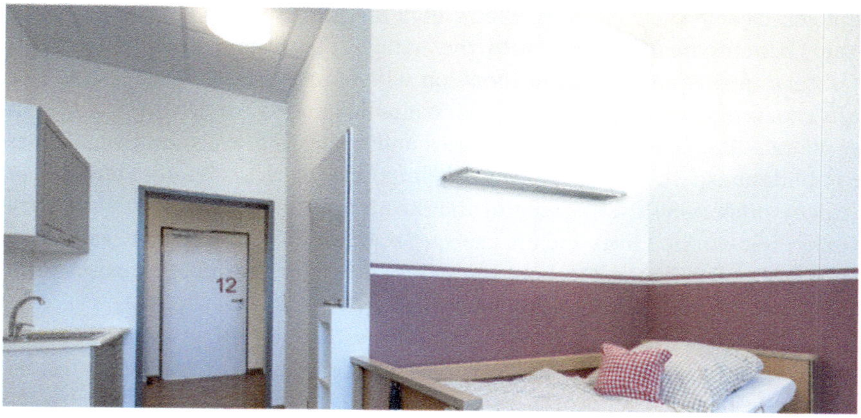

Figure 12.17 A living room in a retirement home in Neubrandenburg, Germany. Source: Reproduced with permission from Ref. [66], Herbert Waldmann GmbH & Co. KG.

Table 12.18 Summarised results from the lighting study in a retirement home in Neubrandenburg 2019.

Effects on the residents	– Cortisol level double to triple higher and comparable with the cortisol level of 17–20 years old persons – Significant reduction of daytime sleepiness; instead of drowsiness, subjects feel energetic – Emotional stabilising effect through the lighting environmentassociated with less nervousness with higher light quality – Significantly higher feeling of safety when walking or driving with the wheelchair and less fear of falls – Light perception and lighting intensity at much higherrates – Higher level of alertness; more comfortable atmosphere
Positive effects also for the staff	– Increased level of alertness, attention and activity – Staff also reported a stabilising effect on the sleep–wake rhythm – Adaptation of lighting technology is a fundamental upgrading of the workplace – On an individual level, it corresponds to an increased quality of life and quality of work

Source: Adapted from Waldmann [66] https://www.waldmann.com/last accessed January 19, 2023.

In the expert standard '*Fall prophylaxis in nursing care*', the fall rate for inpatient care for the elderly is given as approx. 45% and for outpatient care as 35% [66].

12.6.3 Proposal for Lighting Design for Elderly People's Homes and Nursing Homes

The above-mentioned test results from research institutions over a longer period of time in elderly people's homes and in the homes of elderly people with ADRD symptoms clearly show positive effects in terms of stabilisation of sleep–wake rhythm, better synchronisation with the natural timer, more mobility, activity, better sleep quality and reduced depression with appropriate lighting design. The results can be of great humane as well as economic importance by reducing the fall rate, the rate of collapse, nursing expenses and medication consumption. That is one important aspect. The second, equally important aspect is lighting technology for *elderly professionals*, especially in the 50–67 age range, as they are still very active in their working lives and are expected to cope with daily challenges. This is the primary goal of lighting technology in this case, which should be realised with affordable and technologically feasible means.

In lighting science and in luminaire technology, *LED technology* currently offers many theoretical possibilities, but these must also be weighed up economically and financially. In the early years of the HCL discussion, and to some extent even today, HCL was associated with *luminaire technology with variable colour temperatures*, so-called '*tuneable white technology*'. This, however, does not represent the entire range of possible solutions of today's lighting technology for the circadian synchronisation of the human biological rhythm and for the expression of the wake–rest rhythm with a high amplitude. From a technological point of view, several luminaire configurations are imaginable for this purpose (see also Section 11.4.1 of this book):

- The configuration with four LED channels like RGB LEDs and with warm white LEDs: This is a good lighting variant, but from today's point of view, it is cost- and effort-intensive. For a fast and widespread HCL lighting technology, a more cost-effective luminaire configuration is needed.
- A variant with warm white and cool white LEDs with dimmable luminous flux and variable colour temperature is already a frequently tested solution. The disadvantage of this variant is that the radiation component around 470–510 nm cannot be varied. Since elderly people from about 50 years of age up to an advanced age of about 85–90 years lose a lot of eye lens transmission in the blue wavelength range up to 450–510 nm, the compensation in this wavelength range by this luminaire variant is not effective.
- The variant with a warm white, a neutral (around 4200 K) or cold white LED group (around 6000 K) and with a coloured LED group with a peak wavelength around 470–480 nm to vary circadian effectiveness systematically. Such a luminaire was optimised by the authors of the present book and provided the spectra for CCTs between 3000 and 6000 K in Figure 12.18.

Figure 12.18 shows highly adjustable radiation components that lie around 470–500 nm. Table 12.19 contains the photometric, colorimetric, and circadian-effective parameters for the assumption that a *vertical* illuminance of $E_v = 500$ lx is present at the eye of the room user.

The following can be seen from Figure 12.18:

(a) The spectrum at 3116 K does not contain any radiation component of the blue LED from 470 to 480 nm. This variant for 3116 K or at approx. 3000 K can be

Table 12.19 Photometric, colorimetric, and circadian-effective characteristics for the spectra in Figure 12.18.

Name	3K1	3K5	4K	4K5	5K	5K5	5K8	6K
CCT in K	3116	3500	4000	4500	5001	5501	5801	6001
Duv	3.00E-03	−1.99E-03	−3.01E-03	−3.01E-03	−3.01E-03	−2.13E-03	−1.01E-03	−2.38E-04
CIE CRI R_a	82.5	84.1	85.4	84.7	81.9	80.0	80.0	80.0
IES TM-30-15 R_f	84.6	81.6	81.4	80.0	79.5	77.7	77.6	77.5
ES TM-30-15 R_g	92.4	88.5	89.2	89.2	89.7	89.3	89.7	90.1
Vertical illuminance (E_v in lux)	500	500	500	500	500	500	500	500
a_{mel}	0.44	0.61	0.72	0.82	0.91	0.98	1.01	1.03
CS (version 2018)	0.41	0.33	0.39	0.44	0.47	0.49	0.50	0.50
$E_{v,mel,D65}$	246.2	336.7	399.2	451.9	503.5	543.3	559.6	569.9
Melanopic illuminance	269.8	369.5	438.7	497.2	554.6	598.9	617.32	628.80

Source: TU Darmstadt.

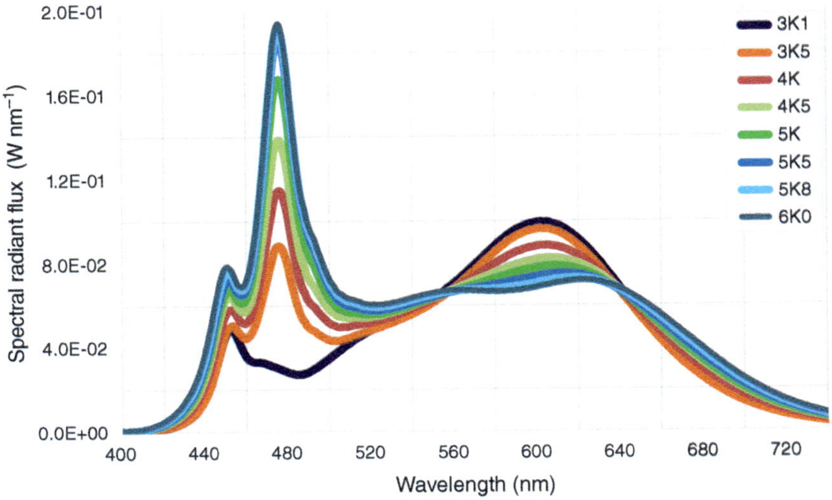

Figure 12.18 Spectral radiant flux of an HCL-oriented luminaire with three LED channels. Legend: see Table 12.19. Source: TU Darmstadt.

used for the evening with appropriate illuminance. In the evening from 8 p.m. onwards, it is often recommended not to exceed a CS value of 0.1. At 3116 K, this means a vertical illuminance of less than 100 lx, even much better at 50–60 lx, depending on the activities in the evening.

(b) For working people with an age of up to about 50 years, Table 12.19 applies with the colour temperature range up to 5000 K and with a horizontal illuminance around 1000–1250 lx, with the spectra in Figure 12.18. This corresponds to a vertical illuminance of 500–600 lx in Table 12.19.

(c) For professionals in the age range of 50–67 years, there is an increasing yellowing of the lens of the eye (see Figure 12.15). The spectra with a slightly higher blue component (i.e. a_{mel}- value) in the range 5000–5800 K can be used with a vertical illuminance around 500 lx.

(d) For the persons from approx. 70 years and also in elderly people's homes, the spectra for 5800–6000 K should be used. Since Figuiero et al. found in the three studies in the elderly people's homes [54, 55, 65] positive results with a vertical illuminance at the eye of more than 400 lx (approximately with a horizontal illuminance around 800–850 lx), these values should be adopted.

Other characteristic values in Table 12.13 can still be used.

References

1 Khanh, T.Q. and Bodrogi, P. (2018). Nutzerpräferenzen in der heutigen Innenraumbeleuchtung (User preferences in today's interior lighting). *Z. Licht* 8-2018: 64–71. ISSN: 0024-2861.

2 Luckiesh, M. and Moss, F.K. (1934). A visual thresholdometer. *J. Opt. Soc. Am.* 24 (11): 305–307.
3 Blackwell, H.R. (1946). Contrast thresholds of the human eye. *J. Opt. Soc. Am.* 36 (11): 624–643.
4 Rea, M.S. and Ouelette, M.J. (1991). Relative visual performance: a basis for application. *Light. Res. Technol.* 23 (3): 135–144.
5 Koedam, M. and Opstelten, J.J. (1971). Measurement and computer-aided optimization of spectral power distributions. *Light. Res. Technol.* 3: 205.
6 Kaplan, S. (1987). Aesthetics, affect and cognition: environmental preference from an evolutionary perspective. *Environ. Behav.* 19: 2–32.
7 Flynn, J.E., Spencer, T.J., Martyniuk, O., and Hendrick, C. (1973). Interim study of procedures for investigating the effect of light on impression und behaviour. *J. Illum. Eng. Soc.* 3 (1): 87–94.
8 Flynn, J.E., Hendrick, C., Spencer, T., and Martyniuk, O. (1979). A guide to methodology procedures for measuring subjective impressions in lighting. *J. Illum. Eng. Soc.* 8 (2): 95–110.
9 Rowlands, E., Loe, D.L., McIntosh, R.M. et al. (1985). Lighting adequacy and quality in office interiors by consideration of subjective assessment and physical measurement. *CIE J.* 4: 23–37.
10 Veitch, J.A. and Newsham, G.R. (1998). Lighting quality and energy-efficiency effects on task performance, mood, health, satisfaction and comfort. *J. Illum. Eng. Soc.* 27: 107–129.
11 DIN 5035 – Teil 1 (1990). *Beleuchtung mit künstlichem Licht: Begriffe und allgemeine Anforderungen. Normenausschuss für Lichttechnik. Lighting with Artificial Light: Terms and General Requirements.* Lighting Standards Committee.
12 Dehoff, P. (2000). Zum Stand der Innenraumbeleuchtung – Ein Überblick (On the state of interior lighting - An overview). Proceedings of the Conference *Licht 2000*, Goslar, Germany. 40–51, LiTG, Deutsche Lichttechnische Gesellschaft e.V. (German Lighting Society).
13 Dehoff, P. (2001). Zum Stand der Innenraumbeleuchtung – ein Überblick (On the state of interior lighting – an overview.). *Z. Licht* 8: 602–609. http://www.cyberlux.de/deutsch/articles/planner/gesetze/dehoff/Dehoff.pdf.
14 Rea, M.S. (2000). *The IESNA Lighting Handbook: Reference and Application.* New York: Illuminating Engineering Society of North America.
15 Kramer, H. (1999). Glaubenfrage Lichtqualität (Faith question "light quality"). *Prof. Light. Des.* 8 (4): 36–41.
16 Kramer, H. (1998). Die acht Gebote guter Lichtgestaltung (The eight commandments of good lighting design). *Licht* 98: 438–446. Tagungsband.
17 Ganslandt, R. and Hofmann, H. (2013). *Handbuch der Lichtplanung (Handbook of Lighting Design).* Springer-Verlag.
18 Veitch, J.A. and Newsham, G.R. (2000). Preferred luminous conditions in open-plan offices: research and practice recommendations. *Light. Res. Technol.* 32 (4): 199–212.
19 Veitch, J.A. (2001). Psychological processes influencing lighting quality. *J. Illum. Eng. Soc.* 30 (1): 124–140.

20 Veitch, J.A. and Newsham, G.R. (1998). Determinants of lighting quality I: state of the science. *J. Illum. Eng. Soc.* 27 (1): 92–106.

21 SHC (Solar Heating Cooling Programme/International Energy Agency) (2020). Literature review of user needs, toward user requirements. IEA SHC Task 61/EBC Annex 77, 2020-27-09, T61.A.1. https://doi.org/10.18777/ieashc-task61-2020-0001.

22 Kruisselbrink, T., Dangol, R., and Rosemann, A. (2018). Photometric measurements of lighting quality: an overview. *Build. Environ.* 138: 42–52.

23 Dehoff, P. and Tralau, B. (2017). *Lichtqualität – ein Prozess statt einer Kennzahl. Methodik zum Erfassen der Anforderungen an eine Lichtlösung und zu ihrer Bewertung zur Bestimmung ihrer Qualität (Lighting Quality – A Process Instead of a Figure of Merit. Methodology for Recording the Requirements of a Lighting Solution and Evaluating them to Determine its Quality)*, 1e. Deutsche Lichttechnische Gesellschaft e. V. (LiTG). ISBN: 978-3-927787-58-2.

24 Fördergemeinschaft Gutes Licht (2018). licht. wissen 21: Leitfaden (Guide to) Human Centric Lighting (HCL). licht.de.

25 ANSI-IES, IES TM-30-15 (2015). IES method for evaluating light source color rendition.

26 CIE (2017). Color Fidelity Index for accurate scientific use. *CIE Publication* 224: 2017.

27 Rea, M.S. and Figueiro, M.G. (2016). Light as a circadian stimulus for architectural lighting. *Light. Res. Technol.* 0: 1–14.

28 DIN 5034-1-2011-07 (2011). *Tageslicht in Innenräumen – Teil 1: Allgemeine Anforderungen* (Indoor daylighting – Part 1: General requirements.)

29 DIN EN 17037:2019-03 (2019). Tageslicht in Gebäuden. (Daylight in buildings). Deutsche Fassung EN 17037:201.

30 R.A. Jakobiak, D. Seifert, D. von See et al. (2019). Leitfaden zu DIN EN 17037: Tageslicht in Gebäuden. (Guide to DIN EN 17037: Daylight in buildings). Consortium of Peter Andres, VFF, daylighting.de, LiTG, Velux, Fitlicht, FVLR, TU Berlin, BF. https://www.fvlr.de/downloads/Leitfaden-zu-DIN-EN-17037-Tageslicht-in-Gebaeuden(1).pdf.

31 Moosmann, C. (2015). Visueller Komfort und Tageslicht am Büroarbeitspalt – eine Felduntersuchung in neun Gebäuden. Promotionsarbeit (Visual comfort and daylight at the office workspace - a field study in nine buildings). PhD thesis. KIT Scientific Publishing, ISBN 978-3-7315-0312-52014.

32 Velux Deutschland GmbH (2020). Neue Richtwerte für Tageslichtversorgung (New guideline values for daylight supply). *Z. Licht* 2: 36.

33 Gramm, S., Aydinli, S., und Völker, S. (2014). Tageslichtnutzung in Wohn- und Arbeitsräumen zur Verbesserung der visuellen Behaglichkeit und der Aufenthaltsqualität. Abschlussbericht (Use of daylight in living and working spaces to improve visual comfort and quality of stay). Final report.

34 Giuliani, F., Sokol, N., Lo Verso, V.R. et al. (2021). A study about daylighting knowledge and education in Europe. Results from the first phase of the DAYKE project. *Archit. Sci. Rev.* 64 (1, 2): 169–181.

35 Acosta, I., Leslie, R.P., and Figueiro, M.G. (2017). Analysis of circadian stimulus allowed by daylighting in hospital rooms. *Light. Res. Technol.* 49: 49–61.

36 Schweizer, C., Edwards, R.D., Bayer-Oglesby, L. et al. (2007). Indoor time-microenvironment-activity patterns in seven regions of Europe. *J. Expo. Sci. Environ. Epidemiol.* 17 (2): 170–181.

37 Hubalek, S., Brink, M., and Schierz, C. (2010). Officeworkers' daily exposure to light and it's influence on sleep quality and mood. *Light. Res. Technol.* 42: 33–50.

38 Stefani, O. (2017). Die Wirksamkeit von Tageslichtreplikationen auf die Wissensarbeit (The effectiveness of daylight replications on knowledge work). Dissertation. PhD thesis, TU Dresden.

39 Rosekind, M.R., Gregory, K.B., Mallis, M.M. et al. (2010). The cost of poor sleep: workplace productivity loss and associated costs. *J. Occup. Environ. Med.* 52 (1): 91–98.

40 Thayer, B. (1995). Daylighting and productivity at lockheed. *Sol. Today* 9 (3): 26–29.

41 Romm, J.J. (1999). *Cool Companies. How the Best Businesses Boost Profits and Productivity by Cutting Greenhouse Gas Emissions?* Washington, DC: Island Press.

42 BAuA (2016). *Arbeitszeitreport Deutschland 2016. Dortmund: Bundesanstalt für Arbeitsschutz und Arbeitsmedizin.* (Working Time Report Germany 2016. Dortmund: Federal Institute for Occupational Safety and Health.).

43 Begemann, S.H.A., Aarts, M.P.J., and Tenner, A.D. (1994). *Daylight, Artificial Light and People.* IES Australia.

44 Seidl, M. (1978). Ausreichendes Tageslicht und akzeptable Fensterabmessungen in Wohnräumen (Sufficient daylight and acceptable window dimensions in living spaces). Dissertation. PhD thesis, TU Berlin.

45 LiTG-Publikation 33 (2016). Tageslicht kompakt – Tageslichttechnik und Tageslichtplanung in Gebäuden (Daylight compact – Daylight technology and daylight planning in buildings) 1, 1–58, 978-3-927787-53-7, https://www.baulinks.de/webplugin/2016/1221.php4.

46 Tops, M., Tenner, A.D., Van den Beld, G.J., und Begemann, S.H.A. (1998). The effect of the length of continous presence on the preferred illuminance in offices. CIBSE National Lighting Conference.

47 Juslen, H. (2007). Lighting, Productivity and preferred illuminances–field studies in the industrial environment. Ph.D. thesis. Helsinki University of Technology, Espoo, Finnland.

48 Hubalek, S. (2008). LuxBlick – Messung der täglichen Lichtexposition zur Beurteilung der nicht-visuellen Lichtwirkungen über das Auge (Measurement of daily light exposure to assess non-visual light effects via the eye). PhD thesis. Appendix in ETH E-Collection http://e-collection.ethbib.ethz.ch/view eth: 29804. Shaker Verlag.

49 Gall, D. (2004). *Die Messung Circadianer Strahlungsgrößen (The measurement of circadian radiation quantities).* Ilmenau: TU Ilmenau.

50 DIN SPEC 5031-100:2015-08 (2015). *Strahlungsphysik im optischen Bereich und Lichttechnik – Teil 100: Über das Auge vermittelte, melanopische Wirkung des*

Lichts auf den Menschen – Größen, Formelzeichen und Wirkungsspektren (Radiation Physics in the Optical Field and Light Technology – Part 100: Melanopic Effect of Light on Humans Mediated via the Eye – Magnitudes, Formula Symbols and Effect Spectra). Berlin: Beuth Verlag.

51 Brown, T.M., Brainard, G.C., Cajochen, C. et al. (2022). Recommendations for daytime, evening, and nighttime indoor light exposure to best support physiology, sleep, and wakefulness in healthy adults. *PLoS Biol.* 20 (3): e3001571.

52 Rea, M.S. and Figueiro, M.G. (2016). Light as a circadian stimulus for architectural lighting. *Light. Res. Technol.* 50 (4): 497–510.

53 Smolders, K.C.H.J., de Kort, Y.A.W., and Cluitmans, P.J.M. (2012). A higher illuminance induces alertness even during office hours: findings on subjective measures, task performance and heart rate measures. *Physiol. Behav.* 107 (1): 7–16.

54 Figueiro, M.G., Hunter, C.M., Higgins, P.A. et al. (2015). Tailored lighting intervention for persons with dementia and caregivers living at home. *Sleep Health* 1: 322–330.

55 Figueiro, M.G., Plitnick, B., Lok, A. et al. (2014). Tailored lighting intervention improves measures of sleep, depression and agitation in persons with Alzheimer's disease and related dementia living in long-term care facilities. *Clin. Interv. Aging* 9: 1527–1537.

56 Lucas, R.J., Peirson, S.N., Berson, D.M. et al. (2014). Measuring and using light in the melanopsin age. *Trends Neurosci.* 37 (1): 1–9.

57 Lucas, R.J., Peirson, S.N., Berson, D. et al. (2013). *Irradiance Toolbox User Guide.* Manchester, UK: University of Manchester.

58 Hochschule Luzern, Technik & Architektur (2017–2019). Licht und Kommunikation in der Pflege: Planung und Betrieb dynamischer Lichtdecken. Erfahrungsbericht (Light and communication in care: Planning and operation of dynamic luminous ceilings). Field report.

59 Adrian, W. (1999). On the influence of age on vision under driving conditions. In: *Proceedings of Progress in Automobile Lighting PAL 99*, 1–8. Herbert Utz Verlag.

60 Holladay, L.L. (1926). The fundamentals of glare and visibility. *J. Opt. Soc. Am. A* 12: 271–319.

61 Schweitzer, S., Schinagl, C., Djuras, G. et al. (2016). Investigation of gender- and age-related preferences of men and women regarding lighting conditions for activation and relaxation. *Proc. SPIE* 9954: 99540L.

62 Mishima, K., Hishikawa, Y., and Okawa, M. (1998). Randomized, dim light controlled, crossover test of morning bright light therapy for rest-activity rhythm disorders in patients with vascular dementia and dementia of Alzheimer's type. *Chronobiol. Int.* 15: 647–654.

63 Yamadera, H., Ito, T., Suzuki, H. et al. (2000). Effects of bright light on cognitive and sleep-wake (circadian) rhythm disturbances in Alzheimer-type dementia. *Psychiatry Clin. Neurosci.* 54: 352–353.

64 Van Someren, E.J., Kessler, A., Mirmiran, M., and Swaab, D.F. Indirect bright light improves circadian rest-activity rhythm disturbances in demented patients. *Biol. Psychiatry* 41: 955–963.

65 Figueiro, M., Plitnick, B., and Rea, M. (2016). A selfluminous light table for persons with Alzheimer's disease. *Light. Res. Technol.* 48: 253–259.

66 Pro-persona care GmbH (Deutschland): Referenzbericht & Lichtstudie – Pflegetherapeutische Zentren Altentreptow & Neubrandenburg (Reference Report & Lighting Study - Altentreptow & Neubrandenburg Care Therapy Centres). https://www.waldmann.com (18 January 2021).

67 Deutsches Netzwerk für Qualitätsentwicklung in der Pflege (2006). Expertenstandardin der Pflege. (German Network for Quality Development in Nursing. Expert standard in nursing care). Hochschule Osnabrück.

13

Numerical Relationship Between Non-visual Metrics and Brightness Metrics – Consequences for the Evaluation of HCL Systems and Facilities

13.1 Introduction

In this book on human-centric integrative lighting, some chapters (Chapters 2 and 3) were worked out to provide readers knowledge on human eye physiology, basic quantities of photometry, colorimetry and metrics of non-visual effects of light (Chapters 2 and 3), on visual performance (Chapter 4), brightness, and visual clarity (Chapter 5). Mathematical models based on laboratory experiments regarding brightness in dependence of the signals from the luminance channel (L + M), from the S-cones, and the intrinsically photosensitive retinal ganglion cells (ipRGCs) were also developed and verified.

Since 2005, intensive research activities and discussions have taken place to explore and work out the mechanisms of how the receptor signals (rods, cones, and ipRGCs) can be combined to give rise to the non-visual effects of light. In Section 10.3.2, the Circadian Stimulus (CS) model in version CS_{2018} of Rea and Figuiero was described [1]. In Section 10.7, the melanopic equivalent daylight illuminance (MEDI, in lx) was defined according to the CIE publication [2]. These two metrics are currently proposed for scientific discussion worldwide. They are also subject to further investigations.

From a physiological point of view, these two metrics represent different opinions on how the quantities for non-visual effects can be defined. With the current chapter of the CIE mentioned above, the non-visual effects should be based solely on the signals of the ipRGC layer expressing the effect of the melanopsin pigments so that the MEDI metric can be used for the calculation and evaluation of lighting systems regarding the aspect of their non-visual effects. The CS model of Rea and Figuiero, in the version of 2005 and 2018 (described in [1]), is based on the idea that the non-visual effects result either from the ipRGC channel (if the opponent channel signal $B - Y < 0$, i.e. the blue channel has a weaker signal than the yellow channel, for example in the case of warm white light sources), or a combination of the ipRGC channel and the signals of the combinations of the (L + M), S cone and rod channels (if $B - Y > 0$, see Figure 2.13 in Chapter 2 of this book). The cases $B - Y < 0$ and $B - Y > 0$ correspond to correlated colour temperatures (CCTs) of about CCT <3710 and >3710 K, respectively, see Eq. (10.8).

Human Centric Integrative Lighting: Technology, Perception, Non-Visual Effects, First Edition.
Tran Quoc Khanh, Peter Bodrogi, and Trinh Quang Vinh.

During the time of drafting this book (in German between 2020 and 2021 and in English with updates and extensions between 2021 and 2023), numerous international scientific discussions and analysis work have been carried out to find the right metric for the non-visual effects with datasets of experiments from different research groups on night-time melatonin suppression, as a validation basis. The following important research results were achieved in this period:

1. Improvements of the CS model (Versions 2005 and 2018) in the two years 2020 and 2021 taking into account the exposure time t (in hours) and the field of view and modelling the contributions of the ipRGC-channel, the S cones, the rods, and the (L + M) channel with *improved* terms, for both cases ($B - Y > 0$ and $B - Y < 0$) [3, 4]. This improved formulation has been validated by means of datasets for *melatonin suppression* [4] and will be described in this chapter.
2. Publication on the '*Recommendations for daytime, evening, and night-time indoor light exposure to best support physiology, sleep, and wakefulness in healthy adults*' of a group of sleep researchers and neurophysiologists [5], see Section 9.7.2 of this book.
3. Based on the analysis of experimental data on melatonin suppression during the night and by using MEDI as an input metric for non-visual effects, Giménez et al., in 2022 [6], defined a formula to predict *melatonin suppression*, with exposure time and pupil dilation as further parameters. This new metric has been found by a machine-learning method and is a non-linear transformation of the MEDI metric. In this context, this formula, now denoted as *Gimenez formula*, has to be compared to the CS metric in its 2021 version.

Parallel to the dynamic and intensive development of metrics for non-visual effects of light on humans, the development of *brightness metrics* and *visual clarity metrics* for visual perception have experienced a renaissance with the development of the quasi-monochromatic and phosphor-converted white LEDs with different CCTs and chromaticity coordinates. This new discussion began at the end of the twentieth century by Fotios and Levermore [7], with the development and analysis of new psychophysical methods in 2012 [8] and has been continued by the authors of this book, see Chapter 5 of this book and also [9–11]. This renaissance could be explained by the fact that the brightness of white LED light sources of the same luminance and different spectra (different CCTs, chromaticity and colour saturation enhancements) are perceived at different brightness levels. The root of this perception difference can be argued as the luminance or the signal of the luminance channel (L + M) *alone* is not responsible for the brightness perception, which includes additional contributions of rods, S cones, and the ipRGC channel, see Figure 5.3. This argumentation is demonstrated in the brightness experiments and models in Chapter 5 of this book, see Eqs. (5.3) and (5.6).

With the above considerations in mind, the non-visual effects of light (CS_{2021}-model) can be modelled either with a combination of signals of the ipRGC channel, the S cones, the (L + M) channel, and the rod channel, or *solely* with the ipRGC signals (MEDI, CIE-publication S 026:2018). Similarly, the brightness metric M from Eq. (5.6) can be built by the use of an exponential function of the (L + M) channel signal and the S cone signal. By doing so, the following research questions arise from the point of view of *practical lighting engineering and planning*:

1. Is there a reasonable and useable correlation between the MEDI (CIE S 026:2018 [2]) and the Circadian Light CL_A in the CS models in the versions of 2018 and 2021, as well as *equivalent luminance* L_{eq} according to the Eq. (8.2) of Fotios and Levermore for brightness perception?
2. Is there also a useable correlation between the values of the CS_{2021} metric, the *Gimenez values* for melatonin suppression, and the metric for brightness perception according to the Bodrogi and Khanh model in Eq. (5.6)?
3. How much is the difference between the values of the CS_{2021}-metric, and the Gimenez values for melatonin suppression for the same absolute light source spectra? Is this difference acceptable for the practical use of these metrics?

If there is a useable correlation between the different metrics and the difference between them is acceptable, then a conversation formula can be developed to transform one metric to the other with sufficient accuracy so that a lighting engineer can design and evaluate a lighting system with several metrics recommended today, although these metrics were established from different human physiological viewpoints. This is the main aim of this chapter, which was written at the end of the year 2022 by the authors of this book. Some aspects of this chapter can also be regarded as an extension of Chapter 8. In Sections 13.3 (for the model CS2021) and 13.4 (the formula of Giménez), the model of brightness, the model CS_{2021} of Rea et al. [4], and the formula of Giménez et al. [6] will be presented before the correlations and relationships between the metrics for the non-visual effects and *brightness* will be described based on the calculation of 884 measured light source spectra.

13.2 Brightness Perception and Modelling

Brightness perception was analysed and modelled in Chapter 5 based on experiments of the authors of this book [7–11]. The result of the brightness model is Eq. (5.6) repeated as Eq. (13.1) here (TUD stands for '*TU Darmstadt*').

$$M\,(\text{Brightness, 2020, TUD}) = 2.566\ \left[E_v^{\,0.4167494}\cdot S^{0.1153348}\right] + 37.04 \qquad (13.1)$$

This model (Eq. (13.1)) contains the combination of a term of illuminance, which is compressed with the power function $E_v^{\,0.417}$ and a term with the compressed S-cone signal. The optimisation based on the experimental data also pointed out that, from a mathematical point of view, the contribution of the S cones and of the ipRGC channel was similar and the distinction between them was not decisive. The authors of this book will apply the contribution of the S cones in the subsequent analysis. *Perceived brightness* was also modelled in literature. One such model, the so-called *equivalent luminance* (L_{eq}) *(or equivalent illuminance)* according to Fotios and Levermore [7], divides the signal of the S-cones (S) by the signal of the $V(\lambda)$ function and then calculates a metric with the 0.24 power, see Eq. (13.2).

$$L_{eq} = L_v \cdot (\text{S/V})^{0.24} \qquad (13.2)$$

In Eq. (13.2), the exponent of the photopic luminance (L_v) remains 1.000, i.e. luminance remains uncompressed. To calculate the signals S or V, the relative spectral radiant flux of the light source must be multiplied by the spectral sensitivity function of the S cones or by the $V(\lambda)$ function, respectively, and this product must be integrated into the visible wavelength range. The metrics in Eqs. (13.1) and (13.2) will be used as metrics for *perceived brightness* in the correlation analysis in Section 13.5.

13.3 Circadian Stimulus Models CS_{2018} and CS_{2021}

The conception of Rea et al. covering the models CS_{2005}, CS_{2018}, or CS_{2021} is based on the design of a phototransduction circuit that regards the following mechanisms [1, 4]:

(a) The phototransduction from photoabsorption, signal generation, and signal conversion into a frequency-coded form and the processing of the signals of the different channels (LMS-cones, rods, and ipRGCs) exhibit subadditivity. Additivity is assumed, e.g. in the definition of the illuminance or luminance with the $V(\lambda)$-function or of the MEDI metric, if the spectral sensitivity of the receptor system is multiplied by the spectral radiance or irradiance of the incident radiation and all the effects at each wavelength between 380 and 780 nm can be integrated by a summation to the final effect of the whole polychromatic radiation. No reduction of the signals is expected. In comparison, a possible subadditivity takes place if the effect at a wavelength λ_1 is reduced at an interaction with radiation of a certain wavelength λ_2. The subadditivity can be explained in neurobiology if the neural circuit for the phototransduction contains a spectrally *opponent* channel. In vision science, two spectral opponent channels are known, a (L − M) and a (S − (L + M)) channel. In the context of non-visual effects, the spectrally opponent channel (S − (L + M)) (also denoted as $B - Y$) is taken into consideration. This is an important difference between the CS conception and the conception of the MEDI metric that is defined by the CIE [11] and assumes the *additivity* of the non-visual pathway. Subadditivity, however, was found to be essential in the experiments of Figuiero et al. [12, 13].

(b) The CS conception followed the idea that a non-visual effect consists of two components, a spectral component and an amount component. The spectral component, denoted by the Circadian Light CL_A, which will be described later, expresses the spectral generation of a stimulus at different receptors and channel systems (LMS-cones, rods, and ipRGC) at a certain state of the spectral opponent channel ($B - Y > 0$ or $B - Y < 0$). In the definition of the MEDI metric, no case distinction is made.

(c) The *amount* component in the models from CS_{2005}, CS_{2018} to model CS_{2021} accounts for the exposure time, the characteristic of the viewing field due to the spatial distribution of the ipRGC receptors on the retina and the absolute size of the circadian light value CL_A.

(d) A conversion from CL_A to Circadian Stimulus CS in the model versions CS_{2005} and CS_{2018} was based on the datasets of Thapan et al. [14] and Brainard et al. [15] with quasi-monochromatic stimuli for *nocturnal melatonin suppression*. It was improved and validated in 2021 by datasets of a variety of research groups. Therefore, the CS metric is also valid for the lighting design processes of both evening and night-time illuminations.

13.3.1 The Circadian Stimulus (CS) Models 2005 and 2018

These models are implemented in two steps [1]:

First step: Establishment of Circadian Light CL_A (denoted as $CL_{A\,1.0}$):

$$CL_A = \begin{cases} 1548 \cdot \left[\int M_{c\lambda} E_\lambda d\lambda + \left(a_{b-y} \cdot \left(\int \frac{S_\lambda}{mp_\lambda} E_\lambda d\lambda - k \int \frac{V_\lambda}{mp_\lambda} E_\lambda d\lambda\right) - a_{rod} \cdot \left(1 - e^{\frac{-\int V'_\lambda E_\lambda d\lambda}{\text{RodSat}}}\right)\right)\right] \\ \quad if\left(\int \frac{S_\lambda}{mp_\lambda} E_\lambda d\lambda - k \int \frac{V_\lambda}{mp_\lambda} E_\lambda d\lambda\right) > 0 \\ 1548 \cdot \int M_{c\lambda} E_\lambda d\lambda \\ \quad if\left(\int \frac{S_\lambda}{mp_\lambda} E_\lambda d\lambda - k \int \frac{V_\lambda}{mp_\lambda} E_\lambda d\lambda\right) < 0 \end{cases} \tag{13.3}$$

In Eq. (13.3), the symbols have the following meaning:

- CL_A: *Circadian Light*, where the subscript 'A' designates a numerical equivalence of $CL_A = 1000$ (photopic) lx for CIE illuminant A
- E_λ: light source spectral irradiance
- $M_{c\lambda}$: melanopsin sensitivity (corrected for crystalline lens transmittance, after Wyszecki and Stiles [16])
- $k = 0.2616$
- S_λ: S cone fundamental (Smith and Pokorny [17]).
- $a_{b-y} = 0.7$
- mp_λ: macular pigment transmittance (after Snodderly et al. [18])
- $a_{rod} = 3.3$
- V_λ: photopic luminous efficiency function (ISO/CIE, [19])
- $\text{RodSat} = 6.5\,\text{W m}^{-2}$
- V'_λ: scotopic luminous efficiency function (ISO/CIE, [19])

Second step: Conversion of the CL_A value to the Circadian Stimulus CS_{2018}

The method according to M. Rea et al. transforms the circadian effective light CL_A with the help of Eq. (13.4) into a so-called '*circadian stimulus CS*', which is proportional to *melatonin suppression* in %. For example, a value of CS = 0.4 thus corresponds to a nocturnal melatonin suppression of 40% compared to the state before irradiation, see Figure 10.8 in Chapter 10 of this book.

$$CS = 0.7 - \frac{0.7}{1 + \left(\frac{CL_A}{355.7}\right)^{1.1026}} \tag{13.4}$$

From the mathematical point of view, this logistical function in Eq. (13.4) shows that the CS = 0.7 or the nocturnal melatonin suppression of 70% is reached if the Circadian Light term CL_A is already very high. A higher CS value (higher than 0.7) is *not possible*. This is also a subject of the questions to be discussed later in relation to the formula of Giménez et al. [6] in Section 13.4. This CS value is valid for the exposure time of one hour during the early biological night. Model values were validated by means of a dataset with 13 polychromatic light sources with a correlation coefficient of $r^2 = 0.69$ [4].

13.3.2 The Circadian Stimulus Model 2021

The CS models, CS_{2005} and CS_{2018}, were used over a long time in some research groups and partially also by the USA, as well as by the international lighting industry with recognized advantages and clear deficits [4]. In some research studies, the authors of this book have also used the CS_{2018} model for modelling some results. In a paper from 2021 [4], Rea et al. wrote:

> Over the years, the model has consistently been able to predict nocturnal melatonin suppression of "cool," polychromatic light sources, *but has not been able* to accurately predict suppression from "warm" light sources. Nocturnal melatonin suppression was systematically overestimated for "warm" LED light sources that produced radiant energy throughout the spectrum in a study by Nagare et al. (2019c) Conversely, a second study by Nagare et al. (2019b) revealed that nocturnal melatonin suppression was systematically underestimated for "warm" LED light sources with a discontinuity in spectral energy around 480 nm. The original model used an in vivo estimate of ipRGC-melanopsin spectral sensitivity alone to characterize the spectral sensitivity of the circadian phototransduction circuit to "warm" sources, so this simple assumption for determining CL_A for "warm" sources was clearly inadequate.

After the analysis of the authors of this book, the borderline between '*cool*' and '*warm*' white polychromatic light sources in this aspect turned out to be about 3400–3710 K. Rea et al. also wrote in [4]:

> The 2005 two-state CL_A formulation … did not include a physiologically based threshold term for the ipRGC-melanopsin response. In the revised formulation ($CL_{A\,2.0}$ …), the ipRGC-melanopsin response is directly modulated by a threshold term involving both rods and cones that, through the AII amacrine neuron, elevates the threshold response of the M_1 ipRGCs to light….

Accordingly, the two steps for building the improved CS model 2021 were as follows:

First step: Establishment of a new formula for Circadian Light CL_A (denoted as $CL_{A\,2.0}$):

$$CL_A 2.0 = \begin{cases} 1548 \cdot [\left(\int M_{c\lambda}E_\lambda d\lambda - a_{rod1} \cdot \left(\frac{\int V'_\lambda E_\lambda d\lambda}{\int V_{c\lambda}E_\lambda d\lambda + g_1 \int S_{c\lambda}E_\lambda d\lambda}\right) \cdot \left(1 - e^{\frac{-\int V'_\lambda E_\lambda d\lambda}{RodSat}}\right)\right) \\ \quad + (a_{b-y} \cdot (\int S_{c\lambda}E_\lambda d\lambda - k \int V_{c\lambda}E_\lambda d\lambda) \\ \quad - a_{rod2} \cdot \left(\frac{\int V'_\lambda E_\lambda d\lambda}{\int V_{c\lambda}E_\lambda d\lambda + g_2 \int S_{c\lambda}E_\lambda d\lambda}\right) \cdot \left(1 - e^{\frac{-\int V'_\lambda E_\lambda d\lambda}{RodSat}}\right))] \text{ if } b - y > 0 \\ 1548 \cdot \left\{\int M_{c\lambda}E_\lambda d\lambda - a_{rod1} \cdot \left(\frac{\int V'_\lambda E_\lambda d\lambda}{\int V_{c\lambda}E_\lambda d\lambda + g_1 \int S_{c\lambda}E_\lambda d\lambda}\right) \cdot \left(1 - e^{\frac{-\int V'_\lambda E_\lambda d\lambda}{RodSat}}\right)\right\} \text{ if } b - y \le 0 \end{cases} \tag{13.5}$$

$$b - y = \int S_{c\lambda}E_\lambda d\lambda - k \int V_{c\lambda}E_\lambda d\lambda \tag{13.6}$$

$$V_{c\lambda} = \frac{\left(\frac{V_\lambda}{mp_\lambda}\right)}{\max\left(\frac{V_\lambda}{mp_\lambda}\right)} \text{ and } S_{c\lambda} = \frac{\left(\frac{S_\lambda}{mp_\lambda}\right)}{\max\left(\frac{S_\lambda}{mp_\lambda}\right)} \tag{13.7}$$

with:

- $k = 0.2616$
- E_λ: light source spectral irradiance
- $a_{b-y} = 0.21$
- $M_{c\lambda}$: melanopsin sensitivity (corrected for crystalline lens transmittance) (Wyszecki and Stiles [16]) $a_{rod1} = 2.30$
- S_λ: S-cone fundamental (Smith and Pokorny [17])
- $a_{rod2} = 1.60$
- mp_λ: macular pigment transmittance (Snodderly et al. [18])
- $g_1 = 1.00$
- V_λ: photopic luminous efficiency function (ISO/CIE, [19])
- $g_2 = 0.16$
- V'_λ: scotopic luminous efficiency function (ISO/CIE, [19])
- $RodSat = 6.50\ W\ m^{-2}$

Second step: Conversion of $CL_{A\,2.0}$ value to Circadian Stimulus CS_{2021}

In comparison to the model versions CS_{2005} and CS_{2018}, the basic structure of version 2021 with the logistic function remains unchanged with the exponent 1.1026 and the half-saturation constant 355.7 (with the CS value of 35% as the *half* of the maximally possible melatonin suppression, 70%). Two new factors have been implemented, *exposure time t* (in hours) from 0.5 and 3 hours and factor *f* describing the spatial distribution of circadian light exposure.

$$\mathrm{CS}_{t,f} = 0.7 \cdot \left[1 - \frac{1}{1 + \left(\frac{\mathrm{CL}_{\mathrm{A},2021} \cdot t \cdot f}{355.7}\right)^{1.1026}}\right] \tag{13.8}$$

Regarding the factor f in three different viewing modes, Rea et al. wrote [4]:

> (a) full visual field, as with a Ganzfeld, $f = 2.0$ (b) central visual field, as with a discrete light box on a desk, $f = 1.0$ and (c) superior visual field, as from ceiling mounted down-light fixtures, $f = 0.5$. These three levels of f were developed to make a relative estimate of how many circadian phototransduction circuits were activated in the four studies used for testing the model....

The above definition of the viewing conditions is, however, not precise for lighting applications in the real world. For numerical analysis purposes in this book, f will be set to 1.0.

13.4 The Formula of Giménez et al. for Nocturnal Melatonin Suppression

The numerical study of Giménez et al. [6] pursued similar aims like the CS models with the following research conception:

1. The authors of this study aimed at building a metric with MEDI (melanopic illuminance) as the starting parameter and extended the analysis with the contributions of the LMS cones based on a correlation analysis of 29 different datasets of nocturnal melatonin suppression published in scientific papers.
2. The co-parameters were *exposure time* and the *state of the pupil* of the test subjects during the experiments (with or without pupil dilation). The metric to be found was intended to be a metric for the prediction of *nocturnal melatonin suppression* similar to the above-mentioned CS_{2021} model of Rea et al.

The data analysis was based on the *random forest* (RF) method, a machine-learning approach to solve classification and regression problems. The model was built in two steps. In the first step, the MEDI illuminance, the photopic illuminance, the *Rhodopic* EDI (for rods), the L-opic EDI, M-opic EDI, and the S-opic EDI were subjects to separate correlation analyses at different exposure times for narrowband and polychromatic light spectra. From 21 to 10 000 lx, the MEDI metric exhibited the best correlation coefficients. The S-cone EDI outperformed the MEDI metric only in the range below 21 lx.

With the MEDI illuminance as starting parameter, further components like LMS-cone signals, exposure time (duration), and pupil dilation were added to the set of input parameters, and a logistic four-parameter function was built and compared to the available datasets. The accuracy of the regression analysis was expressed by the root mean square of error (RMSE). The optimal RF model was the model with the lowest RMSE value and the lowest number of predictors.

Additionally, the coefficient r^2 was also used. The results of this analysis can be summarized as follows:

(a) An additional contribution of L- and M-cones did not improve model quality in comparison to the combination with S-cones as well as ipRGC *alone*.
(b) The combination of ipRGC and S-cones resulted in a higher correlation coefficient and a smaller RMSE value in comparison to ipRGC *alone* (MEDI). The difference was rather small so the research group of Giménez decided to ignore the S-signal part in their model.
(c) The model with a logistic function, therefore, includes MEDI, exposure time, and pupil dilation, see Eq. (13.9).

$$\text{Suppression}_{\text{melatonin}} = \frac{0-100}{1+\left(\frac{\log_{10}(\text{EDI}_{\text{melanopic}} \cdot 10^6)}{9.002-0.008 \cdot \Delta_{t\ \text{exposure}} - 0.462 \cdot \text{dil}_{\text{pupil}}}\right)^{7.496}} + 100 \tag{13.9}$$

In Eq. (13.9), the symbols have the following meaning:

- $\text{suppression}_{\text{melatonin}}$ = melatonin suppression (in %)
- $\text{EDI}_{\text{melanopic}}$ = melanopic EDI (lx)
- $\Delta_{t\ \text{exposure}}$ = exposure duration (in minutes)
- $\text{dil}_{\text{pupil}}$ = pupil dilation applied: 0 = no, 1 = yes

In comparison to the CS model of Rea et al., the value of melatonin suppression ranges up to 100% at infinite illuminance. According to the opinion of the authors of this book, the most important *difference* between the CS model in version CS_{2021} (or CS_{2018}) and the model of Giménez et al. is the *value scaling aspect*. (Note that both models were built with regression methods for nocturnal melatonin suppression. Both models were validated by means of partly similar datasets of well-known research groups.)

13.5 Numerical Analysis of the Relationship Between Brightness and Non-visual Metrics

13.5.1 Introduction

The brightness metrics and the metrics for the *non-visual effects of light* described above can be categorized into the following groups:

(a) A group of *linear* metrics: Fotios's model for *equivalent luminance* without signal compression (Eq. (13.2)); Circadian Light CL_A for the model versions 2018 and 2021; and MEDI (which is also an input parameter of the logistic function of Giménez et al.).
(b) A group of non-linear metrics like CS (CS_{2018} and CS_{2021}); the metric according to Giménez et al. (output parameter of the logistic function in Eq. (13.9));

and the brightness model M of the authors of this book, see Chapter 5 and Eq. (13.1).

(c) Because the CS model includes the case distinction between '*warm*' and '*cold*' white light ($B-Y>0$ or $B-Y<0$, or empirically if CCT ≥ 3710 K or CCT < 3710 K), the correlation with experimental data can depend on the kind of light spectra actually used.

(d) Because the CS model (in the versions CS_{2005} and CS_{2018}) has been used by the USA and international lighting science and industry for a long time, an analysis of the differences and correlations between the values of the versions CS_{2018} and CS_{2021}, as well as the '*warm*' and '*cold*' light categories, is necessary.

13.5.2 Method of Correlation Analysis

For this correlation analysis, several measured and simulated light source spectra were analysed, see Table 13.1. This set of spectra includes real measured light sources from incandescent lamps (thermal radiators, 28 light sources), compact and linear fluorescent lamps (252 light sources), different LED types (419 light sources), and 185 measured daylight spectra. These 884 light source spectra are shown in Figure 10.35, see Chapter 10 of this book.

In order to calculate the values of the *brightness* metrics, the CS values, and the MEDI and Giménez values, the 884 spectra in Table 13.1 were converted into a set of spectra at three fixed photopic vertical illuminance levels: 125, 500, and 750 lx, which cover the range of illumination levels in the practical indoor lighting applications. Then, the brightness and the non-visual parameters were computed at each one of these three illuminance levels.

13.5.3 Relation Between the Linear Brightness Metrics and the Non-visual-Effect Parameters

Table 13.2 shows the correlation coefficients between the linear brightness metrics and the non-visual-effect parameters for all 884 light source spectra in Table 13.1 (regardless of their CCTs, i.e. including both the warm white and the cold white spectra).

Table 13.1 Light sources and their photometric and colorimetric quantities in the correlation analysis.

No.	Light source types and their parameters CCT = 2201–17 815 K; $-1.467\times10^{-2}<$ Duv $<1.529\times10^{-2}$; $0.0797<z<0.4792$; $80<$ CIE $R_a<100$; $0.2784<$ MDER <1.46
1.	Conventional incandescent lamps + filtered incandescent lamps, 28 spectra
2.	Fluorescent tubes + compact fluorescent lamps, 252 spectra
3.	LED lamps + LED luminaires, 419 spectra
4.	Daylight (CIE-model + measurements), 185 spectra

Table 13.2 Correlation between the linear brightness metrics and the non-visual-effect parameters for all spectra in Table 13.1 (correlated colour temperatures between 2201 and 17 815 K).

r^2	MEDI	$L_{\text{eq of Fotios}}$	$CL_{A,2018}(CL_{A\ 1.0})$	$CL_{A,2021}(CL_{A\ 2.0})$
MEDI	1.00	0.94	0.72	0.97
$L_{\text{eq of Fotios}}$	0.94	1.00	0.59	0.89
$CL_{A,2018}(CL_{A\ 1.0})$	0.72	0.59	1.00	0.85
$CL_{A,2021}(CL_{A\ 2.0})$	0.97	0.89	0.85	1.00

The following can be seen in Table 13.2:

- The correlation between MEDI, $CL_{A,2021}$ ($CL_{A\ 2.0}$), and *equivalent luminance* according to Fotios is high, with the following r^2 values:
  - 0.89 ($L_{\text{eq, Fotios}}$ vs. $CL_{A,2021}$ ($CL_{A\ 2.0}$));
  - 0.94 (MEDI vs. $L_{\text{eq, Fotios}}$); and
  - 0.97 ($CL_{A,2021}$ ($CL_{A\ 2.0}$) vs. MEDI).
- The correlation coefficient between $CL_{A,2018}$ ($CL_{A\ 1.0}$) and MEDI or $L_{\text{eq, Fotios}}$ is much lower (r^2 equals 0.72 or 0.59).
- For all spectra between 2201 and 17 815 K, the correlation coefficient between $CL_{A,2018}$ ($CL_{A\ 1.0}$) and $CL_{A,2021}$ ($CL_{A\ 2.0}$) equals 0.85.

The $CL_{A,2021}$ ($CL_{A\ 2.0}$) values tend to correlate very well with MEDI (see Figure 13.1, RMSE = 21.25) and the equivalent luminance of Fotios, much better than with $CL_{A,2018}$ ($CL_{A\ 1.0}$). For all spectra between 2201 K and 17 815 K, a linear

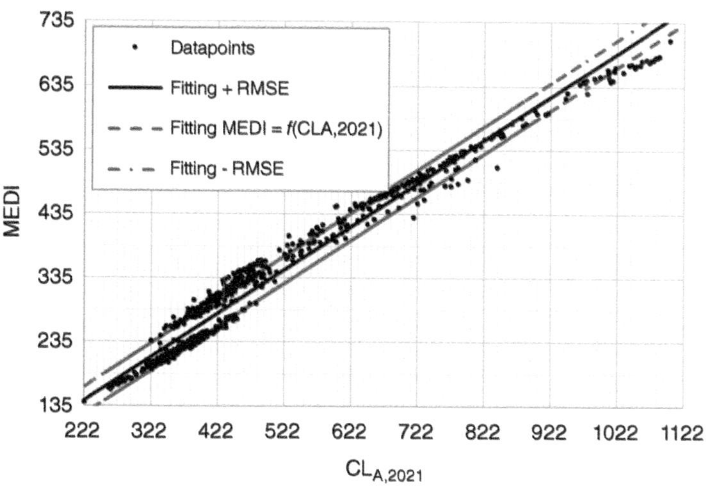

Figure 13.1 Correlation between MEDI and $CL_{A,2021}$ ($\sim CL_{A\ 2.0}$) for all spectra in Table 13.1 between 2201 and 17 815 K.

relationship was found between MEDI and $CL_{A,2021}$ ($CL_{A\ 2.0}$), see Eq. (13.10) and Figure 13.1.

$$MEDI = 0.6792 \cdot CL_{A,2021} - 8.3139 \tag{13.10}$$

If only the spectra in the '*warm white*' range (CCT < 3710 K, $B-Y<0$) are regarded, a different picture emerges. These correlations are shown in Table 13.3.

The following can be seen in Table 13.3:

- $CL_{A,2018}(CL_{A\ 1.0})$ correlates poorly with $CL_{A,2021}(CL_{A\ 2.0})$ ($r^2 = 0.41$), with $L_{eq,\ Fotios}$ and with MEDI.
- $L_{eq,\ Fotios}$ correlates also moderately, rather poorly with MEDI and $CL_{A,2021}(CL_{A\ 2.0})$.
- MEDI and Circadian Light $CL_{A,2021}(CL_{A\ 2.0})$ exhibit a good correlation with $r^2 = 0.81$.

If, finally, the CCT range between 3710 and 17 815 K (neutral white and cold white lights) is regarded, then the correlation coefficients are very good between all brightness metrics and non-visual metrics, see Table 13.4. The values of $CL_{A,2021}(CL_{A\ 2.0})$ correlate very well with MEDI, $CL_{A,2018}(CL_{A\ 1.0})$, and $L_{eq,\ Fotios}$, see Figure 13.2.

A formula was derived for the range 3710 K < CCT < 17 815 K with an RMSE-value of only 11.85, see Eq. (13.11).

$$MEDI = 0.5924 \cdot CL_{A,2021}(\sim CL_{A,2.0}) + 55.5 \tag{13.11}$$

From the practical point of view of lighting engineering, it must be emphasized that most buildings and rooms in the private and professional sectors in

Table 13.3 Correlation between the linear brightness metrics and the non-visual-effect parameters for the correlated colour temperature range 2200 K < CCT < 3710 K (warm white lights).

r^2	MEDI	$L_{eq\ of\ Fotios}$	$CL_{A,2018}(CL_{A\ 1.0})$	$CL_{A,2021}(CL_{A\ 2.0})$
MEDI	1.00	0.50	0.07	0.81
$L_{eq\ of\ Fotios}$	0.50	1.00	0.00	0.36
$CL_{A,2018}(CL_{A\ 1.0})$	0.07	0.00	1.00	0.41
$CL_{A,2021}(CL_{A\ 2.0})$	0.81	0.36	0.41	1.00

Table 13.4 Correlation between the linear brightness metrics and the non-visual-effect parameters for the correlated colour temperature range 3710 K < CCT < 17 815 K (neutral and cold white lights).

r^2	MEDI	$L_{eq\ of\ Fotios}$	$CL_{A,2018}(CL_{A\ 1.0})$	$CL_{A,2021}(CL_{A\ 2.0})$
MEDI	1.00	0.92	0.97	0.99
$L_{eq\ of\ Fotios}$	0.92	1.00	0.97	0.96
$CL_{A,2018}(CL_{A\ 1.0})$	0.97	0.97	1.00	1.00
$CL_{A,2021}(CL_{A\ 2.0})$	0.99	0.96	1.00	1.00

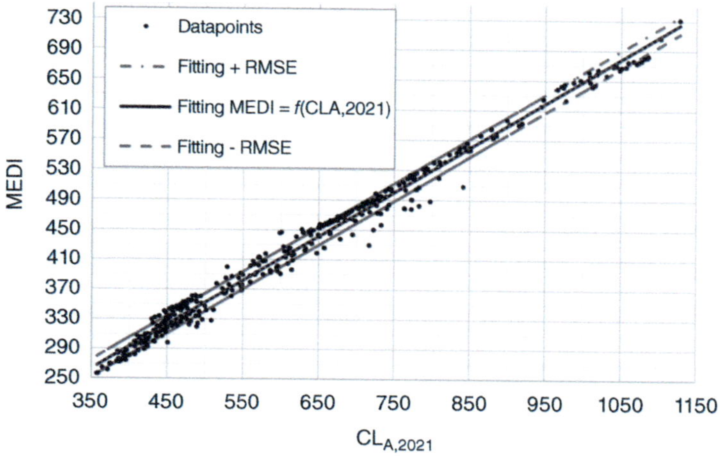

Figure 13.2 Correlation between MEDI and $CL_{A,2021}$ for the spectra of Table 13.1 between 3710 and 17 815 K.

the daytime with daylight incidence (i.e. with windows) shall have illumination situations with CCTs being more than 3710 K so that the conversion between MEDI and $CL_{A,2021}(CL_{A\ 2.0})$, according to Eq. (13.11), can be applied with high accuracy.

The above results lead to the conclusion that the correlation between MEDI and the new version $CL_{A,2021}(CL_{A\ 2.0})$ is good for the warm white range (e.g. for evening-time applications) and very good for the range between 3710 and 17 815 K for indoor lighting in private and professional buildings (e.g. for offices, schools, and supermarkets), as well as for outdoor daylight. The improvement of Circadian Light $CL_{A,2018}$ ($CL_{A\ 1.0}$) to the new version $CL_{A,2021}(CL_{A\ 2.0})$ is essential.

13.5.4 Relation Between Non-linear Brightness Metrics and Non-visual-Effect Parameters

The correlation analysis for all spectra between 2201 and 17 815 K in Table 13.1 provides the results in Table 13.5 in this case. For this analysis, the exposure duration of one hour was selected in the case of CS_{2021} and for the formula of Giménez et al. (denoted as $\text{Sups.Gim.}_{t=1\text{h,pul.d.}=0}$), and no pupil dilation was taken into consideration according to usual illuminating applications in practice.

The following can be seen in Table 13.5:

- CS_{2018} shows moderate correlations to $\text{Sups.Gim.}_{t=1\text{h,pul.d.}=0}$ and brightness $M_{2020,\text{TUD}}$ (r^2 equals 0.51 and 0.53, respectively).
- $CS_{2021,t=1\text{h},f=1}$ correlates very well with $\text{Sups.Gim.}_{t=1\text{h,pul.d.}=0}$ and $M_{2020,\text{TUD}}$ (r^2 equals 0.97 in both cases).
- The correlation between the brightness metric $M_{2020,\text{ TUD}}$ and $CS_{2021,t=1\text{h},f=1}$ and the Giménez metric is very good with $r^2 = 0.97$ and 0.99, respectively.

Figure 13.3 shows the relationship between the Giménez values (one hour exposure time, no pupil dilation) and the $CS_{2021,t=1\text{h},f=1}$ values (exposure time one

Table 13.5 Correlation between the non-linear brightness metrics $M_{2020,TUD}$ and the CS model versions 2018 and 2021, as well as the Giménez metric for all spectra in Table 13.1 in the range 2201 K < CCT < 17 815 K.

r^2	CS_{2018}	$CS_{2021,t=1h,f=1}$	$M_{2020,TUD}$	$Sups.Gim._{t=1h,pul.d.=0}$
CS_{2018}	1.00	0.68	0.53	0.51
$CS_{2021,t=1h,f=1}$	0.68	1.00	0.97	0.97
$M_{2020,TUD}$	0.53	0.97	1.00	0.99
$Sups.Gim._{t=1h,pul.d.=0}$	0.51	0.97	0.99	1.00

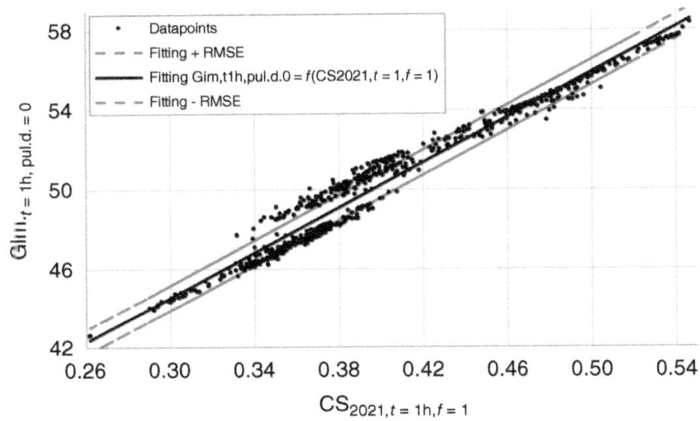

Figure 13.3 Relationship between the Giménez value and $CS_{2021,t=1h,f=1}$ for all spectra in Table 13.1 between 2201 and 17 815 K.

hour, $f = 1$) for a practically relevant data range of $CS_{2021,t=1h,f=1}$ between 0.26 and 0.54. This relationship was modelled by the formula in Eq. (13.12) with the RMSE value of 0.6376 and r^2 of 0.97.

$$Sups.Gim._{t=1h,pul.d.=0}\,(\text{in }\%) = 56.699 \cdot CS_{2021,t=1h,f=1} + 27.5 \tag{13.12}$$

Table 13.6 shows the correlation coefficients in the case of the warm white range CCT < 3710 K. As can be seen from Table 13.6, the CS_{2018} values do not correlate with $CS_{2021,t=1h,f=1}$, $Sups.Gim._{t=1h,pul.d.=0}$ (of the Giménez-metric), and the brightness values according to the $M_{2020,TUD}$-formula. The values $Sups.Gim._{t=1h,pul.d.=0}$ of the Giménez metric correlate well with the brightness metric values $M_{2020,TUD}$ ($r^2 = 0.96$) and with $CS_{2021,t=1h,f=1}$ ($r^2 = 0.85$).

As a summary of all results in this chapter, it can be stated that CS_{2021} represents an essential improvement in comparison to CS_{2018}, which did not work in the warm white range (CCT < 3710 K). The correlation between the brightness metric $M_{2020,\,TUD}$, the Giménez metric (based on the *MEDI* metric), and CS_{2021} based on $CL_{A,2021}$ ($CL_{A\ 2.0}$) is good or even very high. Consequently, these three metrics can be converted by linear formulae with acceptable accuracy from the point of view of lighting engineering.

Table 13.6 Correlation between the non-linear brightness metric $M_{2020,TUD}$ and the CS model versions 2018 and 2021, as well as the Giménez metric for the warm white spectra in Table 13.1 (2201 K < CCT < 3710 K).

r^2	CS_{2018}	$CS_{2021,t=1h,f=1}$	$M_{2020,TUD}$	$Sups.Gim._{t=1h,pul.d.=0}$
CS_{2018}	1.00	0.29	0.01	0.04
$CS_{2021,t=1h,f=1}$	0.29	1.00	0.79	0.85
$M_{2020,TUD}$	0.01	0.79	1.00	0.96
$Sups.Gim._{t=1h,pul.d.=0}$	0.04	0.85	0.96	1.00

References

1 Rea, M. and Figuiero, M.G. (2016). Light as a circadian stimulus for architectural lighting. *Light. Res. Technol.* 50 (4): 497–510.

2 CIE (2018). CIE S 026:2018: CIE System for Metrology of Optical Radiation for ipRGC-Influenced Responses to Light. https://doi.org/10.25039/S026.2018.

3 Rea, M.S., Nagare, R., and Figueiro, M.G. (2020). Predictions of melatonin suppression during the early biological night and their implications for residential light exposures prior to sleeping. *Sci. Rep.* 10 (1): 1–12.

4 Rea, M.S., Nagare, R., and Figueiro, M.G. (2021). Modeling circadian phototransduction: quantitative predictions of psychophysical data. *Front. Neurosci.* 15: 615322.

5 Brown, T.M., Brainard, G.C., Cajochen, C. et al. (2022). Recommendations for daytime, evening, and nighttime indoor light exposure to best support physiology, sleep, and wakefulness in healthy adults. *PLoS Biol.* 20 (3): e3001571.

6 Giménez, M.C., Stefani, O., Cajochen, C. et al. (2022). Predicting melatonin suppression by light in humans: unifying photoreceptor-based equivalent daylight illuminances, spectral composition, timing and duration of light exposure. *J. Pineal. Res.* 72: e12786. https://doi.org/10.1111/jpi.12786.

7 Fotios, S.A. and Levermore, G.J. (1998). Chromatic effect on apparent brightness in interior spaces, II: SWS lumens model. *Light. Res. Technol.* 30 (3): 103–106.

8 Fotios, S. and Atli, D. (2012). Comparing judgments of visual clarity and spatial, brightness through an analysis of studies using the category rating procedure. *Leukos* 8 (4): 261–281.

9 Bodrogi, P. and Khanh, T.Q. (2017). Visual clarity and brightness in indoor and outdoor lighting: experiments and modelling. In: *Proceedings of the CIE Midterm Meeting 2017* (CIE, 2017).

10 Bodrogi, P., Guo, X., and Khanh, T.Q. (2018). Brightness and visual clarity rather than luminance in indoor lighting assessment. Light 2018, Creative Lighting Solutions, 23rd European Lighting Congress, 2018. Davos, Switzerland.

11 Bodrogi, P. and Khanh, T.Q. (2019). Beschreibung von Helligkeit und Sehklarheit für die Innenraumbeleuchtung (Describing brightness and visual clarity for

interior lighting). *Z. Licht* https://lichtnet.de/artikel/beschreibung-von-helligkeit-und-sehklarheit-fuer-die-innenraumbeleuchtung/.

12 Figueiro, M.G., Bullough, J.D., Bierman, A., and Rea, M.S. (2005). Demonstration of additivity failure in human circadian phototransduction. *Neuro. Endocrinol. Lett.* 26: 493–498. 019(2): 80–84.

13 Figueiro, M.G., Bierman, A., and Rea, M.S. (2008). Retinal mechanisms determine the subadditive response to polychromatic light by the human circadian system. *Neurosci. Lett.* 438: 242–245. https://doi.org/10.1016/j.neulet.2008.04.055.

14 Thapan, K., Arendt, J., and Skene, D.J. (2001). An action spectrum for melatonin suppression: evidence for a novel non-rod, non-cone photoreceptor system in humans. *J. Physiol.* 535 (1): 261–267.

15 Brainard, G.C., Hanifin, J.P., Greeson, J.M. et al. (2001). Action spectrum for melatonin regulation in humans: evidence for a novel circadian photoreceptor. *J. Neurosci.* 21 (16): 6405–6412.

16 Wyszecki, G. and Stiles, W.S. (1982). *Color Science: Concepts and Methods, Quantitative Data and Formulae*, 2e, 950. New York: John Wiley and Sons.

17 Smith, V.C. and Pokorny, J. (1975). Spectral sensitivity of the foveal cone photopigments between 400 and 500 nm. *Vis. Res.* 15 (2): 161–171.

18 Snodderly, D.M., Auran, J.D., and Delori, F.C. (1984). The macular pigment. II. Spatial distribution in primate retinas. *Invest Ophthalmol Vis Sci.* 25 (6): 674–85. PMID: 6724837.

19 ISO/CIE 23539:2023 (2023). CIE TC 2-93 Photometry – The CIE system of physical photometry. ISO/CIE. doi: 10.25039/IS0.CIE.23539.2023.

14

Summary and Outlook

14.1 Summary

The task of research in the context of *Human-Centric Lighting* (HCL) or *Integrative Lighting*, in the various fields of neurobiology, cell research, psychology, sleep research, human medicine, and occupational science, is to gain new, systematic knowledge about the effects of light (or better of *visible optical radiation*) on humans. These effects are expressed in physiological terms (e.g. pupil diameter, hormone formation, EEG, ECG, or body core temperature), in psychological and emotional terms (e.g. mood, evaluation, preference, semantics, scaling, and classification), in light and colour perception terms (brightness, shadiness, homogeneity, hue, and colour saturation) as well as in a combinatorial form of various factors (alertness, drowsiness, and concentration). The direct, indirect, and long-term consequences of these effects are work performance, motivation, identification, happiness, joy, room atmosphere, social bonding, and ultimately *health*.

The task of building architecture, lighting architecture, and building services engineering is to dimension building installations and technical lighting installations (luminaires, daylight systems, and objects in the room) in order to enable the optimal working and living conditions of the room users in an individuality-dependent and humane manner. Bridge builders between strict science and research on the one hand and interior design, on the other hand, are the lighting designers or lighting engineers, who implement the knowledge of visual and non-visual lighting effects beyond the standard specifications into the *object planning* on the basis of the selected lighting products.

This book was written for lighting scientists, architects and lighting designers, product developers, occupational physicians, psychologists, and students. The writing of this book was a long scientific–technical journey for the authors. This knowledge and its practicable consequences can be summarised as follows:

1. Both visual and non-visual light effects are contributed by the five photoreceptors (LMS-cones, rods, and ipRGCs). The contribution of the individual receptor types depends on the light intensity, the spectrum, the time duration of the light stimulus, and the adaptation level.

Human Centric Integrative Lighting: Technology, Perception, Non-Visual Effects, First Edition.
Tran Quoc Khanh, Peter Bodrogi, and Trinh Quang Vinh.

2. The signals formed by the receptors are projected to the different brain regions, which in turn communicate with each other. At night, melatonin suppression has a close link to wakefulness, slow eye movement, and EEG activity, see Table 9.6. Research of non-visual effects for night-time has progressed to this point. The non-visual effects in the nocturnal period such as the action spectrum of melatonin suppression (its spectral dependence), the '*dose–response curve*', the 'Circadian Stimulus' (CS) as a parameter for melatonin suppression at different exposure durations, the light effect on the *phase shift* at known light conditions (time, timing, amount of light, and light spectrum) are already well researched. For the daytime as well as for the evening, the *colour quality*, and emotional effects, as far as lighting technology is concerned, have also been well researched by colour and lighting research since about 2005 until today. The formulas for brightness and visual clarity (in Chapter 5), which have been verified in various visual experiments, state that brightness and visual clarity depend on illuminance, S-cone signal, ganglion cell signal, and saturation (Eqs. (5.6), (5.7), and (7.1)). According to the formulae in Chapter 7, colour preference and scene preference depend on illuminance, colour temperature (respectively the S-cone signal), and colour saturation. The *light quality models* in Chapter 7 have been confirmed experimentally several times in real rooms (e.g. in various museum rooms, in shop scenes) and give great importance to illuminance and colour temperature in combination. The quality of light has thus been presented in detail in this book.
3. The transition from the quality of *light* to the quality of *illumination* can be achieved by dimensioning the luminance distribution on the walls and on the ceiling in the right relation, both relative to each other and in absolute terms. The means to achieve this is to vary the luminous intensity distribution of the luminaires in the room in order to change the indirect and direct components depending on the weather and the time of day, see Chapter 9. A ratio of the indirect component to the total illuminance on the working plane of higher than 0.6 (in the range between 0.6 and 0.8) provides very good subjective spatial assessment and facial recognition for visual communication, and a good modelling effect. This is combined with a relatively high vertical illuminance at the eye for an increased non-visual lighting effect and also related, as consequence, to the reduction of glare after the UGR-glare formula.
4. The analysis of the effect of illuminance on work performance/visual performance (Chapter 4), on lighting quality (Chapter 7), and on non-visual effects (Sections 9.6 and 9.7) shows the consistent picture that the range of optimal horizontal illuminance at the working plane is around 1000–1400 lx (to 1250 lx). Parallel to this, the core range of preferred colour temperature is between 4200 and 5000 K. This range can be extended to 4000–5500 K. The colour-rendering index CIE CRI R_a should preferably be better than 86, see Chapter 6.
5. Not only the illuminance but also the time of light exposure is decisive for the effect of light on humans. Bright light in the evening causes a longer sleep latency and a general (self-reported) poorer sleep quality. Bright light in the morning – brighter than 1000 lx – allows for better sleep quality and sleep

maintenance compared to 'moderate and dim light'. In general, a higher light level throughout the day also brings positive sleep quality. Consequently, light levels higher than 1000 lx should be preferred in the morning and throughout the day and avoided in the evening. This consequence for a higher illuminance level higher than 1000 lx, derived from research on non-visual light effects (sleep research), is consistent with research on the psychology of light and colour research (Chapters 6 and 7) and work performance/visual performance (Chapter 4) and leads to a detailed lighting design concept presented in Chapter 12.

6. Thus, in Chapter 12, a comprehensive specification for good light quality (or lighting quality) was proposed as an important result of this book, see Table 12.14 with the draft recommendation for a preferred and circadian-effective HCL lighting system for daytime applications. The content of this table is consistent with the results in recent literature, see Table 12.15.
7. In the context of modern, intelligent interior lighting, daylight plays a major role. In order to couple the highest possible amount of daylight into the interior, the window area proportion to the façade proportion should be higher than 40%, even better higher than 60%. For a 'relatively good' or 'good' quality of sleep, exposures (vertically on the person's face) between 7394 and 21 815 lxh are desirable.
8. Even if the findings and experiences on HCL conception consider the three components of visual performance, psychological–emotional aspects, and non-visual effects as *separate* components (and this book does so in various chapters for analytical reasons only), in truth these three components are closely interconnected – physiologically and cognitively. Chapter 3 also reported that the non-visual and visual processing channels in the brain are linked. This is illustrated in the case of the amygdala in the brain to the extent that the mechanisms of alertness and emotion are closely related. Higher alertness and positive emotions lead to higher cognitive *work performance* and concentration, as well as a better mood.
9. This close and almost inseparable bond among the above three HCL components was confirmed in Chapter 8 and in the special Chapter 13, which describes the relationship of visual metric 'brightness' to the two currently discussed non-visual metrics CS (Circadian Stimulus in the version of 2021) and MEDI (after CIE 2017), by a correlation analysis of their metrics. For the most part, the metrics of visual performance (brightness) correlate with the metrics of the non-visual light effects (see Figures 8.9 and 8.13). It must also be considered and acknowledged that the classical experiments in the 1970s by Boyce or Lindner (see Chapter 4), among others, on work performance pursued the same questions as today's non-visual experiments such as PVT, reaction time, or '*executive performance*'. Research on non-visual light effects such as wakefulness did not begin with the discovery of the ipRGCs, but decades earlier.
10. The metrics on non-visual effects such as a_{mel}, CS, 'melanopic illuminance' and D65-equivalent illuminance according to the CIE (see Chapters 2, 9, 10, and 13) are currently to be regarded more as suggestions. Their significance

and existence have triggered academic and practical discussions for some time. Actually, D65-equivalent illuminance, 'melanopic illuminance', and a_{mel} factors represent those aspects and mechanisms that have a *purely* melanopsin-containing background. Neurology and neurobiology, on the other hand, have shown in recent years that the ipRGCs not only receive the signals from the melanopsin-containing pigments (so-called *intrinsic* effects) but also *extrinsically* via the signals of other photoreceptor types such as the S-cones via a special amacrine cell, see Chapter 3.

11. The CS metric was generated from the action spectrum data for melatonin suppression at the beginning and was improved several times until the version in 2021 on the basis of experimental data of several international research groups and in its current form (model version 2021) takes into account the melatonin suppression in the night-time at different exposure durations and viewing field angles. This is a metric for the effect of 'melatonin suppression' that therefore applies to *night-time*. It considers not only the contributions of the ipRGCs but also those of the LMS-cones and the rods.
12. For the assessment of light effects during the daytime, such as alertness or concentration, the CS parameter is not suitable. The detailed correlation analyses in Chapter 8 showed that the brightness according to the Fotios formula and the a_{mel} factor for the melanopic effect correlate very well with each other, see Figure 8.9. The same statement applies to the very good correlation between CS2021 and MEDI or Giménez-formula value and *brightness* according to the Fotios formula and the model value of the authors of this present book in Chapter 13, which uses the luminance channel (L + M- channel) and the S-channel as signal sources in the brain for the characterisation of brightness, see Figure 8.13. In the future, mathematical constructs similar to brightness or the CS2021 parameter should be developed, which represent a purely mathematical combination of the signals from the ipRGCs, the LMS-cones, and the rods. This combination should be formed on the basis of suitable experiments with regard to a special light effect such as daytime wakefulness.
13. Nevertheless, both the CS and the other metrics such as the D65-equivalent illuminance or 'melanopic illuminance' are currently used and already implemented in international and national lighting technology recommendations. Experience shows that this implementation can be cemented over a longer period of time, if we think of the experiences with the $V(\lambda)$-function since 1924, which is rather problematic for methodological and actually also for practical reasons. Nevertheless, the $V(\lambda)$-function and all photometric parameters calculated from it, such as luminous flux, luminance, luminous intensity, and illuminance, still seem to be recognised over a longer period of time. This could also be the case for the three non-visual parameters CS, CS2021, D65 – equivalent illuminance, or 'melanopic illuminance'. In this book, the authors found a mathematical relationship between these three parameters based on an analysis of more than 500 typical light spectra (see Chapter 2, especially Figure 12.8 and in the special Chapter 13) to facilitate the work

of lighting designers and lighting engineers and the discussion among the researchers.

14. The parameters of brightness and the parameter MEDI and CS can initially be used as transitional parameters for the non-visual lighting effects. The measurements of these parameters should be efficient, sufficiently accurate, and practice-oriented without any special, expensive measuring instrument. In Chapter 10, methods were presented for measuring the CS and MEDI value (in Section 10.7) relatively accurately and practically with a portable colorimeter using the display value of illuminance (in lx) and colour temperature (in Kelvin) or the signal of the z-channel. This is a manageable method for today's lighting designers. Accordingly, in Chapter 10, results of *personal dosimetry* indoors and outdoors with daylight using a portable colorimeter were presented and evaluated.
15. From today's point of view, modern lighting in the sense of HCL or 'integrative lighting' cannot exploit its full potential without data communication, without networking the luminaires with each other or with a central processing unit such as a cloud, or a server for a whole building complex. The concept of 'Smart Lighting' with the use of colour sensors was described in detail in Chapter 11. An important gain in knowledge is that smart luminaires can be developed today and in the future in six stages with different variation possibilities. HCL does not necessarily have to be equipped with a luminaire with variable colour temperatures. The non-visual lighting effects can be realised through *four variations* in the following order: timing/timing of light exposure; variation of illumination levels; variation of spectra/colour temperature; and variation of light distribution in the visual field. In this sense, HCL at the beginning of the new era of modern lighting technology with the right light (right light level) at the right time (light in the early morning, light in the evening, and light in the night shift before the minimum of the body's core temperature) has already been very useful and cost-effective to facilitate the introduction of practicable lighting solutions in the service of their users. Such luminaires do not necessarily have to consist of different LED channels of different spectra.

14.2 Outlook

The following considerations, therefore, arise at the end of this book:

(a) Strictly speaking, night shift and alternating shift systems are work-organisational and economic necessities. If they cannot be avoided, one should try to conceptualise sensible lighting technology as humanely as possible under all circumstances. In contrast to recent review papers on shift work, the authors of this book assume that today's knowledge is sufficient for the time being to try out good lighting concepts in the field, with proper dimensioning of light exposure (lighting conditions) well before the minimum body core temperature, and with light protection measures on the way home. Such concepts can soon be worked out and tested in the field under practical conditions. The knowledge of research for the nocturnal hours gained in the laboratory or with a few test subjects is already advanced.

(b) Lighting quality research on work performance, as well as psychological, emotional, and colour quality aspects, is also very advanced today. This knowledge and the current understanding of non-visual lighting effects lead to an optimum daytime lighting level in the range of horizontal illuminance between 1000 and 1400 lx (or 850–1400 lx) with a static or dynamic lighting scheme as shown in the figures in Section 12.5, e.g. Figure 12.10. According to the most recent studies in literature, only few or no comprehensible effects of dynamic light have been experimentally determined to date. This is also due to the fact that the selected light levels and the schemes of dynamic light control have not yet been optimised in research work to date. Further, carefully prepared field studies with lighting levels above 850 lx or above 1000 lx during the day are therefore necessary.

In general, laboratory studies cannot replace field studies. The literature studies in the years 2017–2020 on the investigations of light effects in the laboratory during the day showed a very heterogeneous picture. In addition to the fundamental studies in the laboratories of neurology, psychology, and sleep research, many more well-prepared light-related field studies with interdisciplinary participation should be conducted. Nevertheless, the knowledge of today for night-time and daytime already forms a good basis for good lighting concepts with already available luminaire technologies.

Index

Human Centric Integrative Lighting: Technology, Perception, Non-Visual Effects, First Edition.
Tran Quoc Khanh, Peter Bodrogi, and Trinh Quang Vinh.

d

m

n